Theory of
Rotating Stars

PRINCETON SERIES IN ASTROPHYSICS

Edited by Jeremiah P. Ostriker

Jean-Louis Tassoul

THEORY OF
ROTATING STARS

PRINCETON UNIVERSITY PRESS
Princeton, New Jersey

297719

He [man] is ignorant in his life's first age, but he never ceases to learn as he goes forward, for he has the advantage not only of his own experience but also of his predecessors', because he always keeps in his memory the knowledge he has once acquired, and that of the ancients is always at hand in the books they have left. And since he keeps his knowledge, he can also easily increase it, so that men today are in a certain sense in the same condition in which those ancient philosophers would be if they could have prolonged their old age until now, adding to the knowledge they had what their studies might have won for them by the grace of so many centuries. Hence it is that by a special prerogative not only does each man advance from day to day in the sciences, but all men together make a continual progress as the universe grows old, because the same thing happens in the succession of men as in the different ages of an individual man. So that the whole series of men during the course of so many centuries should be considered as one self-same man, always in existence and continually learning.[1]

Blaise Pascal

[1] Preface to the "Treatise on the Vacuum" (1663), translated by Richard Scofield. Reprinted by permission from *Great Books of the Western World*, Vol. 33. Copyright 1952 by Encyclopaedia Britannica, Inc.

Contents

Preface

Few graduate students will need to be convinced that one of the major weaknesses in the existing network of scientific communication is the forbidding gap between the typical review paper, addressed by specialists to other specialists already familiar with the discipline, and the more reassuring undergraduate textbook in which unresolved problems are usually not touched upon. This is especially true in the field of stellar rotation, which has become one of the most differentiated of all astrophysical disciplines in terms of research specialties. Inasmuch as there appears to be no volume available that gathers information from scattered sources and treats the theory of rotating stars concisely but with a broad scope, this monograph is meant to fill the gap. Of course, the major ideas and results used in the text are, as is usual in scientific work, those of many individuals. Although I have endeavored to compare theory with observation, the text is basically theoretical and offers, I hope, a compromise between the wildest phenomenological arguments and the most abstruse mathematical developments. A detailed study of the whole problem, however, would involve several volumes of considerably larger size than this one; accordingly, the presentation of a concise picture of rotating stars has necessarily entailed the sacrifice of a considerable amount of detail. Furthermore, because the subject matter has so many separate branches, I have tried to concentrate almost entirely on topics of long standing that may be treated in Newtonian mechanics, rather than on those many greener fields (such as rotating neutron stars, pulsars, and binary X-ray sources) which are in the process of rapid and diverse growth within the framework of general relativity. Much for the same reasons, Dicke's work on the solar oblateness and the solar inner rotation has not been discussed (but references are given) in the book.

The basic plan of the book is as follows. After an introductory chapter concerned with a short historical sketch of stellar rotation, the main observational data on which the subsequent theory is based are presented in Chapter 2. The basic Newtonian principles and equations for studying rotating stars are set up in Chapter 3. The theoretical discussion actually starts with Chapter 4, and is divided into two parts. The first part, Chapters 4–8, provides the theoretical background necessary for the understanding of the structure, stability, and evolution of rotating stars; in the second part, Chapters 9–16, we present a systematic exposition of theoretical results which bear some interest on specific groups of rotating stars.

In Chapter 4, we deal with some simple mechanical properties that are relevant to the subject matter. (Sections 4.2 and 4.3 are of special importance, for most of the theoretical edifice rests on properties discussed in these pages.) In Chapters 5 and 6, we describe the main techniques by which one can determine the structure of a rotating star and discuss its stability with respect to infinitesimal disturbances. (Most parts of Chapters 7 and 8 can be studied directly after Chapter 4; nevertheless, we recommend reading Section 6.4, which is essential for the understanding of further stability considerations.) Problems relating to how a radiating star actually moves about its center of mass are discussed in Chapters 7 and 8. Chapter 7 specifically studies the problem of the angular momentum distribution within a star, whereas Chapter 8 provides a critical discussion of meridional currents in a rotationally distorted star. Were it not for the fact that a satisfactory meridional flow pattern in a nonmagnetic star has yet to be obtained, Chapters 7 and 8 would have been written so as to form a single unit.

Because a consensus about the causes of the Sun's differential rotation does not seem to exist at this time, Chapter 9 is devoted to various theories that attempt to explain the phenomenon. Apparently, we are faced here with another major unresolved problem: the interaction between convection and rotation in a star; but, then, it may be argued that we should perhaps try to understand convection without rotation in the first place! (The Sun is also discussed in Sections 2.2 and 15.5.) In Chapter 10, we are concerned with the differences that exist between uniformly rotating bodies and differentially rotating bodies that greatly deviate from spherical symmetry. This almost classical problem is illustrated by means of the polytropes; Section 10.5 contains a short summary. In Chapters 11–13, we deal with the major phases of stellar evolution as they are currently understood. Broadly speaking, Chapter 11 is devoted to the pre-main-sequence phases, as well as to one of the major open questions in the whole field of stellar rotation: will a single contracting star—initially endowed with some angular momentum—eventually separate into a pair of detached fragments to form a double star? In Chapter 12, we are concerned with the study of main-sequence stars; it concludes with a short discussion of the post-main-sequence phases of stellar evolution. (The interplay of theory and observation is particularly apparent in Sections 12.3–12.6.) Chapter 13 is entirely devoted to the basic characteristics of rotating white dwarfs. Finally, in Chapters 14–16, we treat various topics that could not have been incorporated into the previous chapters without seriously detracting from systematic exposition. These three chapters are devoted successively to some specific considerations about stability, large-scale magnetic fields in rotating stars, and axial rotation in close binaries.

References have been listed at the end of each chapter for elaboration of the material discussed in the corresponding sections or for further pursuit of topics not touched upon in the text. (All contributing authors are also listed in the Index of Names.) The book was written during the years 1973–1976; and it was updated in October 1976, using all material available to me at that time. (Later papers were included only if I had received some advance intimation of their contents.) The comments in the Bibliographical Notes are purely personal, but I felt it necessary to provide the advanced graduate student and research worker in related fields with some guide to the sheer bulk of literature on stellar rotation. Particular attention was paid to the many (and often insoluble) questions of priorities and independent discoveries. For any omission or understatement of credit to authors to whom more was due, I offer a sincere apology in advance.

Naturally, the writing of such a book as this would not have been possible without a great deal of help from others in the form of suggestions, criticism, encouragement, and financial support. The latter was provided by the National Research Council of Canada, under Grant A-5484, I am also indebted to a number of colleagues who read individual chapters and made helpful comments, including in particular Dr. H. A. Abt, Mr. Y. Charland, Dr. R. A. Chevalier, Dr. B. R. Durney, Dr. G. Fontaine, Dr. J. D. Landstreet, Dr. A. Maeder, Dr. L. Mestel, Dr. G. Michaud, Dr. A. Slettebak, and Dr. R. C. Smith. The help of these and other individuals is gratefully acknowledged, but of course they are in no way responsible for any errors of fact or judgment that the book may contain. Finally, I wish to record my indebtedness to my wife, Monique, who worked at every stage of the project, watched the gestation of the final draft, and criticized it mercilessly. Without her help, encouragement, and continuing tolerance of that consuming preoccupation, this book could not have been written, and I dedicate it to her in friendship and love.

Département de Physique,   J.L.T.
Université de Montréal,   *November 1976*
Montréal, Québec

A Short List
of Abbreviations

This list only includes the cryptic abbreviations that occur in the main text, footnotes, tables, and figure captions. Full titles or abbreviations recommended by the International Astronomical Union are thoroughly used in the Bibliographical Notes.

A.Ap.	Astronomy and Astrophysics
Adv.A.Ap.	Advances in Astronomy and Astrophysics
A.J.	The Astronomical Journal
A.N.	Astronomische Nachrichten
Ann.Ap.	Annales d'Astrophysique
Ap.J.	The Astrophysical Journal
M.N.	Monthly Notices of the Royal Astronomical Society
P.A.S.P.	Publications of the Astronomical Society of the Pacific
Z.f.Ap.	Zeitschrift für Astrophysik

Theory of
Rotating Stars

1

An Historical Overview

1.1. DISCOVERY OF THE SOLAR ROTATION

The study of stellar rotation began about the year 1610, when sunspots were observed for the first time through a refracting telescope. The first public announcement of an observation came in June 1611 from Johannes Goldschmidt (1587–1615)—a native of East Friesland, Germany—who is generally known by his Latinized name Fabricius. From his observations he correctly inferred the spots to be parts of the Sun itself, thus proving axial rotation. He does not appear to have appreciated the importance of this conclusion, and pursued the matter no further.

According to his own statement, Galileo Galilei (1564–1642) observed the sunspots toward the end of 1610. He made no formal announcement of the discovery until May 1612, by which time such observations had also been made by Thomas Harriot (1560–1621) in England and by the Jesuit Father Christoph Scheiner (1575–1650) in Germany. A controversy about the nature of sunspots made Scheiner a bitter enemy of Galileo and developed into a quarrel regarding their respective claims to discovery.[1] Since Scheiner had been warned by his ecclesiastical superiors not to believe in the reality of sunspots (because Aristotle's works did not mention it), he announced his discovery in three letters written under the pseudonym of Apelles. He explained the spots as being small planets revolving around the Sun and appearing as dark objects whenever they passed between the Sun and the observer. These views opposed those of Fabricius and Galileo, who claimed that the spots must be on or close to the solar surface.

In the following year, 1613, Galileo replied with the publication of his *Istoria e Dimostrazioni intorno alle Macchie Solari e loro Accidenti*. In these three letters he refuted Scheiner's conclusions and, for the first time, publicly declared his adherence to the heliocentric theory of the solar system—thus initiating the whole sad episode of his clashes with the Roman Inquisition. As far as solar rotation is concerned, one of his

---

[1] For a dispassionate analysis of priority disputes in this and other instances, see: Merton, R. K., "Priorities in Scientific Discovery: A Chapter in the Sociology of Science" in *American Sociological Review* **22**, 635, 1957; reprinted in *The Sociology of Science* (Barber, B., and Hirsch, W., eds.), pp. 447–485, New York: The Free Press of Glencoe, 1962.

arguments is so simple—and at the same time so convincing—that it may be worthwhile to reproduce it here. Galileo noticed that a spot took about fourteen days to cross the solar disc, and that this time was the same whether the spot passed through the center of the solar disc or along a shorter path at some distance from the center. However, he also noticed that the rate of motion of a spot was by no means uniform, but that the motion always appeared much slower when near the solar limb than when near the center. This he recognized as an effect of foreshortening, which would result if and only if the spots were near the solar surface.

Scheiner's views were thus crushingly refuted by Galileo. Eventually, the Jesuit father's own observations led him to realize that the Sun rotates with an apparent period of about 27 days. To him also belongs the credit of determining with considerably more accuracy than Galileo the position of the Sun's equatorial plane and the duration of its rotation. In particular, he showed that different sunspots gave different periods of rotation and, furthermore, that the spots farther from the solar equator moved with a slower velocity. Scheiner published his collected observations in 1630 in a volume entitled *Rosa Ursina sive Sol*. (Dedicated to the Duke of Orsini, the title of the book derives from the badge of the Orsini family, which was a rose and a bear.) This was truly the first monograph on solar physics.

For more than two centuries the problem of solar rotation was practically ignored, and it is not until the 1850s that any significant advance was made. Then, a long series of observations of the apparent motion of sunspots was undertaken by Richard Carrington (1826–1875), a wealthy English amateur, and by Gustav Spörer (1822–1895), a German astronomer. They confirmed, independently, that the outer visible envelope of the Sun does not rotate like a solid body, i.e., its period of rotation varies as a function of heliocentric latitude. They showed that the rotation period is minimum at the equator and increases gradually toward the poles. After correcting for the annual motion of the Earth around the Sun, Carrington derived a mean period of 24.96 days at the solar equator.

Several attempts have been made to represent analytically the dependence of rotational speed on latitude by means of an empirical formula. From his own observations made during the period 1853–1861, Carrington derived the following expression for the daily angle of solar rotation:

$$\xi(\text{deg/day}) = 14°\!.42 - 2°\!.75 \sin^{7/4} \phi, \tag{1}$$

where ϕ is the heliocentric latitude. Somewhat later, the French astronomer Hervé Faye (1814–1902) found that the formula

$$\xi(\text{deg/day}) = 14°\!.37 - 3°\!.10 \sin^2 \phi \tag{2}$$

more satisfactorily represented the dependence of rotation on heliocentric latitude. Other formulae have been suggested, but Faye's empirical law has remained in use until now.

The next step was taken by Hermann Vogel (1841–1907) in Potsdam. In 1871, by means of a new spectroscope devised by Johann Zöllner (1834–1882), Vogel showed that the solar rotation can be detected from the relative Doppler shift of the spectral lines at opposite edges of the solar disc, one of which is approaching and the other receding.[2] Accurate and extensive observations were made visually by Nils Dunér (1839–1914) in Lund and Upsala, and then by Jakob Halm (1866–1944) in Edinburgh. Dunér and Halm used a relative method of measurement based on the difference in behavior of the solar and terrestrial lines. They concluded that Faye's law adequately represents the spectroscopic observations also, but their coverage of latitude was double that of the sunspot measurements. After these early visual observations made during the period 1887–1906, photography almost completely superseded the human eye. The first spectrographic determinations of solar rotation were undertaken at the turn of the century by Walter S. Adams (1876–1956) and George E. Hale (1868–1938) at Mount Wilson Solar Observatory, California.

1.2. EARLY MEASUREMENTS OF STELLAR ROTATION

In 1877, Sir William de Wiveleslie Abney (1843–1920) suggested that axial rotation might be responsible for the great widths of certain stellar absorption lines. He correctly pointed out that a spectrum is actually a composite of light from all portions of the star's disc. Because of the Doppler effect, wavelengths in the light of the receding edge of the star would be shifted toward the red, those from the approaching edge toward the violet. Abney concluded: "There would be a total broadening of the line, consisting of a sort of double penumbra and a black nucleus.... More than this, rotation might account for the disappearance of some of the finer lines of the spectrum.... I'm convinced that from a good

---

[2] In 1842, the Austrian physicist Christian Doppler (1803–1853) noted that the apparent pitch of sound is altered by the relative motion between the source and the observer. Thence, he suggested that the *color* of a star might be changed according to its velocity of approach or recession relative to the observer. The principle was subsequently verified experimentally for sound by the Dutch meteorologist Christopher Buys-Ballot (1817-1890) in 1845. A more useful application for the optical case was given by the French physicist Armand-Hippolyte Fizeau (1819-1896) in 1848; he showed that stellar radial velocities are much too small to cause an appreciable color change, but that the Doppler effect could be detected by very small changes in the wavelengths of individual spectral lines. The experiment was successfully performed for the first time in 1868 by Sir William Huggins (1824–1910), who was able to measure the minute Doppler shifts of the hydrogen lines in the spectrum of Sirius.

photograph much might be determined" (M.N. **37**, 278, 1877). These suggestions were severely criticized by Vogel because some stellar spectra contained both broad and narrow lines. Concerning the use of photography, Vogel went further in saying that "even on the most successful photograph only relative measures of lines would be possible in regard to their width, so that no conclusions could be obtained in regard to rotation; the widths of lines on different photographic plates would depend upon the length of exposure, the sensitivity of the plate, and the length of development" (A.N. **90**, 71, 1877).

Abney's suggestion found little favor among his contemporaries. The reason may be due to the enormous weight accorded the opinion of Vogel, who, for many years, dominated the field of stellar spectroscopy. In any case, only a few years after his emphatic predictions, Vogel introduced photography into stellar spectroscopy and completely reversed his stand in 1898, expressing himself in favor of the hypothesis that rotation *does* produce a measurable broadening of stellar lines.

As often happens in the history of science, the discovery of axial rotation in stars was purely accidental. In 1909, convincing evidence of rotation in the eclipsing and spectroscopic binary δ Librae was obtained by Frank Schlesinger (1871–1943), then at Allegheny Observatory, Pittsburgh. He noticed that just before and just after light minimum, the radial velocities he had measured on his spectrograms departed from their expected values. A positive excess was observed just before mid-eclipse, while after mid-eclipse the departure was found to be negative. Schlesinger concluded that this occurrence could be produced if the brightest star rotates around an axis, so that, at the time of partial eclipse, the remaining portion of its apparent disc is not symmetrical with respect to its axis of rotation. One year later, Schlesinger observed a similar phenomenon in the eclipsing system λ Tauri. In 1924, at the University of Michigan, Richard A. Rossiter (1886–1977) positively established the effect in β Lyrae and gave a complete curve of the residuals in velocity during the eclipse, while Dean B. McLaughlin (1901–1965) investigated the effect in the binary star β Persei. These were the first accurate measurements of the axial rotation of stars.

Another approach to the determination of stellar rotation was provided by spectroscopic binaries not known to be variable in light. In 1919, at Mount Wilson, Adams and Alfred H. Joy (1882–1973) studied the binary W Ursae Majoris of spectral type F8 and having a period of 0.334 day. They observed that "the unusual character of the spectral lines is due partly to the rapid change in velocity during even our shortest exposures but mainly to the rotational effect in each star, which may cause a difference of velocity in the line of sight of as much as 240 km/sec between the two limbs of the star" (Ap. J. **49**, 190, 1919). Ten years later, a systematic study of rotational line-broadening in spectroscopic binaries was undertaken jointly by Grigori Abramovich Shajn (1892–1956) in the Soviet Union and Otto Struve (1897–1963) in the United States. (Their collaboration took place by mail.) Shajn and Struve positively established that in spectroscopic binaries of short period, at least, line broadening is essentially a result of rotation.

The next step was to extend these measurements to *single* stars. Indeed, the possibility existed that rapid rotation occurred only in binary stars, perhaps because tidal forces produced synchronization of axial rotation and orbital revolution. Since the Sun has a very slow axial rotation (about 2 km/sec at its equator), most astrophysicists at the time believed that the rotation of all other single stars was probably also small. During 1930–1934, a systematic study of stellar rotation was undertaken by Struve, in collaboration with Christian T. Elvey (1899–1972) and Miss Christine Westgate, at the Yerkes Observatory of the University of Chicago. The measurements were made by fitting the observed contour of a spectral line to a computed contour obtained by applying different amounts of Doppler broadening to an intrinsically narrow line-contour having the same equivalent width as the observed line. They showed that the measured values of the rotational component of the velocity along the line of sight fell in the range 0–250 km/sec, and may occasionally be as large as 400 km/sec or even more. A correlation of rotational velocity with spectral type was originally discovered by Struve and Elvey in 1931: the O-, B-, A-, and early F-type stars frequently have large rotational velocities, while in late F-type stars and later types rapid rotation occurs only in close spectroscopic binaries. They also found that supergiants of early and late types, and normal giants of type F and later, never show conspicuous rotations.

At this juncture the problem was quietly abandoned for fifteen years. Interest in the measurements of axial rotation in stars was revived in 1949 by Arne Slettebak at Ohio State University.

1.3. ROTATING FLUID MASSES

Research into the influence of rotation upon the internal structure and evolution of a star has a long history. Sir Isaac Newton (1643–1727) was the first to realize the importance of the law of gravitation for the explanation of the figures of celestial bodies. He originally discussed the figure of the Earth in the *Philosophiae Naturalis Principia Mathematica* (Book III, Propositions 18–20, 1687) on the hypothesis that it might be treated as a homogeneous, slightly oblate spheroid rotating with constant angular velocity. Assuming that such a spheroid is a figure of equilibrium, Newton asserted that two perpendicular columns of fluid—one axial and the other equatorial—bored straight down to the Earth's center must have equal weight. From this condition, he was able to derive the causal relationship $f = (5/4)m$ between the ellipticity f of a meridional section and the ratio m of the centrifugal force at the equator to the (average) gravitational attraction on the surface.

Further progress was made by Christiaan Huygens (1629–1695) in his *Discours de la cause de la pesanteur*, which was published at Leiden in 1690. To him we owe a necessary condition for the relative equilibrium of a rotating mass of fluid—that the resultant force of the attraction and the centrifugal force at any point of the free surface must be normal to the surface at that point. Huygens never accepted that adjacent particles of matter attract each other, but he did admit the existence of an attracting force always directed to a fixed point. Accordingly, in one of his models, he assumed that the Earth's gravity is a single, central force varying inversely as the square of the distance from its center. Thence, in making use of Newton's principle of balancing columns, Huygens derived the relation $f = (1/2)m$ for his model, when departure from sphericity is small. As we now know, this is equivalent to the hypothesis that the density of the Earth is infinite at the center; and it should be contrasted with Newton's work in which it is assumed that the Earth is homogeneous in structure. In actual practice, the measured quantity f for the Earth is comprised between the values derived by Newton and Huygens for their two extreme models.

While Newton's ideas led ultimately to contemporary mechanics, at this stage they did not gain immediate acceptance in Europe. During the first half of the eighteenth century, most continental scientists were strong advocates of a vortex theory that had been devised by René Descartes (1596–1650) in 1644. The Cartesians rejected Newton's ideas mainly because attraction, regarded as a cause, was unintelligible; and they somehow inferred from their systems of vortices that the Earth should be flatter at the equator. In the 1730s, to settle the dispute between the Cartesians and the Newtonians, geodetic expeditions were sent to different parts of the Earth to measure the length of an arc of meridian. The expedition to Lapland returned first and confirmed that the Earth is indeed flatter at the poles. Pierre Louis Moreau de Maupertuis (1698–1759)—its leader and the first Frenchman with the courage openly to declare himself a Newtonian—was called "the great flattener." At this time the vortex theory was a lost cause.

During 1737–1743, the most significant advances in the subject were made by Alexis-Claude Clairaut (1713–1765) in Paris and by Colin

Maclaurin (1698–1746) in Edinburgh. Again, to appreciate the importance of their discoveries, it must be borne in mind that the science of hydrostatics was then in an imperfect stage of its development. Actually, this early work was made *without* a clear understanding of the concept of internal pressure. Hence, whenever Clairaut and Maclaurin established a proposition, they had to rely upon Newton's principle of balancing columns, Huygens's principle of the plumb line, or both. As a matter of fact, necessary *and sufficient* conditions of hydrostatic equilibrium remained unknown until 1755, when Leonhard Euler (1707–1783) definitively established the general equations of motion for an inviscid fluid.[3]

As we have seen, Newton assumed without demonstration an oblate spheroid as a possible figure of equilibrium for a slowly rotating mass of fluid. In 1737, Clairaut obtained an expression for the attraction of a homogeneous spheroid at any point of its surface, when the body does not greatly depart from a sphere. To first order in the ellipticity, he then showed that Huygens's principle of the plumb line obtains at every point on the free surface of Newton's model.

This result was generalized in 1740 by Maclaurin, who completely solved the problem of the attraction of a homogeneous spheroid when departure from sphericity cannot be considered small. Again using Newton's principle of balancing columns, he proved that *any* oblate homogeneous spheroid is a possible figure of relative equilibrium; he also was successful in relating the angular velocity of rotation to the ellipticity of a meridional section. Finally, he demonstrated that at every point the direction of the plumb line is perpendicular to the free surface. The basic concept of level surfaces was also introduced by Maclaurin, who observed that at all depths the resultant body force is always normal to a surface belonging to a family of homothetic spheroids.

We now come to the brilliant treatise of Clairaut, *Théorie de la figure de la terre*, published in 1743. As far as our subject is concerned, the most interesting part of this book deals with the figures of equilibrium of slowly rotating, centrally condensed bodies—a problem hitherto practically untouched. Clairaut considered a self-gravitating configuration composed of quasi-spherical strata of varying densities. Given this assumption, he showed that the level surfaces necessarily coincide with the surfaces of equal density. On applying Huygens's principle of the plumb line *to each*

---

[3] It is not until 1822 that Baron Augustin-Louis Cauchy (1789–1857) introduced the concept of stress tensor, thus initiating the theory of viscous fluids. The principle of the conservation of energy was first established in 1842 by the Heilbronn physician, Julius Robert Mayer (1814–1878), and again independently by Ludwig August Colding (1815–1888), Hermann von Helmholtz (1821–1894), James Prescott Joule (1818–1889), and many others.

level surface, he then proved that oblate spheroids are compatible with relative equilibrium, and he derived an equation relating the varying ellipticity of the strata with their density. Finally, to first order in the ellipticity, he showed that the diminution of the surface effective gravity in passing from the poles to the equator varies as the square of the cosine of the latitude.

While all the ideas necessary for the general theory of rotating bodies were proposed by these pioneers, most of our knowledge on this subject derives from the contributions of the Marquis Pierre-Simon de Laplace (1749–1827) and of Adrien-Marie Legendre (1752–1833) over the period 1773–1793. To Legendre we owe the concept of gravitational potential and the general theory of the attraction of a homogeneous ellipsoid. This work eventually led him to introduce the polynomials that now bear his name. It was also Legendre's suggestion to identify the level surfaces of a uniformly rotating body with the condition that the sum of the gravitational and centrifugal potentials is a constant over these surfaces. From this condition he was able to bring Clairaut's theory to its present form and to *prove* that the strata must be oblate spheroids when departure from sphericity is small. As for Laplace, his studies of the problem led him to introduce the spherical harmonics and the differential equation with which his name has since been associated. The well-known proposition that the surfaces of equal pressure, the surfaces of equal density, and the level surfaces all coincide in a uniformly rotating figure of equilibrium was first demonstrated by Laplace. In 1825, as a by-product of his studies of Clairaut's equation, Laplace also introduced the concept of barotropes; and he explicitly solved and discussed the spherical polytrope of index $n = 1$!

All the foregoing results were included by Laplace in his impressive *Traité de mécanique céleste* (1799–1825), though he often forgot to give appropriate credit to the important contributions of his colleague Legendre. The differential equation that relates the gravitational potential at an internal point of a body to the density at that point was originally derived by Siméon-Denis Poisson (1781–1840) in 1829, and it has also become a permanent part of the theory of self-gravitating bodies.

The next important discovery was made by Karl Jacobi (1804–1851), who pointed out, in 1834, that axially symmetric configurations are not the only admissible figures of equilibrium. Jacobi gave decisive evidence of a homogeneous ellipsoid with three unequal axes as a possible form of relative equilibrium for rotating bodies. The same year, Joseph Liouville (1809–1882) published the analytical proof of this statement; and, in 1843, he gave the first description of these ellipsoids in terms of their angular momentum.[4] It was already known that, as the angular momentum increases from zero to infinity, the Maclaurin spheroids range from a sphere to an infinitely flat disc. As for the Jacobi ellipsoids, which range from an axially symmetric configuration to an infinitely long needle, Liouville found that they represent possible figures of equilibrium only for angular momenta exceeding a certain amount.

No further progress was made until the problem was taken up, independently, by Aleksandr Mikhailovich Liapunov (1857–1918) and Henri Poincaré (1854–1912). To Poincaré we owe the general theory of the equilibrium and stability of ellipsoidal forms, and the concept of figures of bifurcation. In particular, he showed the existence of pear-shaped figures that branch off from the sequence of Jacobi ellipsoids. As a matter of fact, the pear-shaped figures had already been discovered during the previous year, 1884, by Liapunov. Unfortunately, the results of the Russian mathematician remained largely unknown to the Western scientists until his papers were translated into French, at the request of Poincaré himself. The Liapunov-Poincaré figures gave rise to extensive investigations, for it was then conjectured that they may eventually come apart in two detached masses rotating about each other. This theory of the origin of double stars was originally proposed by Lord Kelvin (William Thomson, 1824–1907) and Peter Guthrie Tait (1831–1901) in 1883; Sir James Hopwood Jeans (1877–1946) was its most ardent upholder during the 1920s. A critical review of this theory will be found in Section 11.3.

As for centrally condensed bodies that greatly deviate from spherical symmetry, important discoveries were also made during the period 1885–1935, although they attracted much less attention. Indeed, the stratification of a rapidly rotating, inhomogeneous body is never known in advance and is to be found from the conditions of relative equilibrium. By using a method of successive approximations, Liapunov was the first to develop an exact theory of inhomogeneous bodies in which the level surfaces differ but little from homocentric ellipsoids. This work is undoubtedly a masterpiece of functional analysis, but it remains very difficult to use in actual practice. For this reason, other methods were devised by Leon Lichtenstein (1878–1933) in Berlin and Leipzig, and by Rolin Wavre (1896–1949) in Geneva. Most of these techniques have been now superseded by numerical computations; nevertheless, they led to the discovery of a few fundamental

---

[4] The first detailed study of the Maclaurin-Jacobi sequences was made in 1842 by Otto Meyer at Koenigsberg. The classical papers in this subject are: Jacobi, K., *Ann. Phys. und Chemie* (*Poggendorff*) **33**, 229, 1834 (*Werke* **2**, 19, 1882); Liouville, J., *J. Ecole Polytech. Paris* **14** (23), 289, 1834; Meyer, C. O., *J. f. reine und angew, Math.* (*Crelle*) **24**, 44, 1842; Liouville, J., *C. R. Acad. Sci. Paris* **16**, 216, 1843.

theorems that form the basis of the theory of rapidly rotating, centrally condensed bodies. Particular mention should also be made of the French mathematician Pierre Dive, who made a thorough investigation of differential rotation. This kind of research was to be out of fashion in the mid-thirties, and most of these classical results have been largely ignored over the past forty years. They will be discussed anew in Chapter 4.

1.4. THE GOLDEN AGE OF STELLAR ROTATION

Although the theory of rotating stars has greatly advanced during the past fifteen years, the basic *concepts* underlying most of the contemporary researches were developed during the period 1919–1941. Let us briefly summarize these early studies; all of them will receive considerable attention in the main body of this book.

At the turn of the century, it was already clear that the stars were gaseous and centrally condensed configurations. In 1919, Jeans conceived the idea that sequences of uniformly rotating, centrally condensed bodies may not mirror the classical Maclaurin-Jacobi sequences. To illustrate the problem, he constructed various sequences of uniformly rotating polytropes, the polytropic index n being a measure of central condensation. He observed that for values of n larger than $n_c \approx 0.8$ the Maclaurin-Jacobi pattern does not prevail; that is to say, when $n \geqslant n_c$ (giving a central condensation larger than about three), the sequences of uniformly rotating polytropes do not admit of a point of bifurcation, but they instead terminate by the balancing of the gravitational and centrifugal accelerations at the equator. Jeans therefore concluded that if one imagines a slowly contracting sequence of such figures, equatorial break-up would eventually occur; in other words, any further contraction would result in matter being shed from the equator in a continuous stream, centrifugal force outweighing gravity at the equator. This semiquantitative picture led Struve to suggest, in 1931, that the Be stars are uniformly rotating B-type stars on the verge of equatorial break-up, with surrounding gaseous rings.

During the same period, as more was discovered about the physical processes that take place within a star, it also became clear that energy is transported by radiation—not by convection—in the main bulk of the vast majority of stars. In 1923, this result led Edward Arthur Milne (1896–1950) at Oxford to build the first detailed model for a slowly rotating star in pure radiative equilibrium. Ten years later, the technique was generalized and applied to slightly distorted polytropes by the Indian-born astrophysicist Subrahmanyan Chandrasekhar, then a visiting fellow in Copenhagen. Most of the recent stellar models that do not greatly deviate from spherical symmetry rely upon these two pioneering studies.

The next important step in the elucidation of stellar rotation was made in Upsala by Hugo von Zeipel (1873–1959). In 1924, he proved the following theorem: if a chemically homogeneous star, rotating as a rigid body with angular velocity Ω is in static radiative equilibrium, the rate of liberation of energy ϵ_{Nuc} at any point in the interior must be given by

$$\epsilon_{Nuc} = constant\left(1 - \frac{\Omega^2}{2\pi G\rho}\right), \tag{3}$$

where G and ρ are, respectively, the constant of gravitation and the density at that point. Clearly, since the actual energy sources of a star do not fulfill condition (3), at least one of the foregoing assumptions must be relaxed. This fact was first discussed, independently, by Heinrich Vogt (1890–1968) in Heidelberg and by Sir Arthur Stanley Eddington (1882–1944) in Cambridge. In 1925, they pointed out that strict radiative equilibrium must necessarily break down in a uniformly rotating star. This, they argued, will upset the constancy of temperature and pressure over the level surfaces; a pressure gradient will thus ensue, causing a flow of matter primarily in planes passing through the axis of rotation. A quantitative study of the physical properties of this meridional circulation was originally made by Eddington in 1929.

Finally, no survey of this Golden Age of stellar rotation would be complete without special mention of the Oslo school of fluid mechanics. To Vilhelm Bjerknes (1862–1951) we owe a systematic study of physical hydrodynamics. In particular, he laid the foundations of the theory of barotropes and baroclines, and he systematized the problem of the small oscillations of a continuous medium about a state of equilibrium. Although these questions were mainly discussed in the specific context of meteorology, they paved the road to many astrophysical applications. In this respect, let us mention the pioneering work of Svein Rosseland and Gunnar Randers. Of particular importance is the stability analysis of a rotating compressible fluid by Halvor Solberg (1895–1974) and Einar Høiland (1907–1974). A first criterion was obtained by Solberg, in 1936, for the case of axisymmetric disturbances in a rotating star with constant entropy. Under these conditions, he proved that dynamical instability occurs if

$$\frac{d}{d\varpi}(\Omega^2\varpi^4) < 0, \tag{4}$$

where the angular velocity Ω is some specified function of the distance ϖ from the rotation axis. In 1941, this result was generalized by Høiland, who considered the case where Ω depends on the coordinate along the rotation axis as well. Thence, he proved that dynamical instability occurs whenever the angular momentum per unit mass $\Omega\varpi^2$ decreases from the poles to the equator on a surface of constant entropy per unit mass. Elegant derivations of this criterion were subsequently presented by Ragnar Fjørtoft in Oslo and by Jörgen Holmboe in Los Angeles. Because these and related questions of stability now form an important chapter in the theory of rotating stars, they will be discussed further in Section 7.3.

BIBLIOGRAPHICAL NOTES

Section 1.1. Historical accounts of the problem will be found in:

1. Berry, A. B., *A Short History of Astronomy*, London: John Murray, 1898 (New York: Dover Public., Inc., 1961).
2. Bray, R. J., and Loughhead, R. E., *Sunspots*, London: Chapman and Hall, 1964.
3. Meadows, A. J., *Early Solar Physics*, Oxford: Pergamon Press, 1970.

This last reference also contains English translations of many classical papers in the subject. A useful source paper for all solar rotation measurements made during the period 1611–1933 is:

4. DeLury, R. E., *J. Roy. Astron. Soc. Canada* **33** (1939): 345–378.

Section 1.2. The account in this section is largely based on:

5. Struve, O., *Popular Astronomy* **53** (1945): 201–218, 259–276.

See also:

6. Struve, O., and Zebergs, V., *Astronomy of the 20th Century*, New York: The Macmillan Co., 1962.

Section 1.3. All relevant publications from the time of Newton to that of Laplace were critically reviewed by Todhunter:

7. Todhunter, I., *A History of the Mathematical Theories of Attraction and the Figure of the Earth*, London: McMillan and Co., 1873 (New York: Dover Public., Inc., 1962).

See also reference 28 (pp. 1–14) of Chapter 3. A penetrating discussion of the genesis of Newtonian mechanics and hydrodynamics will be found in:

8. Truesdell, C., *Essays in the History of Mechanics*, Berlin: Springer-Verlag, 1968.

Another interesting reading is:

9. Aiton, E. J., *The Vortex Theory of Planetary Motions*, London: Macdonald and Co., 1972.

The period 1885–1935 has been adequately covered by Jardetzky:

10. Jardetzky, W. S., *Theories of Figures of Celestial Bodies*, New York: Interscience Publishers, 1958.

Section 1.4. Early theoretical works on stellar rotation are discussed in:

11. Milne, E. A., "Thermodynamics of the Stars" in *Handbuch der Astrophysik* **3** (1), pp. 235–255, Berlin: Springer-Verlag, 1930.

12. Krat, V. A., *Equilibrium Figures of Celestial Bodies* (in Russian), Moscow: G.I.T.T.L., 1950.

13. Vogt, H., *Aufbau und Entwicklung der Sterne*, Zweite Aufläge, pp. 135–155, Leipzig: Akademische Verlagsgesellschaft, 1957.

2

The Observational Data

2.1. INTRODUCTION

In Sections 1.1 and 1.2 we discussed the early measurements of the axial rotation of the Sun and the stars. Remarkably, most present-day methods of reduction still largely rely upon these pioneering studies. Our intent here, then, is to present a general survey of the recent observational data pertaining to stellar rotation, and the various assumptions underlying their reduction. The Sun, the single stars, and the binaries are examined in turn. For the present, we shall not aim at completeness, but limit ourselves, rather, to the main observational features that we will need in the subsequent chapters. Specific groups of stars will be further discussed in Chapters 11–16, together with their theoretical interpretations.

2.2. THE SUN

The solar rotation rate may be deduced from measurements of the longitudinal motions of semipermanent features across the solar disc (such as sunspots, faculae, dark filaments, or even coronal activity centers), or from spectrographic observations of Doppler displacements of selected spectral lines near the solar limb. From observations of sunspots made prior to 1630 by Scheiner, it was apparent that spots near the solar equator had shorter periods of revolution than those at higher heliocentric latitudes. Spectrographic measurements have since provided an independent verification of the solar differential rotation. For example, from the observations of long-lived sunspots made at Greenwich during 1878–1944, Newton and Nunn obtained a daily displacement $\xi(\phi)$ in the form

$$\xi(\text{deg/day}) = 14°38 - 2°77 \sin^2 \phi, \qquad (1)$$

where ϕ is the heliocentric latitude. (Compare this with Carrington's and Faye's original results discussed in Section 1.1.) The empirical formula (1) corresponds to a sidereal period of 25.03 days at the equator, and to a mean synodic period of 26.87 days at the equator.

Each of the two methods—displacement of a tracer and relative Doppler shift of spectral lines—for deriving rotation rates has its own limitations, although few of these limitations are common to both. Actually, the determination of solar rotation from tracers requires that these semipermanent features be both randomly distributed throughout the fluid and undergo no appreciable proper motion with respect to the medium in which they are embedded. In practice, no tracers have been shown to possess both characteristics; moreover, they tend to occur in a limited range of latitudes. By the spectrographic method, rotation rates can be found over a wider range of latitudes. But, then, the accuracy is limited by the presence of inhomogeneities of the photospheric velocity field, and by macroscopic motions within coronal and prominences features, so that the scatter between repeated measurements is large.

Figure 2.1 assembles the rotation rate measurements determined from different solar tracers. The observations refer to the sunspots, the filaments and prominences, the photospheric magnetic field, and the polar faculae. This figure also includes the rotation rate of localized features in the electron corona (K-corona) and of coronal regions of bright green emission (green corona). The decrease of velocity with increasing latitude is clear. On the average, the tracers all seem to have *roughly* the same daily displacement at the same latitude.

The difficulty of measuring the solar rotation is well illustrated by the sunspots. The two curves enclosing the gray area in Figure 2.1 are given for the sunspots' daily displacement. Actually, as Ward has shown, the average longitudinal motion of a sunspot is a function of its size and shape, with large groups moving up to 2 percent less rapidly than smaller ones. The lower curve was derived by Newton and Nunn from long-lived sunspots that are predominantly in large circular groups. The upper curve depicts Ward's results. He used a random selection of all spots, and his solar rotation rate is 1 percent higher than that calculated by Newton and Nunn, about $0°.15$ per day.

At higher latitudes, however, the difference becomes larger. Around $\phi = 60°-65°$, it seems well established that the coronal rotation rate is at least $12°$ per day, while the filament rate is about $12°.2$ per day, and the photospheric rate found from polar faculae is about $11°$ per day. It should be noted that there may be a fundamental difference in the nature of the tracers of the solar rotation. Indeed, the faculae are very short-lived, with lifetimes of 40 to 50 hours at most, while some of the coronal features and filaments persist for many months. It is thus not clear whether they represent real differences of rotation at various levels in the solar atmosphere or reflect a characteristic behavior of the tracers themselves—such as unequal proper motion or asymmetric growth.

Velocity measurements from Doppler shifts of Fraunhofer lines or emission lines are also a good means of obtaining the differential rotation of the Sun, but there are complications because of the limitations in the accuracy of lineshift measurements, and because of the "noise" introduced by local velocity fields in the solar atmosphere. An extensive series of

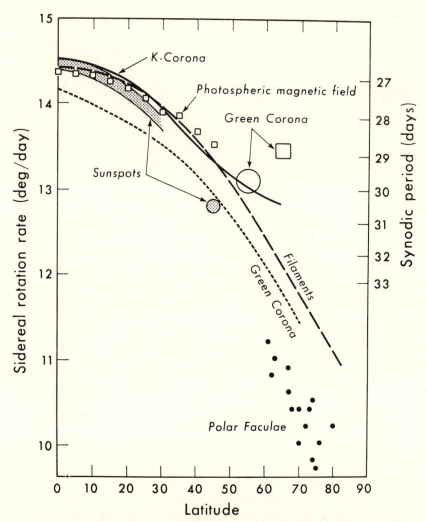

Figure 2.1. Comparison of the solar differential rotation obtained by different tracers. *All sunspots* (top of gray area: Ward, F., *Ap. J.* **145**, 416, 1966); *Long-lived sunspots* (bottom of gray area: Newton, H. W., and Nunn, M. L., *M.N.* **111**, 413, 1951); *High latitude spots at* $\phi = 45°$ (Waldmeier, M., *Z. f. Ap.* **43**, 29, 1957; Kopecký, M., Kvičala, J., and Ptáček, J., *Bull. Astron. Inst. Czech.* **8**, 106, 1957); *Polar faculae* (Müller, R., *Z. f. Ap.* **35**, 61, 1954; Waldmeier, M., *Z. f. Ap.* **38**, 37, 1955); *Filaments* (D'Azambuja, M., and D'Azambuja, L., *Ann. Obs. Meudon* **6**, No 7, 1948; Bruzek, A., *Z. f. Ap.* **51**, 75. 1961); *K-Corona at* $1.125 R_{\odot}$. (Hansen, R. T., Hansen, S. F., and Loomis, H. G., *Solar Phys.* **10**, 135, 1969); *Green corona* (dotted curve: Trellis, M., *Ann. Ap. Suppl.* No 5, 1957; circle: Waldmeier, M., *Z. f. Ap.* **27**, 24, 1950; square: Cooper, R. H., and Billings, D. E., *Z. f. Ap.* **55**, 24, 1962); *Photospheric magnetic field* (Wilcox, J. M., and Howard, R., *Solar Phys.* **13**, 251, 1970). Source: this last paper.

measurements in the line Fe I 5250.216 was made by Howard and Harvey over the period 1966–1968. Averaging over the entire period, they obtained an equatorial rotation rate of $13°76$ per day in the reversing layer. Also, they derived the following expression for the average daily angle of solar rotation:

$$\xi(\text{deg/day}) = 13°76 - 1°74 \sin^2 \phi - 2°19 \sin^4 \phi. \qquad (2)$$

This result indicates a smaller decrease of angular velocity with latitude than found by the tracers. Measurements in the chromosphere and in the lower corona were also made by Henze and Dupree in the extreme ultraviolet radiation. Spectroheliograms were taken in the line Mg X 625 and in a 3.2 Å section of the Lyman continuum at 897 Å. Figure 2.2 provides a comparison between these and other results. Clearly, the average of their extreme ultraviolet results agrees better with the spectrographic observations than with the tracer measurements. According to Simon and Noyes, however, the brightest points in the extreme ultraviolet spectroheliograms rotate approximately like the sunspots. Recent observations by Antonucci and Svalgaard also indicate that the rotation rate in the corona varies with the age of coronal features: the long-lived features rotate almost uniformly, whereas the short-lived features rotate much as sunspots do.

In summary, measurements of the position of various tracers show that *on the average* the solar surface exhibits differential rotation, with the equatorial regions rotating most rapidly. Spectrographic measurements yield results essentially similar to those derived from the tracers, but with peak-to-peak fluctuations of about 25 percent and a *mean* rotation rate significantly lower than that given by various tracers. Also, observations in the chromosphere and in the lower corona clearly show that there is little variation of the rotation rate with height above the photosphere; that is to say, there is approximate "corotation" up to two solar radii above the solar surface.

To conclude, let us also briefly discuss the various attempts to observe a mean north-south motion (i.e., a mean meridional circulation). Tuominen originally found a motion of sunspots toward the equator at low latitudes and a poleward motion at latitudes higher than $\phi \approx 16°$. In Ward's opinion, however, these results are not statistically significant at the 5 percent level; that is to say, his detailed analysis of the Greenwich sunspot data seems to indicate only the existence of a mean meridional circulation (longitudinally averaged north-south) in either direction, with a speed less than one meter per second. Similarly, the spectrographic measurements by Howard and Harvey show no evidence in the mean of photospheric meridional motions. According to Plaskett, however, in the

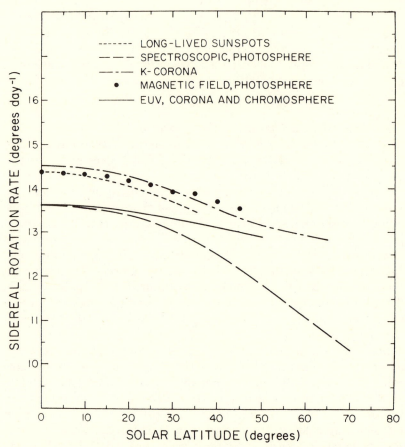

Figure 2.2. Comparison of the solar differential rotation obtained by different methods. *Long-lived sunspots* (Newton, H. W., and Nunn, M. L., *M.N.* **111**, 413, 1951); *Spectroscopic* (Howard, R., *Solar Phys.* **16**, 21, 1971); *K-Corona at 1.125*R$_\odot$ (Hansen, R. T., Hansen, S. F., and Loomis, H. G., *Solar Phys.* **10**, 135, 1969); *Magnetic field* (Wilcox, J. M., and Howard, R., *Solar Phys.* **13**, 251, 1970); *Extreme ultra-violet* (Henze, W., Jr., and Dupree, A. K., *Solar Phys.* **33**, 425, 1973). Source: this last paper.

northern and southern hemispheres of an equatorial zone of constant angular velocity, there occurs at a latitude $\phi \approx 10°$ a dominantly meridional current. (This result is based on 445 sight-line velocities measured from spectra photographed at Oxford during 1966 and 1968.) In the same work, Plaskett also reports a pole-equator difference in temperature of about 38°K. Again, this result disagrees with all other measurements, which indicate at most a temperature difference of a few degrees. We shall not go into these disputes.

Another subject of interest in the theoretical models is the possible existence of large-scale, eddy-like motions on the solar surface. For example, Ward has interpreted his statistical analysis of sunspot motions as an indication of global velocity fields in the photosphere, with spots (and, hence, angular momentum) being transported from higher latitudes toward the equator. Other evidences of such non-axisymmetric velocity fields in the photosphere have been reported by Plaskett and Howard from their Doppler velocity measurements. According to Howard, these large-scale features (with dimensions of the order of one solar radius and velocities in the range 50–75 m/sec) are observed to live for at least several days in general and to rotate approximately at the solar rotation rate. Finally, let us also mention the observations by Bumba, Howard, and Smith, which exhibit an apparent regularity in longitude of large-scale magnetic regions on the solar surface. As they first pointed out, this suggests the presence of giant regular structures, with linear dimension of about 10^5 km or larger. These structures could well be the manifestation of giant convective cells in the hydrogen convective zone, having much larger horizontal size and larger duration than granules and supergranules. These and related matters will be further discussed in Chapter 9.

2.3. SINGLE STARS

If a star is rotating about an axis, the observed radiation originating at the receding areas of the stellar disc is shifted toward the red, whereas the radiation coming from the approaching areas is shifted toward the violet. Thus, the effect of axial rotation on a stellar spectrum is to broaden *all* spectral lines, the amount of broadening depending upon the degree of axial rotation and the inclination of the rotation axis to the line of sight. This is the basic difference between rotational broadening and, say, the Stark effect, which is important only in the hydrogen and helium lines. Unfortunately, some stars—notably the supergiants—have a large amount of turbulence in their atmospheres, and these random motions also broaden the spectral lines because of the Doppler effect. However, for certain classes of stars—especially the supergiants and O-type stars— rotational broadening will give very different line widths (due to inclination effects and different rotational velocities), but turbulence may give the same broadening for every star in the class. Indeed, supergiants show moderate broadening but nearly identical widths for all stars at the same absolute magnitude (cf. §12.6).

For single stars, a study of line broadening interpreted as due to axial rotation does not provide the true equatorial velocities v_e, but their projections $v_e \sin i$ upon the line of sight (see Figure 2.3). When $i = 90°$, the

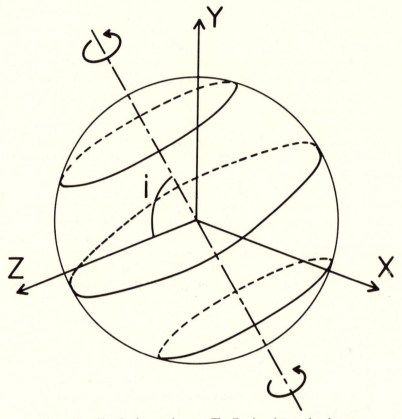

Figure 2.3.   Sketch of a rotating star. The Z-axis points to the observer.

axis of the star is at right angles to the line of sight, and we really measure v_e; on the contrary, when $i = 0°$, the axis is directed along the line of sight, and the spectral lines are sharp. If we except the eclipsing binaries and the Ap stars (in which spectrum, magnetic field, or light variations give independent information on the rotational period), the angle i is usually unknown. For a large, homogeneous group of stars, assumptions can be made about the orientation in space of the rotational axes; and a statistical procedure then provides a mean value of the true equatorial velocity for the group.

THE APPARENT ROTATIONAL VELOCITIES

A graphical method for obtaining the $v_e \sin i$'s was originally suggested by Shajn and Struve in 1929, and was first employed for single stars by

Elvey. The procedure was subsequently justified by Carroll, who derived the integral equation governing the rotational broadening of a stellar line.

Consider a rectangular frame of reference (X,Y,Z) with its origin at the center of the star. The Z-axis points to the observer, and the Y-axis lies in the plane passing through the axis of rotation and the line of sight (see Figure 2.3). To illustrate the method, we shall assume that the steady star is uniformly rotating, and that its observed disc is a circle of radius R. For convenience, we shall let $x = X/R$ and $y = Y/R$.

Let $I(x,y; \lambda - \lambda_c)$ be the intensity of the radiation originating at the point (x,y) on the disc of a nonrotating star, in a spectral line at a distance $\lambda - \lambda_c$ from the central wavelength λ_c. By virtue of the Doppler effect, if the star is rotating the wavelength λ_c must be replaced by $\lambda_c - \lambda_c v_z/c$, where v_z is the radial velocity at the point (x,y) and c is the speed of light ($v_z \ll c$). Since $v_z = -x v_e \sin i$, the intensity of the radiation originating at the point (x,y) on the disc of a uniformly rotating star thus becomes

$$I\left(x,y; \lambda - \lambda_c - \lambda_c x \frac{v_e}{c} \sin i\right). \tag{3}$$

Let now $I_0(x,y)$ be the intensity of the continuum radiation originating at the point (x,y) on the disc. Then, the ratio of the energy emitted by the star in a spectral line to the energy emitted in the continuum is

$$\overline{r}(\lambda - \lambda_c) = \frac{\int_{-1}^{+1} dx \int_0^{(1-x^2)^{1/2}} I\left(x,y; \lambda - \lambda_c - \lambda_c x \frac{v_e}{c} \sin i\right) dy}{\int_{-1}^{+1} dx \int_0^{(1-x^2)^{1/2}} I_0(x,y)\, dy}. \tag{4}$$

This expression defines, in general, the line profile in the spectrum of a rotating star.

Assume now that the chosen line profile is constant over the entire disc. We have, therefore,

$$I(x,y; \lambda - \lambda_c) = r(\lambda - \lambda_c) I_0(x,y). \tag{5}$$

By combining equations (4) and (5), we find

$$\overline{r}(\lambda - \lambda_c) = \int_{-1}^{+1} r\left(\lambda - \lambda_c - \lambda_c x \frac{v_e}{c} \sin i\right) A(x)\, dx, \tag{6}$$

where $A(x)$ is the rotational broadening function:

$$A(x) = \frac{\int_0^{(1-x^2)^{1/2}} I_0(x,y)\, dy}{\int_{-1}^{+1} dx \int_0^{(1-x^2)^{1/2}} I_0(x,y)\, dy}. \tag{7}$$

As is usual, we now assume a linear law of darkening in the continuous radiation over the stellar disc:

$$I_0 = constant \, (1 + \beta \cos \vartheta), \tag{8}$$

where β is the coefficient of limb darkening, and ϑ is the angular distance from the center of the disc. Since $\sin \vartheta = (x^2 + y^2)^{1/2}$, we obtain

$$I_0(x, y) = constant \, \{1 + \beta [1 - (x^2 + y^2)]^{1/2}\}; \tag{9}$$

and equation (7) reduces to

$$A(x) = \frac{3}{3 + 2\beta} \left[\frac{2}{\pi} (1 - x^2)^{1/2} + \frac{\beta}{2} (1 - x^2) \right]. \tag{10}$$

Clearly, the rotational broadening function can be visualized as the line profile that would be observed in a rotating star if the intrinsic line profile produced by each element on the disc of the star were extremely narrow. When $\beta = 0$, this function has the shape of a half-ellipse with a maximum width $2v_e \sin i$. If the intrinsic line width is comparable with the width of the rotationally broadened line, however, the theoretical profile must then be determined by means of equations (6) and (10).

It is now apparent from our previous results that rotational broadening may be distinguished from other broadening effects by the following characteristics: (i) rotation uniformly broadens all the lines of stellar origin in a spectrum, (ii) the equivalent line widths are independent of the amount of rotational broadening, and (iii) the amount of broadening is proportional to the wavelength, as it should be in the case of a phenomenon caused by the Doppler effect.

In actual practice, we assume a series of values for $v_e \sin i$, compute the broadened profiles from equations (6) and (10), and plot them in angstrom units on a transparent paper. By directly comparing an observed profile with a series of computed profiles of the same line, we then obtain an estimate of $v_e \sin i$ for the star. Following Slettebak, Figure 2.4 illustrates the observed contour of the line He I 4026 in ι Herculis and the rotationally broadened line contours derived from it for values of $v_e \sin i$ ranging from 234 to 560 km/sec, on the assumption of a spherical, undarkened star.

The Shajn-Struve method has been applied to determine the values of $v_e \sin i$ for many thousand stars. However, it should be remembered that most observations were reduced in assuming uniform rotation and spherical symmetry. In addition to limb darkening (cf. equation [8]), the reduction of rotationally broadened line profiles should take into account the following factors: (i) differential rotation, (ii) departure from spherical

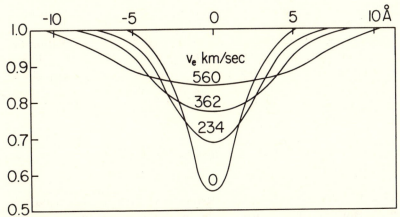

Figure 2.4.   The He I 4026 line contour observed in ɩ Herculis (0 km/sec) and three rotational contours derived from it, on the assumption of a spherical, undarkened star. Source:
Slettebak, A., *Ap. J.* **110**, 498, 1949. (By permission of The University of Chicago Press.)

symmetry, (iii) gravity darkening (i.e., the variation in surface brightness
due to the increase in effective gravity from the equator to the poles;
cf. §7.2), and (iv) variation of absorption line strength with latitude and
with angular distance from the limb. Although these and related matters
will be discussed in Section 12.3, let us mention here that Slettebak and
his associates have presented an atlas of rotationally broadened spectra
based on model-atmosphere calculations. It is found that rotational velocities derived from these calculations are smaller than those based on the
Shajn-Struve method by about 5 percent for the A- and F-type stars, and
by about 15 percent for the B-type stars. Thus, the possibility of appreciable
systematic errors in the $v_e \sin i$ values so far published should not be
excluded.

THE TRUE EQUATORIAL VELOCITIES

As we noted before, observations provide the *apparent* rotational velocities
$v_e \sin i$ only, whereas it is the *true* equatorial velocities v_e in which we are
interested. Assume now that we have a representative sample of stars
belonging to one spectral type, and that we know the probability distribution $\varphi(v_e \sin i)$ of the apparent velocities among the group. Then,
*what is the probability distribution $f(v_e)$ of the true equatorial velocities
among these stars?* The following analysis is due to Chandrasekhar and
Münch, and it is based on the assumption that the axes of rotation are
distributed at random in space. This hypothesis is amply justified, since

the observations reveal a lack of correlation between the measured v_e sin i's and the galactic coordinates of the stars.

By virtue of our assumption, the probability that the inclination i lies in the interval $(i, i + di)$ is known to be sin $i\, di$. Similarly, the probability of occurrence of v_e in the interval $(v_e, v_e + dv_e)$ is $f(v_e)\, dv_e$. Then, since v_e and i are independent quantities, the probability that they occur jointly in their respective intervals $(v_e, v_e + dv_e)$ and $(i, i + di)$ is

$$f(v_e) \sin i\, dv_e\, di. \tag{11}$$

Consider now the apparent rotational velocities

$$v = v_e \sin i, \tag{12}$$

with a probability distribution governed by the supposedly known frequency function $\varphi(v)$. Clearly, the probability that the apparent velocity v falls in the *fixed* interval $(v, v + dv)$ is given by the double integral

$$\varphi(v)\, dv = \iint\limits_{v \leqslant v_e \sin i \leqslant v + dv} f(v_e) \sin i\, dv_e\, di. \tag{13}$$

Now, for a fixed value of v_e, we have

$$di = \frac{dv}{v_e \cos i}. \tag{14}$$

Hence, equation (13) reduces to

$$\varphi(v)\, dv = dv \int\limits_{v_e \sin i = v} f(v_e)\, \frac{\sin i}{v_e \cos i}\, dv_e. \tag{15}$$

Using equation (12) to eliminate the inclination i in the integrand of equation (15), we find[1]

$$\varphi(v) = v \int_v^\infty \frac{f(v_e)}{v_e(v_e^2 - v^2)^{1/2}}\, dv_e. \tag{16}$$

This is the required integral equation that relates the probability distribution of v with that of v_e.

Equation (16) can be reduced to Abel's integral equation and solved for the unknown function $f(v_e)$. We obtain

$$f(v_e) = -\frac{2}{\pi} v_e^2 \frac{d}{dv_e} v_e \int_{v_e}^\infty \frac{\varphi(v)}{v^2(v^2 - v_e^2)^{1/2}}\, dv. \tag{17}$$

---

[1] The same integral equation occurs in the discussion of the mass function of binary stars (Kuiper, G. P., *P.A.S.P.* **47**, 15, 1935), and in the discussion of the true and apparent distributions of the stars in a globular cluster (Eddington, A. S., *M.N.* **76**, 572, 1916).

In actual practice, however, this formal solution is not very useful since it requires differentiation of an empirically determined function $\varphi(v_e \sin i)$, which is usually known in the form of a histogram. For very much the same reason, a numerical integration of equation (16) can hardly be expected to give trustworthy results. It is preferable, instead, to assume a simple analytic form for $f(v_e)$ involving one or more parameters, and to derive the corresponding distribution $\varphi(v)$ by means of equation (16). The adjustable parameters are then determined by fitting this function to the empirical distribution of the $v_e \sin i$'s for the group.

In any case, it is very often advisable to use only the moments of the observed function $\varphi(v)$, and thence derive the moments of the true frequency function $f(v_e)$. To this end, we multiply equation (16) by v^n and integrate over the whole range of v; after inverting the order of the integrations on the r.h.s., we find

$$\int_0^\infty \varphi(v)v^n \, dv = \int_0^\infty f(v_e) \frac{dv_e}{v_e} \int_0^{v_e} \frac{v^{n+1}}{(v_e{}^2 - v^2)^{1/2}} \, dv. \qquad (18)$$

Next, letting $\tau = v/v_e$ in the integral over v, we obtain

$$\int_0^\infty \varphi(v)v^n \, dv = \int_0^\infty f(v_e)v_e{}^n \, dv_e \int_0^1 \frac{\tau^{n+1}}{(1 - \tau^2)^{1/2}} \, d\tau. \qquad (19)$$

Hence, it follows that

$$\langle v^n \rangle = \frac{\sqrt{\pi}}{2} \frac{\Gamma\left(\dfrac{n}{2} + 1\right)}{\Gamma\left(\dfrac{n}{2} + \dfrac{3}{2}\right)} \langle v_e{}^n \rangle, \qquad (20)$$

where brackets indicate an average value. In particular, we can write

$$\langle v_e \rangle = \frac{4}{\pi} \langle v \rangle, \quad \langle v_e{}^2 \rangle = \frac{3}{2} \langle v^2 \rangle, \quad \langle v_e{}^3 \rangle = \frac{16}{3\pi} \langle v^3 \rangle. \qquad (21)$$

Similarly, the mean square deviation and the skewness of the true distribution $f(v_e)$ can be derived from the moments of the apparent distribution $\varphi(v)$ according to the following formulae:

$$\langle (v_e - \langle v_e \rangle)^2 \rangle = \frac{3}{2} \langle v^2 \rangle - \frac{16}{\pi^2} \langle v \rangle^2, \qquad (22)$$

and

$$\langle (v_e - \langle v_e \rangle)^3 \rangle = \frac{16}{3\pi} \langle v^3 \rangle - \frac{18}{\pi} \langle v^2 \rangle \langle v \rangle + \frac{128}{\pi^3} \langle v \rangle^3. \qquad (23)$$

THE OBSERVATIONAL MATERIAL

As far back as 1931, Struve and Elvey recognized that rapid rotation occurs almost exclusively in spectral types O, B, A, and early F, although it is quite common to find rapidly rotating stars of later types in close binaries. A large amount of observational material has since accumulated. Nevertheless, because of differences in the calibration of the $v_e \sin i$'s as well as in the manner of sampling, it is still impossible to combine all the observational material into a single statistical study. These difficulties are

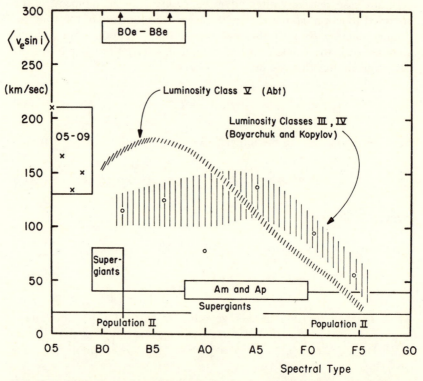

Figure 2.5. Mean observed rotational velocities for a number of different classes of stars compared with normal main-sequence stars. *Luminosity Class V* (Abt, H. A., and Hunter, J. H., Jr., *Ap. J.* **136**, 381, 1962; Slettebak, A., *Ap. J.* **110**, 498, 1949; Slettebak, A., *Ap. J.* **119**, 146, 1954; Slettebak, A., *Ap. J.* **121**, 653, 1955; Slettebak, A., and Howard, R. F., *Ap. J.* **121**, 102, 1955; Herbig, G. H., and Spalding, J. F., Jr., *Ap. J.* **121**, 118, 1955); *Luminosity Classes III and IV* (Boyarchuk, A. A., and Kopylov, I. M., *Astron. Zh.* **35**, 804, 1958); *O stars* (Slettebak, A., *Ap. J.* **124**, 173, 1956); *Be stars* (Boyarchuk, A. A., and Kopylov, I. M., *Public. Crimean Astroph. Obs.* **31**, 44, 1964; Slettebak, A., *Ap. J.* **145**, 121, 1966); *Ap- and Am-stars* (Abt, H. A., Chaffee, F. H., and Suffolk, G., *Ap. J.* **175**, 779, 1972; Abt, H. A., and Moyd, K. I., *Ap. J.* **182**, 809, 1973). Source: Slettebak, A., in *Stellar Rotation* (Slettebak, A., ed.), p. 5, New York: Gordon and Breach, 1970.

reflected in the subsequent table and figures, which are taken from different sources.

The main results pertaining to stellar rotation have been assembled by Slettebak and are summarized in Figure 2.5. In this figure, the narrow, cross-hatched band exhibits the general features of rotation in single, normal, main-sequence stars. Actually, the data have been grouped in order to smooth out irregularities in the distribution of $\langle v_e \sin i \rangle$ as a function of spectral type. Figure 2.6 illustrates the same relationship, but now the maxima and minima have been retained. (In both figures, the curve for main-sequence stars represents a *mean* relation only, and individual stars may have velocities anywhere from zero to 300–500 km/sec.) These irregularities were originally found by Boyarchuk and Kopylov. Following van den Heuvel, the maximum around the spectral type A3 is probably due to the exclusion of the Am- and Ap-stars from the $v_e \sin i$ statistics. Similarly, as van den Heuvel pointed out, the apparent underabundance of rapidly rotating stars in the B0–B4 region may be explained by the omission of the Be stars from the statistics. In any case, whether one believes in the reality of these dips or not, a definite picture emerges

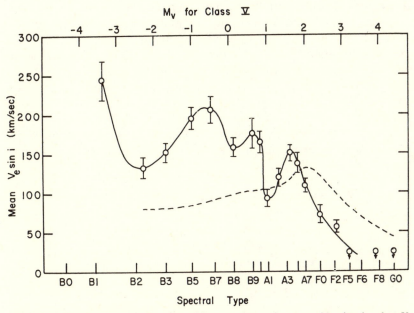

Figure 2.6. The distribution of $v_e \sin i$ with spectral type for stars of luminosity class V (*individual means with probable error bars, and the solid line*) and class III (*dotted line*), as derived by Slettebak and Howard. Source: Abt, H. A., in *Stellar Rotation* (Slettebak, A., ed.), p. 196, New York: Gordon and Breach, 1970.

from the observational material: *the rotational velocities of main-sequence stars increase from very low values in the late-type stars to some maximum in the B-type stars.* According to Slettebak, the exact behavior of early B-type and O-type stars is still uncertain, for we do not know the relative importance of rotation and macroturbulence as line-broadening factors in these early-type stars.

As has been mentioned, *no* correlation exists between the observed rotational velocity of a star and its location in the galactic coordinate system. Hence, by virtue of equation (21), the mean equatorial velocity $\langle v_e \rangle$ for a group of stars is larger than the corresponding $\langle v_e \sin i \rangle$ value by the constant factor $4/\pi$ (or 1.273). The variations of $\langle v_e \rangle$ along the main sequence, and the rather sudden drop in $\langle v_e \rangle$ about the spectral type F5 along this sequence are thus real phenomena. However, as was originally shown by McNally and Walker, *an entirely different picture emerges when one considers the mean periods of rotation rather than the mean equatorial velocities.* To illustrate this point, let us again assume that the main-sequence stars are uniformly rotating. Table 2.1 lists typical values of the masses, mean radii and equatorial velocities for these stars. The angular velocities $\Omega = v_e/R$ and the axial periods $P = 2\pi R/v_e$ are also tabulated. We observe that the periods reach a minimum value of about 0.56 days near the spectral type A5, and increase rather rapidly on both sides so that the G0- and O5-type stars have approximately the same average period of rotation. *The large observed values $\langle v_e \rangle$ for the upper main-sequence stars are thus entirely due to the large radius of these stars.* At the present time, however, this result remains purely qualitative, for we know that solid-body rotation does not adequately approximate the actual rotation law of all main-sequence stars (cf. §12.3)

Let us next briefly examine the other classes of stars depicted by Slettebak on Figure 2.5. The open circles represent mean rotational

TABLE 2.1

Average rotational velocities of main-sequence stars

Spectrum (class V)	M ($M_\odot$)	R ($R_\odot$)	v_e (km/sec)	Ω (10^{-5}/sec)	P (days)
O5	39.5	17.2	190	1.5	4.85
B0	17.0	7.6	200	3.8	1.91
B5	7.0	4.0	210	7.6	0.96
A0	3.6	2.6	190	10.0	0.73
A5	2.2	1.7	160	13.0	0.56
F0	1.75	1.3	95	10.0	0.73
F5	1.4	1.2	25	3.0	2.42
G0	1.05	1.04	12	1.6	4.55

SOURCE: McNally, D., *The Observatory* **85**, 166, 1965.

velocities for stars belonging to the luminosity classes III and IV; they are connected by a broad cross-hatched band, thus suggesting uncertainties in the mean rotational velocities for the giant stars. According to Slettebak, the very low point at A0 can probably be interpreted in terms of selection effects. In any case, the broad band indicates that the early-type giants rotate more slowly than the main-sequence stars of corresponding spectral types, while for the late A- and F-types the giants rotate more rapidly than their main-sequence counterparts. On the contrary, the supergiants of all spectral types never show conspicuous rotations. As a rule, the apparent rotational velocities of population II stars are also small. Supergiants and population II stars are shown schematically near the bottom of Figure 2.5. Observe also that the mean rotational velocities of the peculiar A-type stars and metallic-line stars are considerably smaller than the means for normal stars of corresponding spectral types. Finally, going to the other extreme, we note that the Be stars rotate most rapidly, and individual rotational velocities of 500 km/sec have been observed by Slettebak. These stars are shown separately on Figure 2.5, with arrows indicating that their mean rotational velocities are in reality larger than shown. Rotation in white dwarfs, variable stars, and magnetic stars will be discussed in Chapters 12 through 15.

To conclude this overall picture, let us now summarize the main observations of axial rotation in stars belonging to galactic clusters and associations. From the pioneering work of Struve and his collaborators it was already apparent that rapidly rotating stars are more frequent in the Pleiades than in the Hyades or in the Galaxy at large. According to a statistical study by Abt, in most clusters the brighter stars have unusually large *or* unusually small mean rotational velocities, although the fainter stars in these clusters have the same $\langle v_e \sin i \rangle$ values as the field stars. A more refined statistical analysis by Bernacca and Perinotto shows that cluster and field main-sequence stars between spectral types B5 and A8 have the same $\langle v_e \rangle$ values, whereas for spectral types earlier than B5 and later than A8, field stars rotate more slowly than their cluster's counterparts. The dependence of mean rotational velocity on luminosity or spectral type is shown in Figure 2.7 for main-sequence field stars and for main-sequence members of various clusters and associations. The Ap- and Am-stars have been excluded because they form a group of atypically slow rotators, but all spectroscopic binaries have been included in the statistics (cf. §2.4). Another way of exhibiting the differences between individual clusters and associations is shown in Figure 2.8. This diagram depicts the frequency histograms for the B5–B9.5 stars of luminosity classes V, IV, and III, excluding only the peculiar stars. As van den Heuvel pointed out, in portrayals of this kind there is a suggestion of groups of

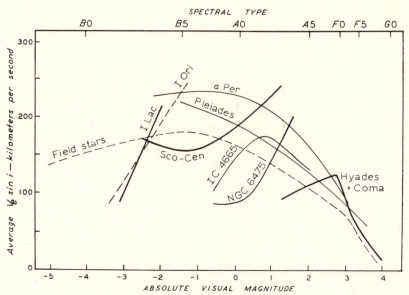

Figure 2.7. Mean observed rotational velocities for nine stellar systems compared with normal main-sequence stars. *I Lacerta* (Abt, H. A., and Hunter, J. H., Jr., *Ap. J.* **136**, 381, 1962); *I Orion* (McNamara, D. H., and Larsson, H. J., *Ap. J.* **135**, 748, 1962; McNamara, D. H., *Ap. J.* **137**, 316, 1963; Sharpless, S., *A. J.* **79**, 1073, 1974); *Sco-Cen* (Slettebak, A., *Ap. J.* **151**, 1043, 1968); α *Persei* (Kraft, R. P., *Ap. J.* **148**, 129, 1967); *Pleiades* (Anderson, C. M., Stoeckly, R., and Kraft, R. P., *Ap. J.* **143**, 299, 1966); *IC* 4665 (Abt, H. A., and Chaffee, F. H., *Ap. J.* **148**, 459, 1967); *NGC* 6475 (Abt, H. A., and Jewsbury, C. P., *Ap. J.* **156**, 983, 1969); *Hyades + Coma* (Kraft, R. P., *Ap. J.* **142**, 681, 1965). Source: Hack, M., *Sky and Telescope* **40**, 143, 1970.

slow (0–135 km/sec) and fast (180–315 km/sec) rotators in each cluster or association. Finally, in some clusters (such as the Pleiades, Sco-Cen, and NGC 2516) the fastest rotations seem to occur near the projected center of the group, although this concentration is not apparent among other clusters (such as *h* and χ Persei). Plausible causes of these differences observed between individual stellar groups will be set forth in Section 12.5.

2.4. DOUBLE STARS

As we have seen in Section 1.2, the first convincing evidence of stellar rotation came from the spectrographic study of small distortions in the velocity curves of eclipsing binaries. This effect was originally discovered by Schlesinger in 1909. Figure 2.9 illustrates the velocity curve of the spectroscopic and eclipsing binary U Cephei. Open circles represent the velocities of the bright B8 star; crosses designate the velocities of the

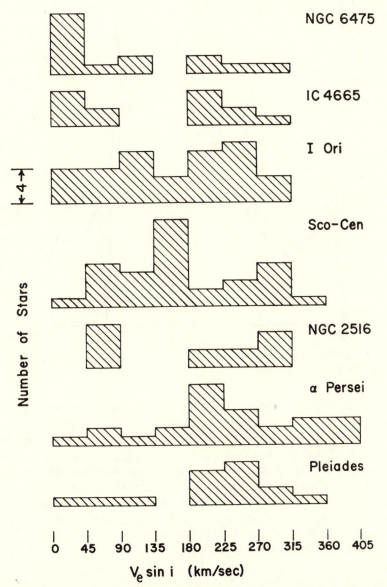

Figure 2.8. Frequencies of values of $v_e \sin i$ in seven clusters and associations. All Ap stars have been excluded. (Not illustrated: NGC 2232, IC 2602, IC 2391; see: Levato, H., *P.A.S.P.* **86**, 940, 1974.) Source: Abt, H. A., in *Stellar Rotation* (Slettebak, A., ed.), p. 195, New York: Gordon and Breach, 1970.

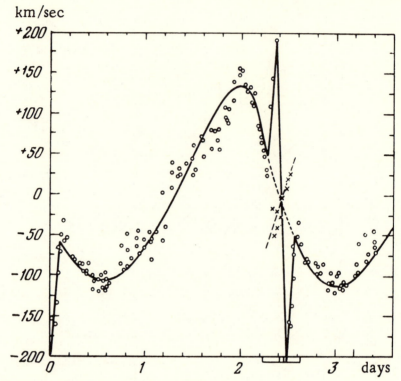

km/sec

Figure 2.9.   The radial velocity curve of U Cephei. Source: Struve, O., *Ap. J.* **99**, 222, 1944.
(By permission of The University of Chicago Press.)

subgiant G2 star, which can be photographed only during the total
eclipse. Figure 2.10 shows part of the relative orbit of the bright component
when it disappears behind its darker, though larger, companion. The
relevant portion of the velocity curve is schematically depicted at the
bottom of this figure. The arrows indicate the directions of the orbital
and axial motions. In this model, the bright star thus rotates in the same
sense in which it moves in its orbit, and the axis of rotation is normal to
the orbital plane. Now, during the early stage of the eclipse, the dark
companion will encroach upon the part of the disc that is approaching
the observer. This area is therefore covered, and we receive only the
radiation originating at the receding part of the disc. Accordingly, the
spectral lines, which before the eclipse were symmetrically broadened by
axial rotation, become asymmetrical. Since redshifts predominate, the
measured points thus fall above the predicted velocity curve. Similarly,
after the end of the total eclipse, the obscuring body covers these parts

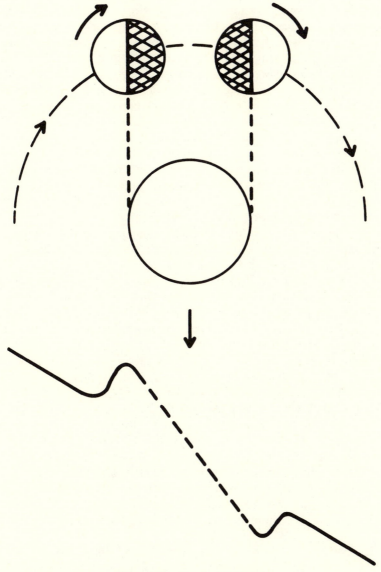

Figure 2.10.   Sketch of U Cephei.

of the disc that are receding, while the approaching portions of the disc remain uncovered. As a result of the axial rotation of the bright component, the spectral lines become again asymmetrical, and an excess of negative radial velocities is then observed.

The distortions in the velocity curve of the binary star U Cephei thus show ample evidence of the axial rotation of the primary component. The same phenomenon has been observed in many dozens of systems, although the disturbances are usually smaller in amount (such as in β Lyrae and in β Persei). In any case, they are always present when the bright component has an appreciable axial rotation, and when the eclipse is favorable for providing the necessary asymmetry in the measured lines. *In all close binaries that were investigated, the direction of axial rotation is always the same as that of orbital revolution.* This result is particularly important in regard to the formation of binaries (cf. §§11.2–11.3).

When studying double stars we are naturally led to ask the following questions: to what extent is the axial rotation of each component affected by its companion? In particular, is there any synchronization between the rotation and the revolution of the component stars?

According to Slettebak, no significant differences between the mean rotational velocities for components of visual binaries and single stars exist, either for main-sequence stars or for giant stars. A different picture emerges, however, when we look for evidence of rotational coupling in the sense of the more rapidly rotating secondary components being associated with the more rapidly rotating primary components. Using a sample of 50 binaries with both components on the main sequence and with spectral types earlier than F0, Steinitz and Pyper found a linear correlation coefficient equal to 0.46. (This value should be contrasted with the value 0.001 for pairs formed by matching to each primary a secondary at random.) Omitting binaries with components of types Am and Ap from the sample analyzed by Steinitz and Pyper, Bernacca found a correlation coefficient equal to 0.75. *It is thus apparent that some kind of coupling between the axial rotations of the components of visual binaries does in fact exist.* Finally, as was demonstrated by Weis, there is also a tendency for the rotation axes of visual binaries to be aligned perpendicular to the orbital planes in such systems. This tendency appears strong among the F-type stars, while the evidence for A-type stars is rather weak. As we shall presently see, more definite conclusions can be made for close binaries.

Figures 2.11 and 2.12 summarize the main results of a statistical study by Olson, who measured the rotational broadening of spectral lines in close binaries. The apparent rotational velocities were determined in this manner for 40 component stars in 29 early-type binaries. Most of the systems are eclipsing binaries with periods between one and six days. If we except the very few spectroscopic binaries, the orbital inclinations are therefore known. The true equatorial velocities were obtained by assuming that the rotational axes are normal to the orbital planes. Olson found no

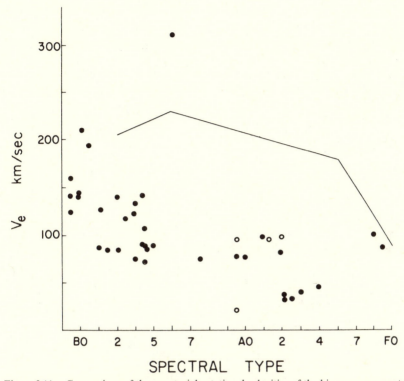

Figure 2.11. Comparison of the equatorial rotational velocities of the binary components and the single luminosity class V stars. The latter are shown by the broken line. Source: Olson, E. C., *P.A.S.P.* **80**, 185, 1968.

tendency for different spectral lines to give systematically different rotational velocities.

In Figure 2.11, the velocity v_e is plotted against spectral type, with the mean equatorial velocity for single main-sequence stars shown by the broken line. *We find a definite tendency of the components of close binaries to rotate more slowly than single stars of the same spectral type.* The influence of the other component is thus clearly perceptible in close binaries of short periods. This effect was first noticed by Kreiken in 1935.

The question of synchronism between rotation and orbital motion is a somewhat more difficult one. Figure 2.12 compares the observed equatorial velocities v_e with the computed equatorial velocities $v(\mathrm{syn})(=2\pi R/P_0)$ that would be synchronous with the orbital period P_0, the 45-degree line showing the cases of perfect synchronism. *We observe that most points do scatter along this line, indicating that synchronization obtains in most close*

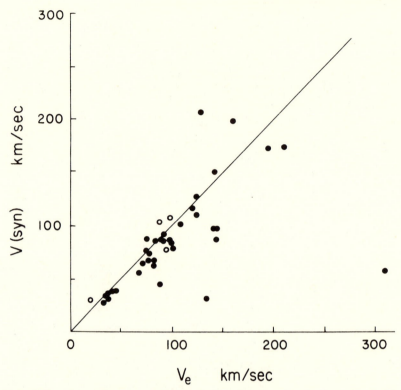

Figure 2.12. The synchronous rotational velocities plotted against the observed velocities (corrected for sin *i* by assuming that the rotational axes are normal to the orbital planes). Open circles are spectroscopic binaries, and the line is the locus of synchronous rotation. Source: Olson, E. C., *P.A.S.P.* **80**, 185, 1968.

binaries of short periods, either perfectly or approximately. Definite evidence of synchronism (or quasi-synchronism) in the majority of short-period binaries was originally found by Swings, in 1936, from a less homogeneous material.

These results nevertheless call for a few remarks. *First,* the inference of synchronism in close binaries of short periods is based on the assumption that the rotation axes are perpendicular to the orbital planes. This has never been properly demonstrated. *Second,* in systems of relatively long periods—larger than four or five days—the synchronization usually breaks down, and we observe components that rotate more rapidly than they would if the two periods were synchronized. *Third,* as was recently shown by Levato, the periods below which binary components are still rotating in synchronism are systematically larger for the more evolved stars than for the less evolved ones; that is to say, synchronization is a function of age, and it appears to occur gradually during the main-sequence lifetime, as well as during the contraction phase prior to the zero-age main sequence. *Fourth*, even among the short-period binaries, there are a few systems for which the axial and orbital periods are definitely unequal. (They are not illustrated in Figure 2.12.) A conspicuous example is U Cephei, whose orbital period is 2.5 days. In this case, if the bright primary should rotate in synchronism with its orbital motion, we would expect an equatorial velocity of about 60 km/sec, which is much less than the observed velocity of about 200–300 km/sec. Similarly, the rotational velocity of the subgiant secondary is 75 km/sec, which is below the 100 km/sec expected from synchronization. Actually, there are a few other close binaries (such as RZ Scuti) in which rotation and orbital motion are definitely out of step. Various theoretical speculations about these problems will be presented in Section 16.3.

BIBLIOGRAPHICAL NOTES

No attempt at a complete bibliography has been made in this chapter.

Section 2.2. A fairly complete survey of the problem will be found in:

1. Belvedere, G., Godoli, G., Motta, S., and Paternò, L., "Solar Rotation: Phenomenology and Superficial Models" in *Atti della XV Riunione della Societa Astronomica Italiana*, pp. S45–S70, Firenze: Baccini e Chiappi, 1972.
2. Gilman, P. A., "Solar Rotation" in *Annual Review of Astronomy and Astrophysics* **12**, pp. 47–70, Palo Alto, California: Annual Reviews, Inc., 1974.
3. Howard, R., "The Rotation of the Sun" in *Scientific American* **232**, 4 (1975): 106–114.

The discussion in this section is largely derived from:

4. Hansen, R. T., Hansen, S. F., and Loomis, H. G., *Solar Phys.* **10** (1969): 135–149.
5. Wilcox, J. M., and Howard, R., *ibid.* **13** (1970): 251–260.

The following papers are also quoted in the text:

6. Howard, R., and Harvey, J., *ibid.* **12** (1970): 23–51.
7. Simon, G. W., and Noyes, R. W., *ibid.* **26** (1972): 8–14.
8. Henze, W., Jr., and Dupree, A. K., *ibid.* **33** (1973): 425–429.
9. Antonucci, E., and Svalgaard, L., *ibid.* **34** (1974): 3–10.

Further contributions are due to:

10. Glackin, D. L., *ibid.* **36** (1974): 51–60.
11. Stenflo, J. O., *ibid.* **36** (1974): 495–515.

12. Deubner, F.-L., and Vazquez, M., *ibid.* **43** (1975): 87–90.
13. Schröter, E. H., and Wöhl, H., *ibid.* **42** (1975): 3–16.
14. Wagner, W. J., *Astroph. J. Letters* **198** (1975): L141–L144.
15. Wolff, C. L., *Solar Phys.* **41** (1975): 297–300.
16. Chistyakov, V. F., *Bull. Astron. Inst. Czechoslovakia* **27** (1976): 84–91.

The most recent papers devoted to the question of meridional circulation are:

17. Plaskett, H. H., *Monthly Notices Roy. Astron. Soc. London* **163** (1973): 183–207.
18. Ward, F., *Solar Phys.* **30** (1973): 527–537.
19. Tuominen, J., *ibid.* **34** (1974): 15–16.

See also reference 13. Evidences of eddy-like motions on the solar surface have been reported in:

20. Ward, F., *Pure and Applied Geophys.* **58** (1964): 157–186.
21. Ward, F., *Astroph. J.* **141** (1965): 534–547.

See also reference 17 (and papers quoted therein) and

22. Howard, R., *Solar Phys.* **16** (1971): 21–36.

Previous work by Bumba and his associates can be traced to:

23. Bumba, V., *ibid.* **14** (1970): 80–88.

Section 2.3. The following general references may be noted:

24. Unsöld, A., *Physik der Sternatmosphären*, Zweite Aufläge, pp. 508–518, Berlin: Springer-Verlag, 1955.
25. Huang, S. S., and Struve, O., "Stellar Rotation and Atmospheric Turbulence" in *Stellar Atmospheres* (Greenstein, J. L., ed.), pp. 321–369, Chicago: The University of Chicago Press, 1960.

Many papers covering various observational aspects of stellar rotation will be found in:

26. Slettebak, A., ed., *Stellar Rotation*, New York: Gordon and Breach, 1970.

The classical papers on the rotation broadening of stellar lines are those of:

27. Shajn, G., and Struve, O., *Monthly Notices Roy. Astron. Soc. London* **89** (1929): 222–239.
28. Carroll, J. A., *ibid.* **93** (1933): 478–507.

A system of standard stars for rotational velocity determinations will be found in:

29. Slettebak, A., Collins, G. W., Boyce, P. B., White, N. M., and Parkinson, T. D., *Astroph. J. Supplements* **29** (1975): 137–160.

A Fourier-transform technique has been also suggested by Carroll (reference 28); preliminary applications of this method will be found in references 40 and 115–117 of Chapter 12, below. Equation (16) was originally discussed by:

30. Chandrasekhar, S., and Münch, G., *Astroph. J.* **111** (1950): 142–156.

Further developments along these lines are due to:

31. Brown, A., *ibid.* **111** (1950): 366–374.
32. Böhm, K. H., *Zeit. f. Astroph.* **30** (1952): 117–133.

See also the paper by Bernacca in reference 26 (pp. 227–248). The main bulk of the observations can now be found in catalogs:

33. Boyarchuk, A. A., and Kopylov, I. M., *Public. Crimean Astroph. Observatory* **31** (1964): 44–99.
34. Bernacca, P. L., and Perinotto, M., *Contrib. Oss. Astrofis. Asiago, Univ. Padova,* No 239 (1970), No 250 (1971).
35. Uesugi, A., and Fukuda, I., *Mem. Fac. Sci. Kyoto Univ., Ser. Phys., Astroph., Geophys., Chem.,* **33** (1970): 205–250.

See also the captions for Figures 2.5 and 2.7. Statistical studies of stellar rotation are contained in:

36. Huang, S. S., *Astroph. J.* **118** (1953): 285–303.
37. Boyarchuk, A. A., and Kopylov, I. M., *Astron. Zh.* **35** (1958): 804–809.
38. Treanor, P. J., *Monthly Notices Roy. Astron. Soc. London* **121** (1960): 503–518.
39. van den Heuvel, E.P.J., *The Observatory* **85** (1965): 241–245.
40. Walker, E. N., *ibid.* **85** (1965): 162–165.
41. van den Heuvel, E.P.J., *Bull. Astron. Inst. Netherlands* **19** (1968): 309–325.

The present section is largely based on the survey papers by Slettebak and Abt in reference 26 (pp. 3–8, 193–203). See also:

42. Hack, M., "Stellar Rotation" in *Atti della XV Riunione della Societa Astronomica Italiana,* pp. S91–S103, Firenze: Baccini e Chiappi, 1972.
43. Bernacca, P. L., and Perinotto, M., *Astron. and Astroph.* **33** (1974): 443–450.

The references to McNally and Walker are to their papers:

44. McNally, D., *The Observatory* **85** (1965): 166–169.
45. Walker, N., *ibid.* **85** (1965): 245–250.

Section 2.4. A brief but interesting survey will be found in:

46. Struve, O., and Huang, S. S., "Spectroscopic Binaries" in *Handbuch der Physik* (Flügge, S., ed.), **50**, pp. 243–273, Berlin: Springer-Verlag, 1958.

Rotation in visual binaries was first discussed in:

47. Slettebak, A., *Astroph. J.* **138** (1963): 118–139.

See also the paper by Steinitz and Pyper in reference 26 (pp. 165–177). Further works are due to:

48. Bernacca, P. L., *Astroph. J.* **177** (1972): 161–175.
49. Weis, E. W., *ibid.* **190** (1974): 331–337.

Synchronism in close binaries is considered in:

50. Plaut, L., *Publ. Astron. Soc. Pacific* **71** (1959): 167–168.
51. Koch, R. H., Olson, E. C., and Yoss, K. M., *Astroph. J.* **141** (1965): 955–964.
52. Olson, E. C., *Publ. Astron. Soc. Pacific* **80** (1968): 185–191.

See also the papers by Plavec and van den Heuvel in reference 26 (pp. 133–146, 178–186). Further investigations of the same subject are due to:

53. Abt, H. A., and Hudson, K. I., *Astroph. J.* **163** (1971): 333–336.
54. Nariai, K., *Publ. Astron. Soc. Japan* **23** (1971): 529–538.
55. Levato, H., *Astron. and Astroph.* **35** (1974): 259–265.
56. Levato, H., *Astroph. J.* **203** (1976): 680–688.

The following papers are also quoted in the text:

57. Kreiken, E. A., *Zeit. f. Astroph.* **10** (1935): 199–208.
58. Swings, P., *ibid.* **12** (1936): 40–46.

3

Stellar Hydrodynamics

3.1. INTRODUCTION

As we may infer from the observations, most stars remain in a state of hydrostatic equilibrium under the action of their own gravitation and centrifugal force of axial rotation. However, detailed study of the Sun has demonstrated that such a balance of forces is only approximate. Indeed, the solar surface shows traces of internal motions, both around the axis of rotation and in the meridional planes (cf. §2.2). Departure from strict equilibrium is also apparent during some phases of stellar evolution; and, in this respect, we may mention the pulsating stars, the novae and supernovae, and the flare stars. All these problems are the domain of *stellar hydrodynamics*, i.e., ordinary hydrodynamics suitably enlarged to include all the atomic and nuclear processes that take place within a star.

Broadly speaking, stellar hydrodynamics proceeds on the hypothesis that a star is practically continuous in structure. That is to say, the properties of *any* small portion of the star are the same as those of the matter in bulk, on a macroscopic scale that is large compared with the distances between the constitutive particles. This hypothesis implies that it is possible to attach a meaning to average values (such as density, velocity, pressure) at each point of the star, and that in general these mean values are continuous functions of position in time. In practice, to define an average quantity we must consider domains that are small compared with the total volume of the star, but much larger than the mean free path of the particles; for then the random motions of these particles have no effect on mean values. As an example, whenever we speak of the velocity of a "mass element" (or "fluid particle") we always mean the average velocity of a large number of particles contained within a volume of finite extent, although this volume must be regarded as a point.

This chapter will derive the general equations of stellar hydrodynamics on the basis of this continuum model. Discontinuous phenomena (such as shock waves) will not be discussed here. Sections 3.5–3.6 also present some general properties that are relevant to the study of stellar rotation. The virial equations are discussed *in fine*. Convection and turbulence have been omitted; they will be treated in subsequent chapters. The reader will need basic knowledge of the internal structure of *spherical* stellar models, and some experience in vector analysis and partial differential equations will prove helpful.[1]

3.2. EULERIAN AND LAGRANGIAN VARIABLES

The mathematical description of stellar interiors from the continuum point of view allows two distinct methods of approach. We may either specify the state of the configuration at all points of space, at all instants, or we may describe the history of every mass element of the star. To outline the methods, let us introduce the Cartesian coordinates $\mathbf{x} = (x_1, x_2, x_3)$ in an inertial frame of reference. Cylindrical and spherical coordinates are discussed in Appendix B; their use will be dictated by the symmetry of the problem.

The first approach, usually called the Eulerian method, provides a complete description of a star with the spatial coordinates $\mathbf{x}$ and the time t regarded as independent variables. In this approach, the coordinates $\mathbf{x}$ refer to a fixed location in space, and not to some moving mass element of the star. From a purely kinematical point of view, the state of motion is described by the velocity field

$$\mathbf{v} = \mathbf{v}(\mathbf{x}, t); \tag{1}$$

we thus introduce *ab initio* the velocity of the fluid particle that happens to be at the location $\mathbf{x}$ at the instant t. The state of the system is then completely determined by the additional knowledge of the density ρ, the pressure p, the temperature T, and the gravitational potential V; and all these quantities must be regarded as functions of $\mathbf{x}$ and t.

The second method, the Lagrangian description, consists of labeling each fluid particle by its original position $\mathbf{X} = (X_1, X_2, X_3)$ at some arbitrarily chosen initial instant $t = 0$ (say). To be specific, consider a typical mass element that, at time $t = 0$, occupies the location $\mathbf{X}$; and suppose it has moved, at the subsequent instant t, to the new location $\mathbf{x}$. The vector $\mathbf{x}$ is thus determined as a function of the independent variables $\mathbf{X}$ and t; and the flow may be represented by an equation of the form

$$\mathbf{x} = \mathbf{x}(\mathbf{X}, t). \tag{2}$$

For a fixed value of the vector $\mathbf{X}$, equation (2) specifies the path of a mass element initially at the location $\mathbf{x} = \mathbf{X}$; similarly, at a given instant t, equation (2) determines the transformation of the region initially occupied by the whole mass of the star into its actual position at time t. By

---

[1] See, e.g., Rektorys, K., ed., *Survey of Applicable Mathematics*, Cambridge, Mass.: The M.I.T. Press, 1969.

virtue of its definition, the velocity of a fluid parcel is

$$\mathbf{v}(\mathbf{X},t) = \frac{\partial \mathbf{x}}{\partial t}, \tag{3}$$

where the partial derivative indicates that the differentiation must be carried out for a given mass element, i.e., holding $\mathbf{X}$ constant. And the functions ρ, p, T, and V must be viewed as depending on the variables $\mathbf{X}$ and t.

The coordinates $\mathbf{x}$ used in equation (1) are called Eulerian (or spatial) variables; in contrast, the coordinates $\mathbf{X}$ that label individual mass elements are called Lagrangian (or material) variables.[2] By virtue of equation (2), any scalar, vector, or tensor quantity Q describing the motion and which is a function of the spatial variables $(\mathbf{x},t)$ is also a function of the material variables $(\mathbf{X},t)$, and vice versa. However, $Q(\mathbf{x},t)$ denotes the value of Q felt by a fluid particle, instantaneously, at the location $\mathbf{x}$; while $Q(\mathbf{X},t)$ is the value of Q experienced, at time t, by a mass element initially at the location $\mathbf{X}$.

Throughout this book, *when using spatial variables*, we shall let

$$\frac{\partial Q}{\partial t} = \frac{\partial Q}{\partial t}\bigg]_{\mathbf{x}\ constant} \tag{4}$$

and

$$\frac{DQ}{Dt} = \frac{\partial Q}{\partial t}\bigg]_{\mathbf{X}\ constant} \tag{5}$$

which, physically, are quite different quantities.[3] The *spatial* derivative (4) defines the rate of change of Q apparent to an observer located at the fixed point $\mathbf{x}$; equation (5) defines the *material* derivative of Q, for it measures the rate of change of Q as we follow a mass element along its path. Using equations (2)–(5), we can write

$$\frac{DQ}{Dt} = \frac{\partial Q}{\partial t} + (\mathbf{v}.\mathrm{grad})Q. \tag{6}$$

Equation (6) merely expresses the time rate of change of an arbitrary function $Q(\mathbf{x},t)$ to an observer traveling with the fluid particle, located at the point $\mathbf{x}$, at some instant t. Part of this change is due to any local variation at the fixed point $\mathbf{x}$, and part is due to the mass element moving into a different position where Q has another value.

---

[2] Both representations were actually introduced by Euler in the 1750s. See: Truesdell, C., in *Euleri Opera Omnia*, II, **12**, cxix–cxxv, Lausanne, 1954.

[3] Hereafter, the symbol d/dt will always designate the derivative of a function that depends on time only.

The difference between Eulerian and Lagrangian variables is particularly apparent when we consider the acceleration $\mathbf{a}$ of a fluid particle. In spatial variables, we have

$$\mathbf{a}(\mathbf{x},t) = \frac{D\mathbf{v}}{Dt} = \frac{\partial \mathbf{v}}{\partial t} + (\mathbf{v}.\mathrm{grad})\mathbf{v}, \tag{7}$$

where $\mathbf{v}$ depends on $\mathbf{x}$ and t. By virtue of equation (3), the same vector becomes

$$\mathbf{a}(\mathbf{X},t) = \frac{\partial \mathbf{v}}{\partial t} = \frac{\partial^2 \mathbf{x}}{\partial t^2}, \tag{8}$$

when $\mathbf{v}$ is expressed in terms of the material variables $\mathbf{X}$ and t.

THE REYNOLDS THEOREM

In the sequel we shall make use of one further result. Let $v = v(t)$ denote an arbitrary *material* volume that is moving within a star, i.e., a volume that is at all instants composed of the same constitutive particles, and consider the time rate of change

$$\frac{d}{dt} \int_v Q(\mathbf{x},t)\, d\mathbf{x} \tag{9}$$

(cf. note 3). If we make use of the continuous transformation (2), the moving volume $v(t)$ in the spatial variables $\mathbf{x}$ may be replaced by the fixed volume $v_0 = v(0)$ in the material variables $\mathbf{X}$. Hence, we can write

$$\int_v Q(\mathbf{x},t)\, d\mathbf{x} = \int_{v_0} Q(\mathbf{X},t)J\, d\mathbf{X}, \tag{10}$$

where

$$J(\mathbf{X},t) = \frac{\partial(x_1,x_2,x_3)}{\partial(X_1,X_2,X_3)} \tag{11}$$

denotes the Jacobian of the transformation that relates the element of volume $d\mathbf{x}$ to the corresponding element $d\mathbf{X}$ in the material variables. Clearly, we now have

$$\frac{d}{dt} \int_{v_0} QJ\, d\mathbf{X} = \int_{v_0} \left(\frac{\partial Q}{\partial t} J + Q \frac{\partial J}{\partial t} \right) d\mathbf{X}. \tag{12}$$

On the other hand, by making use of equations (2), (3), and (11), we obtain at once the identity

$$\frac{\partial J}{\partial t} = J \,\mathrm{div}\, \mathbf{v}. \tag{13}$$

If we next combine equations (12) and (13), it follows that

$$\frac{d}{dt} \int_{v_0} QJ \, d\mathbf{X} = \int_{v_0} J\left(\frac{\partial Q}{\partial t} + Q \operatorname{div} \mathbf{v}\right) d\mathbf{X}. \tag{14}$$

From equations (5), (10), and (14), we obtain

$$\frac{d}{dt} \int_{v} Q \, d\mathbf{x} = \int_{v} \left(\frac{DQ}{Dt} + Q \operatorname{div} \mathbf{v}\right) d\mathbf{x}. \tag{15}$$

By virtue of equation (6), we may also recast this last equation in the alternate form

$$\frac{d}{dt} \int_{v} Q \, d\mathbf{x} = \int_{v} \left[\frac{\partial Q}{\partial t} + \operatorname{div}(Q\mathbf{v})\right] d\mathbf{x}. \tag{16}$$

Finally, in applying Gauss's theorem, we can write

$$\frac{d}{dt} \int_{v} Q \, d\mathbf{x} = \frac{\partial}{\partial t} \int_{v_t} Q \, d\mathbf{x} + \oint_{s} Q\mathbf{v}.\mathbf{n} \, dS, \tag{17}$$

where $\mathbf{n}$ is the outer normal to the surface $s = s(t)$ bounding the material volume. From equation (17), we observe that the rate of change (9) is the sum of two distinct contributions: (i) the rate of change over a fixed volume v_t coinciding, at the instant t, with the material volume $v(t)$; and (ii) the flux of the quantity Q out of the bounding surface $s(t)$. Note that equations (15)–(17) express a purely kinematical property that is quite independent of any physical law.

3.3. DYNAMICAL PRINCIPLES

Within the framework of classical mechanics, the general equations of stellar hydrodynamics embody three fundamental principles: (i) the conservation of mass, (ii) the balance of linear momentum, and (iii) the conservation of energy. Let us formulate these principles in succession.

CONSERVATION OF MASS

Consider an arbitrary material volume $v = v(t)$. We now postulate the following principle: *the mass enclosed within any material volume $v(t)$ cannot change as we follow the volume in its motion.* In other words:

$$\frac{d}{dt} \int_{v} \rho(\mathbf{x},t) \, d\mathbf{x} = 0. \tag{18}$$

From equation (16), it follows that

$$\int_v \left[\frac{\partial \rho}{\partial t} + \text{div}(\rho \mathbf{v}) \right] d\mathbf{x} = 0. \tag{19}$$

Since this equation must hold for any material volume, the integrand must vanish. Thus, we must have, at every point,

$$\frac{\partial \rho}{\partial t} + \text{div}(\rho \mathbf{v}) = 0, \tag{20}$$

or, in view of equation (6),

$$\frac{D\rho}{Dt} + \rho \, \text{div} \, \mathbf{v} = 0. \tag{21}$$

This is the spatial form of the equation of continuity, and it is originally due to Euler.

To obtain the corresponding equation in Lagrangian variables, let us multiply equation (21) by $J(\mathbf{X},t)$. By virtue of equation (13), we find

$$\frac{\partial}{\partial t}(\rho J) = 0, \qquad \text{or} \qquad \rho J = \rho_0, \tag{22}$$

where $\rho_0 = \rho(\mathbf{X},0)$ is the initial density distribution.

As a simple application of the conservation of mass, let us recast the transport theorem (16) into a more convenient form. In replacing Q by ρQ in this equation, and in making use of equation (20), we obtain at once

$$\frac{d}{dt} \int_v \rho Q \, d\mathbf{x} = \int_v \rho \frac{DQ}{Dt} \, d\mathbf{x}. \tag{23}$$

BALANCE OF LINEAR MOMENTUM

The fundamental principle of the dynamics of stellar interiors can be stated as follows: *the rate of change of linear momentum of a material volume $v(t)$ equals the resultant force acting on the volume*. In the present case, only two kinds of forces need to be reckoned with:

(i) the *body* forces which are proportional to the mass enclosed within the volume $v(t)$. In the absence of magnetic field (cf. §3.6), the only body force per unit mass $\mathbf{f}(\mathbf{x},t)$ is due to the gravitational attraction of the star. We then have

$$\mathbf{f} = -\text{grad} \, V, \tag{24}$$

and the gravitational potential $V(\mathbf{x},t)$ is given by

$$V(\mathbf{x},t) = -G \int_{\mathscr{V}} \frac{\rho(\mathbf{x}',t)}{|\mathbf{x} - \mathbf{x}'|} \, d\mathbf{x}', \qquad (25)$$

where G is the constant of gravitation. The volume integral must be performed over the whole domain $\mathscr{V}$ occupied by the star.

(ii) the *surface* forces that are proportional to the amount of surface acted upon (such as pressure and viscous stresses). Let $\mathbf{t}$ be the surface vector force per unit area acting on the boundary $s(t)$ of the material volume $v(t)$. By hypothesis, the value of the stress vector $\mathbf{t}$ depends on position and time, as well as on the orientation of the surface element dS, i.e.,

$$\mathbf{t} = \mathbf{t}(\mathbf{x},t; \mathbf{n}). \qquad (26)$$

Therefore, if we make use of equations (24) and (26), the balance of linear momentum may be expressed by the following statement:

$$\frac{d}{dt} \int_{v} \rho \mathbf{v} \, d\mathbf{x} = \int_{v} \rho \mathbf{f} \, d\mathbf{x} + \oint_{s} \mathbf{t} \, dS. \qquad (27)$$

By virtue of the transport theorem (23), this equation can also be written in the form

$$\int_{v} \rho \frac{D\mathbf{v}}{Dt} \, d\mathbf{x} = \int_{v} \rho \mathbf{f} \, d\mathbf{x} + \oint_{s} \mathbf{t} \, dS. \qquad (28)$$

Without loss of generality, in equation (28) integration over the moving volume $v(t)$ can be now replaced by integration over a fixed volume.

An immediate result can be obtained from equation (28). Let L^3 be the volume of v. If we now divide both sides of this equation by L^2 and let v tend toward zero, we obtain

$$\lim_{v \to 0} \frac{1}{L^2} \oint_{s} \mathbf{t} \, dS = 0. \qquad (29)$$

Thus, at each point of the fluid, the stress forces are in equilibrium. Consider next the tetrahedron OABC illustrated in Figure 3.1. Let the surface ABC (with outer normal $\mathbf{n}$) have a surface area Σ. The outer normals to the other surfaces are $-\mathbf{i}_1$, $-\mathbf{i}_2$, and $-\mathbf{i}_3$; and the areas are $n_1\Sigma$, $n_2\Sigma$, and $n_3\Sigma$, respectively. Since Σ is of order L^2, equation (29) may be applied to the tetrahedron to give

$$\mathbf{t}(\mathbf{n}) + n_1\mathbf{t}(-\mathbf{i}_1) + n_2\mathbf{t}(-\mathbf{i}_2) + n_3\mathbf{t}(-\mathbf{i}_3) = 0, \qquad (30)$$

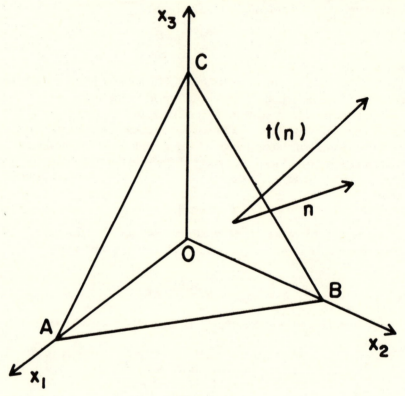

Figure 3.1.

where we have abbreviated $t(n) = t(x,t; n)$. Now, it also follows from the principle of action and reaction that

$$t(-n) = -t(n). \tag{31}$$

Hence, we have

$$t(n) = n_1 t(i_1) + n_2 t(i_2) + n_3 t(i_3). \tag{32}$$

The stress vector t may therefore be expressed in terms of the three stress vectors acting on the coordinates planes. Accordingly, we readily verify that

$$t_i = n_k T_{ki}(x,t) \qquad (i = 1, 2, 3) \tag{33}$$

or, in vectorial notation,

$$t = n.T(x,t). \tag{34}$$

(*Hereafter, a summation over repeated Roman indices is always to be understood.*) The nine scalars T_{ki} given here represent the nine components of the stress tensor **T**. The first index indicates the plane upon which the stress acts, and the second the direction along which the stress acts.

If we now combine equations (28) and (34), and apply Gauss's theorem, we obtain

$$\int_v \rho \frac{D\mathbf{v}}{Dt} \, d\mathbf{x} = \int_v (\rho\mathbf{f} + \text{div } \mathbf{T}) \, d\mathbf{x}. \tag{35}$$

Since the material volume $v(t)$ is arbitrary, it follows at once that

$$\frac{D\mathbf{v}}{Dt} = \mathbf{f} + \frac{1}{\rho} \text{div } \mathbf{T}. \tag{36}$$

This vectorial equation was originally derived by Cauchy. To proceed any further a suitable expression for the stress tensor **T** must be found, taking into account both the thermal motion of the constitutive particles and the radiation field.

THE IDEAL FLUID MODEL.   For most problems in stellar hydrodynamics, the force exerted by the surrounding fluid on a mass element is normal to the surface area dS on which it acts. We then *choose* the components of the stress vector in the form

$$t_i = -n_i p(\mathbf{x}, t). \tag{37}$$

This equation defines the *total* pressure that in the present context is the sum of the thermal pressure p_g and the radiation pressure p_r (cf., e.g., equation [71]). The minus sign in equation (37) results from our convention that the stress vector **t** represents the force exerted *by* the surroundings *on* the surface area. Obviously, the vector **t** is normal to the element dS, and its magnitude is independent of the orientation of this surface area. From equations (33) and (37), the stress tensor is diagonal:

$$T_{ik} = -p \, \delta_{ik} \tag{38}$$

($\delta_{ik} = 1$ if $i = k$, $\delta_{ik} = 0$ if $i \neq k$). Hence, by virtue of equations (24) and (38), the equations of motion (36) can now be brought to the very simple form

$$\frac{D\mathbf{v}}{Dt} = -\text{grad } V - \frac{1}{\rho} \text{grad } p, \tag{39}$$

which is originally due to Euler.

For further use, let us also derive the Lagrangian form of the equations of motion. From equations (8) and (39), it follows that

$$\frac{\partial^2 x_i}{\partial t^2} = -\frac{\partial V}{\partial x_i} - \frac{1}{\rho} \frac{\partial p}{\partial x_i}. \tag{40}$$

In multiplying next equation (40) by $\partial x_i / \partial X_k$, we obtain

$$\frac{\partial x_i}{\partial X_k} \frac{\partial^2 x_i}{\partial t^2} = -\frac{\partial V}{\partial X_k} - \frac{1}{\rho} \frac{\partial p}{\partial X_k}. \tag{41}$$

These are the required equations in the material variables X and t.

THE VISCOUS FLUID MODEL.　In differentially rotating stars, viscous friction is not always entirely negligible; that is to say, the stress vector t is then inclined over the normal n to the surface upon which it acts.

　Let us first neglect the radiation field, and write

$$T_{ik} = -p\,\delta_{ik} + \tau_{ik}, \tag{42}$$

where the τ_{ik}'s are the components of the viscous stress tensor τ (compare with equation [38]). A *plausible* form for the tensor τ can be obtained as follows. Consider the deformation tensor D:

$$D_{ik} = \frac{1}{2}\left(\frac{\partial v_i}{\partial x_k} + \frac{\partial v_k}{\partial x_i}\right). \tag{43}$$

Clearly, this symmetric tensor is a measure of the relative motion between the various parts of the fluid, for it vanishes when the material is locally at rest ($v \equiv 0$) or in a state of rigid rotation with constant angular velocity $\Omega(v = \Omega \times x)$. Following Stokes, we next assume that the τ_{ik}'s are linear and homogeneous functions of the D_{ik}'s. Physically, such a causal relationship is consistent with the idea that internal friction occurs only when adjacent mass elements move with different velocities. Further, if there is no preferred direction in space, it can be shown that the relation between τ and D involves *two* arbitrary coefficients only. Using standard notation, we eventually obtain

$$\tau_{ik} = 2\mu(D_{ik} - \tfrac{1}{3}\delta_{ik}\,D_{ss}) + \mu_{9}\,\delta_{ik}\,D_{ss}, \tag{44}$$

where the coefficients of *shear* viscosity μ and *bulk* viscosity μ_9 both depend on local thermodynamic properties only.[4] It is sometimes convenient to let $\mu = \rho v$ which then defines the coefficient of *kinematical* viscosity.

　The interaction between matter and radiation is somewhat more involved. Following Thomas, Hazlehurst, and Sargent, for all situations usually encountered in stellar interiors this coupling can be described by a radiative stress tensor that is similar in structure to the material

---

[4] No exact relation exists that connects the transport coefficients μ and μ_9 to the usual thermodynamic variables. Their values must be derived from a statistical study of the irreversible transfer of momentum between the constitutive particles, when a macroscopic velocity gradient prevails.

tensor **T**. Actually, the radiative stresses may be separated into a radiation pressure p_r and tangential stresses with a coefficient of *radiative* viscosity μ_r defined by

$$\mu_r = \frac{4aT^4}{15c\kappa\rho},\tag{45}$$

where a, c, and κ are, respectively, the radiation pressure constant, the speed of light, and the opacity coefficient (which is itself a known function of ρ, T, and the chemical composition of the star).

Inserting next the foregoing results into equation (36), we obtain the Navier-Stokes equations

$$\frac{D\mathbf{v}}{Dt} = -\operatorname{grad} V - \frac{1}{\rho}\operatorname{grad} p + \frac{1}{\rho}\operatorname{div}\boldsymbol{\tau},\tag{46}$$

where the *total* tensor $\boldsymbol{\tau}$ has the components[5]

$$\tau_{ik} = (\mu + \mu_r)\left(\frac{\partial v_i}{\partial x_k} + \frac{\partial v_k}{\partial x_i} - \frac{2}{3}\delta_{ik}\frac{\partial v_s}{\partial x_s}\right) + (\mu_\vartheta + \tfrac{5}{3}\mu_r)\,\delta_{ik}\frac{\partial v_s}{\partial x_s}.\tag{47}$$

In the present case, since the presence of a velocity gradient implies a departure from local thermodynamic equilibrium, we must specify what is meant by the symbol p occurring in equations (42) and (46). Thereafter, we shall take p to be the same *total* pressure as would hold in local thermodynamic equilibrium, i.e., $p = p_g + p_r$ as given by an equation of state (cf., e.g., equation [71]). This pressure is then determined by a knowledge of ρ, T, and the chemical composition; but it will no longer be the total pressure in the usual sense. Indeed, from equation (47), the average of the forces per unit area exerted upon the three coordinates planes is $\bar{p} = p - (\mu_\vartheta + \tfrac{5}{3}\mu_r)\operatorname{div}\mathbf{v}$. Note that the difference between the total pressure p and the mean pressure $\bar{p}$ is proportional to $\operatorname{div}\mathbf{v}$, i.e., the rate of volume change of an infinitesimal fluid particle.

BOUNDARY CONDITIONS. Whether we choose to use equations (39) or (46), appropriate boundary conditions must be imposed on the free surface of the star. *First*, the gravitational attraction must be continuous across the surface $\mathscr{S}$. By virtue of equation (25), this condition is always satisfied.

---

[5] As was properly shown by Hazlehurst and Sargent, however, the components of the tensor τ also include the following terms:

$$-\frac{1}{c^2}(v_i\mathscr{F}_k + v_k\mathscr{F}_i + \delta_{ik}v_s\mathscr{F}_s),$$

where the $\mathscr{F}_k$'s are the components of the radiative flux vector $\mathscr{F}_r$ defined in equation (67). These terms are usually neglected; they will be discussed further in Section 7.5.

Second, the stress vector acting on the stellar surface must vanish; hence, we require

$$n_k(-p\,\delta_{ik} + \tau_{ik}) = 0, \qquad \text{on } \mathscr{S}, \tag{48}$$

where the n_k's are the components of the outer normal to the surface $\mathscr{S}$. For an ideal fluid, these *three* conditions reduce to the *single* condition

$$p = 0, \qquad \text{on } \mathscr{S}. \tag{49}$$

This reduction in the number of boundary conditions derives from the fact that the second-order derivatives occurring in equation (46) are not present in the ideal fluid model.

ROTATING FRAME OF REFERENCE.   In some applications, it is convenient to describe the motions as they appear to an observer at rest in a frame rotating with the constant angular velocity $\mathbf{\Omega}$. We can write

$$\mathbf{v}(\mathbf{x},t) = \mathbf{u}(\mathbf{x},t) + \mathbf{\Omega} \times \mathbf{x}, \tag{50}$$

where the velocity $\mathbf{u}$ refers to the moving axes. Similarly, the material acceleration (7) has the form

$$\mathbf{a}(\mathbf{x},t) = \frac{\partial \mathbf{u}}{\partial t} + (\mathbf{u}.\mathrm{grad})\mathbf{u} + 2\mathbf{\Omega} \times \mathbf{u} - \mathrm{grad}(\tfrac{1}{2}|\mathbf{\Omega} \times \mathbf{x}|^2). \tag{51}$$

where $2\mathbf{\Omega} \times \mathbf{u}$ and $\tfrac{1}{2}|\mathbf{\Omega} \times \mathbf{x}|^2$ represent, respectively, the Coriolis acceleration and the centrifugal potential. Since the tensor τ is invariant to a uniform rotation, the Navier-Stokes equations (46) then become

$$\frac{D\mathbf{u}}{Dt} + 2\mathbf{\Omega} \times \mathbf{u} = -\mathrm{grad}(V - \tfrac{1}{2}|\mathbf{\Omega} \times \mathbf{x}|^2) - \frac{1}{\rho}\,\mathrm{grad}\,p + \frac{1}{\rho}\,\mathrm{div}\,\tau, \tag{52}$$

where the material derivative $D\mathbf{u}/Dt$ is now measured in the rotating frame of reference.

3.4. CONSERVATION OF ENERGY

Equations (20), (25), and (46), together with an appropriate equation of state, provide *six* relations among the *seven* unknown quantities p, ρ, T, V, and $\mathbf{v}$. It is thus necessary to augment these equations with an equation based on the principles of thermodynamics and by the use of certain constitutive relations. Diffusion processes will be neglected altogether. Ohmic dissipation is discussed in Section 3.6.

Let U denote the total internal energy per unit mass. We have

$$U = U_g + \frac{1}{\rho}\,aT^4, \tag{53}$$

where U_g denotes the thermal energy (plus any ionization or excitation energy) associated with the particles; and aT^4 is the energy density of black body radiation. The total energy content of a material volume $v(t)$ is, therefore,

$$\int_v \rho(\tfrac{1}{2}|\mathbf{v}|^2 + U + E_s)\, d\mathbf{x}, \tag{54}$$

where E_s designates the subatomic energy associated with the proper mass of the particles.

Conservation of energy can now be stated as follows: *the rate of change of the total energy of a material volume $v(t)$ is equal to the rate at which work is being done on the volume plus the rate at which heat and radiation is conducted into the volume.* This principle is expressed by the statement

$$\frac{d}{dt}\int_v \rho(\tfrac{1}{2}|\mathbf{v}|^2 + U + E_s)\, d\mathbf{x}$$

$$= \int_v \rho \mathbf{f}.\mathbf{v}\, d\mathbf{x} + \oint_s \mathbf{t}.\mathbf{v}\, dS - \oint_s (\mathscr{F} + \mathscr{F}_r).\mathbf{n}\, dS, \tag{55}$$

where $\mathscr{F}$ and $\mathscr{F}_r$ are, respectively, the heat and radiative flux vectors (cf. equations [66] and [67]). The minus sign results from the fact that if these vectors are directed inwards, the quantity $(\mathscr{F} + \mathscr{F}_r).\mathbf{n}$ will be negative. Now, since the volume $v(t)$ is arbitrary, it follows from equations (23), (34), and (55) that

$$\frac{D}{Dt}(\tfrac{1}{2}|\mathbf{v}^2| + U + E_s) = \mathbf{f}.\mathbf{v} + \frac{1}{\rho}\operatorname{div}(\mathbf{T}.\mathbf{v}) - \frac{1}{\rho}\operatorname{div}(\mathscr{F} + \mathscr{F}_r). \tag{56}$$

This is the total energy equation.

The foregoing conservation law is not very convenient. For this reason, we multiply equation (36) by $\rho\mathbf{v}$ and integrate over the volume $v(t)$. By making use of equations (23) and (34), we can then write

$$\frac{d}{dt}\int_v \rho(\tfrac{1}{2}|\mathbf{v}|^2)\, d\mathbf{x} = \int_v \rho \mathbf{f}.\mathbf{v}\, d\mathbf{x} + \oint_s \mathbf{t}.\mathbf{v}\, dS - \int_v \mathbf{T}:\mathbf{D}\, d\mathbf{x}, \tag{57}$$

where $\mathbf{T}:\mathbf{D}$ denotes the product $T_{ik}D_{ik}$. Combining next equations (55) and (57), we obtain

$$\frac{d}{dt}\int_v \rho(U + E_s)\, d\mathbf{x} = \int_v \mathbf{T}:\mathbf{D}\, d\mathbf{x} - \oint_s (\mathscr{F} + \mathscr{F}_r).\mathbf{n}\, dS. \tag{58}$$

Thus, at every point, we must have

$$\rho\frac{DU}{Dt} = \mathbf{T}:\mathbf{D} + \rho\epsilon_{Nuc} - \operatorname{div}(\mathscr{F} + \mathscr{F}_r), \tag{59}$$

where we have introduced the rate of energy released by the thermonuclear reactions

$$\epsilon_{Nuc} = -\frac{DE_s}{Dt}. \tag{60}$$

If we now make use of equations (42), (43), and (47), it is a simple exercise of algebra to prove that

$$\mathbf{T:D} = \Phi_v - p \operatorname{div} \mathbf{v}, \tag{61}$$

where

$$\Phi_v = \tfrac{1}{2}(\mu + \mu_r)\left(\frac{\partial v_i}{\partial x_k} + \frac{\partial v_k}{\partial x_i} - \frac{2}{3}\delta_{ik}\frac{\partial v_s}{\partial x_s}\right)^2 + (\mu_\vartheta + \tfrac{5}{3}\mu_r)\left(\frac{\partial v_s}{\partial x_s}\right)^2. \tag{62}$$

Substituting the foregoing relations into equation (59), we obtain

$$\rho\frac{DU}{Dt} + p \operatorname{div} \mathbf{v} = \Phi_v + \rho\epsilon_{Nuc} - \operatorname{div}(\mathscr{F} + \mathscr{F}_r). \tag{63}$$

This equation corresponds to the first law of thermodynamics (i.e., *the conservation of thermal energy*), and it will be used instead of the total energy equation.

Now, in assuming quasi-static changes at every point of the star, we may write

$$T\frac{DS}{Dt} = \frac{DU}{Dt} + p\frac{D}{Dt}\left(\frac{1}{\rho}\right), \tag{64}$$

where S is the entropy per unit mass. By virtue of equation (21), a comparison between equations (63) and (64) leads to the result

$$\rho T\frac{DS}{Dt} = \Phi_v + \rho\epsilon_{Nuc} - \operatorname{div}(\mathscr{F} + \mathscr{F}_r), \tag{65}$$

expressing the rate of change of entropy as we follow a fluid particle along its motion. The r.h.s. of this equation is nothing but the amount of "heat" absorbed per unit volume and unit time. The positive function Φ_v represents the rate (per unit volume and unit time) at which heat is generated by viscous friction; and is, accordingly, called the dissipation function. The function ϵ_{Nuc} designates the energy generation rate per unit mass and unit time; it is a known function of ρ, T, and the chemical composition of the star. The term $\operatorname{div}\mathscr{F}$ represents the conduction of heat from neighboring fluid elements. Following Fourier, we shall assume that

$$\mathscr{F} = -\chi \operatorname{grad} T, \tag{66}$$

where χ is the coefficient of *thermal* conductivity.[6] Similarly, if we except the outermost surface regions of a star, the radiative flux vector can be written in the form

$$\mathscr{F}_r = -\chi_r \, \text{grad} \, T, \tag{67}$$

where

$$\chi_r = \frac{4acT^3}{3\kappa\rho} \tag{68}$$

is the coefficient of *radiative* conductivity. Both χ and χ_r are positive since the energy fluxes must be from regions at high temperature to those at low temperature.

ISENTROPIC MOTIONS

It is sometimes convenient to consider motions for which, to a first approximation, the entropy of each fluid particle remains a constant along its path (although the entropy may differ from one pathline to another).[7] We then have

$$\frac{DS}{Dt} = \frac{\partial S}{\partial t} + \mathbf{v}.\text{grad} \, S = 0, \tag{69}$$

and equation (63) reduces to

$$\rho \, \frac{DU}{Dt} + p \, \text{div} \, \mathbf{v} = 0. \tag{70}$$

This equation merely expresses that the rate of change of the total internal energy of a moving mass element is equal to the work done by the surroundings to compress this element.

In the sequel, we shall frequently consider a mixture of black body radiation and a simple perfect gas. Neglecting the ionization and excitation

---

[6] The transport coefficient χ also depends on the local thermodynamic state. It is a measure of the irreversible transfer of thermal energy between the constitutive particles, from regions where the temperature is large to those where it is small.

[7] These motions are sometimes called "adiabatic" (lit. "not passing across") as though the terms adiabatic and isentropic were synonymous. In accordance with the current trend in fluid mechanics, here we shall use the term *isentropic* in its strict etymological sense, i.e., adiabatic *and* reversible. To be specific, an adiabatic motion has $\rho\epsilon_{Nuc} = \text{div}(\mathscr{F} + \mathscr{F}_r)$ everywhere (i.e., "heat" does not flow across a surface within the fluid), whereas an isentropic motion has $DS/Dt = 0$ everywhere. Hence, by virtue of equation (65), an adiabatic motion is isentropic only if the dissipation function Φ_v identically vanishes, i.e., if no heat is irreversibly generated by viscous friction in the system. See, e.g., Thompson, P. A., *Compressible-Fluid Dynamics*, pp. 59–60, New York: McGraw-Hill Book Co., 1972.

energies, we therefore have

$$p = \frac{\mathscr{R}}{\bar{\mu}} \rho T + \frac{1}{3} a T^4, \tag{71}$$

and

$$U = c_v T + \frac{1}{\rho} a T^4, \tag{72}$$

where $\mathscr{R}$ is the perfect gas constant, and $\bar{\mu}$ is the mean molecular weight. We also define the constant γ to be the ratio c_p/c_v of the specific heats, at constant pressure and constant volume, per unit mass. Then, in view of equation (21), the condition of isentropy (70) becomes

$$\frac{Dp}{Dt} = \Gamma_1 \frac{p}{\rho} \frac{D\rho}{Dt}, \tag{73}$$

where

$$\Gamma_1 = \beta + \frac{(4 - 3\beta)^2(\gamma - 1)}{\beta + 12(\gamma - 1)(1 - \beta)}, \tag{74}$$

and $\beta = p_g/p$. This equation is originally due to Eddington. As was shown by Chandrasekhar, we can also write

$$\frac{1}{T} \frac{DT}{Dt} = (\Gamma_3 - 1) \frac{1}{\rho} \frac{D\rho}{Dt} = \frac{\Gamma_2 - 1}{\Gamma_2} \frac{1}{p} \frac{Dp}{Dt}, \tag{75}$$

where Γ_2 and Γ_3 are related to Γ_1 by the following relations:

$$\Gamma_3 - 1 = \frac{\Gamma_1 - \beta}{4 - 3\beta} = \Gamma_1 \frac{\Gamma_2 - 1}{\Gamma_2}. \tag{76}$$

Finally, the variation of the ratio β is given by

$$\frac{1}{\beta} \frac{D\beta}{Dt} = (\Gamma_3 - \Gamma_1) \frac{1}{\rho} \frac{D\rho}{Dt}. \tag{77}$$

Note that the Γ's reduce to the usual adiabatic exponent γ in the limit $p_r \ll p_g$; and they reduce to 4/3 for black body radiation alone ($p_g \ll p_r$). Hence, for a mixture of a perfect gas and black body radiation, the generalized adiabatic exponents are intermediate in value between 4/3 and γ.

3.5.   CIRCULATION AND VORTICITY

In many instances, it is particularly instructive to describe a fluid motion in terms of the vorticity field

$$\boldsymbol{\omega} = \text{curl } \mathbf{v}, \tag{78}$$

which represents the local and instantaneous rate of rotation of the fluid. By definition, a line that is everywhere tangent to the local vorticity vector $\boldsymbol{\omega}(\mathbf{x},t)$ is called a *vortex line*; the family of such lines at any instant is defined by the equations

$$\frac{dx_1}{\omega_1} = \frac{dx_2}{\omega_2} = \frac{dx_3}{\omega_3}. \tag{79}$$

Also, the tube formed by all the vortex lines passing through a closed curve is termed a *vortex tube*.

In view of equation (78), we always have

$$\operatorname{div} \boldsymbol{\omega} = 0. \tag{80}$$

Accordingly, by virtue of Gauss's theorem, the flux of vorticity across an arbitrary closed surface S is equal to zero:

$$\oint_S \boldsymbol{\omega}.\mathbf{dS} = 0. \tag{81}$$

Applying this result to the volume consisting of a section of vortex tube capped by two surfaces Σ_1 and Σ_2, we obtain at once

$$\int_{\Sigma_1} \boldsymbol{\omega}.\mathbf{dS}_1 = \int_{\Sigma_2} \boldsymbol{\omega}.\mathbf{dS}_2. \tag{82}$$

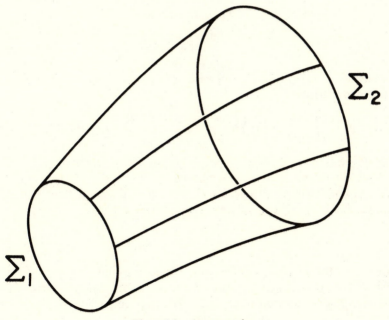

Figure 3.2.   A vortex tube.

Thus, *the flux of vorticity across any section of a vortex tube is the same*; and it is called the tube strength. In particular, if the cross-sectional area is infinitesimal, the quantity **ω.dS** remains constant along the tube. Hence, *vortex lines cannot begin or end in the fluid; they are either closed curves or terminate on the boundary.*

Let us now derive some kinematical identities expressing the rate of change of vorticity in an arbitrary continuous motion. Using a formula well known in vector analysis,

$$\tfrac{1}{2}\,\text{grad}|\mathbf{v}|^2 = \mathbf{v} \times \text{curl } \mathbf{v} + \mathbf{v}.\text{grad } \mathbf{v}, \tag{83}$$

we can take the curl of equation (7) to obtain

$$\text{curl } \mathbf{a} = \frac{\partial \boldsymbol{\omega}}{\partial t} + \text{curl}(\boldsymbol{\omega} \times \mathbf{v}). \tag{84}$$

Now, by virtue of equation (80), we have

$$\text{curl}(\boldsymbol{\omega} \times \mathbf{v}) = \mathbf{v}.\text{grad } \boldsymbol{\omega} - \boldsymbol{\omega}.\text{grad } \mathbf{v} + \boldsymbol{\omega} \text{ div } \mathbf{v}; \tag{85}$$

it thus follows that

$$\text{curl } \mathbf{a} = \frac{D\boldsymbol{\omega}}{Dt} - \boldsymbol{\omega}.\text{grad } \mathbf{v} + \boldsymbol{\omega} \text{ div } \mathbf{v}. \tag{86}$$

In combining equations (21) and (86), we find

$$\frac{D}{Dt}\left(\frac{\boldsymbol{\omega}}{\rho}\right) = \frac{\boldsymbol{\omega}}{\rho}.\text{grad } \mathbf{v} + \frac{1}{\rho}\text{curl } \mathbf{a}. \tag{87}$$

Following Ertel, by a straightforward calculation based on equation (87), we can also write

$$\frac{D}{Dt}\left(\frac{\boldsymbol{\omega}}{\rho}.\text{grad } Q\right) = \frac{\boldsymbol{\omega}}{\rho}.\text{grad } \frac{DQ}{Dt} + \frac{1}{\rho}\text{curl } \mathbf{a}.\text{grad } Q, \tag{88}$$

where $Q(\mathbf{x},t)$ is an arbitrary scalar, vector, or tensor function.

Consider next a *material* contour $c(t)$, i.e., a closed curve drawn in the star, and which is at all instants composed of the same fluid particles. Let this be the velocity circulation along the contour $c(t)$:

$$\oint_c \mathbf{v}.\mathbf{dc} \tag{89}$$

where **dc** is the line element of the curve. Since the contour moves with the fluid, we may calculate the rate of change of the velocity circulation. Assume that the motion of the curve $c(t)$ is given parametrically by the equation $\mathbf{c} = \mathbf{x}(u,t)$, where u locates individual mass elements along the circuit, and t is the time. Then,

$$\frac{d}{dt}\oint_c \mathbf{v}.d\mathbf{c} = \oint_c \frac{D\mathbf{v}}{Dt}.d\mathbf{c} + \oint_c \mathbf{v}.\frac{\partial}{\partial t}\left(\frac{\partial \mathbf{x}}{\partial u}\right)du. \tag{90}$$

In view of equation (3), we have

$$\oint_c \mathbf{v}.\frac{\partial}{\partial t}\left(\frac{\partial \mathbf{x}}{\partial u}\right)du = \oint_c \mathbf{v}.\frac{\partial \mathbf{v}}{\partial u}du = \oint_c d(\tfrac{1}{2}|\mathbf{v}|^2) = 0. \tag{91}$$

Hence, by virtue of definition (7), equation (90) reduces to

$$\frac{d}{dt}\oint_c \mathbf{v}.d\mathbf{c} = \oint_c \mathbf{a}.d\mathbf{c}. \tag{92}$$

This result is purely kinematical, and is thus valid for any motion whatsoever.

To proceed any further, we must now specify the material acceleration occurring in equations (87), (88), and (92). Restricting ourselves to the ideal fluid model ($\mu = \mu_s = \mu_r = 0$), we have

$$\mathbf{a} = -\operatorname{grad} V - \frac{1}{\rho}\operatorname{grad} p. \tag{93}$$

Let us derive another form of this equation. The enthalpy per unit mass is defined by

$$H = U + \frac{p}{\rho}. \tag{94}$$

Since the difference between the values of U for two different mass elements at any instant is given by

$$dU = T\,dS - p\,d\left(\frac{1}{\rho}\right) = T\,dS + \frac{p}{\rho^2}\,d\rho, \tag{95}$$

we obtain

$$dH = T\,dS + \frac{1}{\rho}\,dp. \tag{96}$$

It thus follows that

$$\operatorname{grad} H = T\operatorname{grad} S + \frac{1}{\rho}\operatorname{grad} p, \tag{97}$$

and equation (93) becomes

$$\mathbf{a} = -\operatorname{grad}(V + H) + T\operatorname{grad} S. \tag{98}$$

In taking the curl of equations (93) and (98), we find

$$\operatorname{curl} \mathbf{a} = -\operatorname{grad} \frac{1}{\rho} \times \operatorname{grad} p, \tag{99}$$

or, alternately,

$$\operatorname{curl} \mathbf{a} = \operatorname{grad} T \times \operatorname{grad} S. \tag{100}$$

HOMENTROPIC FLOWS. To clarify the situation, we begin our study of equations (87) and (92) with the special case of motions for which the entropy is uniform over the whole mass of the fluid. We thus require

$$\operatorname{grad} S = 0 \tag{101}$$

at every point of the system; and the motion is called "homentropic." Then, by virtue of equations (99)–(101), it readily follows that the surfaces of equal pressure and those of equal density coincide.[8]

In making use of equations (92), (100), and (101), we obtain

$$\frac{d}{dt} \oint_c \mathbf{v}.\mathbf{dc} = 0. \tag{102}$$

Accordingly, *the velocity circulation around a closed material curve is invariant in a frictionless, homentropic flow.* This is Kelvin's circulation theorem. In applying Stokes's formula, we can alternately write

$$\frac{d}{dt} \int_\Sigma \boldsymbol{\omega}.\mathbf{dS} = 0, \tag{103}$$

where Σ is any open material surface bounded by the circuit $c(t)$. Hence, the flux of vorticity across any material surface does not change with time. By virtue of equation (82), it thus follows that *the strength of a vortex tube is an integral of the equations of motion.*

Another interesting property follows from equation (87), which, under the present assumptions, reduces to

$$\frac{D}{Dt} \left(\frac{\boldsymbol{\omega}}{\rho} \right) = \frac{\boldsymbol{\omega}}{\rho}.\operatorname{grad} \mathbf{v}. \tag{104}$$

---

[8] The distinction between "homentropic" and "isentropic" flows is best illustrated with a perfect gas. The entropy has then the form $S = c_v \log(p/\rho^\gamma) + constant$, and equation (69) reduces to $D/Dt(p/\rho^\gamma) = 0$. For an isentropic motion, the ratio p/ρ^γ remains constant as we follow a given mass element, but it may vary from one element to another. On the contrary, for an homentropic motion, $p/\rho^\gamma = constant$ throughout the whole system; these motions are sometimes called "barotropic" (cf. §§4.2–4.3).

Following Cauchy, by making use of the material variables $\mathbf{X}$ and t, we can integrate this equation at once to obtain

$$\frac{\omega_i}{\rho} = \frac{\omega_{ok}}{\rho_o} \frac{\partial x_i}{\partial X_k}, \tag{105}$$

where $\omega_o = \omega(\mathbf{X},0)$ and $\rho_o = \rho(\mathbf{X},0)$ are the initial values of $\omega(\mathbf{X},t)$ and $\rho(\mathbf{X},t)$. As was shown by Helmholtz, this solution simply means that *the particles that compose a vortex line at one instant will continue to form a vortex line at any subsequent instant*. The proof lies in the fact that a tangent to a vortex line is carried by the fluid so that it always remains tangent to a vortex line. Let dX_i be the components of the vector representing a line element, at the instant $t = 0$, of a vortex line. As we follow its motion, we have

$$dx_i = \frac{\partial x_i}{\partial X_k} dX_k, \tag{106}$$

where the dx_i's are the new components, at time t, of this line element. Now, by hypothesis, we can always write

$$dX_i = \epsilon \frac{\omega_{oi}}{\rho}, \tag{107}$$

where ϵ is some constant. From equations (105)–(107), it follows that

$$dx_i = \epsilon \frac{\omega_{ok}}{\rho_o} \frac{\partial x_i}{\partial X_k} = \epsilon \frac{\omega_i}{\rho}; \tag{108}$$

and the vector with components dx_i is also tangent to a vortex line. This concludes the proof. By virtue of equations (107) and (108), we also note that ω/ρ is proportional to the length of a line element along a vortex line.

NONHOMENTROPIC FLOWS. In this case the situation is quite different. Indeed, from equations (87) and (99), we obtain the following equation

$$\frac{D}{Dt}\left(\frac{\omega}{\rho}\right) = \frac{\omega}{\rho}.\text{grad } \mathbf{v} - \frac{1}{\rho} \text{grad } \frac{1}{\rho} \times \text{grad } p, \tag{109}$$

which was originally discussed by Friedmann. When the last term in equation (109) does not vanish, the vortex lines are not permanently attached to the fluid, and the strength of a vortex tube is not constant in time. *As a result, the possibility of creation and destruction of vortices may arise in an ideal fluid for which the entropy is not constant throughout the whole mass.*

Very much in the same vein, a circulation theorem for frictionless, nonhomentropic flows was established by Bjerknes in 1900. From equations (92) and (93), and Stokes's formula, we obtain

$$\frac{d}{dt}\oint_c \mathbf{v.dc} = \frac{d}{dt}\int_\Sigma \boldsymbol{\omega.dS} = -\oint_c \frac{dp}{\rho}. \tag{110}$$

Since the surfaces $p = constant$ are generally inclined to the surfaces $\rho = constant$, the velocity circulation and the strength of a vortex tube are therefore not constant for all times. A simple geometrical interpretation of equation (110) can be obtained as follows. At a given instant, we plot the image $\bar{c}(t)$ of the curve $c(t)$ in the $(p - \rho)$-plane. We then draw equidistant members of the sets of curves $p = constant$ and $\rho = constant$, and so obtain a series of cells bounded by these curves. The Bjerknes theorem states that *the rate of change of circulation per unit time along a material contour $c(t)$ is proportional to the number of cells surrounded by the curve $\bar{c}(t)$*. Naturally, the number of these cells changes in time, for at a later instant the vortex tubes will consist of different fluid particles.

ISENTROPIC FLOWS. Obviously, the requirement of isentropy does not prevent the creation and destruction of vortices in a frictionless fluid.

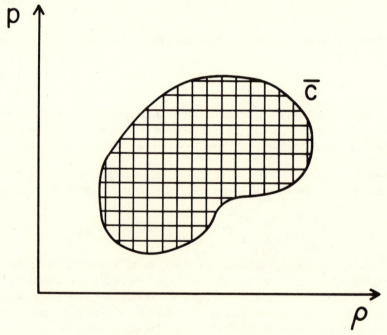

Figure 3.3. The $(p - \rho)$-plane.

Nevertheless, as we shall now see, this condition somewhat simplifies the problem.

First, if we let $Q = S$ in equation (88) and make use of equation (100), we obtain generally

$$\frac{D}{Dt}\left(\frac{\boldsymbol{\omega}}{\rho}.\text{grad } S\right) = \frac{\boldsymbol{\omega}}{\rho}.\text{grad } \frac{DS}{Dt}. \tag{111}$$

Hence, *for isentropic motions, the quantity*

$$\frac{\boldsymbol{\omega}}{\rho}.\text{grad } S \tag{112}$$

remains constant along the path of each fluid particle. Note, however, that this constant generally differs from one mass element to another. This result is due to Truesdell.

Consider now the total energy equation (56). In making use of equations (21), (24), (42), (60), (65), and (94), we can also rewrite this conservation law in the form

$$\frac{D}{Dt}(\tfrac{1}{2}|\mathbf{v}|^2 + H + V) = T\frac{DS}{Dt} + \frac{1}{\rho}\frac{\partial p}{\partial t} + \frac{\partial V}{\partial t}. \tag{113}$$

Thus, for steady isentropic motions, we have

$$\mathbf{v}.\text{grad}(\tfrac{1}{2}|\mathbf{v}|^2 + H + V) = 0. \tag{114}$$

Accordingly, *the "total energy"*

$$\tfrac{1}{2}|\mathbf{v}|^2 + H + V \tag{115}$$

is constant as we follow the steady isentropic motion of a fluid particle. Again, the constant generally depends on the particular choice of pathline. This is Bernoulli's theorem.

Finally, let us write the connection between the gradient of the "total energy," the gradient of the entropy, and the vorticity. For steady motions, in combining equations (7), (78), (93), and (97), we readily obtain

$$\text{grad}(\tfrac{1}{2}|\mathbf{v}|^2 + H + V) = T \text{ grad } S + \mathbf{v} \times \boldsymbol{\omega}. \tag{116}$$

This relation is quite general for steady motions, and it is due to Crocco and Vazsonyi. We observe that the "total energy" (115) is constant over the whole configuration for steady, homentropic flows only, when $\mathbf{v} \times \boldsymbol{\omega}$ identically vanishes at every point.

3.6. STELLAR HYDROMAGNETICS

Stellar material is an electrically conducting fluid. Hence, whenever a magnetic field is prevalent, there will be an interdependence between the field and the flow. *On one hand*, motions of the conducting fluid across the magnetic lines of force cause electric currents to flow; and these currents modify the original magnetic field. *On the other hand*, a body force due to the interaction between the flow of electric currents and the magnetic field appears; and this body force perturbs the original fluid motion.

   The properties of the electromagnetic field and its variation are governed by the Maxwell equations. Since we shall not be concerned with rapid oscillations of the field, we can ignore the displacement currents. As a consequence, we also neglect the accumulation of electric charge; such an approximation is legitimate in the present context, for its variation affects the equation expressing the conservation of electric charge only by terms of order $|\mathbf{v}|^2/c^2$. With these assumptions in mind, the Maxwell equations reduce to

$$\operatorname{curl}\mathbf{H} = 4\pi\mathbf{J}, \tag{117}$$

$$\operatorname{curl}\mathbf{E} = -\frac{\partial\mathbf{H}}{\partial t}, \tag{118}$$

$$\operatorname{div}\mathbf{H} = 0, \tag{119}$$

where the magnetic field $\mathbf{H}$, the electric field $\mathbf{E}$, and the current density $\mathbf{J}$ are measured in e.m.u. The material is supposed to be nonmagnetic, and its permeability is taken equal to unity.

   One more constitutive relation is needed to describe the electric current density. Consider a moving fluid particle. If $\mathbf{v}$ represents its velocity, the total electric field it experiences is $\mathbf{E} + \mathbf{v} \times \mathbf{H}$, i.e., the sum of the field $\mathbf{E}$ arising from sources other than its own motion and the induced field $\mathbf{v} \times \mathbf{H}$. Following Ohm, we then assume that

$$\mathbf{J} = \sigma_e(\mathbf{E} + \mathbf{v} \times \mathbf{H}), \tag{120}$$

where σ_e is the coefficient of electrical conductivity.

   Equations (117)–(120) describe the electromagnetic field. The action of the field on the flow results from the Lorentz force $\mathbf{I}(\mathbf{x},t)$—the body force per unit volume exerted by the magnetic field upon the electric currents. We have

$$\mathbf{I} = \mathbf{J} \times \mathbf{H} = \frac{1}{4\pi}\operatorname{curl}\mathbf{H} \times \mathbf{H}. \tag{121}$$

By using equation (119), we can alternately write the components of the vector $\mathbf{l}$ in the form

$$l_i = -\frac{\partial}{\partial x_i}\left(\frac{|\mathbf{H}|^2}{8\pi}\right) + \frac{\partial}{\partial x_k}\left(\frac{H_k H_i}{4\pi}\right). \tag{122}$$

Thus, the Lorentz force is equivalent to a magnetic pressure $|\mathbf{H}|^2/8\pi$, which is uniform in all directions, together with a tension $|\mathbf{H}|^2/4\pi$ along the magnetic lines of force.

When this electromagnetic force is included into the balance of linear momentum, we obtain the modified Navier-Stokes equations:

$$\frac{D\mathbf{v}}{Dt} = -\operatorname{grad} V - \frac{1}{\rho}\operatorname{grad}\left(p + \frac{|\mathbf{H}|^2}{8\pi}\right) + \frac{1}{4\pi\rho}(\mathbf{H}.\operatorname{grad})\mathbf{H} + \frac{1}{\rho}\operatorname{div}\tau, \tag{123}$$

where p designates the total pressure as given by an equation of state (cf. also equations [25] and [47]). Obviously, the equation of continuity (20) remains unchanged, and the conservation of thermal energy must be modified to include the dissipation (per unit volume) of magnetic energy in the form of heat. For moving conductors, this dissipative contribution is $|\mathbf{J}|^2/\sigma_e$. Accordingly, equation (63) must now be replaced by

$$\rho\frac{DU}{Dt} + p\operatorname{div}\mathbf{v} = \Phi_v + \frac{v_m}{4\pi}|\operatorname{curl}\mathbf{H}|^2 + \rho\epsilon_{Nuc} - \operatorname{div}(\mathscr{F} + \mathscr{F}_r), \tag{124}$$

where

$$v_m = \frac{1}{4\pi\sigma_e} \tag{125}$$

is the coefficient of magnetic diffusivity.

MAGNETIC FIELDS IN A MOVING FLUID.   For most problems, it is convenient to eliminate the vectors $\mathbf{E}$ and $\mathbf{J}$ from equations (117)–(120). We obtain

$$\frac{\partial\mathbf{H}}{\partial t} = \operatorname{curl}(\mathbf{v}\times\mathbf{H}) - \operatorname{curl}(v_m\operatorname{curl}\mathbf{H}). \tag{126}$$

In view of equation (119), this equation becomes

$$\frac{D\mathbf{H}}{Dt} = \mathbf{H}.\operatorname{grad}\mathbf{v} - \mathbf{H}\operatorname{div}\mathbf{v} - \operatorname{curl}(v_m\operatorname{curl}\mathbf{H}). \tag{127}$$

Making use next of equation (21), we find

$$\frac{D}{Dt}\left(\frac{\mathbf{H}}{\rho}\right) = \frac{\mathbf{H}}{\rho}.\operatorname{grad}\mathbf{v} - \frac{1}{\rho}\operatorname{curl}(v_m\operatorname{curl}\mathbf{H}). \tag{128}$$

In particular, when v_m identically vanishes, equation (128) reduces to

$$\frac{D}{Dt}\left(\frac{\mathbf{H}}{\rho}\right) = \frac{\mathbf{H}}{\rho}.\text{grad } \mathbf{v}. \tag{129}$$

We see that the foregoing equation is of exactly the same form as the equation of vorticity (104). The identity is in fact complete since both ω and $\mathbf{H}$ are solenoidal (cf. equations [80] and [119]). Thus, all the theorems discussed in Section 3.5 for frictionless, homentropic flows may be transposed at once for magnetic fields in a perfectly conducting fluid.

First, we define a magnetic tube of force as the surface swept out by the magnetic lines of force passing through a given closed curve. Then, from equation (119), it follows that *the magnetic flux across any section of a magnetic tube of force is the same:*

$$\int_{\Sigma_1} \mathbf{H}.\mathbf{dS}_1 = \int_{\Sigma_2} \mathbf{H}.\mathbf{dS}_2, \tag{130}$$

where Σ_1 and Σ_2 are two arbitrary sections of a magnetic tube (cf. equation [82]). As a consequence, *magnetic lines of force cannot begin or end in the fluid; they are either closed curves or terminate on the boundary.* These properties are quite general.

Second, from equation (129), we observe that *all the fluid particles that initially lie on a magnetic line of force continue to do so in a perfectly conducting medium.* Under the same assumptions, we have

$$\frac{d}{dt}\int_{\Sigma} \mathbf{H}.\mathbf{dS} = 0, \tag{131}$$

where Σ is any open material surface (cf. equation [103]). In other words, *the strength of a magnetic tube of force remains constant as the tube is moved about with the fluid.* Finally, the quantity $\mathbf{H}/\rho$ is proportional to the length of an element along a magnetic line of force. These results were originally noticed by Walén. They are no longer true when the electrical conductivity is less than perfect; then, the magnetic lines of force are not permanently attached to the fluid, and the magnetic field will decay in time.

To conclude, we observe that equation (87) now becomes

$$\frac{D}{Dt}\left(\frac{\omega}{\rho}\right) = \frac{\omega}{\rho}.\text{grad } \mathbf{v} - \frac{1}{\rho}\text{grad }\frac{1}{\rho} \times \text{grad } p + \frac{1}{\rho}\text{curl}\left(\frac{1}{\rho}\mathbf{J} \times \mathbf{H}\right). \tag{132}$$

The last term of this equation does not in general vanish. Hence, the coupling between the field and the flow usually promotes the creation and destruction of vortices.

BOUNDARY CONDITIONS. For illustrative purposes, let us neglect viscous friction ($\mu = \mu_\vartheta = \mu_r = 0$), and assume that the electrical conductivity is infinite ($v_m = 0$). Since the magnetic field usually pervades the star and the surrounding vacuum, the inner and outer fields have to be found and suitably matched on the surface $\mathscr{S}$ of the star. The boundary conditions may be stated as follows. *First*, the gravitational force must be continuous across the surface $\mathscr{S}$. *Second*, the normal component of the material, radiative and magnetic stresses must be continuous at the interface. Thus, instead of the single condition (49), we now impose that

$$\left[\!\left[-\left(p + \frac{|\mathbf{H}|^2}{8\pi} \right) n_i + \frac{H_i}{4\pi}(n_k H_k) \right]\!\right] = 0, \quad \text{on } \mathscr{S}, \tag{133}$$

where brackets designate the jump that the quantity experiences on the boundary. (Note that the pressure p cannot become negative!) *Third*, since the magnetic flux must be continuous across the stellar surface, we must also have

$$[\![n_k H_k]\!] = 0, \quad \text{on } \mathscr{S}. \tag{134}$$

Thus, although the normal component H_n of the magnetic field is continuous, there may be discontinuities in its tangential components $\mathbf{H}_t$, if the star is assumed to be a perfectly conducting medium. Clearly, whenever a jump $[\![\mathbf{H}_t]\!]$ exists, it is parallel to the surface $\mathscr{S}$. This, in turn, implies the existence of a surface current $\mathbf{J}_s$, its density being given by $\mathbf{J}_s = \mathbf{n} \times [\![\mathbf{H}_t]\!]$. For infinite conductivity, surface currents can be avoided only in choosing $\mathbf{H} = 0$ and $p = 0$ on the surface $\mathscr{S}$. Also, no such currents exist when the star has a finite electrical conductivity for, then, the tangential components of the field must also be continuous. This additional condition matches the rise in the order of the differential equation (128) when the magnetic diffusion is included.

3.7. THE VIRIAL EQUATIONS

Consider the ideal fluid model described in Section 3.3. Equation (39) can be rewritten in the form

$$\frac{Dv_k}{Dt} = -\frac{\partial V}{\partial x_k} - \frac{1}{\rho}\frac{\partial p}{\partial x_k}. \tag{135}$$

Multiply now this equation by ρx_i and integrate over the entire volume of the star. By making use of equations (3) and (23), we first have

$$\int_{\mathscr{V}} \rho x_i \frac{Dv_k}{Dt}\, d\mathbf{x} = \frac{d}{dt}\int_{\mathscr{V}} \rho x_i v_k\, d\mathbf{x} - 2K_{ik}, \tag{136}$$

where

$$K_{ik} = \frac{1}{2} \int_{\mathcal{V}} \rho v_i v_k \, d\mathbf{x}. \tag{137}$$

Similarly, by virtue of equation (25), it follows that

$$\int_{\mathcal{V}} \rho x_i \frac{\partial V}{\partial x_k} \, d\mathbf{x} = G \int_{\mathcal{V}} \int_{\mathcal{V}} \rho(\mathbf{x},t) \rho(\mathbf{x}',t) \frac{x_i (x_k - x_k')}{|\mathbf{x} - \mathbf{x}'|^3} \, d\mathbf{x} \, d\mathbf{x}'. \tag{138}$$

Thus, if we let

$$W_{ik} = -\frac{1}{2} G \int_{\mathcal{V}} \int_{\mathcal{V}} \rho(\mathbf{x},t) \rho(\mathbf{x}',t) \frac{(x_i - x_i')(x_k - x_k')}{|\mathbf{x} - \mathbf{x}'|^3} \, d\mathbf{x} \, d\mathbf{x}', \tag{139}$$

equation (138) reduces to

$$\int_{\mathcal{V}} \rho x_i \frac{\partial V}{\partial x_k} \, d\mathbf{x} = -W_{ik}. \tag{140}$$

Finally, the last term can be integrated by parts to give

$$\int_{\mathcal{V}} x_i \frac{\partial p}{\partial x_k} \, d\mathbf{x} = -\delta_{ik} \int_{\mathcal{V}} p \, d\mathbf{x}, \tag{141}$$

in which we made use of Gauss's theorem and condition (49) on the pressure.

Combining equations (136), (140), and (141), we find

$$\frac{d}{dt} \int_{\mathcal{V}} \rho x_i v_k \, d\mathbf{x} = 2K_{ik} + W_{ik} + \delta_{ik} \int_{\mathcal{V}} p \, d\mathbf{x}. \tag{142}$$

Since all tensors on the r.h.s. are symmetric, it thus follows that the l.h.s. must also be symmetric. We can therefore write

$$\frac{d}{dt} \int_{\mathcal{V}} \rho x_i v_k \, d\mathbf{x} = \frac{d}{dt} \int_{\mathcal{V}} \rho x_k v_i \, d\mathbf{x}. \tag{143}$$

This equation embodies the conservation of the total angular momentum. Now, by virtue of equations (23) and (143), we have

$$\frac{d}{dt} \int_{\mathcal{V}} \rho x_i v_k \, d\mathbf{x} = \frac{1}{2} \frac{d^2}{dt^2} \int_{\mathcal{V}} \rho x_i x_k \, d\mathbf{x}; \tag{144}$$

and the virial equations (142) become

$$\frac{1}{2} \frac{d^2 I_{ik}}{dt^2} = 2K_{ik} + W_{ik} + \delta_{ik} \int_{\mathcal{V}} p \, d\mathbf{x}, \tag{145}$$

where

$$I_{ik} = \int_{\mathscr{V}} \rho x_i x_k \, \mathbf{dx}. \tag{146}$$

These are the second-order virial equations in their usual form. The inclusion of viscous stresses and magnetic fields is left as an exercise.

By contracting the indices in equation (145), we obtain

$$\frac{1}{2} \frac{d^2 I}{dt^2} = 2K + W + 3 \int_{\mathscr{V}} p \, \mathbf{dx}, \tag{147}$$

where I, K, and W denote, respectively, the moment of inertia, the kinetic energy, and the gravitational potential energy:

$$I = \int_{\mathscr{V}} \rho |\mathbf{x}|^2 \, \mathbf{dx}, \qquad K = \frac{1}{2} \int_{\mathscr{V}} \rho |\mathbf{v}|^2 \, \mathbf{dx}, \tag{148}$$

and

$$W = \frac{1}{2} \int_{\mathscr{V}} \rho V \, \mathbf{dx} \tag{149}$$

(cf. equations [25] and [139]). For a perfect gas in equilibrium and in a steady state, equation (147) becomes

$$2K + W + 3(\gamma - 1)U_T = 0, \tag{150}$$

where U_T is the total thermal energy of the configuration.

BIBLIOGRAPHICAL NOTES

The literature on classical hydrodynamics is very extensive. The following general references are particularly worth noting:

1. Bjerknes, V., Bjerknes, J., Solberg, H., et Bergeron, T., *Hydrodynamique physique*, Paris: Presses Universitaires de France, 1934.
2. Lamb, H., *Hydrodynamics*, 6th edn., Cambridge: At the Univ. Press, 1932 (New York: Dover Public., Inc., 1945).
3. Landau, L. D., and Lifshitz, E. M., *Fluid Mechanics*, Oxford: Pergamon Press, 1959.

An interesting reference for the foundations of hydrodynamics is:

4. Serrin, J., "Mathematical Principles of Classical Fluid Mechanics" in *Handbuch der Physik* (Flügge, S., ed.), **8** (1), pp. 125–263, Berlin: Springer-Verlag, 1959.

The only book devoted exclusively to the problem of rotation is:

5. Greenspan, H. P., *The Theory of Rotating Fluids*, Cambridge: At the Univ. Press, 1968.

A general survey of hydrodynamics from the standpoint of astrophysics will be found in:

6. Ledoux, P., and Walraven, Th., "Variable Stars" in *Handbuch der Physik* (Flügge, S., ed.), **51**, pp. 353–604, Berlin: Springer-Verlag, 1958.

This paper also contains a brief heuristic discussion of turbulence.

Sections 3.2–3.4. The treatment in the text largely follows references 4 and 6. A particularly clear and concise discussion of the equations of state, energy transfer, opacities, and nuclear reaction rates will be found in:

7. Schwarzschild, M., *Structure and Evolution of the Stars*, pp. 30–95, Princeton: Princeton Univ. Press, 1958 (New York: Dover Public., Inc., 1965).

An exhaustive analysis of these questions occurs in:

8. Cox, J. P., and Giuli, R. T., *Principles of Stellar Structure*, New York: Gordon and Breach, 1968.

This book also contains a detailed study of the Γ's, when both radiation and ionization are taken into account. The radiation field in a moving fluid was considered in:

9. Thomas, L. H., *Quarterly J. Math., Oxford Series*, **1** (1930): 239–251.
10. Hazlehurst, J., and Sargent, W. L. W., *Astroph. J.* **130** (1959): 276–285.

Since hydrodynamics provides no information about the transport coefficients μ, μ_9, χ and σ_e, their expression must be derived from a kinetic theory of the constitutive particles. In view of astrophysical applications, various determinations of these coefficients were made in the following references:

11. Spitzer, L., Jr., *Physics of Fully Ionized Gases*, New York: Interscience Publishers, 1956.
12. Edmonds, F. N., Jr., *Astroph. J.* **125** (1957): 535–549.
13. Hubbard, W. B., *ibid.* **146** (1966): 858–870.
14. Hubbard, W. B., and Lampe, M., *Astroph. J. Supplements* **18** (1969): 297–346.
15. Kovetz, A., and Shaviv, G., *Astron. and Astroph.* **28** (1973): 315–318.
16. Flowers, E., and Itoh, N., *Astroph. J.* **206** (1976): 218–242.

Section 3.5. A particularly complete discussion of circulation and vorticity is due to:

17. Truesdell, C., *The Kinematics of Vorticity*, Bloomington: Indiana Univ. Press, 1954.

See also references 4, 5, and 27. These and related matters are further discussed in:

18. Friedmann, A., *Zeit. f. angew. Math. und Mech.* **4** (1924): 102–107.
19. Bjerknes, V., *Astroph. Norvegica* **2** (1937): 263–339.
20. Crocco, L., *Zeit. f. angew. Math. und Mech.* **17** (1937): 1–7.
21. Ertel, H., *Meteorol. Zeit.* **59** (1942): 277–281, 385–387.
22. Vazsonyi, A., *Quarterly of Applied Math.* **3** (1945): 29–37.
23. Truesdell, C., *Zeit. f. angew. Math. und Phys.* **2** (1951): 109–114.

Section 3.6. A general account of hydromagnetics and of the underlying assumptions will be found in:

24. Landau, L. D., and Lifshitz, E. M., *Electrodynamics of Continuous Media*, Oxford: Pergamon Press, 1960.

See also reference 27. The difficult question of the boundary conditions is discussed at length in:

25. Shercliff, J. A., *A Textbook of Magnetohydrodynamics*, Oxford: Pergamon Press, 1965.

The reference to Walén is:

26. Walén, C., *Arkiv för Matematik, Astronomi och Fysik* **33A**, No 18 (1946): 1–63.

Section 3.7. The classical references on the virial equations are:

27. Chandrasekhar, S., *Hydrodynamic and Hydromagnetic Stability*, Oxford: At the Clarendon Press, 1961.
28. Chandrasekhar, S., *Ellipsoidal Figures of Equilibrium*, New Haven: Yale Univ. Press, 1969.

4

Permanent Rotations

4.1. INTRODUCTION

Let us consider a star for which rotational motion and magnetic field are completely negligible; suppose further that the self-gravitating configuration is isolated from other bodies. Then, as is well known, the system assumes a spherical shape, i.e., the surface upon which the total pressure p vanishes is a sphere. Moreover, the surfaces of constant pressure—the isobaric surfaces—can be described by means of concentric spheres. In consequence, the gravitational potential V, the density ρ, the temperature T, and the luminosity L also possess central symmetry. (Although these assertions may not be evident to a mathematician, they are reasonably obvious on physical grounds.) It is this *geometrical* simplicity that renders the problem of the structure and evolution of radiating stars in hydrostatic equilibrium tractable.

Consider next a single star that rotates about a fixed direction in space with some assigned angular velocity. As we know, in the absence of a magnetic field, the star then becomes an *oblate* figure. We are at once faced with the following questions: what is the geometrical shape of the outer boundary? What is the form of the surfaces upon which the physical parameters of the star remain a constant? To sum up, what is the actual stratification of a rotating star, and how does it depend on the angular velocity distribution? Obviously, the boundary is no longer spherical. Could we describe the isobaric surfaces by means of a set of suitable ellipsoids? Under very special circumstances, this is a possibility; but, for an arbitrary rotating star, ellipsoids do not in general provide an acceptable approximation. At first glance, we could surmise the star to be symmetric with respect to its axis of rotation. This is not always the case, for models having a genuine triplanar symmetry may be constructed. Could we then at least expect the star to be symmetric with respect to a plane perpendicular to its axis of rotation? This is true provided that some stringent conditions are met. As a matter of fact, *we have no a priori knowledge of the actual stratification in a rotating star.* In general, the stratification is unknown and must be derived from the hydrodynamical equations of the problem. This situation is in sharp contrast to the case of a nonrotating star, for which a spherical stratification can be assumed *ab initio.*

At this stage, a further difficulty arises in the theory of stellar rotation. If we except stars that we assume to rotate as a solid body, *friction between the various layers induces a net transfer of angular momentum in the course of time.* Hence, unless simplifying assumptions are made, we cannot expect to adequately approximate stars in a state of differential rotation by means of time-independent models. Specifically, if the time scale of friction is shorter than the lifetime of a rotating star, angular momentum transfer has to be taken into account.

In view of the foregoing remarks, we cannot hope to investigate the structure and evolution of rotating stars to the same degree of generality obtained for stars in hydrostatic equilibrium. In the present chapter we explore some simple *mechanical* properties that are relevant to rotating stars; all of them can be derived without a detailed knowledge of stellar models. Problems pertaining to radiation and thermodynamics will be discussed in Chapter 7.

4.2. ASSUMPTIONS AND DEFINITIONS

Let us first make the following assumptions:

(i) *the star is isolated in space and rotates about a fixed axis with some yet unspecified angular velocity;*

(ii) *the system is stationary when viewed in an inertial frame of reference, and the density of each mass element remains a constant as we follow its motion;*

(iii) *friction may be neglected altogether;*

(iv) *no electromagnetic force is acting on the star.*

By definition, if a configuration satisfies all the above properties it is said to be in a state of permanent rotation for an inertial observer. Assume now the star to rotate about the z-axis, and take the origin of our frame of reference at the center of mass. Thus, in cylindrical coordinates (ϖ, φ, z), the components of the velocity $\mathbf{v}$ have the form

$$v_\varpi = 0, \quad v_\varphi = \Omega\varpi, \quad v_z = 0, \tag{1}$$

in an inertial frame; for the present, the angular velocity Ω is an arbitrary function of the spatial coordinates.

By virtue of the foregoing hypothesis, a first conclusion may be drawn at once:

If a gaseous star is in a state of permanent rotation when viewed in an inertial frame of reference, it necessarily possesses an axial symmetry.

This result is a mere consequence of the conservation of mass. Indeed, we can write

$$\frac{D\rho}{Dt} + \rho \, \text{div} \, \mathbf{v} = 0, \tag{2}$$

or

$$\frac{\partial \rho}{\partial t} + \mathbf{v}.\text{grad} \, \rho + \rho \, \text{div} \, \mathbf{v} = 0 \tag{3}$$

(cf. §3.3: equation [21]). Since the density of a mass element does not vary along its path, we have $D\rho/Dt = 0$; *despite the gaseous nature of the system*, equation (2) thus implies

$$\text{div} \, \mathbf{v} = 0. \tag{4}$$

From equations (1) and (4), we obtain

$$\Omega = constant \quad \text{or} \quad \Omega = \Omega(\varpi, z). \tag{5}$$

Now, in view of assumption (ii), we also have $\partial \rho / \partial t = 0$; hence, equations (3) and (4) give

$$\mathbf{v}.\text{grad} \, \rho = 0. \tag{6}$$

This equation shows that the velocity $\mathbf{v}$ lies in a plane tangent to the surfaces of equal density. Thus, by virtue of equations (1), (5), and (6), we can write

$$\Omega \frac{\partial \rho}{\partial \varphi} = 0; \tag{7}$$

and assuming that Ω differs from zero, we find

$$\rho = \rho(\varpi, z). \tag{8}$$

In making use of equations (5) and (8), we can now easily show that the functions p, V, T, and L also possess axial symmetry.[1]

The foregoing conclusion greatly simplifies the equations of motion (cf. §3.3: equation [39]). We obtain

$$\frac{1}{\rho} \frac{\partial p}{\partial \varpi} = -\frac{\partial V}{\partial \varpi} + \Omega^2 \varpi, \tag{9}$$

and

$$\frac{1}{\rho} \frac{\partial p}{\partial z} = -\frac{\partial V}{\partial z}. \tag{10}$$

---

[1] It is possible to build steady models for which axial symmetry does not exist; however, these configurations are time independent only when viewed from a suitably chosen rotating frame of reference (cf. §10.2).

In general, the gravitational potential and density are related by the Poisson equation

$$\nabla^2 V = 4\pi G\rho, \tag{11}$$

where G represents the constant of gravitation. Instead of equation (11), it is sometimes convenient to use the integral form of the Newtonian potential, i.e.,

$$V(\varpi,z) = -G \int_{\mathscr{V}} \frac{\rho(\mathbf{x}')}{|\mathbf{x} - \mathbf{x}'|}\, d\mathbf{x}', \tag{12}$$

where the triple integral must be evaluated over the (*unknown*) volume $\mathscr{V}$ of the star. By convention, the stellar boundary corresponds to the surface $\mathscr{S}$ upon which the pressure vanishes. The integral form (12) has the advantage that it incorporates *ipso facto* the boundary conditions on the gravitational potential, i.e., the continuity of gravity across the (*unknown*) surface $\mathscr{S}$.

To complete the formulation of the problem, we must now include an appropriate equation of state. In general, the pressure depends on the density, temperature, and chemical composition of the star. At this stage, an explicit relation is not needed, and we write symbolically

$$p = p(\rho,T,\lambda_1,\lambda_2,\ldots), \tag{13}$$

where the λ's denote a set of variables which generally depend on ϖ and z. By definition, we call *barocline* (or baroclinic star) a system for which a *physical* relation of the form (13) holds. Naturally, the temperature is an additional unknown quantity; hence, even if we know the λ's, another thermodynamical equation must be written down to complete our set of equations. For the time being, we shall not need such a relation.

Under special circumstances, it is convenient to complement the equations of motion by means of a *geometrical* relation between just the pressure and density. We then assume that

$$p = p(\rho). \tag{14}$$

Every model for which such a relation exists is called a *barotrope* (or barotropic star). As an example, we can mention the polytropes that, in many instances, serve a useful purpose in that they represent simple systems. Naturally, a zero-temperature white dwarf also satisfies an equation of barotropy; however, physically, it should rather be viewed as a particular limit of a barocline.

The basic distinction between barotropic and baroclinic stars naturally lies in their respective stratification. Obviously, the surfaces of equal density—the isopycnic surfaces—and the isobaric surfaces coincide in a barotrope (lit. "behaving as the pressure"). On the contrary, isopycnic surfaces are in general inclined to and cut the isobaric surfaces in a baroclinic star (lit. "inclined over the pressure"); in the latter case, coincidence sometimes occurs under very specific conditions.

4.3.   BAROTROPES, PSEUDO-BAROTROPES, AND BAROCLINES

Many useful properties can be deduced from equations (9) and (10). For that purpose, let us first define the effective gravity **g** as being the gravitational attraction (per unit mass) corrected for the centrifugal acceleration. In cylindrical coordinates, we thus have

$$g_\varpi = -\frac{\partial V}{\partial \varpi} + \Omega^2 \varpi, \tag{15}$$

and

$$g_z = -\frac{\partial V}{\partial z}. \tag{16}$$

Accordingly, equations (9) and (10) can be rewritten in the form

$$\frac{1}{\rho} \operatorname{grad} p = \mathbf{g}. \tag{17}$$

A first deduction can readily be made:

Given a star in a state of permanent rotation, the effective gravity is everywhere orthogonal to the isobaric surfaces.

This property is valid for both barotropic- and baroclinic-stars.

THE POINCARÉ-WAVRE THEOREM

Let us now assume the star to rotate as a solid body. Equations (15) and (16) then become

$$\mathbf{g} = -\operatorname{grad} \Phi, \tag{18}$$

where, except for an additive constant, we have

$$\Phi(\varpi,z) = V(\varpi,z) - \tfrac{1}{2}\Omega^2\varpi^2. \tag{19}$$

Consider next a differentially rotating star. Under what circumstances can one also derive the effective gravity from a potential? By virtue of equation (16), this is possible if and only if Ω does not depend on z, i.e., when the angular velocity is a constant over cylindrical surfaces centered about the axis of rotation. Instead of equation (19), we obtain

$$\Phi(\varpi,z) = V(\varpi,z) - \int^\varpi \Omega^2(\varpi')\varpi'\,d\varpi'. \tag{20}$$

As we shall presently see, various interesting conclusions can be inferred from the existence of such a potential.

First of all, by virtue of equation (17), we can always write

$$\frac{1}{\rho} dp = g_\varpi \, d\varpi + g_z \, dz. \tag{21}$$

Using equations (18) and (20), we thus find that

$$\frac{1}{\rho} dp = -d\Phi. \tag{22}$$

By definition, for any displacement on a level surface $\Phi = constant$, we have $d\Phi = 0$. Since the foregoing equation shows that $dp = 0$ on the same surface, the isobaric surfaces coincide with the level surfaces. Hence, we have

$$p = p(\Phi) \quad \text{or} \quad \Phi = \Phi(p) \tag{23}$$

and

$$\frac{1}{\rho} = -\frac{d\Phi}{dp}. \tag{24}$$

Therefore, the density is also a constant over a level surface. In conclusion, the surfaces upon which p, ρ, and Φ remain a constant all coincide. As a consequence, when a potential Φ does exist, the vector $\mathbf{g}$ is normal to the isopycnic surfaces.

Reciprocally, let us consider a star for which the isobaric- and isopycnic-surfaces coincide. If we let

$$\Phi(p) = -\int \frac{dp}{\rho(p)}, \tag{25}$$

equation (21) becomes

$$d\Phi = -g_\varpi \, d\varpi - g_z \, dz. \tag{26}$$

The quantity $d\Phi$ is an exact total differential; accordingly, equation (18) must hold true, and $\mathbf{g}$ may be derived from a potential.

Finally, let us suppose that the effective gravity is everywhere normal to the isopycnic surfaces. By virtue of equation (21), we see that the pressure is a constant over such a surface. The coincidence of isobaric- and isopycnic-surfaces is thus established.

If we now collect all the pieces, it is a simple matter to see that we have proved the following theorem:

For a star in a state of permanent rotation, any of the following state-
ments implies the three others: (i) the angular velocity is a constant
over cylinders centered about the axis of rotation, (ii) the effective
gravity can be derived from a potential, (iii) the effective gravity is

normal to the isopycnic surfaces, (iv) the isobaric- and isopycnic-surfaces coincide.

These important results were first demonstrated by Poincaré and Wavre.

The casual student would be tempted to conclude that these four equivalent propositions refer solely to barotropes, and exclude altogether baroclinic stars. This is not quite true. To prove it, consider again a system for which Ω depends on both ϖ and z. If we eliminate the gravitational potential V between equations (9) and (10), we can write

$$\frac{\partial}{\partial z}(\Omega^2 \varpi) = \frac{\partial}{\partial z}\left(\frac{1}{\rho}\right)\frac{\partial p}{\partial \varpi} - \frac{\partial}{\partial \varpi}\left(\frac{1}{\rho}\right)\frac{\partial p}{\partial z}. \tag{27}$$

Returning to equation (1), we obtain

$$2\frac{\partial \Omega}{\partial z}\mathbf{v} = \text{grad}\,\frac{1}{\rho} \times \text{grad}\,p. \tag{28}$$

This equation readily shows that the isobaric- and isopycnic-surfaces all coincide if and only if

$$\frac{\partial \Omega}{\partial z} = 0. \tag{29}$$

This is a direct consequence of the equations of motion and requires no prior knowledge of the equation of state.

Clearly, condition (29) is always satisfied for barotropes in uniform or nonuniform rotation. However, if we consider a rotating barocline for which we *impose* condition (29), equation (28) shows that the isobaric- and isopycnic-surfaces must also coincide in that case—although this is most unnatural and not required *a priori* by the equation of state that relates p, ρ, T, and possibly other parameters! Thus, in general, we can assert that

Whatever the equation of state, baroclinic stars for which $\partial\Omega/\partial z = 0$ are characterized by the following properties: (i) the effective gravity can be derived from a potential, (ii) the effective gravity is normal to the isopycnic surfaces, and (iii) the isobaric- and isopycnic-surfaces always coincide.

Henceforth, these very peculiar baroclines will be called "pseudobarotropes," for they share most of the properties of barotropes. They should be contrasted with *true* baroclines about which we can now make the following statement:

Given a nonuniformly rotating baroclinic star for which the angular velocity depends on both ϖ and z, (i) the effective gravity cannot be derived from a potential, (ii) in no case is the effective gravity normal to

*the isopycnic surfaces, and (iii) the isobaric- and isopycnic-surfaces are
always inclined to each other by a finite angle.*

Further drastic limitations on radiating baroclines and pseudo-barotropes
in a state of permanent rotation will be set forth in Chapter 7.

In summary, if we exclude the peculiar case of pseudo-barotropes, we
can say that barotropic stars are characterized by the two equivalent
statements

$$\frac{\partial \Omega}{\partial z} = 0 \quad \text{and} \quad p = p(\rho). \tag{30}$$

Furthermore, for both barotropes and pseudo-barotropes, the stratification is such that the surfaces upon which p, ρ, and Φ assume a constant
value coincide; also, the lines of force of the effective gravity are orthogonal
to these surfaces. The above properties will greatly simplify the problem
of constructing rotating configurations. On the contrary, matters are not
as simple when we consider genuine baroclines for, then, none of the above
properties remains true. Of particular importance is the fact that the
effective gravity is no longer orthogonal to the isopycnic surfaces.

THE BJERKNES-ROSSELAND RULES

Further conclusions can be deduced from equations (27) or (28). Indeed,
these equations show that when the centrifugal acceleration $\Omega^2 \varpi$ varies
in a direction parallel to the axis of rotation, the isobaric- and isopycnic-
surfaces are inclined to each other by a finite angle. Assume that the
angular velocity is directed along the positive z-axis. Hence, the vectors
$\text{grad}(1/\rho)$ and $\text{grad} \, p$ lie in a plane passing through the z-axis, and the
velocity $\mathbf{v}$ is directed in the positive azimuthal direction. With the foregoing convention, the angle between the vectors $\text{grad}(1/\rho)$ and $\text{grad} \, p$ is
to be measured in the direction away from the axis of rotation.

If $\partial \Omega / \partial z$ is positive at a given point, the right-hand side of equation (28)
must also be positive. Accordingly, at that point, the angle between the
z-axis and $\text{grad} \, p$ is smaller than the angle between $\text{grad}(1/\rho)$ and the same
axis. This means that the isobaric surfaces are *more* oblate than the
isopycnic surfaces. As an example, consider a chemically homogeneous
star that satisfies the equation of state of a perfect gas, i.e., $T \propto p/\rho$; assume
further that the density regularly decreases outward. If $\partial \Omega / \partial z$ is positive
all the way from the poles to the equator, matter on an isobaric surface
is consequently colder at the poles than at the equator, i.e., the isothermal
surfaces are *more* oblate than the isobaric surfaces. When $\partial \Omega / \partial z$ is negative,
these conditions are exactly reversed. Of course, if $\partial \Omega / \partial z$ vanishes throughout the configuration, we come back to the case where the isobaric- and

isopycnic-surfaces coincide; with the present hypothesis, the temperature also assumes a constant value over these surfaces.

THE LICHTENSTEIN THEOREM

A further distinction between barotropes and baroclines occurs with respect to the existence of an equatorial plane of symmetry. Contrary to a current belief, this is far from being a trivial question, for such a plane may not exist! The problem was thoroughly investigated by Lichtenstein and Wavre, who have proved that

> *Rotating stars for which the angular velocity does not depend on z always possess an equatorial plane of symmetry which is perpendicular to the axis of rotation.*

A rigorous mathematical demonstration is outside the scope of the present work. As expected, the case of genuine baroclines is somewhat more involved. Indeed, the angular velocity then depends on both ϖ and z, and a certain condition must be satisfied to insure the existence of a plane of symmetry. Again omitting the proof, we can summarize the result in the following statement:

> *Given a differentially rotating barocline, if the angular velocity is everywhere a single-valued function of the density ρ and the distance ϖ from the axis, then the body is symmetric with respect to an equatorial plane.*

In other words, consider an arbitrary line that is parallel to the axis of rotation; whenever it crosses the body, it intersects each isopycnic surface at two points. If $\Omega(\varpi, z)$ assumes the same value at these two points, $\Omega(\rho, \varpi)$ is a single-valued function; and this insures the existence of a plane of symmetry. The question was originally discussed by Dive. From now on, when dealing with baroclinic models, we shall always *postulate* the existence of an equatorial plane of symmetry.

4.4.   SPHEROIDAL STRATIFICATIONS

At this point, it is of some relevance to ask under what conditions isobaric surfaces can be described by means of a suitable set of concentric spheroids. The question was first raised by Hamy in 1889, and is not as simple as it may at first appear. To illustrate the problem, let us again restrict ourselves to uniformly rotating stars. By virtue of equation (11), on applying the Laplacian operator to both sides of equation (19), we obtain

$$\nabla^2 \Phi = 4\pi G \rho - 2\Omega^2. \tag{31}$$

Moreover, for uniformly rotating bodies, the density is only a function of Φ. Hence, equation (31) can be written symbolically in the form

$$\frac{\partial^2 \Phi}{\partial \varpi^2} + \frac{1}{\varpi} \frac{\partial \Phi}{\partial \varpi} + \frac{\partial^2 \Phi}{\partial z^2} = f(\Phi). \tag{32}$$

Quite generally, we now assume as did Volterra and Pizzetti that the level surfaces $\Phi = constant$ are given by

$$F(\varpi, z, \Phi) = 0, \tag{33}$$

where, for the time being, the above function remains unspecified. We can thus write

$$\frac{\partial F}{\partial \varpi} + F' \frac{\partial \Phi}{\partial \varpi} = 0, \tag{34}$$

where a dash denotes a derivative with respect to Φ; we also have

$$\frac{\partial^2 F}{\partial \varpi^2} + 2 \frac{\partial F'}{\partial \varpi} \frac{\partial \Phi}{\partial \varpi} + F' \frac{\partial^2 \Phi}{\partial \varpi^2} + F'' \left(\frac{\partial \Phi}{\partial \varpi} \right)^2 = 0, \tag{35}$$

and two similar equations in which ϖ is replaced by z. Using these four equations to eliminate the derivatives of Φ in equation (32), we eventually obtain

$$F'^2 \nabla^2 F - 2F' \left(\frac{\partial F}{\partial \varpi} \frac{\partial F'}{\partial \varpi} + \frac{\partial F}{\partial z} \frac{\partial F'}{\partial z} \right)$$

$$+ F'' \left[\left(\frac{\partial F}{\partial \varpi} \right)^2 + \left(\frac{\partial F}{\partial z} \right)^2 \right] + F'^3 f = 0. \tag{36}$$

Clearly, if the assigned stratification provides an acceptable solution, equation (36) must be identically satisfied for all ϖ, z, and Φ related by equation (33).

Choose now a spheroidal stratification of the form

$$F(\varpi, z, \Phi) = \alpha(\Phi)\varpi^2 + z^2 - h(\Phi) = 0, \tag{37}$$

where α and h designate two arbitrary functions of Φ. Equation (36) thus becomes

$$2(2\alpha + 1)(\alpha'\varpi^2 - h')^2 - 8(\alpha'\varpi^2 - h')\alpha\alpha'\varpi^2$$

$$+ 4(\alpha''\varpi^2 - h'')[\alpha(\alpha - 1)\varpi^2 + h] + (\alpha'\varpi^2 - h')^3 f = 0, \tag{38}$$

from which we have already eliminated the variable z. The foregoing equation must be satisfied for all values of ϖ, if spheroids are to provide an acceptable solution.

The last term in equation (38) is of degree six in ϖ, while the remaining terms are of degree four in the same variable. We conclude that

$$\alpha' = 0 \quad \text{or} \quad \alpha = constant. \tag{39}$$

Spheroids (37) must be homothetic and equation (38) reduces to

$$2(2\alpha + 1)h'^2 - 4h''[\alpha(\alpha - 1)\varpi^2 + h] - h'^3 f = 0. \tag{40}$$

We naturally exclude $\alpha = 1$ as a possible solution, for it corresponds to a spherical stratification. Hence, we are left with the choice

$$h'' = 0 \quad \text{or} \quad h' = constant; \tag{41}$$

and, therefore, we must also assume that f is a constant. In view of equations (31) and (32), this implies in turn that ρ must be a constant! In conclusion, the sole acceptable solutions are uniformly rotating spheroids of constant density, with the pressure being a constant over homothetic spheroids.

Similarly, it can be proved that differentially rotating stars for which Ω depends only on ϖ do not allow for spheroidal stratifications. The proof will be omitted here. Nevertheless, we can now enunciate a general proposition:

> *Given a centrally condensed barotrope or pseudo-barotrope in a state of permanent rotation, the isobaric surfaces cannot be described by means of a set of concentric spheroids.*

This result calls for two remarks. First, if we assume that departure from sphericity is small, spheroidal stratifications *can* be used to describe slowly rotating bodies in a first approximation. Many examples will be found at various places in the present work. Second, the foregoing proposition does *not* apply to genuine baroclines. This was originally shown by Dive. Actually, many such models can be constructed by using appropriate spheroidal stratifications. However, realistic stellar models cannot be adequately described by means of these simple solutions. We shall not pursue this matter any further.

4.5. SOME INEQUALITIES

From a purely mechanical point of view, the specification of a particular model in a state of permanent rotation depends on three quantities: (i) the total mass M, (ii) the total angular momentum J, and (iii) an assigned distribution for the angular momentum per unit mass.[2] Consider now a

---

[2] The fact that we neglect friction compels us to prescribe the dependence of the angular momentum per unit mass on the coordinates. Only when viscous effects are taken into account can the rotation field be deduced from the hydrodynamical equations *and* the boundary conditions (cf. §7.5).

fixed value of M and a given rotation field, and construct a sequence of models along which J increases. Besides J, what is the most convenient parameter that characterizes the gross features of a model along the sequence? In the case of a spheroid, we could naturally use the eccentricity of a meridional section; but this is not very suitable when we are dealing with an object for which the boundary is numerically defined. We could also use the equatorial angular velocity Ω_e; again, this is not the best choice, since Ω_e does not always monotonically increase with J. Another possibility is provided by the ratio of the rotational kinetic energy K to the gravitational potential energy W, i.e.,

$$\tau = K/|W|, \tag{42}$$

where

$$K = \frac{1}{2} \int_{\mathscr{V}} \rho \Omega^2 \varpi^2 \, \mathbf{dx}, \tag{43}$$

and

$$W = \frac{1}{2} \int_{\mathscr{V}} \rho V \, \mathbf{dx}. \tag{44}$$

By virtue of the scalar virial theorem, we have

$$2K - |W| + 3 \int_{\mathscr{V}} p \, \mathbf{dx} = 0 \tag{45}$$

(cf. §3.7). As the volume integral over the pressure always remains a non-negative quantity, we can thus write

$$0 \leqslant \tau \leqslant \tfrac{1}{2}. \tag{46}$$

Of course, at this stage we do not know *a priori* whether the entire domain of values for τ (or J) can be covered by suitable models.

To justify our choosing τ as a convenient parameter, let us illustrate the problem by means of uniformly rotating, homogeneous spheroids, i.e., the Maclaurin spheroids. Their existence was proved in the previous section; their actual derivation is straightforward for the gravitational attraction in a homogeneous spheroid is well known. We have

$$\frac{\partial V}{\partial \varpi} = 2\pi G \rho A_1 \varpi \quad \text{and} \quad \frac{\partial V}{\partial z} = 2\pi G \rho A_3 z, \tag{47}$$

where

$$A_1(e) = \frac{(1 - e^2)^{1/2}}{e^3} \sin^{-1} e - \frac{1 - e^2}{e^2}, \tag{48}$$

and

$$A_3(e) = \frac{2}{e^2} - \frac{2(1 - e^2)^{1/2}}{e^3} \sin^{-1} e. \tag{49}$$

The eccentricity e is given by

$$e = (1 - a_3{}^2/a_1{}^2)^{1/2}, \tag{50}$$

where a_1 and a_3 denote the two semi-axes of a meridional section ($a_3 \leqslant a_1$). Both the gravitational attraction and the centrifugal force are linear in the coordinates. Hence, in view of equations (9) and (10), the pressure must be a quadratic form:

$$p(\varpi,z) = 2\pi G\rho^2 a_3{}^2 A_3 \left(1 - \frac{\varpi^2}{a_1{}^2} - \frac{z^2}{a_3{}^2}\right). \tag{51}$$

This result is in agreement with equation (39). We also have

$$\frac{\Omega^2}{2\pi G\rho} = A_1 - (1 - e^2)A_3, \tag{52}$$

$$\tau = \frac{A_1 - (1 - e^2)A_3}{2A_1 + (1 - e^2)A_3}, \tag{53}$$

and

$$\frac{J}{(GM^3\bar{a})^{1/2}} = \frac{\sqrt{6}}{5}(1 - e^2)^{-1/3}[A_1 - (1 - e^2)A_3]^{1/2}, \tag{54}$$

where $\bar{a} = (a_1{}^2 a_3)^{1/3}$. Figures 4.1 and 4.2 illustrate the behavior of e, $f[=(1 - a_3/a_1)]$, J and Ω^2 as a function of τ (see also Appendix D). The eccentricity e and the ellipticity f vary from zero to one, while J increases from zero to infinity. Models can thus be constructed for all values of J; they range from a sphere to an infinitesimally flat disc. When expressed as a function of τ, Ω^2 at first increases, then reaches a maximum, and from there monotonically decreases along the sequence.

As we shall see in Chapters 10–13, the Maclaurin spheroids closely imitate in all essential respects sequences of differentially rotating, centrally condensed barotropes. On the contrary, for stellar models in uniform rotation, the sequences usually terminate well before the limit $\tau = 1/2$ may be reached. In all cases, the angular velocity never increases from zero to infinity. In view of these anticipated results, it would be desirable to derive at this stage an upper limit to the angular velocity without prior knowledge of the detailed structure of a model.

Consider a uniformly rotating star with an otherwise unspecified mass distribution. If we integrate equation (31) over the entire volume $\mathscr{V}$, we find

$$\oint_{\mathscr{S}} \frac{d\Phi}{dn} dS = 4\pi GM - 2\Omega^2\mathscr{V}. \tag{55}$$

(In establishing equation (55), we made use of the following relations:

$$\int_{\mathscr{V}} \nabla^2\Phi \, \mathbf{dx} = \int_{\mathscr{V}} \text{div grad } \Phi \, \mathbf{dx} = \oint_{\mathscr{S}} \text{grad } \Phi.\mathbf{dS} = \oint_{\mathscr{S}} \frac{d\Phi}{dn} dS, \tag{56}$$

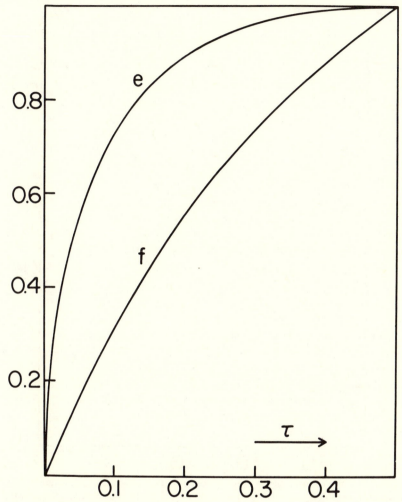

Figure 4.1. The eccentricity e and the ellipticity f along the Maclaurin sequence, as functions of the ratio $\tau = K/|W|$. (See also Appendix D.)

where n denotes the outer normal to the surface.) Clearly, a necessary condition for equilibrium is that the effective gravity **g** be directed inwards at every point on the surface. Thus, we must have

$$\oint_{\mathscr{S}} \frac{d\Phi}{dn}\, dS > 0, \tag{57}$$

or, in view of equation (55),

$$\Omega^2 < 2\pi G\bar{\rho}, \tag{58}$$

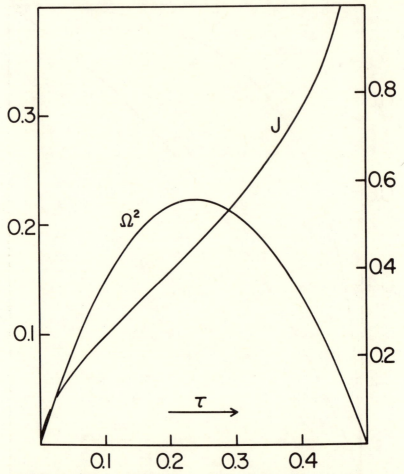

Figure 4.2.   The squared angular velocity Ω^2 and the total angular momentum J along the Maclaurin sequence, as functions of the ratio $\tau = K/|W|$. Units are defined in equations (52) and (54). Ordinates on the left and on the right refer to Ω^2 and J, respectively. (See also Appendix D.)

where $\bar{\rho}$ denotes the mean density $M/\mathscr{V}$. Inequality (58) was originally derived by Poincaré in 1885.

A great deal of effort has been spent refining the foregoing inequality. Under similar assumptions, Crudeli and Nikliborc proved that

$$\Omega^2 < \pi G \rho_c, \tag{59}$$

where ρ_c is the central density—if we assume that ρ monotonically decreases outward. Clearly, equation (59) is no better than equation (58) when $\rho_c/\bar{\rho} > 2$. In 1959, Quilghini obtained the new limit

$$\Omega^2 < \pi G\bar{\rho}, \tag{60}$$

which is a definite improvement but still well above the maximum value attained by Ω^2 for the Maclaurin spheroids, i.e., $\Omega^2 \approx 0.45\pi G\rho$.

Turning next to barotropes and pseudo-barotropes in a state of differential rotation, we must naturally use equation (20). Then, instead of equation (31), we have

$$\nabla^2\Phi = 4\pi G\rho - 2\Omega^2 - \varpi\,\frac{d\Omega^2}{d\varpi}, \tag{61}$$

and inequality (58) becomes

$$\mathscr{V}^{-1}\int_{\mathscr{V}}\left(\Omega^2 + \frac{1}{2}\varpi\,\frac{d\Omega^2}{d\varpi}\right)d\mathbf{x} < 2\pi G\bar{\rho}. \tag{62}$$

The foregoing criterion is due to Wilczynski. Moreover, for any differentially rotating star, it can also be shown that

$$\Omega^2_{\min} + \tfrac{1}{2}g_{\min}(\mathscr{A}/\mathscr{V}) < 2\pi G\bar{\rho} < \Omega^2_{\max} + \tfrac{1}{2}g_{\max}(\mathscr{A}/\mathscr{V}). \tag{63}$$

The quantities $\Omega_{\min}$ and $\Omega_{\max}$ designate, respectively, the minimum and maximum values of Ω on the surface; similarly, $g_{\min}$ and $g_{\max}$ are the extreme values of $|\mathbf{g}|$ on the surface, and $\mathscr{A}$ denotes the area of the surface $\mathscr{S}$. These limitations have been derived by Dive.

4.6. THE OUTER GRAVITATIONAL FIELD

As we know, the gravitational field exterior to a spherical distribution of matter depends solely on the total mass M of the attracting body. Now, on what elements does the outer gravitational field of a rotating star depend? The question was originally raised by Stokes in 1849, and a clear-cut answer was first given by Poincaré in the case of uniformly rotating bodies. The following basic statement can be made, and is due to Wavre:

The outer gravitational potential due to a rotating star is entirely determined by the total mass M, the shape of the boundary $\mathscr{S}$, and the distribution of angular velocity on the surface of the star.

Leaving aside some mathematical subtleties, the proof is straightforward. By virtue of equations (9) and (10), we can always write

$$\frac{1}{\rho}\,dp = -dV + \Omega^2\varpi\,d\varpi, \tag{64}$$

where $\Omega^2(\varpi,z)\varpi$ does not in general derive from a potential. Let us draw a line l on the surface of the star. Along that arbitrary line, we always have $dp = 0$; hence, we obtain

$$dV_s = \Omega_s^2 \varpi \, d\varpi, \tag{65}$$

where a suffix "s" indicates a value on the surface $\mathscr{S}$. Integrate next along the line l from a fixed point P_0 to a variable point P, and let

$$\Gamma = \int_{P_0}^{P} \Omega_s^2 \varpi \, d\varpi. \tag{66}$$

Because V_s is a single-valued function and dV_s is an exact differential, the line integral Γ only depends on the limiting points P_0 and P, and is independent of the path of integration drawn on the surface. We thus have

$$V_s(P) = V_s(P_0) + \Gamma(P_0,P). \tag{67}$$

Now, the outer gravitational potential V_E can always be written in the form[3]

$$V_E(Q) = -\frac{1}{4\pi} \oint_{\mathscr{S}} \frac{1}{D}\left(\frac{dV}{dn}\right)_s dS + \frac{1}{4\pi} \oint_{\mathscr{S}} V_s \frac{d}{dn}\left(\frac{1}{D}\right) dS, \tag{68}$$

where D denotes the distance from an external point Q to the point P. Also, by virtue of equation (11), we obtain

$$M = \frac{1}{4\pi G} \oint_{\mathscr{S}} \left(\frac{dV}{dn}\right)_s dS. \tag{69}$$

Substituting equation (67) into equation (68), we find

$$V_E(Q) = -\frac{1}{4\pi} \oint_{\mathscr{S}} \frac{1}{D}\left(\frac{dV}{dn}\right)_s dS + \frac{1}{4\pi} \oint_{\mathscr{S}} \Gamma(P,P_0) \frac{d}{dn}\left(\frac{1}{D}\right) dS, \tag{70}$$

for $V(P_0)$ is a constant and $1/D$ is an harmonic function. In view of equation (69), $(dV/dn)_s$ is entirely determined by a knowledge of M and $\mathscr{S}$. Thus, the first integral on the r.h.s. of equation (70) is a function of M and $\mathscr{S}$ only. The second term in this equation depends solely on $\mathscr{S}$ and $\Omega_s(\varpi,z)$. This concludes the proof.

In summary, as in the case of a spherical body, no firm conclusion regarding the *internal* stratification of a rotating star can be inferred from the *sole* knowledge of its outer gravitational field. This precludes any investigation of the interior of the Sun—or, for that matter, of a planet—by means of a space probe.

---

[3] See, e.g., Kellogg, O. D., *Foundations of Potential Theory*, pp. 218–223, New York: Ungar Publishing Co., 1929 (New York: Dover Public., Inc., 1954).

BIBLIOGRAPHICAL NOTES

In the past, the properties of self-gravitating bodies in a state of permanent rotation were investigated mainly with a view to their possible applications to geodesy and planetary physics. Accordingly, most authors explicitly stated in their work that the systems under consideration were incompressible. Given the basic assumptions made in Section 4.2, the above hypothesis is entirely redundant. As a matter of fact, many a classical result derived during the period 1740–1940 still retains its usefulness today when applied to gaseous stars. Historic contributions are summarized in reference 7 of Chapter 1. Most of the properties reported in the present chapter are discussed in three monographs:

1. Dive, P., *Rotations internes des astres fluides*, Paris: Albert Blanchard, 1930.
2. Wavre, R., *Figures planétaires et géodésie*, Paris: Gauthier–Villars, 1932.
3. Lichtenstein, L., *Gleichgewichtsfiguren Rotierender Flüssigkeiten*, Berlin: Springer-Verlag, 1933.

Unless otherwise stated below, quotations in the text refer to the foregoing trilogy. Succint accounts of the work by Liapunov, Lichtenstein, Poincaré, and Wavre will be found in reference 10 of Chapter 1. See also:

4. Lebovitz, N. R., "Rotating Fluid Masses" in *Annual Review of Astronomy and Astrophysics* **5**, pp. 465–480, Palo Alto, California: Annual Reviews, Inc., 1967.

Section 4.2. A general discussion of permanent motions (and the requirements they impose on celestial bodies) is given in:

5. Appell, P., *Acta Mathematica* **47** (1926): 15–23.
6. Véronnet, A., *C.R. Acad. Sci. Paris* **185** (1927): 30–32, 249–252.

The following reference may also be noted:

7. Volterra, V., *Rotation des corps dans lesquels existent des mouvements internes*, Paris: Gauthier-Villars, 1938.

Section 4.3. The restriction imposed upon the angular velocity of a barotrope has been originally discussed by Poincaré:

8. Poincaré, H., *Théorie des tourbillons*, pp. 176–178, Paris: Georges Carré, 1893.

The question was revived in a more general context by Véronnet:

9. Véronnet, A., *C.R. Acad. Sci. Paris* **183** (1926): 949–951.

An exhaustive investigation of rotating barotropes was first made by

Wavre. Unfortunately, most of these results have been often ignored or misquoted. The discussion of baroclines is taken from:

10. Bjerknes, V., *Geofysiske Publikasjoner* (*Kristiania*) **2**, No 4 (1921): 44–46.
11. Rosseland, S., *Astroph. J.* **63** (1926): 342–367.

Reference 1 of Chapter 3 also contains a penetrating analysis of the problem. A further work on baroclines is:

12. Walker, E., *Astron. Nachr.* **263** (1937): 253–292.

The tricky question of equatorial symmetry has been adequately discussed by Lichtenstein, Wavre, and Dive only.

Section 4.4. The classical papers on the subject of spheroidal stratifications are:

13. Hamy, M., *Annales de l'Observatoire de Paris, Mémoires* **19** (1889): F1–F54.
14. Hamy, M., *J. Math. pures et appliquées* (4) **6** (1890): 69–143.
15. Volterra, V., *Acta Mathematica* **27** (1903): 105–124.
16. Véronnet, A., *J. Math. pures et appliquées* (6) **8** (1912): 331–463.
17. Pizzetti, P., *Principii della teoria meccanica della figura dei pianeti*, pp. 187–212, Pisa: Enrico Spoerri, 1913.
18. Moulton, E. J., *Trans. American Math. Soc.* **17** (1916): 100–108.
19. Plummer, H. C., *Monthly Notices Roy. Astron. Soc. London* **80** (1919): 26–33.

Major developments and pertinent comments are contained in reference 1. Later discussions of the same problem are those of:

20. Mineo, C., *Rendiconti Accad. Lincei* (6A) **26** (1937): 260–262.
21. Wavre, R., *Comm. Math. Helvetici* **11** (1938): 33–36.
22. Dive, P., *Ann. Scient. Ecole Normale Sup. Paris* (3) **56** (1939): 293–316.
23. Mendes, M., *J. Math. pures et appliquées* (9) **24** (1945): 51–72.
24. Mineo, C., *Rendiconti Accad. Lincei* (8A) **1** (1946): 21–27.
25. Ghosh, N. L., *Bull. Calcutta Math. Soc.* **40** (1948): 229–230; *ibid.* **41** (1949): 92–102, 183–186, 220.
26. Dive, P., *Bull. Sci. Math.* (2) **76** (1952): 38–50.

The analysis in the text follows references 15, 17, and 21. A complete discussion of differentially rotating barotropes necessarily requires a manipulation of the gravitational potential of a spheroid. In this respect, the last paper by Dive is particularly noteworthy. Further properties of baroclinic models with an assigned spheroidal stratification were discussed in:

27. Dive, P., *Bull. Sci. Math.* (2) **63** (1939): 182–192.

Exact baroclinic models have been constructed in the following papers:

28. Ganguly, H. K., *Bull. Calcutta Math. Soc.* **31** (1939): 69–76.
29. Ganguly, H. K., *Zeit. f. Astroph.* **19** (1940): 136–153.
30. Ghosh, N. L., *Bull. Calcutta Math. Soc.* **42** (1950): 101–117, 227– 242, 243–248.
31. Ghosh, N. L., *Proc. Nat. Inst. Sci. India* **17** (1951): 391–401.

Section 4.5. Properties of the Maclaurin spheroids are discussed, e.g., in reference 28 of Chapter 3. A first-hand derivation of the Poincaré criterion is given in:

32. Poincaré, H., *Figures d'équilibre d'une masse fluide*, pp. 11–12, Paris: Gauthier-Villars, 1902.

Inequality (59) was first derived by Crudeli under the hypothesis that the equilibrium structure is convex:

33. Crudeli, U., *Rendiconti Accad. Lincei* (5A) **19**, 1 (1910): 666–668; *ibid.* **19**, 2 (1910): 41–43.

Nikliborc proved that the assumption of convexity was unnecessary:

34. Nikliborc, W., *Math. Zeit.* **30** (1929): 787–793.

This inequality has often been misquoted. Refined limitations on Ω^2 are contained in:

35. Nikliborc, W., *ibid.* **31** (1930): 366–377.
36. Crudeli, U., *Boll. Unione Mat. Italiana* (1) **9** (1930): 257–258.
37. Crudeli, U., *Rendiconti Circolo Mat. Palermo* **57** (1933): 308–310.
38. Crudeli, U., *Mem. Soc. Astron. Italiana* **7** (1933): 131–135.
39. Caldonazzo, B., *Rendiconti Semin. Mat. Fis. Milano* **22** (1952): 212–222.

The reference to Quilghini is to:

40. Quilghini, D., *Boll. Unione Mat. Italiana* (3) **14** (1959): 482–488.
41. Quilghini, D., *Ann. Mat. pura ed applicata* (4) **48** (1959): 281–303.

The first derivation of criterion (62) was made in:

42. Wilczynski, E. J., *Astron. J.* **18** (1897): 57–59.

Section 4.6. The classical paper of Stokes will be found in:

43. Stokes, G. G., *Mathematical and Physical Papers of Sir George Stokes* **2**, pp. 104–130, Cambridge: At the Univ. Press, 1883.

Poincaré's contribution to the problem is contained in reference 32 (pp. 97–112). The analysis in this section follows:

44. Wavre, R., *C.R. Acad. Sci. Paris* **194** (1932): 1447–1449.

This last paper by Wavre is *not* included in reference 2, and covers many situations that are outside the framework of our approximations. References to this and related matters are:

45. Crudeli, U., *Rendiconti Circolo Mat. Palermo* **59** (1935): 336–342.
46. Putnis, A., *Comm. Math. Helvetici* **8** (1935): 181–185.
47. Wavre, R., *C.R. Soc. Phys. Hist. Nat. Genève* **52** (1935): 137–139.
48. Chiara, L., *Mem. Soc. Astron. Italiana* **12** (1939): 105–117.
49. Mineo, C., *Rendiconti Accad. Lincei* (8A) **12** (1952): 635–642.
50. Chiara, L., *Accad. Sci. Lett. Arti Palermo, Parte I: Scienze* (4) **16** (1955): 69–100.

Further contributions by Crudeli and Mineo are quoted in references 45 and 49, respectively. This last paper also contains many pertinent comments on the literature.

5

Stellar Models: Techniques

5.1. INTRODUCTION

In the preceding chapter we discussed some general mechanical properties of stars in a state of permanent rotation. Although, on physical grounds, we must distinguish between barotropes and baroclines, rotation naturally delineates *three* classes of stars: (i) the barotropes, (ii) the pseudo-barotropes, and (iii) the genuine baroclines (cf. §4.3). The first two families are chiefly characterized by their having an angular velocity that is a constant over cylinders centered about the axis of symmetry, i.e., $\partial\Omega/\partial z = 0$; on the contrary, the latter condition never obtains in the case of genuine baroclines. (Clearly, barotropes and pseudo-barotropes include the configurations in uniform rotation.) Let us also remember that, for barotropes, we assume a geometrical $p - \rho$ relation; while, for pseudo-barotropes, p is related to ρ, T, and the chemical composition through an assigned equation of state. If we except this physical difference, barotropes and pseudo-barotropes in a state of permanent rotation do share *all* the same mechanical properties.

Postponing to Chapter 7 the study of some thermodynamical properties of radiating stars, we will now discuss the main techniques by which one can determine the equilibrium structure of barotropes and pseudo-barotropes. Our assumptions are the same as those set forth in Section 4.2. In particular, friction—and the ensuing angular momentum transfer—will not be discussed here. Similarly, we shall ignore for the present the delicate question of meridional circulation, which, for pseudo-barotropes in radiative equilibrium, should necessarily be taken into account. This particular problem will be discussed in Chapter 8.

A great number of techniques have been devised to determine the equilibrium of rotating stars. *The main difficulty lies in the fact that the actual stratification of a centrally condensed star is never known in advance.* Obviously, when departure from spherical symmetry is not large, we can use perturbation techniques and treat the effects of rotation as a small distortion superimposed on a known spherical model. Three such techniques are given in Sections 5.2–5.4. However, when the level surfaces greatly deviate from concentric spheres, other methods must preferably be used. In Section 5.5, we briefly describe the self-consistent field technique devised by Ostriker and his associates; this is, by far, the most efficient computational scheme to treat severely distorted stars in uniform or non-uniform rotation. Other approaches—such as purely numerical schemes or variational techniques—will be discussed in Chapter 10. Finally, a quasi-dynamic method that also embodies a stability analysis with respect to axisymmetric motions will be found in Section 6.8.

5.2.   THE CLAIRAUT-LEGENDRE EXPANSION

To outline the method, let us restrict ourselves to uniformly rotating configurations. Then, by virtue of the equations of motion, we can always write

$$\frac{1}{\rho}\, dp = -d\Phi; \tag{1}$$

ignoring an arbitrary additive constant, we have

$$\Phi = V - \tfrac{1}{2}\Omega^2 \varpi^2 \tag{2}$$

(cf. §4.3: equations [19] and [22]). Moreover, as we know, the level surfaces $\Phi = constant$ coincide with the isobaric- and isopycnic-surfaces. Consequently, if we can find a convenient way to describe the level surfaces, we obtain a complete representation of the rotating system.

Let us now suppose that departure from sphericity is small. Assume further that the density monotonically decreases outward. By definition, we call a the *mean* distance of a given surface $\Phi = constant$ to the center of mass. Hence, to each value of the variable a corresponds one (and only one) level surface; and we have

$$\Phi = \Phi(a), \tag{3}$$

where the mean radius a continuously ranges from the center ($a = 0$) to the surface ($a = a_s$, say). As a consequence, both p and ρ can be thought of as functions of the mean radius a only. Now, a point on a given level surface depends on three parameters. It is convenient to use the spherical coordinates r, θ, and φ. On the other hand, the equation of the level surfaces can be written, in general, as

$$r(a,\theta) = a\left[1 - \sum_{n=1}^{\infty} \epsilon_{2n}(a)P_{2n}(\cos\theta)\right], \tag{4}$$

where the P_n's designate the Legendre polynomials, and the (yet unknown) functions $\epsilon_{2n}(a)$ describe the deviations of the level surfaces from spherical symmetry. Obviously, given the existence of an equatorial plane of symmetry, only the *even* terms should be retained in the expansion. Also, as we assume the rotational distortion to be small, we surmise the function $\epsilon_2(a)$

to be of paramount importance in our description of uniformly rotating bodies.[1]

Consider first the gravitational potential V. At a given interior point $P(r,\theta,\varphi; a)$ which belongs to the level surface of mean radius a, the function $V(P)$ is given by the sum of two terms: (i) the potential $V_i(P)$ due to the attraction of the mass located within the level surface $\Phi = \Phi(a)$, and (ii) the potential $V_e(P)$ due to the attraction of the mass between this surface and the outer boundary $\Phi = \Phi(a_s)$. Thus, we have

$$V(P) = V_i(P) + V_e(P) = -G \int_0^{M_a} \frac{dm'}{|\mathbf{x} - \mathbf{x}'|} - G \int_{M_a}^{M} \frac{dm'}{|\mathbf{x} - \mathbf{x}'|}, \qquad (5)$$

where, in view of our approximations,

$$M_a = 4\pi \int_0^a \rho(a')a'^2 \, da', \qquad (6)$$

and M designates the total mass of the body.

Let $P'(r',\theta',\varphi'; a')$ be a variable point on the level surface of mean radius $a'(a' \lessgtr a)$. Then, the mass element centered at P' is

$$dm' = \rho(a')r'^2(\partial r'/\partial a') \sin \theta' \, d\theta' \, d\varphi' \, da', \qquad (7)$$

where r' is a function of a' through equation (4). The reciprocal of the distance $|\mathbf{x} - \mathbf{x}'|$ between the points P and P' can be expanded in series of Legendre polynomials. We obtain[2]

$$\frac{1}{|\mathbf{x} - \mathbf{x}'|} = \sum_{n=0}^{\infty} \frac{r^n}{r'^{n+1}} P_n(\cos \gamma) \quad (r < r') \qquad (8)$$

and

$$\frac{1}{|\mathbf{x} - \mathbf{x}'|} = \sum_{n=0}^{\infty} \frac{r'^n}{r^{n+1}} P_n(\cos \gamma) \quad (r > r') \qquad (9)$$

where γ designates the angular separation between the two radii vectors, and

$$\cos \gamma = \cos \theta \cos \theta' + \sin \theta \sin \theta' \cos(\varphi - \varphi'). \qquad (10)$$

---

[1] The strict mathematical convergence of the Clairaut-Legendre series has given rise to many discussions. To alleviate the difficulty, other approaches have been used (notably by Liapunov, Lichtenstein, and Wavre); see reference 10 of Chapter 1. However, on the practical side, what matters is the *numerical* convergence of the first few terms of the expansion. The asymptotic nature of the Clairaut-Legendre series has been demonstrated by Lebovitz in the case $n = 1$, i.e., the difference between the exact (unknown) solution and the truncated expansion (17) is of the order of the first neglected term.

[2] See, e.g., Morse, P. M., and Feshbach, H., *Methods of Theoretical Physics* **2**, p. 1274, New York: McGraw-Hill, 1953.

For further references, let us also remember that

$$P_n(\cos \gamma) = P_n(\cos \theta)P_n(\cos \theta')$$

$$+ 2 \sum_{m=1}^{n} \frac{(n-m)!}{(n+m)!} P_n^m(\cos \theta)P_n^m(\cos \theta') \cos m(\varphi - \varphi'), \quad (11)$$

where the P_n^m's denote the Legendre associated functions (see Appendix C); we also have

$$\int_0^\pi \int_0^{2\pi} P_n(\cos \gamma)P_l(\cos \theta') \sin \theta' \, d\theta' \, d\varphi' = \frac{4\pi}{2n+1} P_n(\cos \theta) \delta_{ln}, \quad (12)$$

where $\delta_{ln} = 1$ if $l = n$, and $\delta_{ln} = 0$ if $l \neq n$.

If we now make use of equations (7)–(12), and assume that the distortion of our system is small enough for the squares and cross-products of the ϵ_{2n}'s to be negligible, we find

$$V_i(P) = -\frac{4}{3}\pi G \int_0^a \rho(a') \frac{\partial}{\partial a'} \left(\frac{a'^3}{r} - \sum_{n=1}^{\infty} \frac{3}{4n+1} \frac{a'^{2n+3}}{r^{2n+1}} \epsilon_{2n}P_{2n} \right) da', \quad (13)$$

and

$$V_e(P) = -\frac{4}{3}\pi G \int_a^{a_s} \rho(a') \frac{\partial}{\partial a'} \left(\frac{3}{2}a'^2 - \sum_{n=1}^{\infty} \frac{3}{4n+1} \frac{r^{2n}}{a'^{2n-2}} \epsilon_{2n}P_{2n} \right) da'. \quad (14)$$

As expected, the gravitational potential $V(P)$ does not depend on the azimuthal angle φ.

Returning next to equation (2), we can write the effective potential Φ in the form

$$\Phi = V - \tfrac{1}{2}\Omega^2 r^2 \sin^2 \theta, \quad (15)$$

which, by virture of equation (3), is a function of a only. As we are attempting a first-order theory, we thus have

$$V_i(r,\theta; a) + V_e(r,\theta; a) - \tfrac{1}{3}\Omega^2 a^2[1 - P_2(\cos \theta)] = \textit{function of } a. \quad (16)$$

In the integrals (13) and (14) we can now relate the variables r and a by means of equation (4). Of course, consistent with our approximations, we only retain first-order quantities in the expansion. Given the form of the centrifugal potential in equation (16), we perceive at once that, in the case of a slow uniform rotation, expansion (4) reduces to

$$r(a,\theta) = a[1 - \epsilon(a)P_2(\cos \theta)], \quad (17)$$

in which we write $\epsilon(a)$ for $\epsilon_2(a)$ without confusion. Accordingly, within the framework of our approximations, the functions Φ, p, and ρ assume a constant value over concentric *spheroids* (cf. §4.4). Now, from equation (16), the coefficient of $P_2(\cos \theta)$ must vanish; this gives

$$\frac{5\epsilon(a)}{a} \int_0^a \rho(a')a'^2 \, da' - \frac{1}{a^3} \int_0^a \rho(a') \frac{d}{da} [a'^5 \epsilon(a')] \, da'$$

$$- a^2 \int_a^{a_s} \rho(a') \frac{d}{da'} [\epsilon(a')] \, da' = \frac{5}{12} \frac{\Omega^2}{\pi G} a^2. \quad (18)$$

Multiplying by a^3 and differentiating, we find

$$\left(a^2 \frac{d\epsilon}{da} + 2a\epsilon\right) \int_0^a \rho(a')a'^2 \, da' - a^4 \int_a^{a_s} \rho(a') \frac{d}{da'} [\epsilon(a')] \, da' = \frac{5}{12} \frac{\Omega^2}{\pi G} a^4. \quad (19)$$

Finally, if we divide the foregoing equation by a^4 and differentiate again, we obtain

$$a^2 \frac{d^2\epsilon}{da^2} + 6 \frac{\rho(a)}{\rho_m(a)} \left(a \frac{d\epsilon}{da} + \epsilon\right) = 6\epsilon, \quad (20)$$

where

$$\rho_m(a) = \frac{3}{a^3} \int_0^a \rho(a')a'^2 \, da' \quad (21)$$

is the mean density of the matter comprised within the surface of mean radius a.

If we now let

$$\eta(a) = \frac{a}{\epsilon} \frac{d\epsilon}{da}, \quad (22)$$

equation (20) becomes

$$a \frac{d\eta}{da} + 6 \frac{\rho(a)}{\rho_m(a)} (\eta + 1) + \eta(\eta - 1) = 6. \quad (23)$$

At the origin, we have $a = 0$ and $\rho(0)/\rho_m(0) = 1$. Hence, equation (23) shows that

$$\eta(0) = 0, \quad (24)$$

a condition which is sufficient for a complete specification of the function $\eta(a)$ for any given function $\rho(a)/\rho_m(a)$. Thus, if we consider a spherical configuration for which $\rho(r)/\rho_m(r)$ is known, we can obtain the rotational distortion of the body in solving equation (23). Naturally, as departure from sphericity is small, we can ignore the difference between the variables r and a in the foregoing relations. The function $\epsilon(a)$ can then be obtained from equation (22) together with equation (19), which we evaluate on the surface $a = a_s$.

To conclude, let us mention an interesting property that can be deduced at once from equations (20) and (23). Write

$$m = \frac{3\Omega^2}{4\pi G \rho_m(a_s)} \quad \text{and} \quad f = \tfrac{3}{2}\epsilon(a_s), \tag{25}$$

where m designates the ratio of the centrifugal acceleration at the equator to the (average) gravitational acceleration on the surface, and f denotes the ellipticity of the slightly oblate figure of equilibrium. In view of equation (19), m and f are in a relationship of cause and effect, for we always have

$$\frac{m}{f} = \frac{2}{5}\left[2 + \eta(a_s)\right]. \tag{26}$$

Upper and lower limits for the ratio m/f can be obtained from equation (20) if we restrict ourselves to models in which (in the nonrotating state) the density $\rho(r)$ never exceeds the mean density $\rho_m(r)$. Indeed, any such system is comprised between the homogeneous model ($\rho = \rho_m$ for any value of a) and the Roche model ($\rho/\rho_m = \infty$ for $a = 0$, and $\rho/\rho_m = 0$ for $a > 0$). In the former case, we have $\eta(a) \equiv 0$; and $\eta(a) \equiv 3(a > 0)$ for the latter configuration. Thus, by virtue of equation (26), we can write

$$\frac{4}{5} \leqslant \frac{m}{f} \leqslant 2. \tag{27}$$

Most of the foregoing results were established by Clairaut in 1743 and are still used today.

5.3.   THE CHANDRASEKHAR-MILNE EXPANSION

Again consider a quasi-spherical configuration in slow uniform rotation. Instead of the integral representation (5) for the gravitational potential V, we write down the Poisson equation

$$\nabla^2 V = 4\pi G\rho. \tag{28}$$

In making use of equation (2), we then obtain

$$\nabla^2\Phi = 4\pi G\rho - 2\Omega^2. \tag{29}$$

Remember also that the surfaces upon which Φ, p, and ρ assume a constant value coincide. Hence, once more we can view the density ρ as a sole function of the effective potential Φ.

At this stage, it is convenient to let

$$v = \frac{\Omega^2}{2\pi G\rho_c}, \tag{30}$$

where ρ_c is the central density of the system. From now on, we shall work in spherical coordinates $(r,\mu = \cos\theta,\varphi)$. If R_u denotes the radius of the unperturbed spherical configuration, we can now measure Φ and V in the unit $4\pi G\rho_c R_u^2$. Also, ρ and r are measured in the units ρ_c and R_u, respectively. In the units specified, equation (29) becomes

$$\frac{1}{r^2}\frac{\partial}{\partial r}\left(r^2\frac{\partial\Phi}{\partial r}\right) + \frac{1}{r^2}\frac{\partial}{\partial\mu}\left[(1-\mu^2)\frac{\partial\Phi}{\partial\mu}\right] = \rho - v; \tag{31}$$

we also have

$$\Phi = V - \tfrac{1}{6}vr^2[1 - P_2(\mu)] - V_0, \tag{32}$$

where the constant V_0 designates the value of V at the poles of the configuration.

Suppose next that the rotation is small, so that v may be treated as a small perturbation parameter. On this assumption, the resolution of equation (31) will be carried out consistently to the first order in v. We assume then, for Φ and ρ, a solution of the form

$$\Phi(r,\mu) = V_u(r) + v\left[\Phi_0(r) + \sum_{n=1}^{\infty} A_n\Phi_n(r)P_n(\mu)\right], \tag{33}$$

and

$$\rho(r,\mu) = \rho_u(r) + v\left[\rho_0(r) + \sum_{n=1}^{\infty} B_n\rho_n(r)P_n(\mu)\right], \tag{34}$$

where the A_n's and B_n's are constants unspecified for the present; $V_u(r)$ and $\rho_u(r)$ designate, respectively, the gravitational potential and the density for the corresponding unperturbed spherical configuration.

The relations between the A_n's and B_n's can be obtained by considering ρ as a function of Φ. In expanding ρ in a Taylor's series in a region close to the value $\Phi = V_u$, we find

$$\rho_u(r) = \rho(V_u), \tag{35}$$

$$\rho_0(r) = k\Phi_0(r) \quad \text{and} \quad B_n\rho_n(r) = kA_n\Phi_n(r) \quad (n = 1, 2, \ldots) \tag{36}$$

where

$$k(r) = \frac{d\rho_u}{dV_u} \tag{37}$$

depends only on the nonrotating figure of equilibrium.

Inserting next equations (33) and (34) into equation (31), and neglecting all terms of orders higher than the first in v, we obtain

$$\mathscr{D}_0 V_u = \rho_u, \tag{38}$$

$$\mathscr{D}_0\Phi_0 = k\Phi_0 - 1 \quad \text{and} \quad \mathscr{D}_n\Phi_n = k\Phi_n \quad (n = 1, 2, \ldots) \tag{39}$$

where

$$\mathcal{D}_n = \frac{d^2}{dr^2} + \frac{2}{r}\frac{d}{dr} - \frac{n(n+1)}{r^2}. \tag{40}$$

In establishing the foregoing equations, we made use of relations (35) and (36), and of the equation for the Legendre polynomials. Equation (38) is, of course, the Poisson equation for the unperturbed configuration. Moreover, since $\rho_u = 1$ and $\rho_u' = 0$ at $r = 0$, the solutions of equations (39) must be regular and comply with the conditions

$$\Phi_n(0) = \Phi_n'(0) = 0 \quad (n = 0, 1, 2, \dots) \tag{41}$$

where a prime denotes a derivative with respect to r.

With Φ given by equation (33), we may write the solution for V in the form

$$V(r,\mu) = V_u(r) + c_0 + v\left\{\Phi_0(r) + \sum_{n=1}^{\infty} A_n\Phi_n(r)P_n(\mu)\right.$$

$$\left. + \frac{1}{6}r^2[1 - P_2(\mu)] + c_{1;0}\right\}, \tag{42}$$

where the zero- and first-order contributions to V_0 have been separated:

$$V_0 = c_0 + vc_{1;0}. \tag{43}$$

Now, the potential V_E which is appropriate to the surrounding vacuum satisfies Laplace's equation

$$\nabla^2 V_E = 0. \tag{44}$$

Obviously, the solution of equation (44) which should be associated with solution (42) is

$$V_E(r,\mu) = \frac{\kappa_0}{r} + v\sum_{n=0}^{\infty} \frac{\kappa_{1;n}}{r^{n+1}} P_n(\mu). \tag{45}$$

It remains to determine the various constants that occur in the solutions by requiring the continuity of the gravitational potential and its gradient on the surface of the rotating configuration. For the application of these conditions, it is clearly necessary that we first specify the boundary; let this be the surface:

$$\Xi(\mu) = 1 + v\sum_{n=0}^{\infty} q_n P_n(\mu). \tag{46}$$

The constants q_n introduced in the definition of $\Xi(\mu)$ become determined by the additional requirement that the density vanishes on this surface; by virtue of equations (34)–(36), this requirement gives

$$V_u'(1)q_0 = -\Phi_0(1) \quad \text{and} \quad V_u'(1)q_n = -A_n\Phi_n(1) \quad (n = 1, 2, \dots) \tag{47}$$

Now, the boundary conditions are

$$V(\Xi) = V_E(\Xi) \quad \text{and} \quad V'(\Xi) = V_E'(\Xi). \tag{48}$$

Substituting equations (42) and (45) into the foregoing relations, we eventually obtain

$$c_0 = -V_u'(1), \quad c_{1;0} = -\tfrac{1}{2} - \Phi_0(1) - \Phi_0'(1), \tag{49}$$

$$A_2 = \frac{5}{6} \frac{1}{3\Phi_2(1) + \Phi_2'(1)}, \tag{50}$$

$$\kappa_0 = -V_u'(1), \quad \kappa_{1;0} = -\tfrac{1}{3} - \Phi_0'(1), \tag{51}$$

$$\kappa_{1;2} = \frac{1}{6} \frac{2\Phi_2(1) - \Phi_2'(1)}{3\Phi_2(1) + \Phi_2'(1)}, \tag{52}$$

in which we made use of equations (38) and (47). A careful analysis shows that remaining coefficients must all vanish. This completes the formal solution of the problem.

In summary, we can write

$$\Xi(\mu) = 1 - \frac{v}{V_u'(1)} [\Phi_0(1) + A_2\Phi_2(1)P_2(\mu)], \tag{53}$$

$$V(r,\mu) = V_u(r) + c_0 + v\{\Phi_0(r) + A_2\Phi_2(r)P_2(\mu) \\ + \tfrac{1}{6}r^2[1 - P_2(\mu)] + c_{1;0}\}, \tag{54}$$

$$V_E(r,\mu) = \frac{\kappa_0}{r} + v\left[\frac{\kappa_{1;0}}{r} + \frac{\kappa_{1;2}}{r^3} P_2(\mu)\right], \tag{55}$$

and similar expressions for $\Phi(r,\mu)$ and $\rho(r,\mu)$. Hence, to first order in the parameter v, the rotational distortion of a body can be described by means of the two functions $\Phi_0(r)$ and $\Phi_2(r)$. Both functions must be regular solutions of the second-order equations

$$\mathcal{D}_0\Phi_0 = k\Phi_0 - 1 \quad \text{and} \quad \mathcal{D}_2\Phi_2 = k\Phi_2, \tag{56}$$

and comply with conditions (41). Obviously, their numerical determination requires prior knowledge of the function $k(r)$, i.e., the variation of the density and gravitational potential in the corresponding nonrotating model.

The foregoing method has been widely used to investigate the structure of rotating polytropes and zero-temperature white dwarfs for which departure from sphericity is not very large (cf. §§10.3 and 13.2). As was pointed out by Smith, however, the determination of the structure of rotating models by expansion in the parameter v constitutes a singular perturbation problem. That is to say, as the surface of the rotating configuration is approached, the leading term in expansion (34) approaches zero, so that the second term in this expansion is then not small compared to the first no matter how small v may be. In other words, the perturbation expansion (34) is not uniformly valid over the domain of interest or, more explicitly, the asymptotic series (34) converges but the radius of convergence tends to zero as $r \to 1$. Alternate methods which make use of matched asymptotic expansions and strained coordinates have been proposed by Smith. Both techniques are outside the scope of this book.

5.4. THE QUASI-SPHERICAL APPROXIMATION

Very much in the same vein, other expansions can be used to approximate the structure of centrally condensed bodies. Indeed, in such objects, most of the mass is contained in the central core; hence, since the density remains very small in the outer layers, we can neglect their own gravitational attraction. As was originally shown by Takeda, fairly accurate results can be obtained by means of a double approximation technique: (i) in the inner core where the centrifugal force is always small, we use a first-order expansion in the parameter v, and (ii) we neglect the contribution of the mass in the outer layers, taking the gravitational force as arising solely from the matter in the slightly oblate core. Since, in general, the domains of validity of the two approximation regimes overlap, self-consistent models may readily be constructed. This technique has been generalized to uniformly rotating stars by Roxburgh and his associates. However, as shown recently by Kippenhahn and Thomas, the use of two zones is unnecessary, for the same degree of accuracy can be obtained in choosing an appropriate geometrical representation for the level surfaces. The main advantage of the method lies in the fact that, without much trouble, rotation can be incorporated into the usual programs for stellar evolution.

Let us consider a chemically homogeneous, quasi-spherical star in uniform rotation. Then, as we know, the stratification of this pseudo-barotropic star is such that p, ρ, Φ, and T assume a constant value over the same set of surfaces. By virtue of equation (3), these functions are thus uniquely defined by the mean radius a of the level surfaces $\Phi = constant$.

Assume, for the time being, that the function $\Phi(a)$ is known. By definition, the effective gravity is

$$g = |\mathbf{g}| = \frac{d\Phi}{dn}, \tag{57}$$

where n denotes the outer normal to a level surface. Naturally, as the star rotates, g is not a constant over a surface $\Phi = constant$. For further use, let us define the average values of g and g^{-1} over the surface $\mathscr{S}_a$ corresponding to $\Phi(a)$; we have

$$\langle g \rangle = \frac{1}{\mathscr{A}_a} \oint_{\mathscr{S}_a} g \, dS, \quad \text{and} \quad \langle g^{-1} \rangle = \frac{1}{\mathscr{A}_a} \oint_{\mathscr{S}_a} g^{-1} \, dS, \tag{58}$$

where $\mathscr{A}_a$ designates the area of the surface $\mathscr{S}_a$.

Now, in view of equation (6), the mass element dM_a between the surfaces Φ and $\Phi + d\Phi$ is

$$dM_a = \rho(a) \, d\mathscr{V}_a = 4\pi\rho(a)a^2 \, da. \tag{59}$$

By using equations (57) and (58), we can also write

$$dM_a = \rho \oint_{\mathscr{S}_a} dn \, dS = \rho \, d\Phi \oint_{\mathscr{S}_a} g^{-1} \, dS = \rho \langle g^{-1} \rangle \mathscr{A}_a \, d\Phi. \tag{60}$$

According to equation (59), we first obtain

$$\frac{da}{dM_a} = \frac{1}{4\pi a^2 \rho}. \tag{61}$$

Equation (61) expresses nothing but the conservation of mass; with r replacing a, it reduces to the equation of continuity used in the study of spherical stars. Turning next to equation (1), we can write

$$\frac{dp}{dM_a} = -\rho \frac{d\Phi}{dM_a}; \tag{62}$$

or, by making use of equation (60),

$$\frac{dp}{dM_a} = -\frac{GM_a}{4\pi a^4} f_p, \tag{63}$$

where

$$f_p(a) = \frac{4\pi a^4}{GM_a \mathscr{A}_a} \langle g^{-1} \rangle^{-1}. \tag{64}$$

In the spherical case, we can write $a = r$ and $f_p(a) \equiv 1$.

Assume now the star to be in thermal equilibrium. If L_a denotes the energy that passes per second through a level surface in the outward direction, then the increment suffered by the luminosity L_a between the surfaces Φ and $\Phi + d\Phi$ is given by

$$dL_a = \epsilon_{Nuc} dM_a, \tag{65}$$

where ϵ_{Nuc} designates the energy generation rate per gram and per second. For chemically homogeneous stars, the function ϵ_{Nuc} depends only on ρ and T; and is also a constant over a level surface. Accordingly, we can write

$$\frac{dL_a}{dM_a} = \epsilon_{Nuc}. \tag{66}$$

Consider next the regions where the star is in radiative equilibrium. In those regions, we have

$$\mathscr{F}_a = -\frac{16\sigma T^3}{3\kappa\rho}\frac{dT}{dn} = -\frac{16\sigma T^3}{3\kappa\rho}g\frac{dT}{d\Phi}, \tag{67}$$

where κ denotes the opacity coefficient and σ is the Stefan-Boltzmann constant. Equation (67) merely expresses the fact that the radiative flux $\mathscr{F}_a$ varies in proportion to the effective gravity g on a level surface (cf. §7.2). By definition, the luminosity L_a has the form[3]

$$L_a = \oint_{\mathscr{S}_a}\mathscr{F}_a\,dS = -\frac{16\sigma T^3}{3\kappa}\mathscr{A}_a{}^2\langle g\rangle\langle g^{-1}\rangle\frac{dT}{dM_a}, \tag{68}$$

for κ is also a constant over the surfaces $\Phi = constant$. In view of equations (60) and (68), we find

$$\frac{dT}{dM_a} = -\frac{3\kappa}{16\sigma T^3}\frac{L_a}{(4\pi a^2)^2}f_T, \tag{69}$$

where

$$f_T(a) = \left(\frac{4\pi a^2}{\mathscr{A}_a}\right)^2\langle g\rangle^{-1}\langle g^{-1}\rangle^{-1}. \tag{70}$$

Obviously, we have $f_T(a) \equiv 1$ in the nonrotating case. Finally, in regions where convective equilibrium prevails, one usually *assumes* that

$$\frac{dT}{dM_a} = \left(1 - \frac{1}{\Gamma_2}\right)\frac{T}{p}\frac{dp}{dM_a}, \tag{71}$$

in which the generalized adiabatic exponent Γ_2 also remains a constant over a level surface. The influence of rotation on the onset of convection will be more closely discussed in Section 14.5.

To sum up, equations (61), (63), (66), and (69) [or (71)] now replace the well-known equations of stellar structure in the limit of spherical configurations. Of course, they must be complemented by an appropriate equation of state. The sole difference between rotating and nonrotating stars lies in the fact that we must now evaluate the nondimensional corrective factors $f_p(a)$ and $f_T(a)$. These are purely *geometrical* quantities that depend solely on the shape of the level surfaces. (This can be seen at once by means of a dimensional analysis of equations [64] and [70].) In principle, it would thus be necessary to solve equation (29) simultaneously with the modified stellar structure equations and the appropriate boundary

---

[3] Strictly speaking, it is not possible to consider a radiating pseudo-barotrope without a meridional circulation superimposed on the rotational motion. However, as we shall see in Section 8.2, the *average* flux energy carried by such currents through a level surface is strictly zero. Thus equation (68) holds rigorously.

conditions (48) on the gravitational potential. In practice, we *approximate* the total potential $\Phi(a)$ as being the solution for the Roche model (cf. §5.2). Hence, we choose the level surfaces in the form

$$r(a,\mu) = a\left[1 - \frac{\Omega^2}{3GM}a^3 P_2(\mu)\right],\qquad(72)$$

or

$$\Phi(r,\mu) = -\frac{GM}{r} - \frac{1}{2}\Omega^2 r^2(1-\mu^2).\qquad(73)$$

According to Kippenhahn and Thomas, Figure 5.1 illustrates the behavior

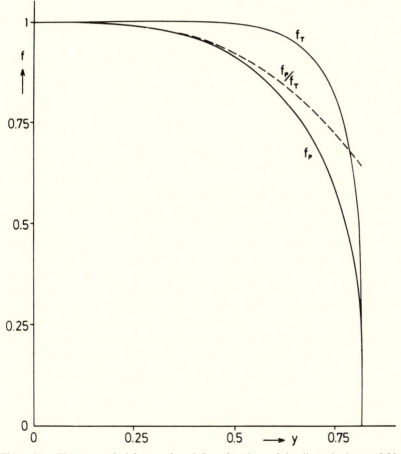

Figure 5.1.   The geometrical factors f_p and f_T as functions of the dimensionless variable $y = (\Omega^2/GM)^{1/3}a$. The ratio of the two functions is plotted in order to show their behavior near the surface. Source: Kippenhahn, R., and Thomas, H. C., in *Stellar Rotation* (Slettebak, A., ed.), p. 24, New York: Gordon and Breach, 1970.

of f_p and f_T as a function of the dimensionless variable $y = (\Omega^2/GM)^{1/3}a$. Although we use the level surfaces of a model that is singular at the origin, the method nevertheless provides realistic solutions with a *finite* central density. Indeed, in the limit of zero rotation, we see that $f_p = f_T = 1$; and we then recover the usual equations for spherical stars. In any event, by our choosing the stratification *a priori*, this technique only applies to pseudo-barotropic stars which do not greatly deviate from spherical symmetry. Since all centrally condensed, *uniformly* rotating stars fall into this class, interesting results can be derived from the method.

5.5. THE SELF-CONSISTENT FIELD METHOD

Perturbation techniques described in Sections 5.2–5.4 are particularly well suited to determine the equilibrium structure of barotropes and pseudo-barotropes for which the level surfaces slightly depart from spheres. Of course, higher-order corrections can be included to allow for larger distortions; but, then, the analytical calculations become hopelessly prohibitive. For that reason, in order to make further progress, the problem should be viewed from another angle. The self-consistent field method (which has been originally applied to barotropic- and pseudo-barotropic stars by Ostriker and his associates) provides such a new approach, for it is especially designed to relax altogether the restrictive assumption of quasi-sphericity.[4] Moreover, since it requires a knowledge of the angular momentum distribution (rather than of the angular velocity distribution), it can treat differentially rotating models without any additional labor.

To illustrate the main features of the technique, let us restrict ourselves to the simple case of differentially rotating barotropes. The total mass M and the total angular momentum J are considered to be fixed quantities. We designate by R the maximum radius of the barotrope under consideration. (Note that for severely distorted models which present an equatorial cusp, R may be greater than the equatorial radius.) Both cylindrical and spherical coordinates will be used with the standard notations. For further use, let m_ϖ be the fractional mass interior to the cylinder of radius ϖ; we have

$$m_\varpi = \frac{2\pi}{M} \int_0^\varpi \int_{-\infty}^{+\infty} \varpi' \rho(\varpi', z') \, d\varpi' \, dz'. \tag{74}$$

Consider now a barotrope on which we impose a rotation with some assigned velocity distribution. Hence, the angular momentum distribution

---

[4] A self-consistent field scheme has been originally devised by Hartree and Fock in connection with their investigation of atomic structures. See, e.g., Hartree, D. R., *The Calculation of Atomic Structures*, pp. 159–162, New York: Wiley and Sons, 1957.

per unit mass is

$$j = j(\varpi) = \Omega(\varpi)\varpi^2. \tag{75}$$

(By virtue of our approximations, remember that the angular velocity cannot depend on z, and must be specified *a priori*; cf. §§4.3 and 4.5.) Following Ostriker and Mark, we can alternately prescribe j as a function of the Lagrangian variable m_ϖ rather than as a function of the radius ϖ; thus, we can define the rotational motion by means of a function of the form

$$j = j(m_\varpi). \tag{76}$$

The above specification is particularly convenient when departure from sphericity is large. Indeed, as we shall see in Chapters 10–13, we can usually expect equilibria in rotating stars for any given angular momentum distribution $j(m_\varpi)$ but not necessarily for any given angular velocity $\Omega(\varpi)$. For the time being, we do not need to choose an explicit form for the function $j(m_\varpi)$. Nevertheless, anticipating somewhat the results of Section 7.3, we can already state that j must always be a monotonically increasing function of m_ϖ.

If we now let

$$U_{\rm rot}(\varpi) = \int_0^\varpi \frac{j^2(m_{\varpi'})}{\varpi'^3}\, d\varpi', \tag{77}$$

and

$$V(\varpi,z) = -G \int_{\mathscr{V}} \frac{\rho(\mathbf{x}')}{|\mathbf{x} - \mathbf{x}'|}\, d\mathbf{x}', \tag{78}$$

the total potential becomes

$$\Phi(\varpi,z) = V(\varpi,z) - U_{\rm rot}(\varpi) \tag{79}$$

(cf. §4.3: equation [20]). As we know, for a barotropic star the isopycnic- and isobaric-surfaces coincide with the level surfaces $\Phi = constant$. Hence, ρ can be viewed as a function of Φ only; by making use of equation (1) and the assigned p–ρ relation, we can write

$$\rho = \rho(\Phi), \tag{80}$$

with the requirement that $\rho(\Phi_s) \equiv 0$, where s designates, as usual, a surface value. Also, by virtue of equations (77) and (78), the functions Φ, ρ, and j are related through potential theory by an integral operator. Consequently, V and $U_{\rm rot}$ at a given point depend for the most part on *mean* properties of the material interior to that point, rather than on *local* properties. Symbolically, we have

$$\Phi = \Phi(\rho;j), \tag{81}$$

where, as we remember, j is a prescribed function of m_ϖ. It is also worth noting that, by making use of the integral representation (78) for the gravitational potential, boundary conditions (48) are automatically satisfied.

The approximation scheme can now be viewed as follows. By using equation (81), a suitably chosen trial density distribution $\rho_0(\varpi,z)$ provides a function $\Phi_0(\varpi,z)$. This function, in turn, leads to an improved density distribution $\rho_1(\varpi,z)$ when equation (80) is taken into account. Similarly, when fed back into equation (81) this new function leads to an improved potential $\Phi_1(\varpi,z)$, etc. Thus, alternately solving equations (80) and (81) as exactly as numerical techniques permit, we eventually obtain a self-consistent solution by this iterative process. The method works remarkably well. According to Ostriker and Mark, for slowly rotating stars, between ten or twenty iterations suffice to give excellent accuracy. For severely distorted barotropes, at the most about one hundred iterations may be needed; however, a detailed investigation indicates that, even in these extreme cases, the relatively slow convergence is due to numerical problems rather than inadequacy in the method.

On the practical side, let us now briefly discuss the main steps of the numerical integration. It is apparent from equation (1) that, given an explicit p–ρ relation, equation (80) relates ρ and Φ through an *algebraic*— not a differential or integral—operator. Hence, the equilibrium step (80) in the iterative process is straightforward. On the other hand, the potential step (81) requires a numerical discretization of two integral operators that we will now discuss.

Suppose that the density $\rho(\varpi, z)$ is known at some step in the iterative scheme. Naturally, that function vanishes on a boundary that is not spherical and, in extreme cases, may be quite far from spheroidal. Expand now the density in approximate polynomial form. It is convenient to define this expansion within a *sphere* of radius R centered about the center of mass of the configuration. We have

$$\rho(x,\mu) \cong \frac{3M}{4\pi R^3} \sum_{l=1}^{N} \sum_{m=l}^{N} A_{lm} x^{2m-2} P_{2l-2}(\mu), \tag{82}$$

where $x = r/R$ and $\mu = \cos\theta$. Note that we use only even-order Legendre polynomials because of the symmetry about the equatorial plane. Correspondingly, since we expand in the range $0 \leqslant x \leqslant 1$, only even powers of the radial coordinate need appear. Moreover, the constants A_{lm} must vanish if $m < l$ in order to maintain continuity at the origin; this first constraint also implies that the density assumes a finite value at the center of mass. In addition, although the true density vanishes on the circumscribed sphere, the approximate density does not vanish identically there; thus, in order to minimize this error, we must add the constraint $\rho(1,\mu) \equiv 0$. Finally, the parameter N, which determines the number of terms taken in the expansion, is chosen to give the requisite accuracy consistent with limitations of time and storage capacity of the computer. Leaving aside further technicalities, we thus notice that the density (at a given step of the integration) is essentially described by means of a triangular matrix A_{lm}.

If we make use next of equations (8)–(12), the gravitational potential (78) becomes

$$V(x,\mu) = -2\pi R^2 G \sum_{n=0}^{\infty} P_n(\mu) \int_{-1}^{+1} d\mu' P_n(\mu') \left(\int_0^x dx' \frac{x'^n}{x^{n+1}} \right.$$

$$\left. + \int_x^1 dx' \frac{x^n}{x'^{n+1}} \right) \rho(x',\mu')x'^2. \tag{83}$$

Substituting equation (82) into the foregoing expansion and integrating over μ' and x', we find

$$V(x,\mu) \cong -\frac{3}{2} \frac{GM}{R} \sum_{l=1}^{N} \sum_{m=l}^{N} \frac{A_{lm} P_{2l-2}(\mu)}{(m-l+1)(4l-3)} \left(x^{2l-2} - \frac{4l-3}{2m+2l-1} x^{2m} \right). \tag{84}$$

Thus, for any trial density distribution (i.e., any given matrix A_{lm}), the gravitational potential can be derived at once. Similarly, for a prescribed function $j(m_\varpi)$, we can thence express the centrifugal potential per unit mass (77) as a polynomial in the normalized variable ϖ/R. As in equations (82) and (84), the density occurs via the matrix A_{lm}. To summarize, given a trial density, we can find the total potential Φ in analytical form. The next step in the iterative process consists of substituting this potential into equation (80) and determining a new density distribution.

This method has been extensively used to describe polytropes and zero-temperature white dwarfs in rapid differential rotation. The same technique has been extended by Mark to describe equilibria for radiating pseudo-barotropes that greatly depart from spherical symmetry. More recently, Jackson has coupled the self-consistent field method with the Henyey scheme to permit the construction of pseudo-barotropic stellar models that fully include the details of nuclear-energy generation and radiative transport. The self-consistent field technique has been also employed by Clement, with the usual method of calculating the gravitational potential by means of double-series expansion being replaced by a two-dimensional finite-difference technique to solve Poisson's equation.

In this method, the gravitational potential is obtained throughout all space, and no boundary conditions are applied on the surface of the configuration. Since the Clement scheme has proved to be a fast, efficient method for finding the gravitational potential in rapidly rotating barotropes, Chambers has also combined this finite-difference method for the solution of Poisson's equation with the Henyey program to produce rapidly rotating stellar models by means of an iterative scheme.

Finally, let us mention that Hubbard, Slattery, and DeVito have recently derived a new perturbation technique for the structure of rotating bodies in mechanical equilibrium. Their method uses an expansion of the density in Legendre polynomials (cf. equation [82]), and can be developed analytically in a manner analogous to the standard Clairaut-Legendre theory (cf. §5.2).

BIBLIOGRAPHICAL NOTES

Section 5.2. The method was originally devised by Clairaut in connection with the internal structure of the Earth. The text follows the more general discussion by Legendre. Reference 7 of Chapter 1 contains a general survey of these important contributions. A modern account of these and other investigations will also be found in:

1. Kopal, Z., *Figures of Equilibrium of Celestial Bodies*, Madison: Univ. of Wisconsin Press, 1960.
2. Zharkov, V. N., Trubitsyn, V. P., and Samsonenko, L. V., *Planetary Interiors*, Tucson, Arizona: Pachart Publ. House, 1977.

Further developments along the same lines are contained in:

3. Lanzano, P., *Icarus* **1** (1962): 121–136.
4. James, R., and Kopal, Z., *ibid.* **1** (1963): 442–454.
5. Kopal, Z., and Lanzano, P., *Astroph. and Space Sci.* **23** (1973): 425–429.
6. Lanzano, P., *ibid.* **20** (1973): 71–83; *ibid.* **29** (1974): 161–178; *ibid.* **37** (1975): 173–181.
7. Kopal, Z., and Kalama Mahanta, M., *ibid.* **30** (1974): 347–360.

See also:

8. Bucerius, H., *Astron. Nachr.* **275** (1947): 49–72.

Refined limitations on the ratio m/f are due to:

9. Chandrasekhar, S., and Roberts, P. H., *Astroph. J.* **138** (1963): 801–808.
10. Seidov, Z. F., *Astrofizika* **5** (1969): 503–505.

See also:

11. Rau, A. R. P., *Monthly Notices Roy. Astron. Soc. London* **168** (1974): 273–286.

A generalized Clairaut-Legendre theory is also given in the following paper:

12. Lebovitz, N. R., *Astroph. J.* **160** (1970): 701–723.

The asymptotic nature of the Clairaut-Legendre expansion is discussed in:

13. Lebovitz, N. R., *Astroph. and Space Sci.* **9** (1970): 398–409.

Section 5.3. The classical papers in this subject are those of:

14. Milne, E. A., *Monthly Notices Roy. Astron. Soc. London* **83** (1923): 118–147.
15. Chandrasekhar, S., *ibid.* **93** (1933): 390–405.

This last paper contains a systematic study of uniformly rotating polytropes and had a profound influence on much subsequent research. However, insufficient attention was paid to the boundary conditions, and this led to some criticisms; fortunately, the ambiguity in the formulation did not at all affect the general structure of the solutions. A rigorous treatment of the boundary conditions was first given (in a more general context) by Krogdahl:

16. Krogdahl, W., *Astroph. J.* **96** (1942): 124–150.

A satisfactory discussion of uniformly rotating polytropes is contained in the following papers:

17. Chandrasekhar, S., and Lebovitz, N. R., *ibid.* **136** (1962): 1082–1104.
18. Kovetz, A., *ibid.* **154** (1968): 999–1003.

The general analysis in this section follows:

19. Sedrakian, D. M., Papoian, V. V., and Chubarian, E. V., *Monthly Notices Roy. Astron. Soc. London* **149** (1970): 25–33.

The above paper also contains a second-order expansion in the parameter v. The equivalence between this and the previous method has been demonstrated in the following papers:

20. Kopal, Z., *ibid.* **98** (1938): 414–424.
21. Chandrasekhar, S., and Krogdahl, W., *Astroph. J.* **96** (1942): 151–154.

See, however, Smith's contribution in references 23 of Chapter 10.

Section 5.4. The two-zone technique has been originally devised by Takeda:

22. Takeda, S., *Mem. College Sci. Kyoto Univ.*, **A**, **17** (1934): 197–217.

The same problem was reconsidered in:

23. Monaghan, J. J., and Roxburgh, I. W., *Monthly Notices Roy. Astron. Soc. London* **131** (1965): 13–21.
24. Monaghan, J. J., *Zeit. f. Astroph.* **67** (1967): 222–226.

Models of uniformly rotating, pseudo-barotropic stars were first constructed by Sweet and Roy:

25. Sweet, P. A., and Roy, A. E., *Monthly Notices Roy. Astron. Soc. London* **113** (1953): 701–715.

Generalizations of Takeda's method to uniformly rotating stars are contained in:

26. Roxburgh, I. W., Griffith, J. S., and Sweet, P. A., *Zeit. f. Astroph.* **61** (1965): 203–221.
27. Faulkner, J., Roxburgh, I. W., and Strittmatter, P. A., *Astroph. J.* **151** (1968): 203–216.

The analysis in this section follows:

28. Kippenhahn, R., and Thomas, H. C., in *Stellar Rotation* (Slettebak, A., ed.), pp. 20–29, New York: Gordon and Breach, 1970.

This paper also contains a discussion of tidally distorted stars. Further progress was made by:

29. Smith, B. L., *Astroph. and Space Sci.* **25** (1973): 195–205.

Other perturbation techniques are discussed in:

30. Epps, H. W., *Proc. Nat. Acad. Sci. U.S.A.* **60** (1968): 51–58.
31. Monaghan, J. J., *Zeit. f. Astroph.* **68** (1968): 461–472.
32. Papaloizou, J. C. B., and Whelan, J. A. J., *Monthly Notices Roy. Astron. Soc. London* **164** (1973): 1–10.

Section 5.5. The analysis in this section is derived from:

33. Ostriker, J. P., and Mark, J. W.-K., *Astroph. J.* **151** (1968): 1075–1088.

The effects of radiation on pseudo-barotropes are included in:

34. Mark, J. W.-K., *ibid.* **154** (1968): 627–643.

The coupling of the Henyey scheme with the self-consistent field method is due to:

35. Jackson, S., *ibid.* **161** (1970): 579–585.

Further technical improvements will be found in:

36. Blinnikov, S. I., *Astron. Zh.* **52** (1975): 243–254.

Related methods are due to:

37. Clement, M. J., *Astroph. J.* **194** (1974): 709–714.
38. Hubbard, W. B., Slattery, W. L., and DeVito, C. L., *ibid.* **199** (1975): 504–516.
39. Chambers, R. H., Ph.D. Thesis (unpublished), Univ. of Toronto, 1976.

6

Small Oscillations and
Stability: Techniques

6.1. INTRODUCTION

In most instances, configurations in a state of permanent rotation provide an adequate description of actual rotating stars. However, it must be borne in mind that our basic underlying assumptions were made on account of mathematical difficulties (cf. §4.2), and that they never *strictly* obtain in nature. For instance, can we surmise *a priori* that the effective gravity **g** always exactly balances the pressure force in a real star? As a matter of fact, not every model in a state of permanent rotation can actually occur in nature: the models must not only obey the general conservation principles of physics but also be *stable*. Thus, an acceptable model must sustain itself against natural fluctuations to which any physical body is subject, i.e., whenever such disturbances arise, they should decrease with time. If the fluctuations do not die down, the system is then said to be *unstable*, for it progressively departs from its initial state of equilibrium.

The mathematical investigation of the stability of a known model proceeds as follows. An arbitrary perturbation is superposed on the system, and the system's subsequent behavior under the natural forces is studied by means of the time-dependent equations of stellar hydrodynamics. For convenience, we then assume the fluctuations to be small enough so that we can *linearize* the nonlinear equations expressing the conservation of mass, linear momentum, and energy. In other words, we neglect all products and powers (higher than the first) of the disturbances and retain only terms that are linear in them. It is usually assumed that the solutions of these linearized equations approximate closely enough the actual solutions to reveal the general trend of the motion in the immediate vicinity of the equilibrium state. In most simple cases, energetic considerations usually provide a physical interpretation of the results. Further insight into the problem necessarily requires an investigation of finite-amplitude motions.

It is the purpose of this chapter to expound the various approximation techniques by which one can discuss the stability of rotating stellar models. Actually, we shall be mainly concerned with the small-perturbation method, which until now has been almost exclusively used to test the stability of configurations in a state of permanent rotation (see, however,

§6.8). To illustrate the problem, we will further restrict the present discussion to *isentropic* perturbations, i.e., motions for which the entropy of each mass element is preserved along its path. Numerical results and their interpretation will be given whenever the need arises. Also, departures from linearity and isentropy will be dealt with for specific questions in subsequent chapters.

6.2. EULERIAN AND LAGRANGIAN VARIATIONS

Broadly speaking, two types of description can be used to analyze the oscillations of a system about a known state of equilibrium: *either* we specify the disturbances noted by an external observer who, at every instant t, views a given volume element at a fixed location in space, *or* we describe the fluctuations within a given mass element, which is then followed along its path in the course of time. Both representations are very much akin to the Eulerian and Lagrangian variables defined in Section 3.2 and, therefore, will be named accordingly.

To be specific, consider a perturbed flow (i.e., the motion of the rotating star which we distort from its equilibrium position) and imagine it is accompanied by a ghostly flow that remains unperturbed (i.e., the known state of permanent rotation). Let $Q(\mathbf{x},t)$ and $Q_0(\mathbf{x},t)$ be the values of any physical quantity (such as pressure p, density ρ, gravitational potential V, or velocity $\mathbf{v}$) in the perturbed and unperturbed flows, respectively.[1] Consider then the vectorial change $\boldsymbol{\xi}(\mathbf{x},t)$ which a mass element, in the perturbed flow, experiences relative to its corresponding image, at time t, in the ghostly flow. The function $\boldsymbol{\xi}(\mathbf{x},t)$ is called the *Lagrangian displacement* and unequivocally describes any departure from a well-defined state of equilibrium.

Now, consider the following quantity:

$$\delta Q = Q(\mathbf{x},t) - Q_0(\mathbf{x},t). \tag{1}$$

By definition, this *Eulerian variation* represents the change noted by an external observer who simultaneously compares the values of Q, in the real and ghostly flows, *at a fixed location* $\mathbf{x}$. On the other hand, we call the increment

$$\Delta Q = Q[\mathbf{x} + \boldsymbol{\xi}(\mathbf{x},t),t] - Q_0(\mathbf{x},t) \tag{2}$$

the *Lagrangian variation* in the quantity Q caused by the disturbances.

---

[1] Given our assumptions (cf. §4.2), the unperturbed flow is stationary and possesses an axial symmetry, i.e., Q_0 solely depends on the cylindrical coordinates ϖ and $z(\partial/\partial t = \partial/\partial\varphi = 0)$. This restriction is by no means essential to the present discussion, which remains valid when Q_0 depends on all space variables and time.

Obviously, it involves the simultaneous values of Q, in both flows, *for the same mass element*; the latter is centered at the point $\mathbf{x} + \boldsymbol{\xi}$ in the perturbed flow, whilst its corresponding image is located at the point $\mathbf{x}$ in the ghostly flow. By virtue of definitions (1) and (2), Eulerian and Lagrangian changes can be related by means of a Taylor expansion; we have

$$\Delta Q = \delta Q + \xi_i \frac{\partial Q}{\partial x_i} + \frac{1}{2}\xi_i\xi_j \frac{\partial^2 Q}{\partial x_i\,\partial x_j} + \cdots . \tag{3}$$

(Henceforth, a summation over repeated Roman indices is to be understood.) Also, from the definition of $\boldsymbol{\xi}$, the Lagrangian variation $\Delta\mathbf{v}$ in the velocity has the form

$$\Delta\mathbf{v}(\mathbf{x},t) = \mathbf{v}(\mathbf{x} + \boldsymbol{\xi},t) - \mathbf{v}_0(\mathbf{x},t) = \frac{D_0\boldsymbol{\xi}}{Dt}, \tag{4}$$

where D_0/Dt is the material derivative as we follow the ghostly flow, i.e.,

$$\frac{D_0}{Dt} = \frac{\partial}{\partial t} + \mathbf{v}_0(\mathbf{x},t).\text{grad}, \tag{5}$$

and $\mathbf{v}_0$ denotes the velocity field of the unperturbed configuration. It is worth noting that equations (1)–(4) are quite general, and do not presuppose any assumption about the smallness of the disturbances.

If we now *approximate* the actual motion by assuming that $\boldsymbol{\xi}$ and the δQ's are so small that their square and products are negligible, we obtain

$$\Delta Q = \delta Q + \xi_i \frac{\partial Q}{\partial x_i}. \tag{6}$$

Thus, to first order in the disturbances, the operators Δ and δ are related by

$$\Delta = \delta + \boldsymbol{\xi}.\text{grad}. \tag{7}$$

Strictly speaking, however, relation (7) is only valid for *infinitesimal* Lagrangian displacements. This approximation will be made in the remainder of the present chapter.

The importance of Lagrangian variations arises because, to first order in the disturbances, the operations of Δ and material derivation commute. (In the present instance, we have

$$\frac{D}{Dt} = \frac{\partial}{\partial t} + \mathbf{v}(\mathbf{x},t).\text{grad} = \frac{\partial}{\partial t} + \left[\mathbf{v}_0(\mathbf{x},t) + \delta\mathbf{v}(\mathbf{x},t)\right].\text{grad}, \tag{8}$$

which should be compared with equation [5].) To establish this useful result, let us first quote some properties of the operators δ and Δ. By virtue of equation (1), the operations of δ and partial derivation *do* commute:

$$\delta \frac{\partial Q}{\partial x_i} = \frac{\partial}{\partial x_i} \delta Q, \quad \text{and} \quad \delta \frac{\partial Q}{\partial t} = \frac{\partial}{\partial t} \delta Q. \tag{9}$$

On the other hand, it is a simple matter to see that the operators Δ and grad (or $\partial/\partial t$) *do not* commute. Indeed, in making use of equations (6) and (9), we find

$$\Delta \frac{\partial Q}{\partial x_i} = \frac{\partial}{\partial x_i} \Delta Q - \frac{\partial \xi_k}{\partial x_i} \frac{\partial Q}{\partial x_k}; \tag{10}$$

similarly, we can write

$$\Delta \frac{\partial Q}{\partial t} = \frac{\partial}{\partial t} \Delta Q - \frac{\partial \xi_k}{\partial t} \frac{\partial Q}{\partial x_k}. \tag{11}$$

In view of equations (4), (8), (10), and (11), we eventually obtain

$$\Delta \frac{DQ}{Dt} = \frac{D}{Dt} \Delta Q, \tag{12}$$

in which we made use of the fact that $D\xi/Dt$ and $D_0\xi/Dt$ only differ by the (negligible) second-order quantity $\delta v.\text{grad } \xi$. As a consequence of equation (12), we also have

$$\delta \frac{DQ}{Dt} = \frac{D}{Dt} \Delta Q - \xi_k \frac{\partial}{\partial x_k} \frac{DQ}{Dt}. \tag{13}$$

6.3. THE PERTURBATION EQUATIONS

For further reference, let us now write down the general equations governing the changes caused by a small arbitrary disturbance of an initial state of permanent rotation. Since confusion can no longer arise, we shall omit from now on the subscript "0" labeling an equilibrium state.

In particular, from the equation expressing mass conservation (cf. §3.3: equation [21]), it follows that

$$\Delta \rho = -\rho \text{ div } \xi, \tag{14}$$

or, by virtue of equation (6),

$$\delta \rho = -\rho \text{ div } \xi - \xi.\text{grad } \rho = -\text{div}(\rho \xi). \tag{15}$$

Also, for isentropic motions, relative changes in pressure and density are proportional (cf. §3.4: equation [73]); that is

$$\Delta p = \Gamma_1 \frac{p}{\rho} \Delta \rho = -\Gamma_1 p \text{ div } \xi, \tag{16}$$

expressing the absence of dissipative mechanisms. For most cases, the generalized adiabatic exponent Γ_1 is a function of position. Similarly, we obtain

$$\delta p = -\Gamma_1 p \operatorname{div} \xi - \xi.\operatorname{grad} p. \tag{17}$$

The discussion of the equations of motion is somewhat more involved. First, as we recall, the vectorial equation governing equilibrium is

$$\frac{1}{\rho} \operatorname{grad} p = \mathbf{g}, \tag{18}$$

where

$$\mathbf{g} = -\operatorname{grad} V + \mathbf{q}; \tag{19}$$

in cylindrical coordinates (ϖ, φ, z), the components of the centrifugal acceleration $\mathbf{q}$ are

$$q_\varpi = \Omega^2(\varpi, z)\varpi, \quad q_\varphi = 0, \quad q_z = 0 \tag{20}$$

(cf. §4.3: equations [15]–[17]). Then, from the vectorial equation of motion

$$\frac{D\mathbf{v}}{Dt} = -\frac{1}{\rho} \operatorname{grad} p - \operatorname{grad} V, \tag{21}$$

it can readily be seen that its small-perturbation counterpart has the form

$$\frac{D_0^2 \xi}{Dt^2} = -\Delta\left(\frac{1}{\rho} \operatorname{grad} p + \operatorname{grad} V\right), \tag{22}$$

where we have used an *inertial* frame of reference;[2] the l.h.s. of equation (22) stems from the obvious relations

$$\Delta \frac{D\mathbf{v}}{Dt} = \frac{D}{Dt} \Delta\mathbf{v} = \frac{D}{Dt}\left(\frac{D_0\xi}{Dt}\right) = \frac{D_0^2\xi}{Dt^2}. \tag{23}$$

By virtue of equations (7), (18), and (19), we can next write

$$\Delta\left(\frac{1}{\rho} \operatorname{grad} p + \operatorname{grad} V\right) = \delta\left(\frac{1}{\rho} \operatorname{grad} p + \operatorname{grad} V\right) + (\xi.\operatorname{grad})\mathbf{q}. \tag{24}$$

In making then use of the first equation (9), we eventually obtain

$$\frac{D_0^2\xi}{Dt^2} = -\frac{1}{\rho} \operatorname{grad} \delta p + \frac{\delta\rho}{\rho^2} \operatorname{grad} p - \operatorname{grad} \delta V - (\xi.\operatorname{grad})\mathbf{q}. \tag{25}$$

Note that both δp and $\delta\rho$ can be related to ξ by means of equations (15) and (17). Equation (25) thus provides a relation between the Lagrangian displacement ξ and the Eulerian change δV in the gravitational potential.

---

[2] Since we shall mostly deal with differentially rotating stellar models, no serious advantage can be gained by working in a rotating frame of reference.

The principal obstacle of the problem is clearly the presence of the variation δV in equation (25). In order to find a convenient expression for δV, consider a mass element $dm' = \rho(\mathbf{x}')\,\mathbf{dx}'$ displaced by an amount $\boldsymbol{\xi}' = \boldsymbol{\xi}(\mathbf{x}',t)$ from its initial position $\mathbf{x}'$. Its contribution to the gravitational potential at some point $\mathbf{x}$ was

$$-G\frac{dm'}{|\mathbf{x} - \mathbf{x}'|}, \tag{26}$$

but it will now be

$$-G\frac{dm'}{|\mathbf{x} - (\mathbf{x}' + \boldsymbol{\xi}')|}. \tag{27}$$

The change of the potential V at the point $\mathbf{x}$ due to that particular mass element is therefore

$$-\boldsymbol{\xi}'.\mathrm{grad}'\left(G\frac{dm'}{|\mathbf{x} - \mathbf{x}'|}\right) = -G\boldsymbol{\xi}'.\mathrm{grad}'\left(\frac{1}{|\mathbf{x} - \mathbf{x}'|}\right)dm', \tag{28}$$

where grad$'$ refers to the variables $\mathbf{x}'$. If we next sum over the entire mass M, the Eulerian variation δV at the location $\mathbf{x}$ and the time t becomes

$$\delta V(\mathbf{x},t) = -G\int_{\mathscr{V}} \boldsymbol{\xi}'.\mathrm{grad}'\left(\frac{1}{|\mathbf{x} - \mathbf{x}'|}\right)dm', \tag{29}$$

or, by using Gauss's theorem,

$$\delta V(\mathbf{x},t) = -G\oint_{\mathscr{S}} \frac{\rho(\mathbf{x}')}{|\mathbf{x} - \mathbf{x}'|}\,\mathbf{n}'.\boldsymbol{\xi}(\mathbf{x}',t)\,dS' - G\int_{\mathscr{V}} \frac{\delta\rho(\mathbf{x}',t)}{|\mathbf{x} - \mathbf{x}'|}\,\mathbf{dx}', \tag{30}$$

where $\delta\rho$ is given in terms of $\boldsymbol{\xi}$ by equation (15). The vector $\mathbf{n}'$ designates the outer normal to the unperturbed boundary $\mathscr{S}$; also, the triple integral must be evaluated over the equilibrium volume $\mathscr{V}$ of the rotating model. Alternatively, δV can be obtained from the Poisson equation

$$\nabla^2\,\delta V = 4\pi G\,\delta\rho = -4\pi G\,\mathrm{div}(\rho\boldsymbol{\xi}). \tag{31}$$

In summary, the small oscillations of a known equilibrium structure are entirely governed by equations (15), (17), (25), and (30) [or (31)], together with some appropriate boundary conditions that we will now discuss. *First*, the total pressure $p + \delta p$ must vanish, at every instant t, on the (*unknown*) distorted surface. Since we are restricting our present analysis to small-amplitude motions, it is sufficient to require the vanishing of the infinitesimal Lagrangian variation Δp on the known surface $\mathscr{S}$. By virtue of equation (16), this condition is automatically satisfied if $\boldsymbol{\xi}$ and its derivatives remain finite everywhere, so this is the only requirement we must place on $\boldsymbol{\xi}$.[3] *Second*, the gravitational force must be continuous across the perturbed boundary. Clearly, the requisite condition is always satisfied when we use the integral representation (30). In contrast, when we derive δV from equation (31), we must also impose the infinitesimal change Δ grad V to be continuous, at every instant t, on the equilibrium surface $\mathscr{S}$. This, in turn, requires the additional determination of the outer gravitational potential of the distorted, time-dependent configuration. As far as rotating models are concerned, the latter approach is not so practicable.

A further point should be made. By convention, the boundary $\mathscr{S}$ is defined by the condition $p = 0$ (cf. §4.2); and nowhere is it assumed that the density vanishes on the equilibrium surface. However, *when ρ does vanish on the surface $\mathscr{S}$, then $\Delta p = \delta p$ on this boundary*. Indeed, in view of equations (7) and (18), we always have

$$\Delta p = \delta p + \boldsymbol{\xi}.\text{grad } p = \delta p + \rho \mathbf{g}.\boldsymbol{\xi}. \tag{32}$$

Since $\rho \mathbf{g}.\boldsymbol{\xi}$ vanishes wherever the density vanishes, the quoted property follows at once. This result is physically obvious for $\rho \mathbf{g}.\boldsymbol{\xi}\ dS$ represents the weight of the matter which is displaced, above or below an area dS drawn on $\mathscr{S}$, as a consequence of the disturbances. Obviously, this quantity becomes negligible when the density decreases to zero near the surface of equilibrium.

Now, having linearized the problem we can always separate the space variables $\mathbf{x}$ and the time t; we thus search for normal-mode solutions of the form

$$\boldsymbol{\xi}(\mathbf{x},t) = \boldsymbol{\xi}(\mathbf{x}; \sigma)e^{i\sigma t}, \tag{33}$$

with similar expressions for the Eulerian and Lagrangian changes. Quite generally, the angular frequency σ can be real, imaginary or complex; and, in this case, we write $\sigma = \sigma_r + i\sigma_i$. The allowed values for σ (and the corresponding eigenfunctions) must be derived next from the perturbation equations and the requisite boundary conditions (in which, formally, we let $\partial/\partial t = i\sigma$, $\partial^2/\partial t^2 = -\sigma^2$). Clearly, instability occurs as soon as we find *one* particular eigenfrequency with a negative imaginary part ($\sigma_i < 0$) for, then, the Lagrangian displacement (33) is multiplied by the growing factor $\exp(|\sigma_i|t)$. In contrast, stability toward small-amplitude fluctuations not only imposes the condition $\sigma_i \geqslant 0$ for *all* modes, but also requires that *every* small arbitrary disturbance can be expanded as a linear com-

---

[3] The present motions should be contrasted with *volume-preserving* disturbances. Indeed, in that case equation (16) reduces to div $\boldsymbol{\xi} = 0$, and the condition $\Delta p = 0$ must be expressed explicitly on the surface $\mathscr{S}$.

bination of the eigenfunctions (since, otherwise, perturbations may exist that would escape our analysis). To date, the completeness of the normal modes has never been demonstrated in general for the problem at hand. In any event, a normal-mode analysis of small isentropic fluctuations is useful on two counts: (i) it gives frequencies of oscillation, some of them providing an approximate description of observed stellar variabilities (but it still remains to find an excitation mechanism to *explain* the phenomenon), and (ii) it exhibits some unstable flow patterns that eventually compel the system to depart from its initial state of equilibrium (but, thence, *finite-amplitude* disturbances must be taken into account).

6.4. THE SPHERICAL STARS: A BRIEF SURVEY

Although a normal-mode analysis usually exhibits infinite sets of eigenfrequencies, only a few of these oscillations present a real interest in the theory of rotating stars. As a preliminary to the general problem, a discussion of spherical stars is in order, for it provides a simple way to visualize the pertinent motions.

In the case of a nonrotating star ($\mathbf{q} \equiv 0$), the space variables are separable in terms of spherical harmonics (see Figure 6.1 and Appendix C). Quite generally, a normal-mode displacement has three distinct components; in spherical coordinates (r,θ,φ), its radial part becomes

$$\xi_r(r,\theta,\varphi,t) = \delta r(r; \sigma,l) Y_l^m(\theta,\varphi) e^{i\sigma t} = \delta r(r; \sigma,l) P_l^m(\cos \theta) e^{im\varphi} e^{i\sigma t} \quad (34)$$

($-l \leqslant m \leqslant + l$), while the tangential components $r\xi_\theta$ and $r \sin \theta \, \xi_\varphi$ are proportional to $\partial Y_l^m/\partial\theta$ and $\partial Y_l^m/\partial\varphi$, respectively. The changes δp, $\delta\rho$, and δV are similar in form to equation (34). Now, by virtue of the central symmetry of a nonrotating star, its eigenfrequencies are independent of m; thus, to each value of l correspond $2l + 1$ distinct displacements. Also, in this case, the squared frequency σ^2 is always real ($\sigma^2 \lessgtr 0$); this fact excludes the possibility of oscillatory growing or damped motions.

To illustrate the problem, we shall consider four typical configurations: (i) *the homogeneous, compressible model*, i.e., a self-gravitating gaseous

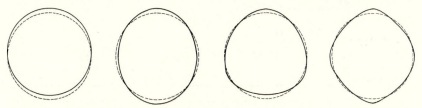

Figure 6.1.   Distortion of a sphere (*dotted lines*) due to the lowest spherical harmonics $Y_l^0(\cos \theta)$, with $l = 1, 2, 3, 4$. (See also Appendix C.)

sphere of constant density, (ii) *the polytropic spheres*, i.e., barotropes for which we assume a geometrical relation of the form $p = K\rho^{1+1/n}$, where K and *n* are constants ($0 \leqslant n \leqslant 5$, $n = 0$ corresponding to model [i]), (iii) *a main-sequence model* of mass $M = 27.8 M_\odot$, which has a core in

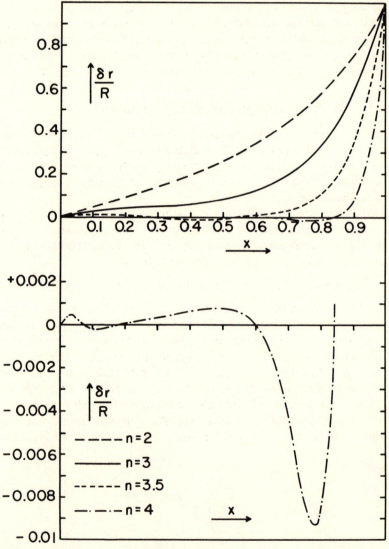

Figure 6.2.   Radial displacement of the f-mode belonging to $l = 2$ for various polytropic spheres ($\gamma = 5/3$). All displacements $\delta r/R$ are normalized to unity at $x = r/R = 1$, where R is the radius. Source: Robe, H., *Ann. Ap.* **31**, 475, 1968.

convective equilibrium covering the two fifths of its radius (chemical composition: $X = 0.75$, $Y = 0.22$, $Z = 0.03$), and (iv) *a composite polytrope* built out of a homogeneous core ($n = 0$) and a radiative envelope ($n = 3$).

Following Cowling, we must distinguish between *three* infinite discrete spectra for the eigenvalues σ^2. Let us now examine in succession these various modes of oscillation.[4]

THE f-MODES. As was shown by Kelvin in 1863, an incompressible fluid sphere allows the frequencies

$$\sigma^2 = \frac{4}{3}\pi G\rho \frac{2l(l-1)}{2l+1} \quad (l = 2, 3, 4, \ldots) \tag{35}$$

and the corresponding displacements

$$\xi(\mathbf{x},t) \propto \text{grad}[r^l P_l^m(\cos\theta)e^{im\varphi}]e^{i\sigma t}. \tag{36}$$

These modes describe surface waves that preserve the volume of the configuration; all of them are stable. (Obviously, the three harmonics belonging to $l = 1$ must be disregarded, since they represent a displacement of the center of mass.) *A similar infinite discrete spectrum exists for a compressible star* (see Figures 6.2 and 6.5, and Tables 6.1–6.2).

TABLE 6.1

Squared frequencies $\sigma^2/\pi G\bar{\rho}$ for polytropic spheres ($l = 2$, $\gamma = 5/3$)

n	0	1	2	3	3.25	3.50	4
$\rho_c/\bar{\rho}$	1.000	3.290	11.40	54.18	88.15	152.9	622.4
p_6	221.4	173.2	151.6	139.8	138.2	137.2	140.3
p_5	161.4	128.2	114.0	107.4	106.9	106.8	116.6
p_4	110.3	89.44	81.51	79.23	79.55	80.39	102.2
p_3	68.16	57.11	54.17	55.29	56.29	57.88	83.82
p_2	34.97	31.32	32.10	35.63	37.17	39.68	67.75
p_1	11.18	12.41	15.41	20.35	22.62	27.91	56.18
f	1.066	1.997	4.151	10.90	14.98	21.55	45.77
g_1	−0.9544	−0.4039	0.7510	6.553	10.63	16.13	36.79
g_2	−0.3050	−0.1844	0.3957	3.771	6.476	11.39	30.67
g_3	−0.1565	−0.1070	0.2452	2.430	4.229	7.655	23.99
g_4	−0.09669	−0.07029	0.1672	1.694	2.965	5.417	20.48
g_5	−0.06609	−0.04989	0.1215	1.248	2.192	4.023	17.02
g_6	−0.04818	−0.03731	0.09238	0.9584	1.686	3.099	13.44

SOURCE: Robe, H., *Ann. Ap.* **31**, 475, 1968.

---

[4] As was shown by Perdang, one should also consider a *fourth* spectrum of oscillation; that is, the roots $\sigma^2 = 0$, which correspond to all displacements ξ consistent with the condition $\xi.\text{grad}V = 0$, leading thus to another equilibrium configuration. When rotation is present, some of these trivial roots give rise to nonzero values (cf. §6.7).

TABLE 6.2

Oscillation periods (in days)
for a main-sequence star of M = 27.8M$_\odot$

	$l = 1$	$l = 2$	$l = 3$	$l = 4$	$l = 8$
g_4	1.799	1.053	0.7580	0.5992	0.3495
g_3	1.423	0.8392	0.6091	0.4861	0.2955
g_2	1.039	0.6220	0.4593	0.3734	0.2436
g_1	0.6453	0.4009	0.3085	0.2613	0.1949
f	∞	0.1395	0.1159	0.1063	0.08775
p_1	0.1023	0.08727	0.07829	0.07263	0.06038
p_2	0.07273	0.06460	0.05944	0.05582	0.04730
p_3	0.05634	0.05142	0.04802	0.04551	0.03923
p_4	0.04601	0.04276	0.04034	0.03849	0.03366

SOURCE: Smeyers, P., D.Sc. Thesis, Univ. of Liège, 1966.

Although no longer volume-preserving, these f-modes nevertheless imitate very closely surface waves on a fluid sphere. In particular, they do not greatly depend on the value of Γ_1. (For the homogeneous, compressible model, solutions [35] and [36] strictly obtain.) An *approximate* expression for the lowest frequency $l = 2$ was derived by Chandrasekhar and Lebovitz. We have

$$\sigma^2 = \frac{4}{5}\frac{|W|}{I} \propto \pi G\bar{\rho}, \qquad (37)$$

where W and I designate, respectively, the gravitational potential energy and the moment of inertia about the center of mass (cf §3.7); formula (37) is exact for the homogeneous, compressible model. Table 6.1 shows that the f-modes are fairly sensitive to the ratio of the central density ρ_c to the mean density $\bar{\rho}$. Figure 6.2, after Robe, illustrates the radial displacement δr of the five harmonics belonging to $l = 2$ for various polytropes. Note the progressive lowering of the amplitudes, near the center, as the ratio $\rho_c/\bar{\rho}$ increases; when $n \gtrsim 3.25$ ($\rho_c/\bar{\rho} \gtrsim 100$), the appearance of nodes along the radius is worth noticing. As was shown by Scuflaire, this unusual behavior (which is also present in the p- and g-modes of centrally condensed bodies) is caused by the fact that the f- and p-modes look like g-modes in the central regions of very condensed models, whereas the f- and g-modes behave like p-modes in the external layers of these models.

THE P-MODES. According to Pekeris, the homogeneous, compressible sphere exhibits the frequencies

$$\sigma^2 = \tfrac{4}{3}\pi G\rho\{\Delta_v + [\Delta_v{}^2 + l(l + 1)]^{1/2}\} \quad (l = 0, 1, 2, \ldots) \qquad (38)$$

where

$$2\Delta_v = [2l + 3 + v(2v + 2l + 5)]\Gamma_1 - 4 \tag{39}$$

and $v = 0, 1, 2, \ldots$ (Γ_1 is taken here as a constant throughout the model). These solutions illustrate the behavior of centrally condensed models for which no analytical solution can be written down (see Tables 6.1–6.2). *Thus, for each value of l, there exists an infinite discrete spectrum.* A close study of the corresponding eigenfunctions shows that the radial function ξ_r dominates over the tangential components of the displacement, and that the pressure experiences substantial fluctuations. In particular, when $l = 0$, these p-modes (or acoustic modes) become purely radial ($\xi_\theta \equiv \xi_\varphi \equiv 0$). Also, the three harmonics belonging to $l = 1$ must be retained in this case, for they do not imply a displacement of the center of mass (albeit their moving of the geometrical center of the system!); this result, which was originally demonstrated by Smeyers, also applies to the g-modes belonging to $l = 1$ (see below).

Nonradial p-modes are always stable (see Figure 6.5 and Tables 6.1–6.2). For moderate central condensations ($\rho_c/\bar{p} \lesssim 100$), the number of nodes regularly increases along the radius as v tends to infinity (see, however, the final comments about the f-modes). Short periods of oscillation are thus associated with short wavelengths, i.e., large l and large v.

Greater interest is attached to the radial spectrum $l = 0$. Equation (38) then reduces to

$$\sigma^2 = \tfrac{4}{3}\pi G\rho\{[3 + v(2v + 5)]\Gamma_1 - 4\}. \tag{40}$$

Figure 6.3 illustrates the functions δr for the polytrope of index $n = 3$. Generally, the radial displacements form a complete set of orthogonal functions, and the number of nodes regularly varies in proportion to v along the radius. The influence of the ratio $\rho_c/\bar{p}$ on the lowest mode is shown on Figure 6.4 (compare with Figure 6.2). Just as for the f-modes, we can *approximate* the lowest frequency by the following expression:

$$\sigma^2 = (3\langle\Gamma_1\rangle - 4)\frac{|W|}{I} \propto (3\langle\Gamma_1\rangle - 4)\pi G\bar{p}, \tag{41}$$

where $\langle\Gamma_1\rangle$ denotes the pressure-weighted average of the adiabatic exponent. Equation (41) is originally due to Ledoux and Pekeris; when Γ_1 is a constant, it rigorously obtains for the homogeneous, compressible sphere.

From equations (40) and (41) we observe that instability sets in by the lowest mode $v = 0$, when $\langle\Gamma_1\rangle$ falls below the critical figure $4/3$. In the simple context of a constant value of the adiabatic exponent ($\Gamma_1 = \gamma = c_p/c_v$), this dynamical instability arises because the total energy E of a

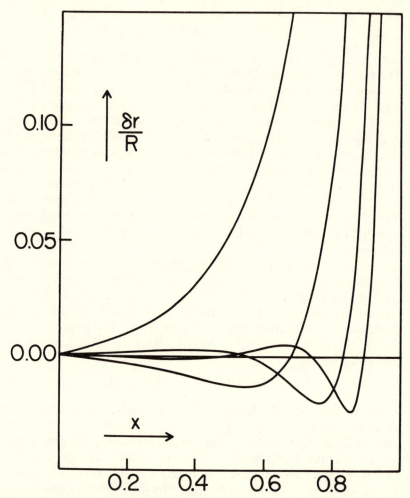

Figure 6.3.   Lowest radial modes of pulsation for the spherical polytrope $n = 3(\gamma = 5/3)$. All displacements $\delta r/R$ converge to unity at $x = r/R = 1$, where R is the radius. Courtesy Dr. H. Robe.

gaseous sphere becomes positive when $\gamma < 4/3$. Indeed, given our assumptions, we have $E = U_T + W$, where U_T designates the total internal energy. Hence, from the scalar virial theorem $W + 3(\gamma - 1)U_T = 0$ (cf. §3.7), we can write $E = -(3\gamma - 4)U_T$, implying a positive value for E when $\gamma < 4/3$. In a real star, however, the situation is somewhat more delicate. In this case, Γ_1 depends on r, and radiation pressure is the main factor to lower efficaciously the average $\langle \Gamma_1 \rangle$. Yet, since the limit $\langle \Gamma_1 \rangle = 4/3$ is never

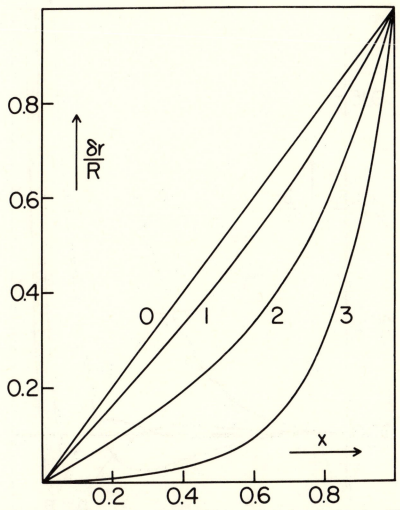

Figure 6.4.   Fundamental mode of radial pulsation for various polytropic spheres ($\gamma = 5/3$). All displacements $\delta r/R$ are normalized to unity at $x = r/R = 1$, where R is the radius. Curves are labeled by the value of the polytropic index to which they belong. Courtesy Dr. H. Robe.

attained for finite masses, instability cannot arise solely on account of radiation pressure (cf. §3.4). In any event, for massive stars, the lowest radial mode is always on the verge of instability ($\sigma^2 \to 0$); hence, small departures from isentropy or Newtonian mechanics may easily enhance the growth of this mode (cf. §§14.2 and 14.4).

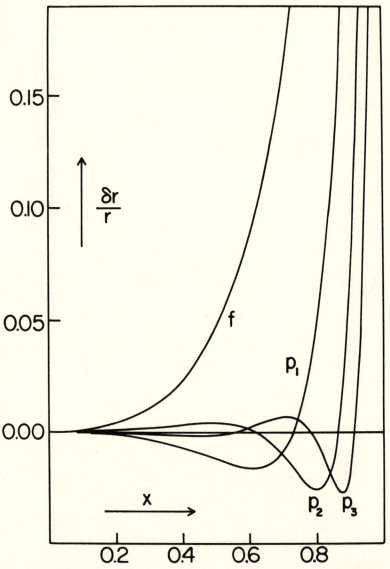

Figure 6.5.   Relative displacements $\delta r/r$ of the f-mode and the lowest p-modes belonging to $l = 4$ for a main-sequence star of $M = 27.8 M_\odot (\gamma = 5/3)$. The functions $\delta r/r$ are normalized to unity at $x = r/R = 1$, where R is the radius. Source: Smeyers, P., D.Sc. Thesis, Univ. of Liège, 1966.

THE g-MODES.   For simplicity, let us first consider the polytropes (with a constant adiabatic exponent Γ_1). In addition to the foregoing modes, Table 6.1 shows the existence of oscillations that are stable or unstable depending on the value of n; when $n = 3/2$ and $\Gamma_1 = 5/3$, these modes become neutral ($\sigma^2 = 0$). Quite generally, in the limiting case $n = 0$ (i.e., the homogeneous, compressible model), we can write

$$\sigma^2 = \tfrac{4}{3}\pi G\rho\{\Delta_v - [\Delta_v^2 + l(l+1)]^{1/2}\} \quad (l = 1, 2, 3, \ldots) \tag{42}$$

where $v = 0, 1, 2, \ldots$ (cf. equation [39]); these modes are always unstable ($\sigma^2 < 0$). In examining the corresponding displacements, we observe that the tangential components $r\xi_\theta$ and $r\sin\theta\xi_\varphi$ now exceed the radial function ξ_r; moreover, the changes in pressure δp are relatively smaller than for the p-modes. Actually, the present modes are caused by the natural tendency of gravity to smooth out material inhomogeneities along the level surfaces $V = constant$. For this reason, they are usually called g-modes (or gravity modes). It is worth noting that $|\sigma^2|$ increases with l, and decreases with v. Moreover, when $\rho_c/\bar{p} \lesssim 100$, the number of nodes in the disturbances regularly augments along the radius as v increases (see, however, the final comments about the f-modes). Accordingly, in an isentropic theory, greater attention must be paid to the unstable g-modes that are characterized by small "horizontal" dimensions (large l) and large "vertical" dimensions (small v). The above remark may prove illusory in practice, however, since dissipation and nonlinearity must eventually hinder the growth rate of these unstable motions.

As was first shown by Cowling, the origin of unstable g-modes in polytropic spheres can be traced back to the fact that the square of the *local* Brunt-Väisälä frequency

$$N^2(r) = \frac{1}{\rho}\frac{dp}{dr}\left(\frac{1}{\rho}\frac{d\rho}{dr} - \frac{1}{\Gamma_1 p}\frac{dp}{dr}\right) \tag{43}$$

is everywhere negative when $\Gamma_1 < 1 + 1/n$, which is also the condition for the temperature lapse rate to be superadiabatic throughout a polytrope.[5] Afterwards, Lebovitz proved the following general statements: (i) if $N^2(r)$

[5] See, e.g., Eckart, C., *Hydrodynamics of Oceans and Atmospheres*, pp. 57–61, Oxford: Pergamon Press, 1960. In the astrophysical literature, the fact that convection occurs if and only if $N^2(r) < 0$ is generally named after Karl Schwarzschild (1873–1916). To the best of my knowledge, this condition was originally derived in 1841 by the American meteorologist, James Pollard Espy; however, it is not until 1862 that Kelvin first published a firm analytical derivation of the criterion for convection, for both dry and wet air (see: Espy, J. P., *The Philosophy of Storms*, Boston: C. C. Little and J. Brown, 1841; Thomson, W., *Mem. Manchester Soc.* 2, 170, 1862 [Lord Kelvin: *Math. and Phys. Papers* 3, 255, 1890]). The Schwarzschild criterion will be discussed further in Sections 7.3 and 14.5.

is thoroughly positive in a spherical star, all squared frequencies σ^2 belonging to nonradial modes are positive (that is, of stable type), and (ii) if $N^2(r)$ is negative in some spherical domain, there must exist at least one nonradial disturbance which grows in time. Finally, Eisenfeld showed that the nonradial modes form a complete set when $N^2(r)$ is everywhere positive and does not identically vanish in some spherical shell. Accordingly, all nonradial disturbances are stable if and only if the temperature gradient is subadiabatic throughout the configuration. *Convective currents are thus the ultimate cause to the existence of unstable g-modes.* It is apparent from Table 6.1 that the g-modes always act in concert; hence, they all become unstable when $N^2(r)$ is negative everywhere. As it was expected on physical grounds, when $N^2(r) \equiv 0$ in the core and $N^2(r) > 0$ in the envelope, the g-modes are thoroughly stable (see Table 6.2); however, in this particular instance, the zeros of the displacements δr remain located in the radiative envelope, and the amplitudes decrease exponentially in the core which is in strict convective equilibrium (see Figure 6.6).

The behavior of the g-modes is particularly intriguing when $N^2(r) < 0$ in the core and $N^2(r) > 0$ in the envelope (see Figure 6.7). As was first shown by Ledoux and Smeyers, for each value of l there exist *two* infinite discrete spectra: the (*stable*) g^+-modes corresponding to gravity oscillations in the radiative zone, and the (*unstable*) g^--modes describing convective currents in the core. According to these authors, the motions are fairly well localized in their respective zones and the corresponding amplitudes decrease exponentially in the contiguous region. This last property can be proved easily for the very high modes ($\sigma^2 \to 0$). Indeed, by letting

$$w(r) = r\rho^{1/2}(r)\,\delta r(r), \tag{44}$$

we can then write the equations of motion in the approximate form

$$\frac{d^2 w}{dr^2} + \frac{l(l+1)}{\sigma^2}\frac{N^2(r)}{r^2}\,w \approx 0, \tag{45}$$

in the vicinity of the spherical surface where $N^2(r)$ changes sign. Consider first the g^+-modes ($\sigma^2 > 0$). From equation (45), we notice that the solutions are oscillatory in the region where $N^2(r) > 0$, and decrease exponentially in the adjacent zone. The reverse situation occurs for the g^--modes ($\sigma^2 < 0$). According to Scuflaire, however, whereas the g^--modes of a star may oscillate only in convective zones, the g^+-modes do *not* decrease exponentially in the exterior convection zones of low mass, evolved stars.

Quite generally, it can also be shown that *the g-modes split into a number of infinite discrete spectra which is equal to the number of spherical shells where $N^2(r)$ keeps a constant sign.* In the limit of very high modes

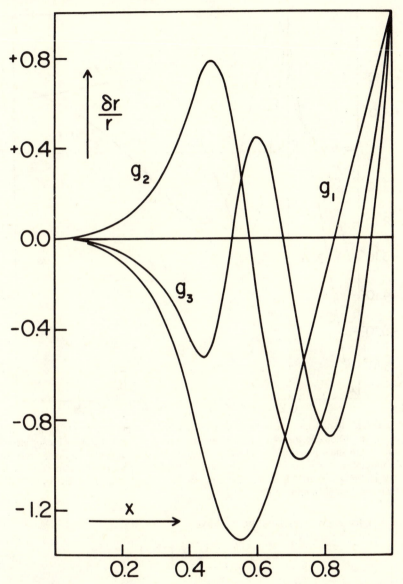

Figure 6.6. Relative displacements $\delta r/r$ of the lowest g-modes belonging to $l = 4$ for a main-sequence star of $M = 27.8M_\odot (\gamma = 5/3)$. The functions $\delta r/r$ are normalized to unity at $x = r/R = 1$, where R is the radius. Source: Smeyers, P., D. Sc. Thesis, Univ. of Liège, 1966.

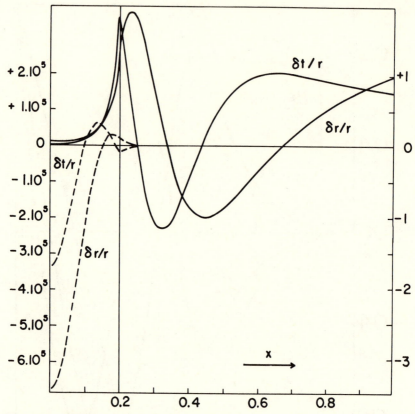

Figure 6.7. Radial ($\delta r/r$) and tangential ($\delta t/r$) relative displacements of the $g_2{}^+$- and $g_2{}^-$-modes belonging to $l = 2$ for a composite model constructed from a homogeneous core ($n = 0$) and a polytropic envelope ($n = 3$), with $\gamma = 5/3$. Ordinates on the left refer to the unstable $g_2{}^-$-mode (*dashed curves*); ordinates on the right refer to the stable $g_2{}^+$-mode (*full-line curves*). Source: Ledoux, P., et Smeyers, P., *C.R. Acad. Sci. Paris* **262B**, 841, 1966. (By permission of Gauthier-Villars, Paris.)

($v \gg l$), for each layer D (say), we have

$$\sigma \approx \pm \frac{[l(l + 1)]^{1/2}}{\pi(v + c)} \int_D \frac{N(r)}{r}\, dr, \tag{46}$$

where the constant c is of order unity and depends on the chosen domain D. Thus, to every spherical layer where the local frequency $N(r)$ is real corresponds one stable g^+-spectrum; on the contrary, in each zone where $N(r)$ is purely imaginary, all squared frequencies are negative. Also, in view of equation (45), high-order disturbances are mainly confined to their corresponding shell, and drop exponentially in the contiguous region(s). To illustrate these results, Goossens and Smeyers have subsequently studied the three g-spectra of composite polytropic models that consist of two convectively stable regions separated by a convectively unstable zone. Apart from the g^--modes associated with the intermediate layer, they also found two types of stable g-modes, one type associated mainly with the core, the other one associated mainly with the outer envelope. However, for these stable modes, they also found accidental "resonances" giving rise to stable modes with an appreciable amplitude in the two stable regions that contribute nearly equally to the eigenvalues of these modes. Because this phenomenon does not appear in the asymptotic analysis leading to equation (46), further analytical work is thus needed to discuss, e.g., the resonant interaction between two convective zones separated by a radiative shell—a case that is obviously more relevant to the study of overshooting and mixing in real stars.

6.5. AN ENERGY PRINCIPLE

As was said before, a rigorous discussion of stability requires a proof of completeness for the normal modes. If this proof is lacking, it is then preferable to discuss the motions themselves, characterized by their initial values. The energy principle of Laval, Mercier, and Pellat provides such an approach.

Assume that the small departures from a known equilibrium structure are described by a Lagrangian displacement $\xi(\mathbf{x},t)$ that satisfies an equation of the form

$$\rho \frac{\partial^2 \xi}{\partial t^2} = \mathbf{L}\xi, \tag{47}$$

where $\mathbf{L}$ designates a linear operator which is both *time-independent* and *symmetric*; we thus have

$$\frac{\partial}{\partial t}\mathbf{L}\xi = \mathbf{L}\frac{\partial \xi}{\partial t}, \tag{48}$$

and

$$\int_{\mathscr{V}} \xi . \mathbf{L}\eta \, d\mathbf{x} = \int_{\mathscr{V}} \eta . \mathbf{L}\xi \, d\mathbf{x}, \tag{49}$$

where $\xi(\mathbf{x},t)$ and $\eta(\mathbf{x},t)$ denote two arbitrary smooth functions. Under these conditions, if

$$\mathscr{W}(t) = \int_{\mathscr{V}} \xi . \mathbf{L}\xi \, d\mathbf{x} \tag{50}$$

is negative for *all* finite displacements $\xi(\mathbf{x},t)$ non-identically zero, the configuration is stable, i.e., the kinetic energy of the perturbed motion

$$\mathcal{K}(t) = \frac{1}{2} \int_{\mathscr{V}} \rho \frac{\partial \boldsymbol{\xi}}{\partial t} \cdot \frac{\partial \boldsymbol{\xi}}{\partial t} \, \mathbf{dx} \qquad (51)$$

cannot grow from its initial value $\mathcal{K}(0)$ at time $t = 0$ (say).

For further reference, let us first quote two important relations. In making use of equations (47)–(51), we can write

$$2\mathcal{K} - \mathcal{W} = \mathscr{E}, \qquad (52)$$

where $\mathscr{E}$ denotes a constant. Similarly, we obtain

$$\frac{1}{2} \frac{d^2 \mathscr{I}}{dt^2} = 2\mathcal{K} + \mathcal{W}, \qquad (53)$$

where

$$\mathscr{I}(t) = \int_{\mathscr{V}} \rho \boldsymbol{\xi} \cdot \boldsymbol{\xi} \, \mathbf{dx}. \qquad (54)$$

Equation (52) merely expresses the conservation of energy of the perturbed system, and equation (53) is the usual scalar virial theorem.

Obviously, if $\mathcal{W}(t)$ is negative for *all* allowable functions $\boldsymbol{\xi}(\mathbf{x}, t)$, equation (52) implies that $\mathcal{K}(t)$ cannot grow from its initial value. The condition is thus sufficient.

To prove that the theorem is also a necessary condition of stability, let us assume there exists a square integrable displacement $\boldsymbol{\eta}(\mathbf{x}, 0)$ such that

$$\mathcal{W}(0) = \sigma_0^2 \int_{\mathscr{V}} \boldsymbol{\eta} \cdot \boldsymbol{\eta} \, \mathbf{dx}, \qquad (55)$$

where σ_0 is a positive constant. Impose next the initial conditions

$$\boldsymbol{\xi} = \boldsymbol{\eta}, \quad \text{and} \quad \frac{\partial \boldsymbol{\xi}}{\partial t} = \sigma_0 \boldsymbol{\eta}, \quad \text{at } t = 0. \qquad (56)$$

With this particular choice, we can determine the value of $\mathscr{E}$, and equation (52) then reduces to

$$2\mathcal{K} - \mathcal{W} = 0. \qquad (57)$$

If we now combine equations (53) and (57), we obtain

$$\frac{d^2 \mathscr{I}}{dt^2} = 8\mathcal{K}. \qquad (58)$$

In making use next of the Schwarz inequality, which, for the present problem, takes the form

$$\left(\frac{d\mathscr{I}}{dt} \right)^2 \leqslant 8 \mathscr{I} \mathcal{K}, \qquad (59)$$

we can write

$$\frac{d^2 \mathscr{I}}{dt^2} \geqslant \frac{1}{\mathscr{I}} \left(\frac{d\mathscr{I}}{dt} \right)^2,$$ (60)

or

$$\frac{d}{dt} \left(\frac{1}{\mathscr{I}} \frac{d\mathscr{I}}{dt} \right) \geqslant 0.$$ (61)

On the other hand, by virtue of our initial conditions (56), we have

$$\frac{1}{\mathscr{I}} \frac{d\mathscr{I}}{dt} = 2\sigma_0, \quad \text{at } t = 0;$$ (62)

hence, in view of equation (61), we obtain

$$\frac{1}{\mathscr{I}} \frac{d\mathscr{I}}{dt} \geqslant 2\sigma_0, \quad \text{at } t \geqslant 0.$$ (63)

This, in turn, implies that

$$\mathscr{I}(t) \geqslant \mathscr{I}(0)e^{2\sigma_0 t}.$$ (64)

Finally, if we derive $\mathscr{K}(t)$ from equations (55), (57), and (59), we obtain

$$\mathscr{K}(t) \geqslant \tfrac{1}{2}\mathscr{W}(0)e^{2\sigma_0 t}.$$ (65)

Thus, by virtue of equation (55), $\mathscr{K}(t)$ must necessarily depart from its initial value.

In summary, the negativeness of $\mathscr{W}(t)$ for all allowable displacements $\xi(\mathbf{x},t)$ is a necessary and sufficient condition of stability. The strength of this principle is that it avoids any assumptions about the existence of normal modes or about their properties when they do exist.

6.6. A VARIATIONAL PRINCIPLE

The small oscillations of a barocline in a state of permanent rotation are generally described by equations (15), (17), (25), and (29), together with the requisite boundary conditions. Unfortunately, in most instances, this problem cannot be solved in closed analytical form; hence, we usually resort to approximation techniques to investigate the motions. We shall now show that the solutions of the problem can be expressed in terms of a variational principle due to Clement, Lynden-Bell, and Ostriker. To outline the general ideas, we shall restrict the present discussion to axisymmetric disturbances.

Returning to equation (25), we can write

$$\rho \left(\frac{\partial}{\partial t} + \mathbf{v}.\mathrm{grad} \right)^2 \xi = \mathbf{K}\xi - \rho(\xi.\mathrm{grad})\mathbf{q};$$ (66)

we have

$$\mathbf{K}\xi = \text{grad}(\Gamma_1 p \text{ div } \xi) + \text{grad}(\xi.\text{grad } p)$$

$$+ \frac{\delta\rho}{\rho} \text{ grad } p - \rho \text{ grad } \delta V, \tag{67}$$

where $\delta\rho$ and δV are given in terms of ξ by equations (15) and (29), respectively. The centrifugal acceleration $\mathbf{q}$ is defined in equation (20), and the components of the velocity $\mathbf{v}$ have the form

$$v_\varpi = 0, \quad v_\varphi = \Omega(\varpi,z)\varpi, \quad v_z = 0, \tag{68}$$

in our inertial frame. By virtue of the assumptions, the equilibrium quantities depend on ϖ and z only.

Up to this point, no assumption has been made about the Lagrangian displacement $\xi(\mathbf{x},t)$. Henceforth, we shall seek for axisymmetric solutions, i.e., motions for which ξ_ϖ, ξ_φ, and ξ_z are independent of φ. In this case, the components of equation (66) become

$$\frac{\partial^2 \xi_\varpi}{\partial t^2} - 2\Omega \frac{\partial \xi_\varphi}{\partial t} = \frac{1}{\rho}(\mathbf{K}\xi)_\varpi - \varpi\left(\xi_\varpi \frac{\partial\Omega^2}{\partial\varpi} + \xi_z \frac{\partial\Omega^2}{\partial z}\right), \tag{69}$$

$$\frac{\partial^2 \xi_\varphi}{\partial t^2} + 2\Omega \frac{\partial \xi_\varpi}{\partial t} = 0, \tag{70}$$

$$\frac{\partial^2 \xi_z}{\partial t^2} = \frac{1}{\rho}(\mathbf{K}\xi)_z, \tag{71}$$

where $(\mathbf{K}\xi)_\varpi$ and $(\mathbf{K}\xi)_z$ solely depend on ϖ and z. Equation (70) expresses the conservation of the z-component of the angular momentum per unit mass, and is a mere consequence of our assuming axisymmetric disturbances. In integrating equation (70) with respect to the time, we can write

$$\frac{\partial \xi_\varphi}{\partial t} + 2\Omega\xi_\varpi = f(\varpi,z), \tag{72}$$

where $f(\varpi,z)$ is an arbitrary function.

Let us now restrict the motions to those for which $f \equiv 0$ (this assumption is discussed below). On this account, after eliminating $\partial\xi_\varphi/\partial t$ from equations (69) and (72), we obtain

$$\frac{\partial^2 \xi_\varpi}{\partial t^2} = \frac{1}{\rho}(\mathbf{K}\xi)_\varpi - 4\Omega^2\xi_\varpi - \varpi\left(\xi_\varpi \frac{\partial\Omega^2}{\partial\varpi} + \xi_z \frac{\partial\Omega^2}{\partial z}\right). \tag{73}$$

Equations (71) and (73) represent an initial-value problem for ξ_ϖ and ξ_z. Hence, once we know ξ_ϖ, we can derive ξ_φ from equation (70); and by an appropriate choice of $\partial\xi_\varphi/\partial t$ at $t = 0$ (say) we can arrange the function f to vanish. Thus, in choosing $f \equiv 0$, we restricted the motions to the class of initial conditions for which this is true. It is easy to see that normal-mode solutions of the form (33) belong to this class.

Since neither equation (71) nor equation (73) involves ξ_φ, it is convenient to redefine $\xi(\varpi,z,t)$ to be a two-dimensional vector with components ξ_ϖ and ξ_z. Accordingly, we may combine equations (71) and (73) into the single vectorial equation

$$\rho \frac{\partial^2 \xi}{\partial t^2} = \mathbf{L}\xi, \tag{74}$$

where $\mathbf{L}$ is a linear operator acting on ξ_ϖ and ξ_z, and whose components are given by

$$(\mathbf{L}\xi)_\varpi = (\mathbf{K}\xi)_\varpi - \rho(\Phi_\varpi \xi_\varpi + \Phi_z \xi_z), \tag{75}$$

and

$$(\mathbf{L}\xi)_z = (\mathbf{K}\xi)_z, \tag{76}$$

where the vector $\mathbf{\Phi}$ has the form

$$\mathbf{\Phi}(\varpi,z) = \frac{1}{\varpi^3} \operatorname{grad}(\Omega^2 \varpi^4). \tag{77}$$

If we make use of equation (67), it can be proved that the time-independent operator $\mathbf{L}$ is *symmetric* (cf. equation [49]). Hence, by virtue of equation (74), we can apply the energy principle of Section 6.5. In particular, if for some admissible choice of ξ

$$\int_V \xi.\mathbf{L}\xi \, \mathbf{dx} > 0, \tag{78}$$

we can assert that the equilibrium configuration is unstable. This necessary and sufficient condition of instability with respect to axisymmetric motions does not invoke the existence of normal modes; and, in this form, the result is due to Lebovitz. As we shall see in Section 6.8, condition (78) expresses the fact that an equilibrium configuration is unstable if its total energy fails to be a minimum.

Let us now restrict the discussion to normal modes. With a dependence of ξ on time of the form (33), equation (74) leads to an eigenvalue problem associated with the equation

$$-\sigma^2 \rho\xi = \mathbf{L}\xi. \tag{79}$$

(In the remainder of this section, ξ solely depends on ϖ and z.) Thus, in making use of the notation defined in equations (50) and (54), we can write

$$\sigma^2 = -\frac{\mathscr{W}}{\mathscr{I}}, \tag{80}$$

where, after performing an integration by parts, we obtain

$$
\mathscr{W} = \int_{\mathscr{V}} \left[-\Gamma_1 p(\mathrm{div}\ \xi)^2 + \xi.\mathrm{grad}(\xi.\mathrm{grad}\ p) + \frac{\delta\rho}{\rho}\ \xi.\mathrm{grad}\ p \right.
$$

$$
\left. - \rho\xi.\mathrm{grad}\ \delta V \right]\mathbf{dx} - \int_{\mathscr{V}} \rho\xi_\varpi(\Phi_\varpi\xi_\varpi + \Phi_z\xi_z)\ \mathbf{dx}, \tag{81}
$$

and

$$
\mathscr{I} = \int_{\mathscr{V}} \rho(\xi_\varpi{}^2 + \xi_z{}^2)\ \mathbf{dx}. \tag{82}
$$

As we shall presently see, equation (80) provides a variational basis for the determination of the allowed values of σ^2. That is; the eigenfrequencies determined from equation (80) are *stationary* with respect to arbitrary variations in ξ.

To prove this property, let $\delta\sigma^2$ be the change in σ^2 when ξ is subject to a small variation $\delta\xi$. The functions ξ and $\delta\xi$ are arbitrary except for the condition that they be consistent with the boundary conditions. According to equation (80), we can thus write

$$
\mathscr{I}\ \delta\sigma^2 = -\delta\mathscr{W} - \sigma^2\ \delta\mathscr{I}. \tag{83}
$$

By virtue of the symmetry of **L**, we obtain

$$
\delta\mathscr{W} = 2\int_{\mathscr{V}} \delta\xi.\mathbf{L}\xi\ \mathbf{dx}; \tag{84}
$$

we also have

$$
\delta\mathscr{I} = 2\int_{\mathscr{V}} \rho\ \delta\xi.\xi\ \mathbf{dx}. \tag{85}
$$

Equation (83) thus becomes

$$
\mathscr{I}\ \delta\sigma^2 = -2\int_{\mathscr{V}} \delta\xi.(\sigma^2\rho\xi + \mathbf{L}\xi)\ \mathbf{dx}. \tag{86}
$$

Clearly, if ξ is a proper solution of equation (79), $\delta\sigma^2$ vanishes. On the other hand, if $\delta\sigma^2 = 0$ for an arbitrary variation $\delta\xi$, then ξ must be the proper displacement corresponding to σ^2. This completes the proof of the variational principle for axisymmetric motions.

From a practical point of view, it is convenient to choose "trial" displacements of the form

$$
\xi(\varpi,z) = \sum_{\alpha=1}^{N} a_\alpha\xi^{(\alpha)}(\varpi,z), \tag{87}
$$

where the $\xi^{(\alpha)}$'s are any fixed functions (e.g., polynomials in ϖ and z), and the coefficients a_α are determined in expressing that σ^2 is an extremum. If we let

$$
A_{\alpha\beta} = \int_{\mathscr{V}} \rho\xi^{(\alpha)}.\xi^{(\beta)}\ \mathbf{dx}, \tag{88}
$$

and

$$B_{\alpha\beta} = \int_{\mathscr{V}} \xi^{(\alpha)}.\mathbf{L}\xi^{(\beta)} \, \mathbf{dx}, \tag{89}$$

equation (80) now reads

$$\sum_{\alpha=1}^{N} \sum_{\beta=1}^{N} (\sigma^2 A_{\alpha\beta} + B_{\alpha\beta}) a_\alpha a_\beta = 0. \tag{90}$$

In varying the a_α's in order to extremize the ratio $\mathscr{W}/\mathscr{I}$, we readily obtain

$$\sum_{\beta=1}^{N} (\sigma^2 A_{\alpha\beta} + B_{\alpha\beta}) a_\beta = 0 \quad (\alpha = 1, 2, \ldots, N) \tag{91}$$

These algebraic equations are linear and homogeneous in the a_α's. Accordingly, their determinant must vanish, and we are led to solve the secular equation

$$|\sigma^2 A_{\alpha\beta} + B_{\alpha\beta}| = 0, \tag{92}$$

which provides approximate values of the eigenfrequencies.

6.7.   THE SECOND-ORDER VIRIAL THEOREM

Instead of using the general equations of hydrodynamics, we can alternatively describe the motions of a self-gravitating system by means of the virial equations to which they are equivalent (cf. §3.7). The main interest of the virial method is that it provides *approximate* representations of small departures from a state of permanent rotation. While the scalar virial theorem allows a simple description of the lowest acoustic mode (41), the second-order virial equations embody motions that, in the limit of zero rotation, correspond to the radial mode (41) and the *five f-modes* (37). Similarly, the third-order virial equations describe motions that reduce to modes belonging to the first- and third-order spherical harmonics in the limiting case of a nonrotating body (cf. §14.5). Higher-order equations naturally provide higher-order modes, but this road becomes rapidly impracticable. The virial method has been extensively used by Chandrasekhar and Lebovitz.

As a preliminary to the discussion of the linearized virial equations, consider first the change δF, which we observe by comparing at time t the values of the integral

$$F(t) = \int_M Q(\mathbf{x},t) \, dm, \tag{93}$$

in the perturbed and unperturbed flows (cf. §6.2). Clearly, a mass element $dm = \rho(\mathbf{x},t) \, \mathbf{dx}$ located at the point $\mathbf{x}$ in the ghostly flow has its corresponding image at the location $\mathbf{x} + \xi(\mathbf{x},t)$ in the actual flow. Hence, by virtue of equations (2) and (6), we have

$$\delta F = \delta \int_M Q \, dm = \int_M \Delta Q \, dm = \int_M \left(\delta Q + \xi_k \frac{\partial Q}{\partial x_k} \right) dm, \qquad (94)$$

embodying the conservation of mass. Very much for the same reason, we can also write

$$\frac{dF}{dt} = \frac{d}{dt} \int_M Q \, dm = \int_M \frac{DQ}{Dt} \, dm = \int_M \left(\frac{\partial Q}{\partial t} + v_k \frac{\partial Q}{\partial x_k} \right) dm \qquad (95)$$

(cf. §3.3: equation [23]). Thus, when using equations (94) and (95), we may perform all integrations over the *equilibrium* volume $\mathscr{V}$. For further use, we shall retain the notation $dm = \rho \, \mathbf{dx}$. Finally, note that the operations of δ and d/dt always commute.

Consider now a barotrope or a pseudo-barotrope in a state of permanent rotation (cf. §4.3). Henceforth, for convenience, a Cartesian system of coordinates (x_1, x_2, x_3) will be used. Accordingly, in an inertial frame of reference, the components of the velocity $\mathbf{v}$ are given by

$$v_1 = -\Omega(\varpi)x_2, \quad v_2 = +\Omega(\varpi)x_1, \quad v_3 = 0, \qquad (96)$$

where $\varpi^2 = x_1{}^2 + x_2{}^2$. Inserting the foregoing expressions in the second-order virial equations, we obtain

$$\langle \Omega^2 \rangle I_{11} + W_{11} + \int_{\mathscr{V}} \frac{p}{\rho} \, dm = 0, \qquad (97)$$

and

$$W_{33} + \int_{\mathscr{V}} \frac{p}{\rho} \, dm = 0, \qquad (98)$$

where we define

$$\langle \Omega^2 \rangle = \int_{\mathscr{V}} \Omega^2(\varpi)\varpi^2 \, dm \Big/ \int_{\mathscr{V}} \varpi^2 \, dm, \qquad (99)$$

which is proportional to the kinetic energy; remaining symbols have their usual meanings (cf. §3.7). Finally, in addition to the "mean squared angular velocity," we will also need the quantity

$$\langle \Omega \rangle = \int_{\mathscr{V}} \Omega(\varpi)\varpi^2 \, dm \Big/ \int_{\mathscr{V}} \varpi^2 \, dm, \qquad (100)$$

which is proportional to the total angular momentum J of the rotating configuration.

Suppose next that the equilibrium is slightly disturbed. Then, from the second-order virial equations

$$\frac{d}{dt} \int_{\mathscr{V}} v_i x_j \, dm - \int_{\mathscr{V}} v_i v_j \, dm = W_{ij} + \delta_{ij} \int_{\mathscr{V}} \frac{p}{\rho} \, dm, \qquad (101)$$

together with equation (94), we readily obtain

$$\frac{d}{dt} \int_{\mathscr{V}} (\Delta v_i x_j + v_i \xi_j)\, dm - \int_{\mathscr{V}} (\Delta v_i v_j + \Delta v_j v_i)\, dm$$

$$= \delta W_{ij} + \delta_{ij} \left(\delta \int_{\mathscr{V}} \frac{p}{\rho}\, dm \right). \quad (102)$$

Now, equations (4), (95), and (96) imply that

$$\frac{d}{dt} \int_{\mathscr{V}} \xi_i x_j\, dm = \int_{\mathscr{V}} \Delta v_i x_j\, dm + \int_{\mathscr{V}} \xi_i v_j\, dm, \quad (103)$$

and

$$\frac{d}{dt} \int_{\mathscr{V}} \xi_i v_j\, dm = \int_{\mathscr{V}} \Delta v_i v_j\, dm + \int_{\mathscr{V}} \xi_i v_k \frac{\partial v_j}{\partial x_k}\, dm. \quad (104)$$

Hence, integrations over $\Delta v_i x_j$ and $\Delta v_i v_j$ appearing in equation (102) may be related at once to integrations over $\xi_i x_j$, $\xi_i v_j$, and $\xi_i v_k \partial v_j / \partial x_k$. Similarly, we can write

$$\delta W_{ij} = \int_{\mathscr{V}} \xi_k \frac{\partial V_{ij}}{\partial x_k}\, dm, \quad (105)$$

where

$$V_{ij} = -G \int_{\mathscr{V}} \frac{(x_i - x_i')(x_j - x_j')}{|\mathbf{x} - \mathbf{x}'|^3}\, dm'. \quad (106)$$

We also have

$$\delta \int_{\mathscr{V}} \frac{p}{\rho}\, dm = - \int_{\mathscr{V}} (\Gamma_1 - 1) \frac{p}{\rho} \frac{\partial \xi_k}{\partial x_k}\, dm, \quad (107)$$

where we have made use of equations (14), (16), and (94). In inserting equations (103)–(107) into equation (102), we finally obtain

$$\frac{d^2}{dt^2} \int_{\mathscr{V}} \xi_i x_j\, dm - 2 \frac{d}{dt} \int_{\mathscr{V}} \xi_i v_j\, dm + \int_{\mathscr{V}} \left(\xi_i \frac{\partial v_j}{\partial x_k} + \xi_j \frac{\partial v_i}{\partial x_k} \right) v_k\, dm$$

$$- \int_{\mathscr{V}} \xi_k \frac{\partial V_{ij}}{\partial x_k}\, dm + \delta_{ij} \int_{\mathscr{V}} (\Gamma_1 - 1) \frac{p}{\rho} \frac{\partial \xi_k}{\partial x_k}\, dm = 0, \quad (108)$$

relating the displacement ξ to quantities that all depend on the known equilibrium structure.

Equation (108) is an *exact* integral relation that must be satisfied in all cases of small-amplitude disturbances; in particular, it must be satisfied by the proper solutions belonging to the normal modes of oscillation (33). Thus, by inserting for ξ a "trial" function with a spatial dependence of the form (87), we should obtain equations to determine the frequencies with some degree of accuracy. Actually, it can be shown that the second-order virial technique is equivalent to a first-order variational approach. Nevertheless, in the present instance, we cannot use "trial" functions with as many parameters as we like; indeed, since equation (108) provides *nine* equations, a "trial" displacement that we may wish to insert must involve *nine* arbitrary coefficients. We will now assume that, in a first approximation, the Lagrangian displacement can adequately be represented by the function

$$\xi_i = L_{i,k} x_k e^{i\sigma t}, \tag{109}$$

where the $L_{i,k}$'s denote nine coefficients that play the role of the variational parameters in a proper variational treatment. Actually, as was shown by Chandrasekhar and Lebovitz, the function (109) corresponds to the *exact* solution in the case of a gaseous configuration for which the density ρ and the angular velocity Ω are constant, i.e., the compressible Maclaurin spheroids (cf. §4.5). For this reason, it is evident that the larger the departure from uniformity in angular velocity and mass distribution, the greater the error in the frequencies will be. In any event, the qualitative features of the results should be trustworthy.

By making use of equation (109), equation (108) provides nine homogeneous equations in the nine unknowns $L_{i,k}$. In the special case where $\Omega \equiv 0$, these equations lead to nine characteristic frequencies: (i) the single root (41) corresponding to a purely radial motion; (ii) the quintuple root (37) belonging to the spherical harmonics Y_2^m (cf. equation [34]); and (iii) the triple root $\sigma^2 = 0$, each of which describes an infinitesimal rotation about the x_1-, x_2-, and x_3-axes, respectively (cf. note 4). Turning next to rotating bodies, we observe that rotation partially removes the degeneracy in the eigenfrequencies; indeed, we now find seven *distinct* squared frequencies and the double root $\sigma^2 = 0$. Actually, equations (108) and (109) lead to frequencies that correspond to zonal, tesseral, and sectorial oscillations, respectively[6]. Let us examine these modes in turn.

THE TESSERAL MODES.   The displacement appropriate to these motions is

$$\xi_1 = L_{1,3} x_3, \quad \xi_2 = L_{2,3} x_3, \quad \text{and} \quad \xi_3 = L_{3,1} x_1 + L_{3,2} x_2, \tag{110}$$

where we have suppressed the time-dependent factor $e^{i\sigma t}$. From equation (110), we note that these oscillations are characterized by a relative shearing of the northern and southern hemispheres; the motions are thus antisymmetric with respect to the equatorial plane. The corresponding secular equation is

$$\sigma^4 [\sigma^2 + \langle \Omega^2 \rangle - (\lambda + \mu)]^2 = 4\sigma^2 \langle \Omega \rangle^2 (\sigma^2 - \mu)^2, \tag{111}$$

---

[6] These second-order oscillations are sometimes called "pulsation modes," "transverse shear modes," and "toroidal (or barlike) modes," respectively. Here we shall use the terms "zonal," "tesseral," and "sectorial" in their strict etymological senses (see Appendix C).

where

$$\lambda = \frac{W_{13;13}}{I_{11}} \quad \text{and} \quad \mu = \frac{W_{31;13}}{I_{33}}; \tag{112}$$

quite generally, we may define

$$W_{ij,kl} = -\int_{\mathscr{V}} x_i \frac{\partial V_{kl}}{\partial x_j} \, dm, \quad \text{and} \quad I_{ij} = \int_{\mathscr{V}} x_i x_j \, dm. \tag{113}$$

By virtue of our assumptions, these frequencies do not depend on the compressibility of the configuration. After eliminating the trivial root $\sigma^2 = 0$, we obtain the *two* cubic equations:

$$\sigma^3 \mp 2\langle\Omega\rangle\sigma^2 + [\langle\Omega^2\rangle - (\lambda + \mu)]\sigma \pm 2\langle\Omega\rangle\mu = 0. \tag{114}$$

The occurrence of two signs means that the six frequencies provided by the above equations have a doublet character, with $+\langle\Omega\rangle$ and $-\langle\Omega\rangle$ playing equivalent roles. All tesseral modes describe *stable* oscillations: $\sigma^2 = 0$, $\sigma^2 \approx \langle\Omega\rangle^2$, and two squared roots that, for a spherical system, reduce to the *f*-modes (37) belonging to the tesseral harmonics Y_2^{+1} and Y_2^{-1} (cf. §10.4).

THE SECTORIAL MODES.   The predominant feature of these oscillations is to transform the rotating model into a genuine triaxial body, while preserving its plane of symmetry; all motions are restricted to planes parallel to the equatorial plane. Suppressing the factor $e^{i\sigma t}$, we thus have

$$\xi_1 = L_{1,1}x_1 + L_{1,2}x_2, \quad \xi_2 = -L_{1,1}x_2 + L_{1,2}x_1, \quad \text{and} \quad \xi_3 = 0. \tag{115}$$

The four corresponding frequencies of oscillation also have a doublet character; we can write

$$\sigma = \pm\langle\Omega\rangle \pm (2\omega + \langle\Omega\rangle^2 - 2\langle\Omega^2\rangle)^{1/2}, \tag{116}$$

where

$$\omega = \frac{W_{12,12}}{I_{11}}. \tag{117}$$

Obviously, under the present assumptions, the sectorial modes are independent of the adiabatic exponent. In the limit of zero rotation, they reduce to the *f*-modes (37) belonging to the sectorial harmonics Y_2^{+2} and Y_2^{-2}. Now, by virtue of equation (116), it follows at once that σ^2 vanishes when

$$\langle\Omega^2\rangle = \omega; \tag{118}$$

moreover, all solutions correspond to stable oscillations if and only if ω is positive and

$$\langle\Omega^2\rangle \leqslant \omega + \tfrac{1}{2}\langle\Omega\rangle^2. \tag{119}$$

As we shall see in Section 10.4, the possible occurrence of the neutral mode $\sigma^2 = 0$ coincides with the appearance of triaxial configurations, when $\langle \Omega^2 \rangle = \omega$ along a sequence of rotating models. Thus, if the limit (118) can be reached along a sequence, this neutral mode simply carries the axially symmetric body over into a neighboring equilibrium state having a genuine triplanar symmetry.

THE ZONAL MODES.   Six squared frequencies are now accounted for. The remaining three values of σ^2 correspond to the displacement

$$\xi_1 = L_{1,1}x_1 + L_{1,2}x_2, \quad \xi_2 = L_{1,1}x_2 - L_{1,2}x_1, \quad \xi_3 = L_{3,3}x_3, \quad (120)$$

where, again, the factor $e^{i\sigma t}$ has been omitted. Clearly, they represent pulsations that preserve both the planar and axial symmetries of the configuration. In addition to the trivial root $\sigma^2 = 0$ corresponding to an infinitesimal rotation about the x_3-axis ($L_{1,1} = L_{3,3} = 0$, $L_{1,2} \neq 0$), we have

$$I_{11}I_{33}\sigma^4 - [(4\langle \Omega \rangle^2 - 3\langle \Omega^2 \rangle)I_{11}I_{33} + (\beta + \alpha)I_{33} + \beta I_{11}]\sigma^2$$
$$+ [\beta(4\langle \Omega \rangle^2 - 3\langle \Omega^2 \rangle)I_{11} + (\beta - \alpha)(\beta + 2\alpha)] = 0, \quad (121)$$

where

$$\alpha = W_{33,11} + \int_{\mathscr{V}} (\Gamma_1 - 1)\frac{p}{\rho}\, dm, \quad (122)$$

and

$$\beta = W_{33,33} + \int_{\mathscr{V}} (\Gamma_1 - 1)\frac{p}{\rho}\, dm. \quad (123)$$

When rotation is absent, the two squared roots (121) can easily be identified: one corresponds to the f-mode belonging to the zonal harmonic $Y_2{}^0$ (cf. equation [37]), and the remaining solution degenerates into the lowest radial mode (41). Let us note that rotation couples the two zonal modes (121), and that they both strongly depend on the compressibility of the system. Further inferences from these axisymmetric motions will be drawn in Section 14.2.

6.8. THE STATIC AND QUASI-DYNAMIC METHODS

The two previous sections provide practical tools for determining the frequencies of oscillation of a rotating star. As we shall see, it is also possible to obtain useful information about the stability of an equilibrium configuration without resorting to such lengthy procedures. The static method leads to a simple stability criterion for axisymmetric motions, and one needs to construct equilibrium models only, without any further calculation. The quasi-dynamic method provides a numerical scheme for solving the equations of equilibrium, and it incorporates a stability test that would otherwise have to be performed separately.

Both systems of analysis rely upon the fact that the total energy—which is extremal for stars in a state of permanent rotation—is a minimum for stable equilibria and fails to be a minimum for unstable equilibria. This can be seen as follows. Take the scalar product of the equations of motion (21) with the velocity $\mathbf{v}$, and integrate over the volume $\mathscr{V}$. After performing an integration by parts and using the fact that the pressure vanishes on the stellar surface, we find

$$\frac{d}{dt} \int_{\mathscr{V}} \tfrac{1}{2}|\mathbf{v}|^2 \, dm = \int_{\mathscr{V}} \frac{p}{\rho} \operatorname{div} \mathbf{v} \, dm - \int_{\mathscr{V}} \mathbf{v}.\operatorname{grad} V \, dm, \qquad (124)$$

where $dm = \rho \, \mathbf{dx}$. Now, for isentropic motions, we have

$$\frac{DU}{Dt} + \frac{p}{\rho} \operatorname{div} \mathbf{v} = 0, \qquad (125)$$

where U is the thermal energy per unit mass (cf. §3.4: equation [70]). By virtue of equation (95), it thus follows that

$$\int_{\mathscr{V}} \frac{p}{\rho} \operatorname{div} \mathbf{v} \, dm = -\int_{\mathscr{V}} \frac{DU}{Dt} \, dm = -\frac{d}{dt} \int_{\mathscr{V}} U \, dm = -\frac{dU_T}{dt}, \qquad (126)$$

where U_T is the total thermal energy. Similarly, we can write

$$\int_{\mathscr{V}} \mathbf{v}.\operatorname{grad} V \, dm = -\frac{dW}{dt}, \qquad (127)$$

where W is the gravitational potential energy. Using equations (126) and (127), we can integrate equation (124) at once to obtain

$$\int_{\mathscr{V}} \tfrac{1}{2}|\mathbf{v}|^2 \, dm + U_T + W = constant. \qquad (128)$$

Suppose now that an axially symmetric motion is superimposed upon the state of equilibrium. Equation (128) then becomes

$$\int_{\mathscr{V}} \tfrac{1}{2}|\mathbf{v}_p|^2 \, dm + E = constant, \qquad (129)$$

where $\mathbf{v}_p$ is the velocity field of the axially symmetric pulsation, and E is the total energy, i.e.,

$$E = \frac{1}{2} \int_{\mathscr{V}} \frac{j^2}{\varpi^2} \, dm + \int_{\mathscr{V}} U \, dm + \frac{1}{2} \int_{\mathscr{V}} V \, dm, \qquad (130)$$

where $j(\varpi, z) = \Omega(\varpi, z)\varpi^2$ designates the angular momentum per unit mass. Clearly, any increase of the kinetic energy of the pulsation must be supplied from the total energy E. Accordingly, this energy must be a minimum for stable, isentropic pulsations of an inviscid star in a state of permanent rotation. This concludes the proof.

Let us now compute the first- and second-variations of E by keeping constant the total mass M and the total angular momentum J. Stable equilibria correspond to the conditions

$$\delta E = 0 \quad \text{and} \quad \delta^2 E > 0. \tag{131}$$

Now, in the case of axisymmetric pulsations, the angular momentum j is preserved for each mass element; hence, we have $Dj/Dt = 0$, and we can write

$$\delta \frac{1}{2} \int_{\mathscr{V}} \frac{j^2}{\varpi^2} \, dm = - \int_{\mathscr{V}} \frac{j^2}{\varpi^3} \xi_\varpi \, dm. \tag{132}$$

Similarly, for isentropic motions, we have $DS/Dt = 0$; equation (125) thus implies that

$$\Delta U = - \frac{p}{\rho} \operatorname{div} \boldsymbol{\xi}. \tag{133}$$

Using equation (94), we find

$$\delta \int_{\mathscr{V}} U \, dm = \int_{\mathscr{V}} \Delta U \, dm = - \int_{\mathscr{V}} \frac{p}{\rho} \operatorname{div} \boldsymbol{\xi} \, dm = \int_{\mathscr{V}} \frac{1}{\rho} \boldsymbol{\xi} . \operatorname{grad} p \, dm. \tag{134}$$

Finally, by virtue of equation (105), the first variation of the gravitational energy is

$$\delta W = \int_{\mathscr{V}} \boldsymbol{\xi} . \operatorname{grad} V \, dm. \tag{135}$$

From equations (132), (134), and (135), it thus follows that

$$\delta E = \int_{\mathscr{V}} \boldsymbol{\xi} . \left(\operatorname{grad} V + \frac{1}{\rho} \operatorname{grad} p - \frac{j^2}{\varpi^3} \mathbf{1}_\varpi \right) dm, \tag{136}$$

where $\mathbf{1}_\varpi$ is the unit vector in the ϖ-direction. As we expected, the condition $\delta E = 0$ defines a state of permanent rotation (cf. equations [18]–[20]). Similarly, it is a simple matter to prove that

$$\delta^2 E = \int_{\mathscr{V}} \boldsymbol{\xi} . \left[\Delta \left(\operatorname{grad} V + \frac{1}{\rho} \operatorname{grad} p \right) + \frac{3j^2}{\varpi^4} \xi_\varpi \mathbf{1}_\varpi \right] dm. \tag{137}$$

In making use of equations (20), (24), (67), (75), and (76), we can next rewrite equation (137) in the compact form

$$\delta^2 E = - \int_{\mathscr{V}} \boldsymbol{\xi} . \mathbf{L} \boldsymbol{\xi} \, d\mathbf{x}. \tag{138}$$

The physical meaning of Lebovitz's condition (78) is now clear: if there exists a virtual displacement ξ such that

$$\int_{\mathscr{V}} \xi.\mathbf{L}\xi \, \mathbf{dx} > 0, \tag{139}$$

we have $\delta^2 E < 0$, and the total energy E fails to be an absolute minimum. In its broad features, the above derivation is due to Fjørtoft, who did not, however, include self-gravitation. The usefulness of the Fjørtoft-Lebovitz criterion will become apparent in Section 7.3.

THE STATIC METHOD. To outline this approach (originally devised by Zeldovich), let us first restrict ourselves to the case of spherical stars. Consider a sequence of equilibrium models satisfying a given equation of state and having a fixed distribution of entropy per unit mass $S(m/M)$, where m/M is the mass enclosed within the sphere of volume $v = 4\pi r^3/3$. Such a family can be parametrized, e.g., by the central density ρ_c of each model. Thus, for a given value of the parameter ρ_c, there corresponds an equilibrium model with mass $M = M(\rho_c)$; and the family can be depicted as a curve of mass versus central density in the $(M-\rho_c)$-plane. The static criterion asserts that *one normal mode of radial pulsation changes stability (i.e., $\sigma^2 = 0$) at each maximum (or minimum) of the curve* $M = M(\rho_c)$, *and that there are no exchanges of stability elsewhere*. Intuitively, this result can be understood as follows. Suppose that the curve $M = M(\rho_c)$ has an extremum at the point $\rho_c = \rho_{cc}$ where $M(\rho_{cc}) = M_c$ (i.e., $dM/d\rho_c = 0$ at $\rho_c = \rho_{cc}$). Thus, in the vicinity of the point $\rho_c = \rho_{cc}$, to each model having a mass arbitrarily close to M_c there exists another model having the same mass but with a different central density. In other words, each of these models may be obtained from a "nearby" configuration by an infinitesimal displacement $\xi(r)$ which does not depend on time. Hence, some squared frequency σ^2 belonging to a radial mode of oscillation must change sign at the point $\rho_c = \rho_{cc}$. This means an exchange of stability at each extremum of the curve $M = M(\rho_c)$. As shown by Thorne, even modes ($v = 0, 2, 4, \ldots$) lose stability at a maximum of the curve $M = M(\rho_c)$, whereas odd modes ($v = 1, 3, 5, \ldots$) lose stability at a minimum of this curve. The static method has been extensively used for discussing the stability of dense stars. It is particularly useful because it avoids detailed calculation of the radial modes of oscillation.

As we shall now prove, a more definite stability criterion can be deduced from the extremization of the total energy E. For a spherical system, we have

$$E = \int_0^{\mathscr{V}} f(m,m',v) \, dv, \tag{140}$$

where $\mathscr{V}$ is the total volume of the star, and

$$f(m,m',v) = Um' - \frac{Gmm'}{r(v)};\tag{141}$$

a prime denotes a derivative with respect to the volume $v = 4\pi r^3/3$. As we know, the functional E is extremal (i.e., $\delta E = 0$) if the mass distribution $m(v)$ satisfies the Euler-Lagrange equation

$$\frac{d}{dv}\left(\frac{\partial f}{\partial m'}\right) - \frac{\partial f}{\partial m} = 0.\tag{142}$$

To discuss the sign of the second variation $\delta^2 E$, let us introduce a function $Z(v)$ satisfying the Jacobi equation

$$\frac{d}{dv}\left(\frac{\partial^2 f}{\partial m'^2}\frac{\partial Z}{\partial v}\right) + \left[\frac{d}{dv}\left(\frac{\partial^2 f}{\partial m \partial m'}\right) - \frac{\partial^2 f}{\partial m^2}\right]Z = 0,\tag{143}$$

with the conditions

$$Z(0) = 0, \quad \text{and} \quad dZ/dv(0) = 1.\tag{144}$$

Then, $\delta^2 E$ is positive if and only if the two following conditions are met: (i) the function $\partial^2 f/\partial m'^2$ is positive, and (ii) the function $Z(v)$ has no zero in the domain $0 < v \leqslant \mathscr{V}$.[7]

In the present context, by virtue of the well known thermodynamical relations (valid for isentropic motions only)

$$\left(\frac{\partial U}{\partial \rho}\right)_s = \frac{p}{\rho^2}, \quad \text{and} \quad \left(\frac{\partial p}{\partial \rho}\right)_s = \Gamma_1\frac{p}{\rho},\tag{145}$$

we find

$$\frac{\partial f}{\partial m} = -\frac{Gm'}{r}, \quad \frac{\partial f}{\partial m'} = \frac{p}{\rho} + U - \frac{Gm}{r},\tag{146}$$

and

$$\frac{\partial^2 f}{\partial m^2} = 0, \quad \frac{\partial^2 f}{\partial m \partial m'} = -\frac{G}{r}, \quad \frac{\partial^2 f}{\partial m'^2} = \Gamma_1\frac{p}{\rho^2}.\tag{147}$$

It is now a simple matter to check that equation (142) reduces to the condition of hydrostatic equilibrium

$$\frac{1}{\rho}\frac{dp}{dr} + \frac{Gm}{r^2} = 0.\tag{148}$$

---

[7] See, e.g., Gelfand, I. M., and Fomin, S. V., *Calculus of Variations*, Englewood Cliffs, N.J.: Prentice-Hall, Inc.,1963.

Also, using relations (147), the Jacobi equation becomes

$$\frac{d}{dr}\left(\frac{1}{4\pi\rho r^2}\frac{\Gamma_1 p}{\rho}\frac{dZ}{dr}\right) + \frac{GZ}{r^2} = 0, \tag{149}$$

and the Legendre condition $\partial^2 f/\partial m'^2 > 0$ is always satisfied.

Consider next a family of spherical stars as described above. For each model, the mass distribution m can thus be viewed as a function of v and ρ_c, i.e., $m = m(v,\rho_c)$. Now take the derivative of equation (142) with respect to the parameter ρ_c; we obtain

$$\frac{d}{dv}\left(\frac{\partial^2 f}{\partial m\,\partial m'}\frac{\partial m}{\partial\rho_c} + \frac{\partial^2 f}{\partial m'^2}\frac{\partial m'}{\partial\rho_c}\right) - \frac{\partial^2 f}{\partial m^2}\frac{\partial m}{\partial\rho_c} - \frac{\partial^2 f}{\partial m\,\partial m'}\frac{\partial m'}{\partial\rho_c} = 0, \tag{150}$$

since ρ_c is an initial condition and as such cannot explicitly appear in the equation of hydrostatic equilibrium. Equation (150) can also be rewritten in the form

$$\frac{d}{dv}\left[\frac{\partial^2 f}{\partial m'^2}\frac{d}{dv}\left(\frac{\partial m}{\partial\rho_c}\right)\right] + \left[\frac{d}{dv}\left(\frac{\partial^2 f}{\partial m\,\partial m'}\right) - \frac{\partial^2 f}{\partial m^2}\right]\frac{\partial m}{\partial\rho_c} = 0. \tag{151}$$

Moreover, because $m(v,\rho_c) \approx \rho_c v$ near the center ($v \approx 0$), we find

$$\left(\frac{\partial m}{\partial\rho_c}\right)_{v=0} = 0, \quad \text{and} \quad \left[\frac{d}{dv}\left(\frac{\partial m}{\partial\rho_c}\right)\right]_{v=0} = 1. \tag{152}$$

A comparison between equations (143)–(144) and equations (151)–(152) shows that the functions Z and $\partial m/\partial\rho_c$ satisfy the same differential equation and are determined by the same conditions. It thus follows that

$$\frac{\partial m}{\partial\rho_c} = Z(v,\rho_c); \tag{153}$$

in particular, at $v = \mathscr{V}$, we have

$$\frac{dM}{d\rho_c} = Z(\mathscr{V},\rho_c). \tag{154}$$

Equation (154) contains the proof of the static criterion. Indeed, whenever $dM/d\rho_c = 0$, the function Z has a zero at the point $v = \mathscr{V}$, and the second variation $\delta^2 E$ cannot be strictly positive for the corresponding model. The physical content of equation (153) is even more interesting. Consider the $(m$–$\rho_c)$-plane and draw the continuous set of curves $m = m(v,\rho_c)$, as v ranges from $v = 0$ to $v = \mathscr{V}$. (Obviously, the uppermost curve $m = m(\mathscr{V},\rho_c)$ is nothing but the curve $M = M(\rho_c)$ used in the static criterion.) Whenever a curve belonging to this set has an extremum (i.e., $\partial m/\partial\rho_c = 0$), the function Z has a zero in the domain $0 < v \leqslant \mathscr{V}$; and, accordingly $\delta^2 E$ fails to be strictly positive. The following statement (due to Calamai) can thus be made: *all models belonging to the family* M = M(ρ_c) *are stable against radial pulsations in the range of central densities where none of the curves* $m = m(v, \rho_c)$ *has an extremum.* In conclusion, while Zeldovich's static criterion locates the (*discrete*) values of ρ_c at which an exchange of stability occurs, the Calamai criterion defines the (*continuous*) range of central densities for which stellar models belonging to the family M = M(ρ_c) are unstable with respect to radial isentropic perturbations. The very same discussion can be made within the framework of general relativity.

Given the foregoing considerations, it is a simple matter to generalize the static approach to the case of barotropes and pseudo-barotropes in a state of permanent rotation. Consider again a family of rotating models characterized by the same equation of state, and with fixed distributions (per unit mass) of entropy $S(m_\varpi/M)$ and angular momentum $j(m_\varpi/M)$, where m_ϖ/M is the fractional mass interior to the cylinder of radius ϖ. Since isentropic axisymmetric pulsations preserve both functions (together with J and M), stability against these motions may be similarly diagnosed by locating the extrema of the curve M = M(ρ_c) in the (M–ρ_c)-plane. The problem was first discussed by Hartle and Thorne in the context of general relativity. A more rigorous analysis was subsequently made by Bisnovatyi-Kogan and Blinnikov, including also numerical examples for rotating polytropes and white dwarfs. As they have pointed out, in spite of its simplicity, the generalized static criterion is not free of shortcomings. *First*, because the method pinpoints exchanges of stability only, its application must be necessarily restricted to modes for which σ^2 is always real, i.e., axisymmetric pulsations (cf. equation [80]). *Second*, as we know from other approaches, rotation couples the various p-, f-, and g-modes discussed in Section 6.4; the question of the correct identification of a neutral mode may thus eventually prove to be a somewhat ambiguous one. Nevertheless, the static criterion may become useful in discussing the influence of rotation upon the onset of axisymmetric convective motions— a problem that requires further research (cf. §14.5).

THE QUASI-DYNAMIC METHOD.   The aim of this approach is to solve the equations describing a state of permanent rotation. As usual, we have

$$0 = -\frac{1}{\rho} \operatorname{grad} p - \operatorname{grad} V + \frac{j^2}{\varpi^3} \mathbf{1}_\varpi, \tag{155}$$

where

$$V(\mathbf{x}) = -G \int_{\mathscr{V}} \frac{\rho(\mathbf{x}')}{|\mathbf{x} - \mathbf{x}'|} \, d\mathbf{x}', \tag{156}$$

and we suppose an equation of state of the form $p = p(\rho,S)$. The present method replaces equation (155) by the quasi-dynamic equation

$$\frac{\partial \mathbf{x}}{\partial \tau} = -\frac{1}{\rho} \operatorname{grad} p - \operatorname{grad} V + \frac{j^2}{\varpi^3} \mathbf{1}_\varpi, \tag{157}$$

where we let $\mathbf{x} = (\varpi,z)$, and τ denotes some continuous parameter. As we shall see, equation (157) can be solved by an iterative scheme that converges toward a solution of equation (155) if and only if the corresponding model is dynamically stable against axisymmetric pulsations. This approach is due to Kovetz, Lebovitz, Shaviv, and Zisman; and it generalizes the usual Rakavy-Shaviv-Zinamon scheme for spherical stars.

Consider a provisional, or "initial," mass distribution and assign to every point $\mathbf{X} = (\varpi_0,z_0)$ an entropy per unit mass $S(\mathbf{X})$ and an angular momentum per unit mass $j(\mathbf{X})$. This specifies all the physical variables appearing in equation (155), but not in such a way that it will be satisfied. Consider next the family of transformations

$$\mathbf{x} = \mathbf{x}(\mathbf{X},\tau), \tag{158}$$

reducing to $\mathbf{x} = \mathbf{X}$ when $\tau = 0$ (say), and specifying the functions S, ρ, and j through the following constraints

$$\frac{\partial S}{\partial \tau} = 0, \quad \frac{\partial j}{\partial \tau} = 0, \quad \frac{\partial \rho}{\partial \tau} = -\rho \operatorname{div} \mathbf{u}, \tag{159}$$

where the quasi-velocity $\mathbf{u}$ designates the quantity $\partial \mathbf{x}/\partial \tau$. Of course, at this stage, the transformation (158) is arbitrary. We now require that it be obtained by solving the initial-value problem for the quasi-dynamic equation (157). And we suppose that the steady-state form of this equation, namely equation (155), has a solution for the assigned distributions of entropy and angular momentum. That is to say, we suppose that there exists a transformation $\mathbf{x} = \mathbf{x}(\mathbf{X})$ providing a solution of equation (155). We will now show that this steady-state solution is an asymptotically stable solution of equation (155), i.e.,

$$\mathbf{x} = \mathbf{x}(\mathbf{X}, \tau) \to \mathbf{x} = \mathbf{x}(\mathbf{X}) \quad \text{as } \tau \to \infty \tag{160}$$

if and only if the steady state solution is dynamically stable in the usual sense.

Denoting the internal energy per unit mass by U, we have

$$E(\tau) = \int_{\mathscr{V}} \left(\frac{1}{2} \frac{j^2}{\varpi^2} + U + \frac{1}{2} V \right) dm, \tag{161}$$

for the total energy of a configuration (not generally in a state of permanent rotation). Differentiating with respect to τ and using conditions (159), the definition of the quasi-velocity $\mathbf{u}$, and the vanishing of the pressure on the surface, we find

$$E'(\tau) = \int_{\mathscr{V}} \mathbf{u} \cdot \left(\operatorname{grad} V + \frac{1}{\rho} \operatorname{grad} p - \frac{j^2}{\varpi^3} \mathbf{1}_\varpi \right) dm \qquad (162)$$

(compare with equation [136]). If, further, equation (157) is satisfisied, equation (162) implies that

$$E'(\tau) = - \int_{\mathscr{V}} |\mathbf{u}|^2 \, dm \leqslant 0, \qquad (163)$$

equality holding only if $\mathbf{u} = 0$, in which case the steady-state equation (155) must hold.

Let next E_1 represent the total energy of the steady-state configuration sought. If E_1 is a minimum compared to all nearby virtual configurations, then this steady-state solution is asymptotically stable by virtue of equation (163). This may be formulated as follows. Suppose the transformation (158) is not determined by the quasi-dynamic equation, but is arbitrary except for the requirement that for some value $\tau = \tau_1$ (say), it yields the steady-state solution. The function $\mathbf{u}(\tau_1)$ is then arbitrary; equation (162) remains valid, however, since it depends only on the auxiliary conditions (159), which continue to apply. It easily follows that

$$E'(\tau_1) = 0, \qquad (164)$$

where a prime denotes a derivative with respect to τ. We also obtain

$$E''(\tau_1) = \int_{\mathscr{V}} \mathbf{u}(\tau_1) \cdot \left[\frac{\partial}{\partial \tau} \left(\operatorname{grad} V + \frac{1}{\rho} \operatorname{grad} p \right) + 3 \frac{j^2}{\varpi^4} \mathbf{1}_\varpi \right]_{\tau = \tau_1} dm \qquad (165)$$

(compare with equation [137]). Now, the derivative with respect to τ appearing in equations (159) and (165) is taken holding $\mathbf{X}$ fixed, and therefore represents a Lagrangian change. We thus have

$$\frac{\partial}{\partial \tau} \left(\operatorname{grad} V + \frac{1}{\rho} \operatorname{grad} p \right) = - \mathbf{K}\mathbf{u}(\tau), \qquad (166)$$

where the operator $\mathbf{K}$ is defined by equation (67); similarly, using definitions (75) and (76) we find

$$E''(\tau_1) = - \int_{\mathscr{V}} \mathbf{u}(\tau_1) \cdot \mathbf{L}\mathbf{u}(\tau_1) \, d\mathbf{x} \qquad (167)$$

(compare with equation [138]). By virtue of the Fjørtoft-Lebovitz criterion (139), equation (167) shows that $E''(\tau_1)$ is positive-definite when the configuration is dynamically stable, and can take on negative values when the configuration is dynamically unstable. This completes the demonstration that the quasi-dynamic method is valid for rotating stellar models.

BIBLIOGRAPHICAL NOTES

Sections 6.1–6.3. In the particular context of stellar hydrodynamics, the problem of small oscillations and their stability originates in the works of Ritter (1878–1883) and Eddington (1919). A detailed survey of these and other investigations is contained in:

1. Ledoux, P., "Stellar Stability" in *Handbuch der Physik* (Flügge, S., ed.), **51**, pp. 605–688, Berlin: Springer-Verlag, 1958.

The most recent addendum to this penetrating paper is:

2. Ledoux, P., "Non-Radial Oscillations" in *Stellar Instability and Evolution* (Ledoux, P., Noels, A., and Rodgers, A. W., eds.), pp. 135–173, Dordrecht: D. Reidel Publ. Co., 1974.

Detailed accounts of various technical matters will also be found in references 1, 27, and 28 of Chapter 3. The first of these three references provides the Lagrangian equations in their perturbed forms, but self-gravitation is not included. See also references 40 and 46.

Section 6.4. Reference 2 contains an overall picture of the question. The homogeneous, compressible model has been discussed in:

3. Sterne, T. E., *Monthly Notices Roy. Astron. Soc. London* **97** (1937): 582–593.
4. Pekeris, C. L., *Astroph. J.* **88** (1938): 189–199.

Further investigations that also include the f-modes are due to:

5. Robe, H., *Bull. Acad. Roy. Belgique: Classe des Sciences* (5) **51** (1965): 598–603.
6. Smeyers, P., *ibid.* **52** (1966): 1126–1142.

Detailed discussions of the illustrative models may be found in:

7. Ledoux, P., and Smeyers, P., *C.R. Acad. Sci. Paris* **262B** (1966): 841–844.
8. Smeyers, P., *Ann. Astroph.* **29** (1966): 539–548.
9. Smeyers, P., *Bull. Soc. Roy. Sci. Liège* **36** (1967): 357–392.
10. Robe, H., *Ann. Astroph.* **31** (1968): 475–482.
11. Scuflaire, R., *Astron. and Astroph.* **36** (1974): 107–111.

References to Cowling, Lebovitz, and Eisenfeld are to their papers:

12. Cowling, T. G., *Monthly Notices Roy. Astron. Soc. London* **101** (1941): 367–375.
13. Lebovitz, N. R., *Astroph. J.* **142** (1965): 229–242, 1257–1260; *ibid.* **146** (1966): 946–949.
14. Eisenfeld, J., *J. Math. Analysis and Applications* **26** (1969): 357–375.

The existence of g^+- and g^--modes was originally discussed in references 7 and 8; see also:

15. Noels, A., Boury, A., Scuflaire, R., and Gabriel, M., *Astron. and Astroph.* **31** (1974): 185–188.
16. Scuflaire, R., *ibid.* **34** (1974): 449–451.

Equation (46) and its implications are due to:

17. Tassoul, M., and Tassoul, J. L., *Ann. Astroph.* **31** (1968): 251–256.

Further progress will be found in:

18. Goossens, M., and Smeyers, P., *Astroph. and Space Sci.* **26** (1974): 137–151.

Theoretical questions pertaining to the nonradial oscillations of a gaseous sphere are also considered in the following papers:

19. Kaniel, S., and Kovetz, A., *Phys. Fluids* **10** (1967): 1186–1193.
20. Vandakurov, Yu. V., *Astron. Zh.* **44** (1967): 786–797.
21. Iweins, P., et Smeyers, P., *Bull. Acad. Roy. Belgique: Classe des Sciences* (5) **54** (1968): 164–176.
22. Smeyers, P., *Ann. Astroph.* **31** (1968): 159–165.
23. Andrew, A. L., *J. Australian Math. Soc.* **10** (1969): 367–384.
24. Zahn, J. P., *Astron. and Astroph.* **4** (1970): 452–461.
25. Andrew, A. L., *J. Math. Analysis and Applications* **32** (1970): 400–413; *ibid.* **33** (1971): 425–432.
26. Scuflaire, R., *Bull. Acad. Roy. Belgique: Classe des Sciences* (5) **57** (1971): 1126–1136.
27. Grisvard, P., Souffrin, P., and Zerner, M., *Astron. and Astroph.* **17** (1972): 309–311.
28. Denis, J., Denoyelle, J., and Smeyers, P., *Astroph. and Space Sci.* **37** (1975): 221–233.
29. Robe, H., and Ledoux, P., *Bull. Acad. Roy. Belgique: Classe des Sciences* (5) **61** (1975): 198–209.
30. Christensen-Dalsgaard, J., *Monthly Notices Roy. Astron. Soc. London* **174** (1976): 87–89.
31. Sobouti, Y., *Astron. and Astroph.* **55** (1977): 327–337.

The nature of the trivial root $\sigma^2 = 0$ was first investigated by:

32. Perdang, J., *Astroph. and Space Sci.* **1** (1968): 355–371.

Section 6.5. The analysis in this section is derived from:

33. Laval, G., Mercier, C., and Pellat, R., *Nuclear Fusion* **5** (1965): 156–158.

See also:

34. Eisenfeld, J., *J. Math. Analysis and Applications* **31** (1970): 167–181.

35. Kulsrud, R. M., and Mark, J. W.-K., *Astroph. J.* **160** (1970): 471–483.

Section 6.6. The application of variational techniques to stellar hydrodynamics was first made in:

36. Ledoux, P., and Pekeris, C. L., *ibid.* **94** (1941): 124–135.
37. Chandrasekhar, S., *ibid.* **139** (1964): 664–674.

For the inclusion of the *f*-modes, see also:

38. Chandrasekhar, S., and Lebovitz, N. R., *ibid.* **140** (1964): 1517–1528.

The problem of uniformly rotating stars was originally discussed by Clement:

39. Clement, M. J., *ibid.* **140** (1964): 1045–1055.

The general investigation of differentially rotating bodies is due to:

40. Lynden-Bell, D., and Ostriker, J. P., *Monthly Notices Roy. Astron. Soc. London* **136** (1967): 293–310.

These two references also contain a discussion of non-axisymmetric modes. Axisymmetric pulsations are further considered in:

41. Chandrasekhar, S., and Lebovitz, N. R., *Astroph. J.* **152** (1968): 267–291.

The presentation of the subject in the text follows Lebovitz's (reference 12 of Chapter 5). The physical principle underlying the symmetry of the operator **L** is discussed in:

42. Kulsrud, R. M., *Astroph. J.* **152** (1968): 1121–1124.

Other demonstrations of the variational principle derived by Lynden-Bell and Ostriker are contained in the following papers:

43. Unno, W., *Publ. Astron. Soc. Japan* **20** (1968): 356–375.
44. Schutz, B. F., *Astroph. J. Supplements* **24** (1972): 319–342.

Section 6.7. The virial method was originally used in the context of stellar hydrodynamics by Ledoux:

45. Ledoux, P., *Astroph. J.* **102** (1945): 143–153.

The second-order virial equations were first applied to self-gravitating spheroids in:

46. Lebovitz, N. R., *ibid.* **134** (1961): 500–536.

Uniformly rotating bodies are discussed in:

47. Chandrasekhar, S., and Lebovitz, N. R., *ibid.* **135** (1962): 248–260, 659; *ibid.* **136** (1962): 1069–1081.

The analysis in this section follows:

48. Tassoul, J. L., and Ostriker, J. P., *ibid.* **154** (1968): 613–626.

Variational techniques to refine the locations of the neutral mode (118) and the limit of dynamical stability (119) are expounded in:

49. Chandrasekhar, S., and Lebovitz, N. R., *ibid.* **185** (1973): 19–30.
50. Chandrasekhar, S., *ibid.* **187** (1974): 169–174.

See, however, reference 36 of Chapter 10.

Section 6.8. These and related matters were originally discussed in:

51. Thomas, L. H., *Monthly Notices Roy. Astron. Soc. London* **91** (1931): 619–628.
52. Tolman, R. C., *Astroph. J.* **90** (1939): 541–567.

The reference to Fjørtoft is to his paper:

53. Fjørtoft, R., *Geofysiske Publikasjoner (Oslo)* **16**, N°5 (1946): 1–28.

The static method is due to Zeldovich:

54. Zeldovich, Ya. B., *Voprosi Kosmogonii (U.S.S.R.)* **9** (1963): 157–170.

According to Thorne, the method was independently discovered by Wheeler and Bardeen in 1965; see:

55. Thorne, K. S., "Relativistic Stellar Structure and Dynamics" in *High Energy Astrophysics* (De Witt, C., Schatzman, E., and Véron, P., eds.), **3**, pp. 259–441, New York: Gordon and Breach, 1967.

The rigorous proof of the static method is due to Calamai:

56. Calamai, G., *Astroph. and Space Sci.* **8** (1970): 53–58.

The appropriate generalization to rotating bodies will be found in:

57. Hartle, J. B., and Thorne, K. S., *Astroph. J.* **158** (1969): 719–726.
58. Bisnovatyi-Kogan, G. S., and Blinnikov, S. I., *Astron. and Astroph.* **31** (1974): 391–404.
59. Hartle, J. B., *Astroph. J.* **195** (1975): 203–212.

Reference 58 contains many pertinent comments on the literature. The presentation in the text of the quasi-dynamic method largely follows:

60. Kovetz, A., Lebovitz, N. R., and Shaviv, G., *Mém. Soc. Roy. Sci. Liège* (6) **8** (1975): 25–30.
61. Kovetz, A., Shaviv, G., and Zisman, S., *Astroph. J.* **206** (1976): 809–814.

7

The Angular Momentum Distribution

7.1. INTRODUCTION

As we recall from Chapter 4, many fundamental properties of a configuration in a state of permanent rotation can be derived in a straightforward
manner from the condition of mechanical equilibrium. Although these
results have a direct bearing on several aspects of the theory of stellar
rotation, they provide no clue as to what the angular momentum distribution is in a star. Also, because we have hitherto circumvented the
use of the condition of energy conservation when constructing equilibrium
models, we do not know as yet whether we can apply these results, without
modification, to an actual radiating star. For instance, is there any constraint imposed by the condition of strict radiative equilibrium on the
angular momentum distribution within a star? Finally, to what extent
is it necessary to modify the properties derived in Chapter 4 when viscous
friction and electromagnetic forces are taken into account? To sum up,
*is it actually possible to build a realistic stellar model by assuming the
material velocity to be wholly one of pure rotation?* In view of their importance, we shall devote this chapter, in its entirety, to the study of
these questions.

In principle, the instantaneous angular momentum distribution within
a star should be calculable from initial conditions. Obviously, this is an
impossible task at the present level of knowledge of the subject, even
were these initial conditions known. The alternate procedure, which is
now widely used, is to *choose* a steady angular momentum distribution
by ruling out those rotating configurations that are (dynamically or
thermally) unstable, as well as those that do not comply with the simultaneous conditions of mechanical and thermal equilibrium. A restriction
of the latter type for regions in strict radiative equilibrium is set forth
in Section 7.2, while necessary (but not sufficient) conditions of stability
are demonstrated in Sections 7.3 and 7.4. Thence, by ruling out the models
that do not satisfy these conditions of equilibrium and stability, we are
led to conclude that a chemically homogeneous star cannot be in strict
radiative equilibrium while subject to steady rotation. For this reason,
it becomes necessary to consider radiative zones with a large-scale circulation superimposed on a pure rotational motion. Postponing to Chapter 8 the study of these meridional currents, we conclude the chapter

with some considerations about the role of viscosity and magnetic fields in a rotating star with negligible circulation. Practical ways to determine the steady (or quasi-steady) rotation laws in a viscous star are also discussed. Time-dependent velocity fields will be examined further in Section 13.3.

7.2. THE VON ZEIPEL PARADOX

In this section we shall be primarily concerned with stellar models (or regions thereof) which are in strict radiative equilibrium. Hence, at every point, the radiative flux vector $\mathscr{F}$ and the nuclear reaction rate ϵ_{Nuc} are related by

$$\operatorname{div} \mathscr{F} = \rho\epsilon_{Nuc};\tag{1}$$

we have

$$\mathscr{F} = -\frac{4acT^3}{3\kappa\rho}\operatorname{grad} T,\tag{2}$$

where a, c, and κ are, respectively, the radiation pressure constant, the speed of light, and the opacity coefficient. To illustrate the problem, let us also assume that the total pressure p is related to the density ρ and temperature T by the equation of state

$$p = \frac{\mathscr{R}}{\bar{\mu}}\rho T + \frac{1}{3}aT^4,\tag{3}$$

where $\bar{\mu}$ is the (constant) mean molecular weight, and $\mathscr{R}$ is the perfect gas constant.

Assume now that the configurations are in a state of permanent rotation (cf. §4.2), and that their angular velocity Ω is a function of the distance ϖ from the axis of rotation only. Since these pseudo-barotropes possess axial symmetry, it is convenient to use cylindrical coordinates (ϖ,φ,z). The condition of mechanical equilibrium then becomes

$$\frac{1}{\rho}\operatorname{grad} p = -\operatorname{grad}\Phi,\tag{4}$$

where

$$\Phi(\varpi,z) = V(\varpi,z) - \int^{\varpi}\Omega^2(\varpi')\varpi'\,d\varpi'.\tag{5}$$

Finally, the gravitational potential is related to the density by the Poisson equation

$$\nabla^2 V = 4\pi G\rho,\tag{6}$$

where G is the constant of gravitation.

Now, because we restrict our analysis to chemically homogeneous pseudo-barotropes, the isobaric-, isopycnic-, and isothermal-surfaces (i.e.,

the surfaces $p = constant$, $\rho = constant$, and $T = constant$, respectively)
all coincide with the level surfaces $\Phi = constant$ (cf. §4.3).[1] Thence, if we
assume that $\epsilon_{Nuc} = \epsilon_{Nuc}(\rho, T)$ and $\kappa = \kappa(\rho, T)$, both the energy generation
rate and the opacity coefficient depend on Φ only. It thus follows that

$$\mathscr{F} = -\frac{4ac}{3}\frac{T^3}{\kappa\rho}\frac{dT}{d\Phi}\,\mathrm{grad}\,\Phi, \tag{7}$$

or

$$\mathscr{F} = f(\Phi)\,\mathrm{grad}\,\Phi, \tag{8}$$

where

$$f(\Phi) = -\frac{4ac}{3}\frac{T^3}{\kappa\rho}\frac{dT}{d\Phi}. \tag{9}$$

Now, $f(\Phi)$ is a constant over each level surface, the constant depending
on the chosen level surface. Accordingly, because the surface of a pseudo-
barotrope *is* a level surface, we may thus assert that

*The variation of brightness over the surface of a pseudo-barotrope
corresponds exactly to the variation of effective gravity.*

This is von Zeipel's law of gravity darkening, and it will be discussed
further in Section 12.3.

Let us consider next the divergence of equation (8). We readily find

$$\mathrm{div}\,\mathscr{F} = f'(\Phi)\left(\frac{d\Phi}{dn}\right)^2 + f(\Phi)\,\nabla^2\Phi, \tag{10}$$

where dn is along the outward normal to a level surface, and a prime
denotes a derivative with respect to Φ. We also have

$$\left(\frac{d\Phi}{dn}\right)^2 = \left(\frac{\partial\Phi}{\partial\varpi}\right)^2 + \left(\frac{\partial\Phi}{\partial z}\right)^2 = g^2. \tag{11}$$

Clearly, $d\Phi/dn$ is the magnitude of the effective gravity $\mathbf{g}$, i.e., the gravitational attraction per unit mass modified by the centrifugal acceleration.
From equations (5) and (6), we next obtain

$$\nabla^2\Phi = 4\pi G\rho - \frac{1}{\varpi}\frac{d}{d\varpi}(\Omega^2\varpi^2). \tag{12}$$

By making use of equations (10)–(12), we can thus rewrite equation (1)
in the form

$$f'(\Phi)g^2 + f(\Phi)\left[4\pi G\rho - \frac{1}{\varpi}\frac{d}{d\varpi}(\Omega^2\varpi^2)\right] = \rho\epsilon_{Nuc}. \tag{13}$$

---

[1] The same result holds true if $\bar{\mu}$ depends on p and ρ only (cf. equation [3]).

This is the condition of *strict* radiative equilibrium for a pseudo-barotrope in a state of permanent rotation.

The case of a uniformly rotating star is particularly straightforward, and it was originally discussed by von Zeipel. Equation (13) then reduces to

$$f'(\Phi)g^2 + f(\Phi)(4\pi G\rho - 2\Omega^2) = \rho\epsilon_{Nuc}. \tag{14}$$

As we know, g is *not* constant over a level surface of a rotating body, because the distance from one level surface to the next is not the same for every point on it. Accordingly, since ρ, ϵ_{Nuc}, and Ω are all constant on level surfaces, the coefficient of g in equation (14) must vanish separately. We have, therefore,

$$f'(\Phi) = 0 \quad \text{or} \quad f(\Phi) = constant. \tag{15}$$

Hence, equation (14) assumes the form

$$\epsilon_{Nuc} \propto \left(1 - \frac{\Omega^2}{2\pi G\rho}\right), \tag{16}$$

and equation (8) reduces to

$$|\mathscr{F}| \propto g. \tag{17}$$

Obviously, condition (16) is never fulfilled in an actual star. It thus follows that rigid rotation is impossible for a pseudo-barotrope in static radiative equilibrium.

Before discussing in detail the implications of equation (16), let us briefly consider the general conservative law $\Omega = \Omega(\varpi)$. In this case, as was shown by Rosseland and Vogt, it is intuitively evident that the law $\Omega = \Omega(\varpi)$ is incompatible with condition (13). Indeed, while Ω will be constant over cylinders centered about the axis of rotation, g will be constant over certain oblate surfaces not very different from the level surfaces. Therefore, by virtue of equation (13), conditions (15) and (17) still obtain, but we must impose the additional condition

$$\frac{1}{\varpi}\frac{d}{d\varpi}(\Omega^2\varpi^2) = constant. \tag{18}$$

After integrating, we find

$$\Omega^2 = c_1 + \frac{c_2}{\varpi^2}, \tag{19}$$

where c_1 and c_2 denote two arbitrary constants. If $c_2 = 0$, we simply recover the case of a uniformly rotating pseudo-barotrope. Similarly, if $c_2 \neq 0$, the rotation law (19) must be also disregarded, for it also leads to an impossible constraint on ϵ_{Nuc}, i.e., condition (16) with Ω^2 being replaced by c_1. Thus, the above argument shows that differentially rotating pseudo-barotropes cannot remain in static radiative equilibrium. This is von Zeipel's paradox. A rigorous proof of this general property of pseudo-barotropes has been given by Roxburgh.

In summary, since neither condition (16) nor condition (19) are satisfied in actual stars, we can thus make the following statement:

Pseudo-barotropic models in a state of permanent rotation cannot be used to describe rotating stars in strict radiative equilibrium.

As a matter of fact, it is not the usual energy generation rates that prevent the rotation laws $\Omega = constant$ and $\Omega = \Omega(\varpi)$ from being realized, but rather the condition of strict radiative equilibrium. Indeed, in the limit of zero rotation, g is a constant over each spherical surface $\Phi \equiv V = constant$, and there is no requirement that some terms in equation (13) should vanish independently of the remaining terms. Therefore, equation (13) must be regarded as an indication that, for nonspherical stars, at least one of the assumptions leading to equations (16) and (19) must be relaxed.

The von Zeipel paradox can be solved in two different ways: *either* we allow Ω to depend on both ϖ and z *or* we assume that strict radiative equilibrium breaks down in a rotationally distorted star. According to Vogt and Eddington, the latter solution will disrupt the constancy of temperature and pressure over the level surfaces, and the ensuing pressure gradient between the equator and the poles will cause a flow of matter along the level surfaces. As we shall see in Chapter 8, this flow caused by the thermal imbalance on the level surfaces must necessarily continue and take the form of a permanent meridional circulation. In the remainder of this chapter, we shall mainly discuss the former solution; that is to say, genuine baroclines for which Ω necessarily depends on both ϖ and z (cf. §4.3).

7.3. THE SOLBERG AND HØILAND CRITERIA

Consider a star in a state of permanent rotation for an inertial observer (cf. §4.2), and assume that the body rotates with some prescribed angular velocity $\Omega = \Omega(\varpi,z)$. Then, under what conditions is this star stable with respect to small isentropic perturbations? In other words, what restrictions must we place on the rotation law $\Omega = \Omega(\varpi,z)$ to avoid the existence of monotonically increasing motions with a time-scale of order $(\pi G \bar{\rho})^{-1/2}$, i.e., a few rotation periods? No definitive answer can be given at the present time. Nevertheless, interesting results can be obtained for axisymmetric motions by means of the Fjørtoft-Lebovitz criterion derived in Sections 6.6 and 6.8. As we recall, if for some admissible choice of virtual displacement ξ

$$\int_{\mathscr{V}} \xi.\mathbf{L}\xi \, \mathbf{dx} > 0, \tag{20}$$

the equilibrium configuration is dynamically unstable with respect to axisymmetric motions, i.e., the total energy of the star fails to be an absolute minimum. In the present instances, the linear operator $\mathbf{L}$ acts on the two-dimensional vector ξ with components ξ_ϖ and ξ_z (cf. §6.6), and is given by

$$\mathbf{L}\xi = \frac{\delta\rho}{\rho} \operatorname{grad} p - \operatorname{grad} \delta p - \rho \operatorname{grad} \delta V$$

$$- \frac{\rho}{\varpi^3} \xi.\operatorname{grad}(\Omega^2\varpi^4)\mathbf{1}_\varpi, \tag{21}$$

where $\mathbf{1}_\varpi$ is the unit vector in the ϖ-direction. The Eulerian changes $\delta\rho$ and δp are given by the relations

$$\delta\rho = -\rho \operatorname{div} \xi - \xi.\operatorname{grad} \rho, \tag{22}$$

$$\delta p = -\Gamma_1 p \operatorname{div} \xi - \xi.\operatorname{grad} p \tag{23}$$

(cf. §6.3); similarly, by assuming that the density vanishes on the equilibrium surface $\mathscr{S}$, we have

$$\delta V = -G \int_{\mathscr{V}} \frac{\delta\rho(\mathbf{x}',t)}{|\mathbf{x} - \mathbf{x}'|} \, \mathbf{dx}', \tag{24}$$

where $\delta\rho$ is given in terms of ξ by equation (22).

To proceed any further, we must now insert equations (21)–(24) into equation (20). In particular, integrating the second pressure term in equation (21) by parts and making use of the fact that the Lagrangian change Δp must vanish on $\mathscr{S}$, we find

$$\int_{\mathscr{V}} \xi.\operatorname{grad} \delta p \, \mathbf{dx} = -\int_{\mathscr{V}} \delta p \operatorname{div} \xi \, \mathbf{dx}, \tag{25}$$

since $\Delta p \equiv \delta p$ on the boundary $\mathscr{S}$ where, by assumption, ρ vanishes (cf. §6.3: equation [32]). Rearranging the various terms in equation (20), we eventually obtain

$$\int_{\mathscr{V}} \xi.\mathbf{L}\xi \, \mathbf{dx} = -\int_{\mathscr{V}} \xi\mathbf{M}\xi \, dm - \int_{\mathscr{V}} \left[\frac{(\delta p)^2}{\Gamma_1 p\rho} + \xi.\operatorname{grad} \delta V \right] dm, \tag{26}$$

where $dm = \rho(\mathbf{x}) \, \mathbf{dx}$, and the tensor $\mathbf{M}$ has the form

$$\mathbf{M} = \frac{1}{\rho} \operatorname{grad} p \left(\frac{1}{\rho} \operatorname{grad} \rho - \frac{1}{\Gamma_1 p} \operatorname{grad} p \right) + \frac{1}{\varpi^3} \operatorname{grad} \varpi \operatorname{grad}(\Omega^2\varpi^4). \tag{27}$$

For further use, let us next define the following vectors:

$$\boldsymbol{\Phi} = \frac{1}{\varpi^3} \, \mathrm{grad}(\Omega^2\varpi^4) = 2\frac{\Omega}{\varpi} \, \mathrm{grad} \, j, \tag{28}$$

$$\boldsymbol{\Phi}_0 = \mathrm{grad} \, \varpi = \mathbf{1}_\varpi, \tag{29}$$

$$\boldsymbol{\Psi} = \frac{1}{\Gamma_1 p} \, \mathrm{grad} \, p - \frac{1}{\rho} \, \mathrm{grad} \, \rho = \frac{1}{c_p} \frac{\gamma - 1}{\Gamma_3 - 1} \, \mathrm{grad} \, S, \tag{30}$$

$$\boldsymbol{\Psi}_0 = -\frac{1}{\rho} \, \mathrm{grad} \, p = -\mathbf{g}, \tag{31}$$

where $j = \Omega\varpi^2$ is the angular momentum per unit mass, and S is the entropy per unit mass. (The second equality [30] is generally valid for a mixture of gas and radiation, and it may be derived from the relations given in Section 3.4; in the present context, c_p is the specific heat at constant pressure for the mixture.) With these definitions in mind, we can rewrite the tensor $\mathbf{M}$ in the compact form

$$\mathbf{M} = \boldsymbol{\Phi}_0\boldsymbol{\Phi} + \boldsymbol{\Psi}_0\boldsymbol{\Psi}. \tag{32}$$

Since the vectors (28)–(31) and the tensor (32) play an essential role in our subsequent discussion, we shall briefly summarize their main properties. First of all, by virtue of the condition of mechanical equilibrium, these four vectors are not independent. We generally have

$$\frac{1}{\rho} \, \mathrm{grad} \, p + \mathrm{grad} \, V = \Omega^2\varpi\mathbf{1}_\varpi. \tag{33}$$

(Equation [4] does not apply, since Ω now depends on z as well.) Taking the curl of equation (33), we find

$$\mathrm{grad} \, \frac{1}{\rho} \times \mathrm{grad} \, p = \frac{1}{\varpi^3} \, \mathrm{grad}(\Omega^2\varpi^4) \times \mathbf{1}_\varpi. \tag{34}$$

By virtue of equations (28)–(31), it thus follows that

$$\boldsymbol{\Phi} \times \boldsymbol{\Phi}_0 + \boldsymbol{\Psi} \times \boldsymbol{\Psi}_0 = 0, \tag{35}$$

and, accordingly, the rotation $\boldsymbol{\Phi} \to \boldsymbol{\Phi}_0$ is always opposite to the rotation $\boldsymbol{\Psi} \to \boldsymbol{\Psi}_0$. Equation (35) is also the condition that makes the tensor $\mathbf{M}$ symmetric. Indeed, since the sets $(\boldsymbol{\Phi}_0, \boldsymbol{\Psi}_0)$ and $(\boldsymbol{\Phi}, \boldsymbol{\Psi})$ can be interchanged, we obtain

$$\boldsymbol{\Phi}\boldsymbol{\Phi}_0 + \boldsymbol{\Psi}\boldsymbol{\Psi}_0 = \boldsymbol{\Phi}_0\boldsymbol{\Phi} + \boldsymbol{\Psi}_0\boldsymbol{\Psi}, \tag{36}$$

and the curves $\boldsymbol{\xi}\mathbf{M}\boldsymbol{\xi} = constant$ represent a family of concentric conics. At each point of the star, we can thus find their two (orthogonal) principal axes (ϖ', z'), so that

$$\xi \mathbf{M} \xi = \alpha_\varpi \xi_{\varpi'}^2 + \alpha_z \xi_{z'}^2, \tag{37}$$

where $\xi_{\varpi'}$ and $\xi_{z'}$ are the components of ξ along the two principal axes. Since the trace and the determinant of $\mathbf{M}$ are invariant with respect to a rotation of the axes, we also have

$$\alpha_\varpi + \alpha_z = \text{trace } \mathbf{M} = \boldsymbol{\Phi}.\boldsymbol{\Phi}_0 + \boldsymbol{\Psi}.\boldsymbol{\Psi}_0, \tag{38}$$

and

$$\alpha_\varpi \alpha_z = \det \mathbf{M} = (\boldsymbol{\Phi}_0 \times \boldsymbol{\Psi}_0).(\boldsymbol{\Phi} \times \boldsymbol{\Psi}). \tag{39}$$

Finally, note that the vectors $\boldsymbol{\Phi}_0$ and $\boldsymbol{\Psi}_0$ are always directed along the outer normal to the surfaces $\varpi = constant$ and $p = constant$, respectively. Similarly, the vectors $\boldsymbol{\Phi}$ and $\boldsymbol{\Psi}$ are orthogonal to the surfaces $j = constant$ and $S = constant$, respectively; but we do not know *a priori* whether they are directed along the inner or outer normal to the surfaces to which they are orthogonal.

To discuss the implications of the Fjørtoft-Lebovitz criterion, let us now assume that the Eulerian changes δp and δV can be neglected in the configuration. For isentropic motions, the hypothesis $\delta p \equiv 0$ is equivalent to using the so-called parcel method, and it is valid whenever the characteristic time of the disturbances exceeds the travel time of a sound wave in the perturbed domain. (This amounts to filtering out the p-modes discussed in Section 6.4.) The hypothesis $\delta V \equiv 0$ implies that we restrict our analysis to disturbances having many nodes (cf. equation [24]), i.e., perturbations with wavelengths much shorter than the mean radius of the star. Given these assumptions, the stability of an equilibrium depends thus on the character of the quadratic form $\xi \mathbf{M} \xi$. By virtue of equations (20) and (26), the system is dynamically stable with respect to axisymmetric disturbances if and only if $\xi \mathbf{M} \xi$ is positive definite. Indeed, if this condition is not satisfied, it is always possible to choose a virtual displacement ξ that will decrease the total energy of the system, thus indicating an unstable state of equilibrium. From equations (37)–(39), we observe that $\xi \mathbf{M} \xi$ is positive definite if and only if α_ϖ and α_z are both positive. Hence, the conditions of stability are

$$\text{trace } \mathbf{M} > 0, \quad \text{and} \quad \det \mathbf{M} > 0, \tag{40}$$

or, returning to the original variables,

$$\frac{1}{\varpi^3} \frac{\partial j^2}{\partial \varpi} + \frac{1}{c_p} \frac{\gamma - 1}{\Gamma_3 - 1} (-\mathbf{g}).\text{grad } S > 0, \tag{41}$$

and

$$-g_z \left(\frac{\partial j^2}{\partial \varpi} \frac{\partial S}{\partial z} - \frac{\partial j^2}{\partial z} \frac{\partial S}{\partial \varpi} \right) > 0. \tag{42}$$

In this general form, the foregoing stability conditions were originally obtained by Høiland.

In the limit of zero rotation ($j \equiv 0$), conditions (41) and (42) reduce to the single inequality

$$(-\mathbf{g}).\text{grad } S > 0, \tag{43}$$

thus implying that the entropy always increases outward in a stable spherical system. This is also the condition for the temperature lapse rate to be subadiabatic throughout the system, or, equivalently, the condition for all g-modes to be of stable type (cf. §6.4: equation [43]). Hence, whenever condition (43) is not satisfied, convective currents must appear, which tend to mix the stellar material in such a way as to reduce the temperature gradient. Of course, the present derivation is based on the conditions $\delta p \equiv 0$ and $\delta V \equiv 0$. That these assumptions are *not* essential in deriving the Schwarzschild criterion (43) has been fully demonstrated by Lebovitz and Eisenfeld (cf. §6.4).

The case of an homentropic star (i.e., grad $S \equiv 0$) is also straightforward. Then, by virtue of equation (35), the vectors $\boldsymbol{\Phi}$ and $\boldsymbol{\Phi}_0$ are everywhere colinear, and the angular velocity does not depend on z. The configuration degenerates into a pseudo-barotrope, and the stability condition (41) becomes

$$\frac{d}{d\varpi}(\Omega^2\varpi^4) > 0. \tag{44}$$

Therefore, in stable homentropic stars the angular momentum per unit mass must necessarily increase outward. This is the Solberg criterion, and it generalizes to homentropic bodies the well-known Rayleigh criterion for an inviscid, incompressible fluid. As was shown by Randers, the stability condition (44) may be easily explained by the conservation of angular momentum per unit mass of each fluid particle, when it is slightly displaced from its equilibrium position. Indeed, when condition (44) is obtained, any mass element moving outward lacks angular momentum compared to the angular momentum of the material in equilibrium about its new location. This lack of angular momentum means a lack of centrifugal force, and a resulting deceleration of the outward motion. Similarly, if a mass element is impelled inward, the excess of angular momentum moving inward tends to drive the fluid particle out again, thereby stopping the inward motion.

Given the foregoing results, one would be tempted to conclude that, in the general case of a baroclinic star, stability conditions (41) and (42) are equivalent to conditions (43) and (44) simultaneously. This is not quite true, as it will become apparent from the following geometrical discussion due to Høiland and Holmboe. Since we are mainly interested in stability conditions, let us restrict the discussion to the case where trace $\mathbf{M} > 0$

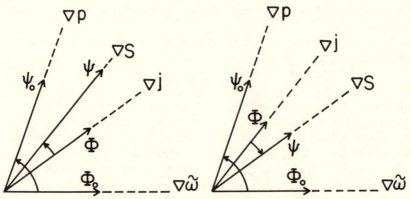

Figure 7.1.   A dynamically stable situation.   Figure 7.2.   A dynamically unstable situation.

(cf. equation [38]). Figures 7.1 and 7.2 depict, at a given point, two plausible orientations of the basic vectors. In Figure 7.1, the vector products $\mathbf{\Phi}_0 \times \mathbf{\Psi}_0$ and $\mathbf{\Phi} \times \mathbf{\Psi}$ both point along the same direction; and, accordingly, the determinant of $\mathbf{M}$ is positive (cf. equation [39]). This implies stability. On the contrary, in Figure 7.2, the vector products $\mathbf{\Phi}_0 \times \mathbf{\Psi}_0$ and $\mathbf{\Phi} \times \mathbf{\Psi}$ have opposite signs. Their scalar product is therefore negative, and det $\mathbf{M}$ is negative. This implies instability. We may summarize the above results in the following proposition:

> *A baroclinic star in permanent rotation is dynamically stable with respect to axisymmetric motions if and only if the two following conditions are satisfied:* (i) *the entropy per unit mass S never decreases outward, and* (ii) *on each surface S = constant, the angular momentum per unit mass* $\Omega\varpi^2$ *increases as we move from the poles to the equator.*

This is the Høiland criterion. Thus, as far as dynamical stability is concerned, stable density stratifications permit certain rotation laws that depend on both ϖ and z. And these nonconservative laws are *not* in conflict with the von Zeipel paradox (see, however, Section 7.5).

Let us conclude with two remarks. *First*, the Høiland criterion is also valid for pseudo-barotropic models. In this case, stable configurations must satisfy both conditions (43) and (44)—except perhaps at the equator ($g_z = 0$) where equation (41) must be used (cf. §14.5). However, because the condition grad $S > 0$ implies stability against convection (i.e., a subadiabatic lapse rate), we are then faced again with the von Zeipel paradox which excludes radiative equilibrium in pseudo-barotropes. *Second*, the Høiland criterion provides necessary and sufficient conditions of stability with respect to the axisymmetric motions which we considered. Since our analysis does not allow for nonaxisymmetric disturbances, and because we have neglected the Eulerian change δV, this criterion must be considered as a necessary condition of stability only. At this time, small departures from axisymmetry in a star are still poorly understood.

7.4. THERMAL IMBALANCE IN BAROCLINES

It has been hitherto assumed that the irreversible processes that always take place in a star are ineffective in modifying its prescribed rotation law. To what extent is this approximation valid for, e.g., a main-sequence star satisfying Høiland's stability criterion? As we recall from Section 3.4, the rate of change of entropy is generally given by

$$\rho T \frac{DS}{Dt} = \Phi_v + \rho \epsilon_{Nuc} + \mathrm{div}[(\chi + \chi_r)\,\mathrm{grad}\ T]. \tag{45}$$

(Magnetic fields are considered in Section 7.5.) The function Φ_v represents the amount of energy which is dissipated by friction, and is proportional to the total coefficient of shear viscosity $\mu + \mu_r$. Following Spitzer and Thomas, we can write

$$\mu + \mu_r = \frac{2m^{1/2}(kT)^{5/2}}{5e^4 \log \Lambda} + \frac{4aT^4}{15c\kappa\rho}, \tag{46}$$

where m is the proton mass, e the electronic charge in e.s.u., and k the Boltzmann constant; the quantity Λ has the form

$$\Lambda = \frac{3}{2e^3}\left(\frac{mk^3 T^3}{\pi\rho}\right)^{1/2}, \tag{47}$$

and is proportional to the cut-off distance for collisions, which we take to be equal to the Debye length. As written, equation (46) applies to a pure hydrogen plasma, which is accurate enough for our purposes. The coefficients μ and μ_r describe the transport of momentum by collision and radiation, respectively, when a velocity gradient prevails. (Transport of angular momentum by the net luminous flux is discussed in Section 7.5.) Similarly, the coefficient of radiative conductivity is

$$\chi_r = \frac{4acT^3}{3\kappa\rho}. \tag{48}$$

Except for degenerate stars, thermal conductivity is completely negligible ($\chi \ll \chi_r$).

In the Sun, we have roughly $\mu \approx \mu_r$, but μ_r becomes progressively more important for more massive stars. Table 7.1 illustrates the magnitude of the coefficients of kinematical viscosity in a main-sequence star with

TABLE 7.1

Coefficients of kinematical viscosity
and the time scale for
viscous diffusion in the envelope
of a 10 M$_\odot$ model[†]

r/R	v_r (cm^2/sec)	v (cm^2/sec)	t_v (years)
0.50	1.25(4)	2.37(2)	4.8(9)
0.60	2.73(4)	2.96(2)	2.2(9)
0.70	6.89(4)	3.87(2)	8.8(8)
0.80	2.20(5)	5.47(2)	2.8(8)
0.90	1.30(6)	9.36(2)	4.7(7)
0.94	4.45(6)	1.36(3)	1.4(7)
0.98	5.24(7)	2.92(3)	1.2(6)

[†] The number in parentheses following each entry is the power of ten by which that entry must be multiplied.

SOURCE: Clement, M. J., *Ap. J.* **156**, 1051, 1969. (By permission of The University of Chicago Press. Copyright 1969 by the University of Chicago.)

$M = 10M_\odot (v_r = \mu_r/\rho, v = \mu/\rho)$; it also includes the time scale t_v for viscous diffusion of momentum in a region with characteristic dimensions equal to one tenth of the radius, i.e., $t_v = 0.01R^2/(v + v_r)$. By comparison, note that $t_v \approx 10^{11}$ years in the Sun, when $v + v_r$ is evaluated at a radius exterior to half of its mass. Actually, the radiative diffusion of thermal energy is always much more rapid than the viscous diffusion of momentum. The relative importance of these two mechanisms is given by the Prandtl number

$$P = \frac{c_v(\mu + \mu_r)}{\chi_r}. \tag{49}$$

Just below the solar convective zone, we have $P \approx 10^{-6}$. Quite generally, if large velocity gradients do not prevail, radiative transfer largely dominates over microscopic transport of momentum.[2] Outside the central regions where nuclear reactions take place, equation (45) then reduces to

$$\rho T \frac{DS}{Dt} = \mathrm{div}(\chi_r \, \mathrm{grad} \, T); \tag{50}$$

and, to the same approximation, the equations of motion remain unaffected by viscosity.

[2] The concept of eddy viscosity (i.e., macroscopic transport of momentum in a turbulent region) will be considered further in Section 8.4.

Consider now a chemically homogeneous barocline that rotates with some prescribed angular velocity $\Omega = \Omega(\varpi,z)$. *Assume further that it satisfies Høiland's stability criterion* (cf. §7.3). As we recall from Section 4.3, the isothermal- and isobaric-surfaces are always inclined to each other at a finite angle. Therefore, since the temperature does not assume a constant value over an isobaric surface, radiative conductivity will act in such a way as to smooth out these temperature differences so as to achieve the coincidence of both families of surfaces. In other words, unless some other mechanism quickly prevents this large-scale circulation, a genuine barocline will slowly readjust its angular momentum distribution, and gradually evolve toward a final state characterized by the condition

$$\frac{\partial \Omega}{\partial z} = 0, \tag{51}$$

while satisfying Høiland's criterion for dynamical stability. This is the main content of a result which was originally discussed by Goldreich and Schubert, and by Fricke.

Following recent developments by Sung and by Smith and Fricke, we shall first derive condition (51) by means of a local stability analysis. For our system, we shall assume a perfect gas with negligible radiation pressure, and we shall ignore viscosity ($P \ll 1$). Additionally, it is convenient at this stage to allow for a gradient of chemical composition. We thus have $p \propto \rho T/\bar{\mu}$, where $\bar{\mu}$ is now a function of position and time. Also, since we are chiefly interested in a dissipative mechanism (i.e., radiative conductivity), we may expect that the most unstable perturbations will be found to have wavelengths that are much smaller than the mean stellar radius. It is therefore expedient to work with a simplified set of equations that approximate the exact equations in a small region of the star. The analysis is restricted to small axisymmetric disturbances, and we assume that their size (L, say) is much smaller than any scale height of the equilibrium model (H, say). Then, the coefficients in the perturbation equations will be independent of ϖ, z, and t, so that the Eulerian changes may be expanded in plane waves of the form

$$\exp[nt + i(k_\varpi \varpi + k_z z)]. \tag{52}$$

Consistent with the above approximations, we may now take $\delta V \equiv 0$. Finally, we assume that (for the wavelengths of interest) the sound frequency is much greater than the rotation frequency.

By virtue of equation (52), the equations of motion reduce to

$$n^2 \boldsymbol{\xi} = -\frac{\delta \rho}{\rho} \boldsymbol{\Psi}_0 - (\boldsymbol{\xi} \cdot \boldsymbol{\Phi}) \boldsymbol{\Phi}_0 - \frac{i}{\rho} \mathbf{k} \delta p, \tag{53}$$

where $\boldsymbol{\xi}$ is a two-dimensional vector with components ξ_ϖ and ξ_z (cf. §6.6: equation [74]). Note that equation (53) already incorporates the conservation of angular momentum of each mass element in its motion. (This property still obtains here for we have neglected viscous friction.) Similarly, because of mass conservation (and equation [4] of Section 6.2), our approximations lead to the further condition $\mathbf{k}.\boldsymbol{\xi} = 0$, thus implying that the wave vector $\mathbf{k}$ is transverse to the displacement $\boldsymbol{\xi}$. Letting next $\boldsymbol{\xi} = \xi\mathbf{a}$ (where $\mathbf{a}$ is the unit vector along the vector $\boldsymbol{\xi}$) and multiplying equation (53) by $\mathbf{a}$, we find

$$n^2\xi = -\frac{\delta\rho}{\rho}(\mathbf{a}.\boldsymbol{\Psi}_0) - \xi(\mathbf{a}.\boldsymbol{\Phi})(\mathbf{a}.\boldsymbol{\Phi}_0) \tag{54}$$

(cf. equations [28], [29], and [31]). Because $\mathbf{k}.\mathbf{a} = 0$, the Eulerian variation δp is therefore eliminated at once from the equations of motion.

Now, the small-perturbation counterpart of equation (50) can be brought to the form

$$n\rho(\delta S + \boldsymbol{\xi}.\mathrm{grad}\ S) = -\chi_r k^2\left[\frac{\delta(T/\bar\mu)}{T/\bar\mu} + \frac{\delta\bar\mu}{\bar\mu}\right], \tag{55}$$

where $k^2 = k_\varpi{}^2 + k_z{}^2$. Quite generally, for a perfect gas, we have

$$S = c_v \log\frac{T}{\bar\mu\rho^{\gamma-1}} + constant. \tag{56}$$

It thus follows that

$$\mathrm{grad}\ S = c_v\left[\frac{1}{T}\mathrm{grad}\ T - (\gamma-1)\frac{1}{\rho}\mathrm{grad}\ \rho - \frac{1}{\bar\mu}\mathrm{grad}\ \bar\mu\right], \tag{57}$$

and

$$\delta S = c_v\left[\frac{\delta(T/\bar\mu)}{T/\bar\mu} - (\gamma-1)\frac{\delta\rho}{\rho}\right] = -c_p\frac{\delta\rho}{\rho}. \tag{58}$$

The second equality (58) stems from the linearized form of the equation of state, which takes the form

$$\frac{\delta\rho}{\rho} + \frac{\delta(T/\bar\mu)}{T/\bar\mu} = 0, \tag{59}$$

because $\delta p/p \approx (L/H)(\delta\rho/\rho)$, and, hence, may be neglected. (Remember that $L \ll H$, by assumption!) Finally, since the rate of diffusion of chemical species is comparable to the (negligible) viscous diffusion rate, we shall also assume that

$$\frac{D\bar\mu}{Dt} = \frac{\partial\bar\mu}{\partial t} + \mathbf{v}.\mathrm{grad}\ \bar\mu = 0. \tag{60}$$

Thence, we obtain

$$\frac{\delta\bar{\mu}}{\bar{\mu}} + \xi.\frac{\operatorname{grad}\bar{\mu}}{\bar{\mu}} = 0. \tag{61}$$

Using equations (57)–(61), we can thus rewrite equation (55) in the form

$$\left(1 + \frac{\epsilon}{n}\right)\frac{\delta\rho}{\rho} = \xi(\mathbf{a}.\boldsymbol{\Psi}) - \frac{\epsilon}{n}\xi\left(\mathbf{a}.\frac{\operatorname{grad}\bar{\mu}}{\bar{\mu}}\right), \tag{62}$$

where

$$\epsilon = \chi_r k^2/\rho c_p, \tag{63}$$

and $\boldsymbol{\Psi} = \operatorname{grad} S/c_p$, with $\operatorname{grad} S$ being defined in equation (57).

It is now a simple matter to eliminate $\delta\rho/\rho$ between equations (54) and (62); we find

$$n^3 + \epsilon n^2 + [(\mathbf{a}.\boldsymbol{\Phi})(\mathbf{a}.\boldsymbol{\Phi}_0) + (\mathbf{a}.\boldsymbol{\Psi})(\mathbf{a}.\boldsymbol{\Psi}_0)]n$$

$$+ \epsilon\left[(\mathbf{a}.\boldsymbol{\Phi})(\mathbf{a}.\boldsymbol{\Phi}_0) - \left(\mathbf{a}.\frac{\operatorname{grad}\bar{\mu}}{\bar{\mu}}\right)(\mathbf{a}.\boldsymbol{\Psi}_0)\right] = 0. \tag{64}$$

As was expected, in the limiting case $\epsilon = 0$, equation (64) provides the requisite dispersion relation for discussing dynamical stability (cf. §7.3). When radiative conductivity is taken into account ($\epsilon > 0$), we shall have an instability when equation (64) possesses a root with positive real part. In the present instances, such a growing mode certainly exists if

$$(\mathbf{a}.\boldsymbol{\Phi})(\mathbf{a}.\boldsymbol{\Phi}_0) - \left(\mathbf{a}.\frac{\operatorname{grad}\bar{\mu}}{\bar{\mu}}\right)(\mathbf{a}.\boldsymbol{\Psi}_0) < 0. \tag{65}$$

This is the required condition for thermal instability with respect to axisymmetric motions, when both radiative conductivity and a gradient of mean molecular weight are taken into account.

Consider first a chemically homogeneous star. Then, by virtue of equation (65), thermal instability occurs whenever a vector $\mathbf{a}$ can be found that will make $(\mathbf{a}.\boldsymbol{\Phi})(\mathbf{a}.\boldsymbol{\Phi}_0)$ negative. After James and Kahn, Figure 7.3 illustrates, at a given point, the vectors $\boldsymbol{\Phi}$ and $\boldsymbol{\Phi}_0$ in the case of a dynamically stable barocline. It is a simple matter to see that all vectors $\mathbf{a}$ that lie in the cross-hatched region make the body thermally unstable at that point. Obviously, the only way to prevent this thermal imbalance in a star is to remove the cross-hatched region at every point. This can be done only if each vector $\boldsymbol{\Phi}$ points in the ϖ-direction, i.e., if condition (51) obtains at every point of the star. Since Høiland's stability criterion does obtain by assumption, a necessary (but not sufficient) condition for thermal stability is, therefore,

$$\frac{\partial\Omega}{\partial z} = 0. \tag{66}$$

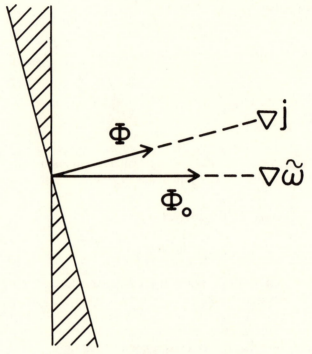

Figure 7.3.   A thermally unstable situation.

We can summarize the above results as follows:

When a chemically homogeneous, inviscid star is in static equilibrium and satisfies Høiland's stability criterion, thermal instability with respect to axisymmetric disturbances always occurs if Ω depends on the coordinate z along the rotation axis.

This is the Goldreich-Schubert-Fricke criterion.[3] It shows that a genuine baroclinic star must necessarily evolve in time in such a way as to restore the condition (66) which is required for thermal stability.

Now, is it possible to maintain a chemically inhomogeneous star in static equilibrium with the steady rotation law $\Omega = \Omega(\varpi,z)$? Condition (65) shows that a stable gradient of chemical composition (i.e., grad $\bar{\mu} < 0$) often has a stabilizing influence on all displacements which are not tangent to the surfaces $\bar{\mu} = constant$. In other words, a suitable stratification of mean molecular weight might well prevent radiative conductivity from smooth-

---

[3] This criterion is the exact analog to the Taylor-Proudman theorem for a slowly rotating, incompressible fluid (see, e.g., reference 27, pp. 83–85, of Chapter 3).

ing out the nonuniformity in density along the isobaric surfaces (cf. equation [59]). Further discussion of this effect necessarily requires the use of a particular model. Similarly, viscous friction that opposes the motion might also inhibit the growth of the thermal instability. However, because the Prandtl number is usually very small in a real star, this stabilizing effect is generally negligible.

Finally, since radiative conductivity permits a general circulation which would tend to restore the condition $\partial\Omega/\partial z = 0$ required for thermal stability, what is the time scale on which Ω evolves? (Obviously, this time scale is much longer than the time scale for dynamical instability which eliminates an adverse angular momentum distribution over a few rotation periods.) Following Fricke, Colgate, and Kippenhahn, the appropriate time scale is of the order of the Kelvin-Helmholtz time. This estimate was subsequently criticized by James and Kahn, who suggested that the appropriate time scale for large-scale redistribution of angular momentum by radiative conductivity is probably related to the circulation time for meridional currents, i.e., to the Eddington-Sweet time (cf. §8.2). This problem requires further study. In any event, it is now apparent that *it is impossible to build a chemically homogeneous stellar model in strict radiative equilibrium with a steady rotation law*. Indeed, when $\partial\Omega/\partial z \neq 0$, radiative conductivity induces a thermal imbalance that, in turn, permits a large-scale circulation. On the other hand, when $\partial\Omega/\partial z = 0$ at every point, strict radiative equilibrium must necessarily break down and cause circulatory currents. Therefore, in both cases, time-independent rotation laws in radiative zones are always approximate. Of course, in zones of efficient convection the transport of energy is not by radiation, and von Zeipel's argument does not apply. In that case, in principle, the steady rotation laws $\Omega = \Omega(\varpi)$ may well be used (see, however, Section 8.5).

7.5. SOME QUASI-PERMANENT SOLUTIONS

Although von Zeipel's paradox does not apply to genuine baroclines, the assumption of strict radiative equilibrium still places restrictions on the function $\Omega(\varpi,z)$ and, for an arbitrary rotation law, circulation currents will generally be set up. Various attempts have been made to obtain the angular velocity distribution in slowly rotating stars which, by assumption, are devoid of meridional circulation. Some authors (notably Schwarzschild and Roxburgh) have sought to determine the function $\Omega(\varpi,z)$ in an inviscid baroclinic star from the condition that equations (1) and (33) are simultaneously satisfied. This procedure is equivalent to assuming that, for some reason, the angular velocity will adjust itself so as to prevent meridional currents, and appears very difficult to justify on physical grounds.

Moreover, as we now know, all these models are thermally unstable. In this section we shall briefly review the role of viscosity in determining the angular velocity distribution, and we shall conclude with some considerations about the role of magnetic fields in equilibrium models.

VISCOUS FRICTION

Consider a slowly evolving, viscous star that rotates, at time t, with the angular velocity $\Omega = \Omega(\varpi,z,t)$. By hypothesis, the motion is so slow that the meridional circulation is negligible. Taking into account the interaction between matter and radiation up to the lowest-order terms in $|\mathbf{v}|/c$, we can write the three components of the Navier-Stokes equations in the form

$$\frac{\partial p}{\partial \varpi} = -\rho\frac{\partial V}{\partial \varpi} + \rho\Omega^2\varpi, \tag{67}$$

$$\rho\frac{\partial}{\partial t}(\Omega\varpi^2) + \frac{1}{c^2}\operatorname{div}(\Omega\varpi^2\mathscr{F}) = \operatorname{div}(\mu\varpi^2\operatorname{grad}\Omega), \tag{68}$$

$$\frac{\partial p}{\partial z} = -\rho\frac{\partial V}{\partial z}, \tag{69}$$

where, for the sake of brevity, μ denotes the total coefficient of shear viscosity (cf. §3.3: equation [46] and note 5). Clearly, the ϖ- and z-components of the equations of motion remain unaffected by dissipation. On the contrary, the φ-component (68) describes the irreversible change in angular momentum due to viscous friction; and it also includes the transport of angular momentum by the net radiative flux. As was originally shown by Jeans, the second term on the l.h.s. of equation (68) expresses the fact that radiation possesses angular momentum due to the rotation of the region that emits it. As usual, to equations (67)–(69) we must add equations (3), (6), and (45). Note that conservation of mass is always satisfied because we have assumed axially symmetric motions with no meridional currents. Finally, these equations must be supplemented by appropriate boundary conditions on the changing surface of the star.

Time-dependent solutions will be presented for zero-temperature white dwarfs in Section 13.3. We shall, for the present, restrict the analysis to permanent or quasi-permanent motions. Then, up to order $|\mathbf{v}|/c$, equation (68) reduces to

$$\operatorname{div}(\mu\varpi^2\operatorname{grad}\Omega) = 0. \tag{70}$$

Since the stress vector must vanish on the stellar surface $\mathscr{S}$, we also have

$$\mathbf{n}.\operatorname{grad}\Omega = 0, \quad \text{on } \mathscr{S}, \tag{71}$$

where **n** is the outer normal to the surface $\mathscr{S}$ (cf. §3.3: equation [48]). To understand the physical meaning of equations (70) and (71), let us first compute the total power transformed into heat by viscous friction. Given our assumptions, the total dissipation function is

$$D(\Omega) = \int_{\mathscr{V}} \Phi_v \, \mathbf{dx} = \int_{\mathscr{V}} \mu \varpi^2 \left[\left(\frac{\partial \Omega}{\partial \varpi} \right)^2 + \left(\frac{\partial \Omega}{\partial z} \right)^2 \right] \mathbf{dx} \qquad (72)$$

(cf. §3.4: equation [62], and Appendix B). Consider now an arbitrary rotation law $\Omega(\varpi,z) + \delta\Omega(\varpi,z,t)$ that is subject only to the constraint that it preserves the surface $\mathscr{S}$ and the volume $\mathscr{V}$ of the configuration. The total dissipation function in the altered motion is

$$D(\Omega + \delta\Omega) = D(\Omega) + D(\delta\Omega) + 2 \int_{\mathscr{V}} \mu \varpi^2 \left(\frac{\partial \Omega}{\partial \varpi} \frac{\partial \delta\Omega}{\partial \varpi} + \frac{\partial \Omega}{\partial z} \frac{\partial \delta\Omega}{\partial z} \right) \mathbf{dx}. \quad (73)$$

After integrating by parts, we can next rewrite the last term of equation (73) in the form

$$2 \int_{\mathscr{S}} \delta\Omega(\mu \varpi^2 \, \text{grad } \Omega).\mathbf{dS} - 2 \int_{\mathscr{V}} \delta\Omega \, \text{div}(\mu \varpi^2 \, \text{grad } \Omega) \, \mathbf{dx}. \qquad (74)$$

Therefore, since the function $\Omega(\varpi,z)$ satisfies equations (70) and (71), the two integrals in expression (74) vanish, and equation (73) reduces to

$$D(\Omega + \delta\Omega) = D(\Omega) + D(\delta\Omega). \qquad (75)$$

Because $D(\delta\Omega)$ is essentially a positive quantity, $D(\Omega + \delta\Omega)$ always exceeds the total dissipation function $D(\Omega)$ in the steady state. Accordingly, *every solution of equations (70) and (71) is characterized by the property that the total power dissipated by friction is an absolute minimum when compared to any other motion consistent with the boundary $\mathscr{S}$ and the volume $\mathscr{V}$.* This result is basically due to Helmholtz, and it was first used for the solar rotation problem by Dedebant and Wehrlé.

Another interesting approach to the problem of a slowly evolving viscous star has been suggested by Balazs. In principle, to find the angular velocity distribution $\Omega(\varpi,z,t)$ in a rotating star, one must solve the equations of motion as an initial value problem. As the time progresses, the rotational kinetic energy will be mainly transformed into heat by internal friction, and angular momentum will be dissipated as well in the radiation field and mass loss from the star. Because the situation is such that the relative rates of kinetic energy and angular momentum dissipation are usually very slow, a quasi-permanent regime can thus establish itself. This suggests that we replace our problem with a simpler one, and search for that steady rotation law $\Omega(\varpi,z)$ that makes the total dissipation function $D(\Omega)$ a minimum, subject to the boundary condition (71) and to the auxiliary conditions that

$$K = \frac{1}{2} \int_{\mathscr{V}} \rho\Omega^2\varpi^2 \, \mathbf{dx} \quad \text{and} \quad J = \int_{\mathscr{V}} \rho\Omega\varpi^2 \, \mathbf{dx} \qquad (76)$$

should be held constant, having the same values as in the actual quasi-permanent regime. (Of course, for this approach to be applicable, it is necessary that the relaxation time for establishing a quasi-permanent state should be much less than the relaxation times for K and J themselves.) Thus, introducing the Lagrange multipliers $-\alpha^2$ and β ($\alpha^2 \geqslant 0$, $\beta \gtrless 0$), which are essentially the reciprocal relaxation times of the quantities we keep artificially fixed, we must minimize the functional

$$D(\Omega) - \alpha^2 K(\Omega) + \beta J(\Omega), \qquad (77)$$

with the boundary condition (71) and the constraints (76). It is a simple matter to prove that the angular velocity $\Omega(\varpi,z)$ which minimizes the functional (77) satisfies the equation

$$\frac{1}{\rho\varpi^2} \operatorname{div}(\nu\rho\varpi^2 \operatorname{grad} \Omega) + \alpha^2\Omega = \beta, \qquad (78)$$

where $\nu = \mu/\rho$ is the coefficient of kinematical viscosity. Balazs did not solve equation (78) in the case of a particular stellar model, but he did set forth a few interesting conclusions that are worth summarizing here. First, in the presence of internal and external dissipative mechanisms (e.g., viscous friction, the outer radiation field, mass loss, . . .), *whatever angular velocity distribution we will find, it will not correspond to a state of rigid rotation*, since $\Omega = constant$ is incompatible with the auxiliary conditions. In other words, when both the kinetic energy and the angular momentum are lost with the passage of time ($\alpha^2 > 0$, $\beta \neq 0$), a quasi-permanent state of differential rotation must be set up. Second, when $\alpha^2 > 0$ and $\beta \neq 0$, the variational problem has no unique solution, and equation (78) necessarily exhibits an infinite set of modes. The angular momentum distribution realized in the quasi-permanent state of differential rotation will therefore depend on the initial conditions at star formation. We thus obtain the interesting result that *under certain conditions, the state of least dissipation is not unique*, and the presence of internal and external dissipative mechanisms is not sufficient to wipe out the influence of the initial state completely. Further research along these lines would be most useful.

SOME PRACTICAL DETERMINATIONS OF Ω (ϖ,z)

As far back as 1932, Gião and Wehrlé solved equation (70) by assuming that μ is a constant throughout the system (cf. note 2). Looking for separable solutions of the form $\Omega(\varpi,z) = f(\varpi)g(z)$ they found

$$\Omega(\varpi,z) = \Omega_0 \cos mz \sum_{n=0}^{\infty} \frac{(m\varpi)^{2n}}{2^{2n}(n + 1)(n!)^2}, \tag{79}$$

where Ω_0 and m are two constants of integration. In the case of a quasi-spherical system (such as the Sun), it is convenient to use spherical co-ordinates $(r,\phi = 90° - \theta,\varphi)$; then, substituting the relations $\varpi = r \cos \phi$ and $z = r \sin \phi$ in equation (79), we obtain

$$\Omega(R,\phi) = \Omega_0 \cos(mR \sin \phi) \sum_{n=0}^{\infty} \frac{(mR \cos \phi)^{2n}}{2^{2n}(n + 1)(n!)^2}, \tag{80}$$

on the mean boundary with mean radius $r = R$. The remarkable feature of equation (80) is that it reproduces with good accuracy the solar rotation law (cf. §2.2), when Ω_0 and mR are fitted at two different heliocentric latitudes. Unfortunately, unless $m = 0$, this rotation law does *not* comply with condition (71). According to Dedebant and Wehrlé, equation (79) should not be viewed as a strictly permanent rotation law, but rather as a quasi-permanent law corresponding to a state of least dissipation. In any event, it is doubtful that this solution adequately describes the inner solar rotation, for the variations of viscosity with depth are not included in the problem.

Another approach, which also takes into account the viscous forces in a restricted form, was suggested by Schwarzschild in 1942. The method consists of the derivation of the function $\Omega(\varpi,z)$, which is a solution of the usual equations for an inviscid star, and which we demand to fulfill the viscous boundary condition (71). In this manner, equation (70) is *not* satisfied, but the formulation at least incorporates the correct behavior of the angular velocity near the surface. Detailed calculations have been made by Clement for slowly rotating baroclinic models on the upper main sequence. Since the time scale for (radiative) viscous diffusion in the envelope of these stars is shorter than their lifetime on the main sequence, the procedure is, therefore, a valid one (see Table 7.1). By treating the rotation as a small perturbation on a spherical model, we then have

$$\left(\frac{\partial \Omega}{\partial r}\right)_{r = R} = 0. \tag{81}$$

Figures 7.4 and 7.5 illustrate, respectively, the curves of constant angular velocity and constant angular momentum per unit mass for a model with a uniformly rotating convective core surrounded by a radiative envelope. (The opacity is due to electron scattering, and the radiation pressure is neglected altogether.) Note that the surfaces $\Omega\varpi^2 = constant$ are approximately cylindrical. Hence, one can only conclude that the thermal stability

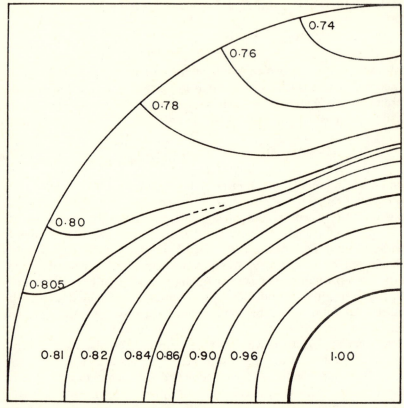

Figure 7.4. Quadrant of a meridional plane showing curves of constant angular velocity for the viscous boundary condition (81). The right-hand edge of the figure is the axis of rotation, and the bottom edge is the equatorial plane. Each curve is labeled by its value of Ω. The quarter-circle in the lower right-hand corner is a section of the rigidly rotating convective core ($\Omega \equiv 1$). Source: Clement, M. J., *Ap. J.* **156**, 1051, 1969. (By permission of The University of Chicago Press. Copyright 1969 by the University of Chicago.)

of the model is open to some question (cf. §7.4). Moreover, because the balance of forces in the azimuthal direction is achieved only near the surface, the model cannot be in strict mechanical equilibrium. Here also the solution must be regarded at best as a quasi-permanent state of differential rotation.

Very much in the same vein, another possible boundary condition, which represents the opposite extreme to equation (81), was suggested by Jeans in 1926. For stationary systems that are free of meridional circulation, equation (68) becomes

$$\frac{1}{c^2}\,\mathscr{F}.\mathrm{grad}(\Omega\varpi^2) = \mathrm{div}(\mu\varpi^2\,\mathrm{grad}\,\Omega), \tag{82}$$

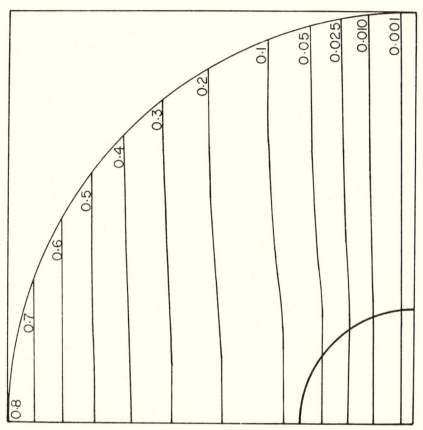

Figure 7.5. Curves of constant angular momentum per unit mass in a meridional plane for the viscous boundary condition (81). Each curve is labeled by its value of $\Omega\varpi^2$. (Compare with Figure 7.4.) Source: Clement, M. J., *Ap. J.* **156**, 1051, 1969. (By permission of The University of Chicago Press. Copyright 1969 by the University of Chicago.)

wherever $\epsilon_{Nuc} = 0$. As it was noted before, the l.h.s. of this equation implies that a flux of radiation is always accompanied by a flux of angular momentum, thus tending to equalize $\Omega\varpi^2$ throughout the star. This radiative braking mechanism dominates over the viscous diffusion of angular velocity, when $T/|\text{grad } T|$ is small compared with $\Omega/|\text{grad }\Omega|$. (If the gaseous viscosity is ignored, this can be seen by writing the coefficient of radiative viscosity in the form

$$\mu_r = \frac{|\mathscr{F}|}{5c^2}\frac{T}{|\text{grad } T|},$$ (83)

which combines equations [2] and [46].) Thus, if the viscous term in equation (82) is ignored, this effect results in the angular velocity falling off as r^{-2} along a given radius near the surface of a slowly rotating radiative envelope; that is,

$$\left[\frac{\partial}{\partial r}(\Omega r^2)\right]_{r=R} = 0. \tag{84}$$

As was shown by Marks and Clement, when the actual balance between radiative braking and viscous diffusion is taken into account, equation (82) requires that

$$\left[\frac{\partial}{\partial r}(\Omega r^{5/3})\right]_{r=R} = 0, \tag{85}$$

at the surface of a slowly rotating star. Note that it is much closer to the radiative braking condition (84) than to the viscous condition (81). This is consistent with Jeans's suggestion that radiative braking is dominant in radiative envelopes. Various baroclinic models satisfying boundary condition (85) have been constructed by Harris and Clement for upper main-sequence stars, assuming the material velocity to be wholly one of pure rotation. The corresponding rotation laws are very similar to the one depicted on Figures 7.4 and 7.5. However, they were severely criticized by Smith, who showed by means of an order-of-magnitude argument that these rotation laws are irrelevant for most observed early-type stars, being destroyed either by meridional circulation or by thermal imbalance in a time which is very short compared to the main-sequence lifetime of such stars.[4]

THE ISOROTATION LAW

Up to now, the possibility of the existence of a general magnetic field in stars has been completely disregarded. Obviously, if a large-scale magnetic field is present, a non-uniform rotation must seriously modify the field. Hence, what condition must we impose on $\Omega(\varpi,z)$ to build a stationary model that is subject to an axisymmetric magnetic field? Consider a completely ionized star in which each mass element revolves about a fixed axis. Let there be an axisymmetric magnetic field whose lines of force lie in meridional planes (i.e., a poloidal field). A steady state is supposed to have been attained. Then, as we recall from Section 3.6, we have

$$\text{curl } \mathbf{H} = 4\pi \mathbf{J}, \tag{86}$$

$$\text{curl } \mathbf{E} = 0, \quad \text{div } \mathbf{H} = 0, \tag{87}$$

---

[4] This remark may be irrelevant, though, since it is based on a local time scale, which might be largely in error (cf. §8.2: equation [50] and note 2).

and

$$\mathbf{J} = \sigma_e(\mathbf{E} + \mathbf{v} \times \mathbf{H}), \tag{88}$$

where the variable coefficient σ_e has, in general, a finite value. Taking the curl of equation (88), we find

$$\mathrm{curl}\,\mathbf{J} = \sigma_e\,\mathrm{curl}(\mathbf{v} \times \mathbf{H}) + \frac{\mathrm{grad}\,\sigma_e}{\sigma_e} \times \mathbf{J}, \tag{89}$$

since curl $\mathbf{E}$ vanishes. Now, by assumption, the magnetic field $\mathbf{H}$ has no φ-component. Accordingly, from equation (86), the current density $\mathbf{J}$ is purely azimuthal, and curl $\mathbf{J}$ has no φ-component. It thus follows that

$$[\mathrm{curl}(\mathbf{v} \times \mathbf{H})]_\varphi = 0, \tag{90}$$

or

$$\frac{\partial}{\partial z}(v_\varphi H_z) + \frac{\partial}{\partial \varpi}(v_\varphi H_\varpi) = 0. \tag{91}$$

Because $v_\varphi = \Omega\varpi$, we can also rewrite this equation in the form

$$H_\varpi \frac{\partial \Omega}{\partial \varpi} + H_z \frac{\partial \Omega}{\partial z} + \Omega\left[\frac{1}{\varpi}\frac{\partial}{\partial \varpi}(\varpi H_\varpi) + \frac{\partial H_z}{\partial z}\right] = 0, \tag{92}$$

or, by virtue of the second equation (87),

$$\mathbf{H}.\mathrm{grad}\,\Omega = 0. \tag{93}$$

The meaning of this result is that Ω is constant over each surface traced out by the complete revolution of a line of force about the axis of rotation. That it to say, *a stationary star subject to a poloidal magnetic field can be divided into axisymmetric shells, each shell rotating bodily with its own angular velocity, and wholly containing its own lines of force.* This is the isorotation law, and it was first discovered by Ferraro. This result is obvious when $\sigma_e = \infty$, since if it were not true the differential rotation would drag the field lines out of the meridional planes, therefore giving an unsteady state. But we see that the theorem also holds true for finite electrical conductivity, i.e., when the magnetic lines of force of a poloidal field are not frozen in the stellar material.

Now, under which circumstances are the rotation laws satisfying condition (93) thermally stable? In general, the relative importance of radiative conductivity and magnetic diffusivity is given by the magnetic Prandtl number

$$\mathsf{P}_m = \frac{c_v \rho v_m}{\chi_r}, \tag{94}$$

where $v_m = 1/4\pi\sigma_e$ (cf. §3.6: equation [125]). For an actual star, this ratio is usually very small, so that we may neglect the diffusion of the field lines across the star. (For the radiative interior of the Sun, we have $P_m \approx 10^{-4}$.) Then, as was shown by Fricke, all isorotation laws are thermally unstable with respect to axisymmetric disturbances (with the possible exception of solid body rotations). The role of finite electrical conductivity was also investigated by Fricke and Schubert. Because of the smallness of magnetic diffusivity, they found that this effect generally provides only a negligible stabilizing effect on a star. The stationariness of baroclinic stars satisfying the isorotation law (93) is thus open to some question. Stellar magnetic fields will be discussed further in Chapter 15.

BIBLIOGRAPHICAL NOTES

General accounts of these and related matters will be found in:

1. Strittmatter, P. A., "Stellar Rotation" in *Annual Review of Astronomy and Astrophysics* 7, pp. 665–684, Palo Alto, California: Annual Reviews, Inc., 1969.
2. Fricke, K., and Kippenhahn, R., "Evolution of Rotating Stars" in *Annual Review of Astronomy and Astrophysics* 10, pp. 45–72, Palo Alto, California: Annual Reviews, Inc., 1972.
3. Roxburgh, I. W., "The Effect of Rotation in Stellar Structure and Evolution" in *Mém. Soc. Roy. Sci. Liège* (6) 8 (1975): 15–24.

The following references on the subject of stability may also be noted here:

4. Spiegel, E. A., and Zahn, J. P., *Comments on Astroph. and Space Phys.* 2 (1970): 178–183.
5. Zahn, J. P., "Rotational Instabilities in Stellar Evolution" in *Stellar Instability and Evolution* (Ledoux, P., Noels, A., and Rodgers, A. W., eds.), pp. 185–195, Dordrecht: D. Reidel Publ. Co., 1974.

Section 7.2. The reference to von Zeipel is to his papers:

6. von Zeipel, H., *Probleme der Astronomie* (Festschrift für H. von Seeliger), pp. 144–152, Berlin: Springer-Verlag, 1924.
7. von Zeipel, H., *Monthly Notices Roy. Astron. Soc. London* 84 (1924): 665–683, 684–701.

The case of differentially rotating stars was first considered by Rosseland (reference 11 of Chapter 4) and Vogt:

8. Vogt, H., *Astron. Nachr.* 255 (1935): 109–112.

The problem has been further discussed by Baker and Kippenhahn (reference 17 of Chapter 8). A rigorous proof of the generalized von Zeipel paradox will be found in:

9. Roxburgh, I. W., *Monthly Notices Roy. Astron. Soc. London* **132** (1966): 201–215.

Section 7.3. This problem actually originates in the work of Helmholtz (1888). The classical references on the subject are those of:

10. Solberg, H., *Procès-Verbaux Ass. Météor., U.G.G.I., 6ᵉ Assemblée Générale (Edinburgh), Mém. et Disc.* **2** (1936): 66–82.

11. Høiland, E., *Archiv f. Math. og Naturv. (Oslo)* **42**, No 5 (1939): 1–69.

12. Høiland, E., *Avhandliger Norske Videnskaps-Akademi i Oslo*, I, *Math.-Naturv. Klasse*, No 11 (1941): 1–24.

13. Kleinschmidt, E., *Ann. Hydrogr. und marine Meteor.* **69** (1941): 305–325.

Criterion (44) was definitely established in reference 10. Solberg's paper also contains a discussion of nonhomentropic flows in the particular context of geophysics. However, conditions (41) and (42) were first derived and discussed in their most general form in 1941 by Høiland, who based his derivation on the Bjerknes theorem (cf. §3.5); they were independently obtained by Kleinschmidt in a somewhat more restricted context. The following papers may also be noted:

14. Randers, G., *Astroph. J.* **95** (1942): 454–460.

15. Holmboe, J., *J. Marine Research* **7** (1948): 163–174.

This last reference amplifies and clarifies the geometrical discussion due to Høiland. The presentation of the subject in the text largely follows Fjørtoft's (reference 53 of Chapter 6). His approach was independently followed by:

16. Fricke, K., and Smith, R. C., *Astron. and Astroph.* **15** (1971): 329–331.

Reference 16 also presents a global proof (with δV neglected) of the Høiland criterion. Non-axisymmetric motions are briefly discussed in references 4 and 5; see also:

17. Sung, C.-H., *Astron. and Astroph.* **33** (1974): 99–104.

18. Sung, C.-H., *Astroph. and Space Sci.* **33** (1975): 127–140.

Section 7.4. The thermal imbalance of genuine baroclines was independently discovered by:

19. Goldreich, P., and Schubert, G., *Astroph. J.* **150** (1967): 571–587.

20. Fricke, K., *Zeit. f. Astroph.* **68** (1968): 317–344.

More refined derivations will be found in:

21. Sung, C.-H., *Astroph. and Space Sci.* **26** (1974): 305–317.

22. Smith, R. C., and Fricke, K., *Monthly Notices Roy. Astron. Soc. London* **172** (1975): 577–584.

The following papers are also quoted in the text:

23. Colgate, S. A., *Astroph. J. Letters* **153** (1968): L81–L83.
24. Kippenhahn, R., *Astron. and Astroph.* **2** (1969): 309–315.
25. James, R. A., and Kahn, F. D., *ibid.* **5** (1970): 232–239; *ibid.* **12** (1971): 332–339.

Pertinent comments will also be found in reference 32 of Chapter 8. Inhomogeneous stars were first discussed in reference 19. See also:

26. Vandakurov, Yu.V., *Astrofizika* **8** (1972): 433–439.

Section 7.5. An early attempt to derive the angular velocity distribution in an inviscid star is due to:

27. Rosseland, S., *Astroph. Norvegica* **2** (1936): 173–191; *ibid.* **2** (1937): 249–262.

Circulation-free baroclinic models have been constructed by:

28. Schwarzschild, M., *Astroph. J.* **106** (1947): 427–456.
29. Porfirev, V. V., *Astron. Zh.* **39** (1962): 710–714.
30. Roxburgh, I. W., *Monthly Notices Roy. Astron. Soc. London* **128** (1964): 157–171, 237–244.
31. Roxburgh, I. W., and Strittmatter, P. A., *ibid.* **133** (1966): 1–14, 345–357.
32. Legg, M. P. C., *Australian J. Phys.* **20** (1967): 651–662.
33. Aikawa, T., *Sci. Reports Tôhoku Univ.* (I) **53** (1970): 21–29, 158.
34. Clement, M. J., in *Stellar Rotation* (Slettebak, A., ed.), pp. 346–351, New York: Gordon and Breach, 1970.
35. Monaghan, J. J., *Monthly Notices Roy. Astron. Soc. London* **154** (1971): 47–57.
36. Durney, B. R., *Astroph. J.* **172** (1972): 479–484.

The results of reference 28 are vitiated by an inappropriate formulation of the problem; for this reason, Schwarzschild's model was reinvestigated by Aikawa. References 30–32 and 36 present slightly distorted models in which the angular velocity is constant over spherical shells, i.e. $\Omega = \Omega(r)$. See also references 21 and 36 of Chapter 8, and note 5 on p. 473. Viscous friction and the principle of least dissipation are discussed in:

37. Gião, A., et Wehrlé, P., *Beitrage z. Physik d. freien Atmosphäre* **19** (1932): 237–245.
38. Dedebant, G., et Wehrlé, P., *La Météorologie*, No 85–86 (1932): 117–121.

39. Dedebant, G., Schereschewsky, P., et Wehrlé, P., *C.R. Acad. Sci. Paris* **199** (1934): 1287–1289.

40. Dedebant, G., Schereschewsky, P., et Wehrlé, P., *La Météorologie*, No 124 (1935): 336–341.

A comprehensive account of this problem will be found in:

41. Fabre, H., *Annales de l'Observatoire de Toulouse* **19** (1949): 16–53.

The reference to Balazs is to his paper:

42. Balazs, N. L., *Phys. Fluids* **5** (1962): 57–68.

Practical means to determine the function $\Omega(\varpi, z)$ were suggested in the following references:

43. Jeans, J. H., *Monthly Notices Roy. Astron. Soc. London* **86** (1926): 444–458.

44. Jeans, J. H., *Astronomy and Cosmogony*, pp. 268–283, Cambridge: At the Univ. Press, 1928 (New York: Dover Public., Inc., 1961).

45. Schwarzschild, M., *Astroph. J.* **95** (1942): 441–453.

Various applications of these ideas were made by:

46. Clement, M. J., *ibid.* **156** (1969): 1051–1068.

47. Marks, D. W., and Clement, M. J., *Astroph. J. Letters* **166** (1971): L27–L29.

48. Harris, W. E., and Clement, M. J., *Astroph. J.* **167** (1971): 321–325.

Comments on these results will be found in:

49. Smith, R. C., *Monthly Notices Roy. Astron. Soc. London* **153** (1971): 33P–35P.

The isorotation law was originally discussed by Ferraro:

50. Ferraro, V. C. A., *ibid.* **97** (1937): 458–472.

See also:

51. Alfvén, H., *Arkiv för Matematik, Astronomi och Fysik* **29A**, No 12 (1943): 1–17.

52. Sweet, P. A., *Monthly Notices Roy. Astron. Soc. London* **109** (1949): 507–516.

An interesting reference on this and related matters is:

53. Cowling, T. G., "Solar Electrodynamics" in *The Sun* (Kuiper, G. P., ed.), pp. 532–591, Chicago: The Univ. of Chicago Press, 1953.

Thermal instability of the isorotation law has been demonstrated in:

54. Fricke, K., *Astron. and Astroph.* **1** (1969): 388–398.

55. Fricke, K., and Schubert, G., *ibid.* **8** (1970): 480–485.

8

Meridional Circulation

8.1. INTRODUCTION

This chapter is primarily concerned with large-scale meridional currents that may occur in a star perturbed away from spherical symmetry. For the sake of simplicity, we shall restrict ourselves to rotationally driven currents only, but similar large-scale motions also exist in tidally or magnetically distorted stars. The possible existence of meridional currents in the radiative regions of rotating stars was first suggested independently by Vogt and by Eddington in order to solve the problem of thermal equilibrium posed by the von Zeipel paradox (cf. §7.2). As they showed, the effect of radiative transfer alone attempting to maintain thermal equilibrium is to produce a slight rise in temperature over some areas of any given level surface and a slight fall over other areas. Hence, in those regions where excessive amounts of heat may be trapped, material will at first rise and then cool by expansion; on the contrary, in the cooling regions, material will sink to be subsequently heated by compression. As a result, a large-scale meridional circulation will commence.

In Section 8.2 we will obtain steady circulation patterns in the radiative zones of a uniformly rotating, inviscid star. Following the usual practice, we thus merely determine the circulation velocity generated by the non-sphericity of a chemically homogeneous star in slow rotation. Section 8.3 critically assesses this approach, which fails to take into account the angular momentum transfer by the currents themselves, and which also neglects the important role played by a gradient in the mean molecular weight; the influence of viscous and magnetic forces, as well as the role of time-dependent currents is also stressed in this section. We conclude the chapter with some considerations about turbulent motions and the possible existence of large-scale currents in the convective zones of a rotating star. Circulatory currents in rotating degenerate stars will be discussed in Section 13.3.

8.2. CIRCULATION IN RADIATIVE ZONES

Let us first confine our attention to the radiative zones of an axially symmetric star for which the viscous and magnetic forces are negligible compared to the centrifugal force arising from the rotational motion of the star. As a further simplifying restriction, we shall assume a perfect gas and neglect the radiation pressure. Then, in an inertial frame of reference, the general equations of the problem reduce to

$$\frac{\partial \rho}{\partial t} + \operatorname{div}(\rho \mathbf{v}) = 0, \tag{1}$$

$$\frac{\partial \mathbf{v}}{\partial t} + \mathbf{v}.\operatorname{grad} \mathbf{v} = -\frac{1}{\rho}\operatorname{grad} p - \operatorname{grad} V, \tag{2}$$

$$\nabla^2 V = 4\pi G\rho, \tag{3}$$

$$\rho c_v\left(\frac{\partial T}{\partial t} + \mathbf{v}.\operatorname{grad} T\right) + p \operatorname{div} \mathbf{v} = \rho \epsilon_{Nuc} - \operatorname{div} \mathscr{F}, \tag{4}$$

$$p = \frac{\mathscr{R}}{\mu}\rho T, \tag{5}$$

where, except in the outermost surface regions, the radiative flux vector is given by

$$\mathscr{F} = -\frac{4acT^3}{3\kappa\rho}\operatorname{grad} T. \tag{6}$$

Remaining symbols have their usual meanings (cf. §§3.3–3.4).

Since we restrict ourselves to axisymmetric motions, it is convenient to use cylindrical coordinates (ϖ,φ,z). The velocity $\mathbf{v}$ may then be written as

$$\mathbf{v} = \Omega\varpi\mathbf{1}_\varphi + \mathbf{u}, \tag{7}$$

where $\Omega(\varpi,z,t)$ is the angular velocity of rotation, $\mathbf{u}$ is the meridional circulation velocity, and $\mathbf{1}_\varphi$ is the unit vector in the azimuthal direction. The φ-component of the equations of motion becomes, therefore,

$$\frac{\partial}{\partial t}(\Omega\varpi^2) + \mathbf{u}.\operatorname{grad}(\Omega\varpi^2) = 0, \tag{8}$$

expressing the conservation of the angular momentum per unit mass $\Omega\varpi^2$ as we follow the motion of a fluid particle. The remaining components of equation (2) may also be recast in the form

$$\frac{\partial \mathbf{u}}{\partial t} + \mathbf{u}.\operatorname{grad} \mathbf{u} = -\frac{1}{\rho}\operatorname{grad} p - \operatorname{grad} V + \Omega^2\varpi\mathbf{1}_\varpi, \tag{9}$$

where $\mathbf{1}_\varpi$ is the unit vector in the ϖ-direction. Finally, because we have assumed axial symmetry, we may also replace $\mathbf{v}$ by $\mathbf{u}$ in equations (1) and (4).

The foregoing equations provide seven relations among the seven unknown functions Ω, $\mathbf{u}$, p, ρ, T, and V. Thus, in principle, the structure of a rotating star with meridional circulation is entirely determined by these equations, together with the usual initial and boundary conditions. However, since a detailed resolution of this problem presents almost insurmountable difficulties, the best one can do at the present time is to make some further simplifying assumptions. Up to now, particular attention has been paid to steady-state solutions ($\partial/\partial t = 0$) for which the meridional circulation speed $|\mathbf{u}|$ is everywhere small compared to the rotational velocity $|\Omega\varpi|$.[1] Under these circumstances, equations (1) and (4) become

$$\text{div}(\rho\mathbf{u}) = 0, \tag{10}$$

$$\rho c_v \mathbf{u}.\text{grad } T + p \text{ div } \mathbf{u} = \rho\epsilon_{Nuc} - \text{div } \mathscr{F}. \tag{11}$$

Similarly, equations (8) and (9) reduce to

$$\mathbf{u}.\text{grad}\,(\Omega\varpi^2) = 0, \tag{12}$$

$$\frac{1}{\rho}\text{ grad } p + \text{grad } V = \Omega^2\varpi\mathbf{1}_\varpi. \tag{13}$$

Equations (3), (5), and (6) remain unaffected. The above system of differential equations is usually solved as follows. One first constructs a model satisfying equation (13) *with some prescribed angular velocity distribution*, and one thence computes the circulation velocity from equations (10) and (11). Unfortunately, these velocity fields are not consistent from the strict point of view of stellar hydrodynamics, for they must also comply with equation (12), which has been hitherto ignored. As was first pointed out by Randers, this equation implies that the steady meridional flow in an inviscid, nonmagnetic star must necessarily follow the surfaces of constant angular momentum per unit mass. Because a steady circulation pattern has closed streamlines, this requirement is an impossible one in a dynamically stable star (cf. §8.3: assumption [iv]). Hence, every "steady-state" solution derived on the basis of the foregoing approximate method must inevitably distort the prescribed rotational field in a time scale $t_c \approx \mathsf{R}/u_c$, where R is the mean radius of the star and u_c is a suitable mean speed of the flow (cf. equation [49]). In other words, the meridional cir-

---

[1] Steady-state models for which the angular velocity adjusts itself so as to prevent meridional currents (i.e., $\mathbf{u} \equiv 0$) have also been discussed in the literature. In this case, equations (4) and (9) reduce to the well-known conditions for radiative and mechanical equilibrium, respectively. As was first shown by Schwarzschild, solutions satisfying both conditions at every point do exist, with Ω depending on both ϖ and z. (Solutions with $\partial\Omega/\partial z = 0$ are excluded by the von Zeipel paradox.) However, as we recall from Section 7.4, all these models are thermally unstable. That is to say, they always generate a circulation that tends to restore the constancy of temperature over the isobaric surfaces. This fact makes these "circulation free" models mainly of academic interest (cf. §7.5).

culation generated by the nonsphericity of a rotating star will in general react back on the driving perturbation; accordingly, a continuous inter-action between the circulation and the driving force will necessarily ensue. In spite of the fact that this interaction is not taken into account, the steady-state approach has been almost exclusively used to describe meridional motions in a star. A critical survey of the main results provided by this method is thus in order. Possible modifications to the theory and the role of a $\bar{\mu}$-gradient are discussed in Section 8.3.

First of all, a general property can be derived at once from equations (10), (11), and (13), when the prescribed angular velocity does not depend on z, i.e., for all pseudo-barotropic models. As we recall from Section 4.3, equation (13) then becomes

$$\frac{1}{\rho} \operatorname{grad} p = -\operatorname{grad} \Phi, \tag{14}$$

where

$$\Phi = V - \int^{\varpi} \Omega(\varpi')\varpi'^2 \, d\varpi'. \tag{15}$$

By making use next of equations (10) and (14), we may also rewrite equation (11) in the form

$$\operatorname{div}\left[\rho \mathbf{u}\left(c_v T + \frac{p}{\rho} + \Phi\right)\right] = \rho \epsilon_{Nuc} - \operatorname{div} \mathscr{F}. \tag{16}$$

Consider now a chemically homogeneous star. Then, as we know, we have $p = p(\Phi)$, $\rho = \rho(\Phi)$, and $T = T(\Phi)$. Integrating the function $\rho \mathbf{u}(c_v T + \Phi + p/\rho)$ over a level surface and making use of equation (10), we generally find

$$\oint_{\mathscr{S}_\Phi} \rho \mathbf{u}\left(c_v T + \frac{p}{\rho} + \Phi\right).d\mathbf{S} = \left(c_v T + \frac{p}{\rho} + \Phi\right)\oint_{\mathscr{S}_\Phi} \rho \mathbf{u}.d\mathbf{S} = 0, \tag{17}$$

where $\mathscr{S}_\Phi$ designates the level surface $\Phi = constant$. (The second equality in equation [17] merely expresses the fact that the total flux of matter across a level surface must vanish.) By virtue of equation (17), we may now rewrite equation (16) in integral form, so that the total luminosity L_Φ across a level surface is given by

$$L_\Phi = \int_{\mathscr{V}_\Phi} \rho \epsilon_{Nuc} \, d\mathbf{x} = \oint_{\mathscr{S}_\Phi} \mathscr{F}.d\mathbf{S}, \tag{18}$$

where $\mathscr{V}_\Phi$ is the total volume enclosed within the surface $\Phi = constant$. This equation can be thought of saying that, though radiative equilibrium does not hold at every point, it does hold *on average*, i.e., averaged over a level surface. *Thus, within the framework of our approximations, the meridional circulation in a pseudo-barotropic star transports no net energy over a level surface.* This result (due to Roxburgh, Griffith, and Sweet) justifies the use of equation (68) of Section 5.4.

Various perturbation techniques for calculating the circulation velocity in slightly distorted stars have been devised. Since all these calculations are very similar in their principles, we shall discuss the problem by means of a simple example. (The present theory is originally due to Sweet, and it easily extends to more general rotation laws.) Consider a chemically homogeneous star that we compel to rotate as a rigid body. Suppose next that departure from sphericity is small. It is thus convenient to use spherical coordinates (r,θ,φ). The condition of mechanical equilibrium (13) then becomes

$$\frac{\partial p}{\partial r} = -\rho \frac{\partial V}{\partial r} + \rho r \Omega^2 \sin^2 \theta, \tag{19}$$

and

$$\frac{\partial p}{\partial \theta} = -\rho \frac{\partial V}{\partial \theta} + \rho r^2 \Omega^2 \sin \theta \cos \theta. \tag{20}$$

Given our assumptions, we may now expand the various quantities as series in the perturbing parameter

$$\lambda = \frac{\Omega^2}{4\pi G \rho_c}, \tag{21}$$

where ρ_c is the central density in the corresponding unperturbed spherical model. For example, we shall let

$$V(r,\theta) = 4\pi G \rho_c k^2 [V_0(r) + \lambda V_1(r,\theta) + \lambda^2 V_2(r,\theta) + \cdots], \tag{22}$$

$$\rho(r,\theta) = \rho_c [\rho_0(r) + \lambda \rho_1(r,\theta) + \lambda^2 \rho_2(r,\theta) + \cdots], \tag{23}$$

$$T(r,\theta) = T_c [T_0(r) + \lambda T_1(r,\theta) + \lambda^2 T_2(r,\theta) + \cdots], \tag{24}$$

where

$$k^2 = 5 \mathscr{R} T_c / 8\pi \bar{\mu} G \rho_c, \tag{25}$$

and T_c is the central temperature in the unperturbed configuration. For convenience, the radial variable r is measured in units of k. The quantities with subscript "0" correspond to the spherically symmetric model. Since the procedure is nothing but an extension of the Chandrasekhar-Milne approach (cf. §5.3), only those equations required to describe the method of computation of the circulation velocity will be considered. Also, for the sake of shortness, we shall explicitly derive the first-order terms only. As we shall see, the second-order contributions are essential to the subsequent discussion; they can be obtained by following the very same procedure.

On eliminating the pressure from equations (19) and (20), we obtain, correct to the first order in λ,

$$\frac{\partial \rho_1}{\partial \theta} \frac{dV_0}{dr} = \frac{d\rho_0}{dr} \frac{\partial V_1}{\partial \theta} - r^2 \frac{d\rho_0}{dr} \sin \theta \cos \theta. \tag{26}$$

Proceeding along the same lines, we may now eliminate $\partial \rho_1/\partial \theta$ between equations (3) and (26) to obtain a second-order equation in the function $\partial V_1/\partial \theta$. If we now write

$$\chi_1(r,\theta) = \frac{\partial V_1}{\partial \theta} = r^2 f(r) \sin \theta \cos \theta, \tag{27}$$

it is a simple matter to prove that the radial function $f(r)$ satisfies the nonhomogeneous equation

$$\frac{d^2 f}{dr^2} + \frac{6}{r} \frac{df}{dr} - \frac{1}{g_0} \frac{d\rho_0}{dr} (f - 1) = 0, \tag{28}$$

where $g_0 = dV_0/dr$ is the gravitational acceleration in the (known) unperturbed configuration. Physically acceptable solutions of equation (28) must be finite at $r = 0$, and converge to zero as $r \to \infty$. Now, once the function $f(r)$ has been obtained for a given spherical model, the first-order terms $\partial \rho_1/\partial \theta$ and $\partial T_1/\partial \theta$ become also known (cf. equations [5], [20], and [26]).

The next step is to find the velocity components u_r and u_θ in terms of these known functions. For this purpose, we let

$$u_r(r,\theta) = \lambda u_{1r}(r,\theta) + \lambda^2 u_{2r}(r,\theta) + \cdots, \tag{29}$$

and a similar expression for $u_\theta(r,\theta)$; we also write

$$\mathscr{F}(r,\theta) = \mathscr{F}_0(r) + \lambda \mathscr{F}_1(r,\theta) + \lambda^2 \mathscr{F}_2(r,\theta) + \cdots \tag{30}$$

Let us now turn to equation (11) which may be recast in the form

$$\rho T \mathbf{u}.\text{grad } S = \rho \epsilon_{Nuc} - \text{div } \mathscr{F}, \tag{31}$$

where $S = c_v \log(p/\rho^\gamma) + constant$ is the entropy per unit mass. Thus, outside the central regions where nuclear burning takes place, we obtain, correct to the first order in λ,

$$\rho_0 T_0 \frac{dS_0}{dr} u_{1r} = -\text{div } \mathscr{F}_1, \tag{32}$$

where S_0 is the entropy per unit mass in the spherical model. Equation (32) implies that the rotation causes at each point a net influx of energy of amount $(-\text{div } \mathscr{F}_1)$, and that the rate of circulation must adjust itself automatically so that this amount of energy is convected away. By differentiating equations (32) and (6) with respect to θ, we can thus write

$$\frac{\partial u_{1r}}{\partial \theta} = Q(r,\theta), \tag{33}$$

where $Q(r,\theta)$ is presently a known function. Hence, we have

$$u_{1r}(r,\theta) = \int_0^\theta Q(r,\theta')\, d\theta' + Q_1(r), \tag{34}$$

where $Q_1(r)$ is a function that can be determined from the condition that the total flux of matter over any spherical surface $r = constant$ must vanish. It then follows that

$$u_{1r}(r,\theta) = \int_0^\theta Q(r,\theta')\, d\theta' - \frac{1}{2} \int_0^\pi \sin\theta_1 \int_0^{\theta_1} Q(r,\theta)\, d\theta\, d\theta_1. \tag{35}$$

The second-order term $u_{2r}(r,\theta)$ can be obtained by using a similar approach. Following Maheswaran, we eventually find

$$u_r(r,\theta) = \frac{LR^2}{GM^2} \left\{ \lambda\alpha_0(r)P_2(\theta) + \lambda^2[\alpha_2(r)P_2(\theta) + \alpha_4(r)P_4(\theta)] + \cdots \right\}, \tag{36}$$

where the $P_n(\theta)$'s are the Legendre polynomials, and the $\alpha_n(r)$'s are some functions which depend on the chosen spherical model. The constants L, M, and R are, respectively, the total luminosity, the total mass, and the zero-order radius of the star. The θ-component of the circulation velocity is given at once by the equation of continuity.

To be specific, we shall now determine the rotationally driven circulation in a Cowling point-source model with constant mean molecular weight. The model consists of a convective core and a radiative envelope, with all the energy generation in the core (i.e., $\epsilon_{Nuc} = 0$ in the radiative zone). First-order circulation velocities were originally calculated by Sweet for a Cowling model in slow uniform rotation. However, as we know from the pioneering work of Öpik, this first-order calculation is inadequate for a uniformly rotating star, and the second-order "corrections" must necessarily be taken into account. Up to this order, the radial component of the velocity $\mathbf{u}$ is

$$u_r(r,\theta) = \frac{\bar{L}R^2}{\bar{M}^2}\, Y(r,\theta), \tag{37}$$

where barred quantities are expressed in solar units. The function $Y(r,\theta)$ has the form

$$Y(r,\theta) = \left\{ \lambda\beta_0(r)P_2(\theta) + \lambda^2 \frac{\bar{\rho}}{\rho_0}[\beta_2(r)P_2(\theta) + \beta_4(r)P_4(\theta)] \right\} \bigg/ \left(n - \frac{3}{2}\right), \tag{38}$$

where the $\beta_n(r)$'s are regular functions of r. As usual, n is the effective polytropic index,

$$n = d(\log \rho_0)/d(\log T_0), \tag{39}$$

and $\bar{\rho}$ designates the mean density of the configuration. For the Cowling model, the ratio α of the centrifugal force to the gravitational attraction at the equator is given by

$$\alpha = \frac{\Omega^2 R^3}{GM} = 1.11 \cdot 10^2 \lambda. \tag{40}$$

Table 8.1 gives the values of the function $Y(r,\theta)$ for the case $\alpha = 0.1$. These numerical results exhibit two important features: (i) the circulation pattern breaks up into two regions with motions in opposite senses, and (ii) the magnitude of the circulation velocity over the bulk of the radiative zone is

TABLE 8.1

Values of $Y(r,\theta)$, in cm/sec, for a uniformly rotating
Cowling model with $\alpha = 0.1$[†]

r/k	$\theta = 0°$	$\theta = 30°$	$\theta = 60°$	$\theta = 90°$
1.20	$+1.60(-5)$	$+1.00(-5)$	$-2.00(-6)$	$-8.00(-6)$
1.60	$+2.16(-6)$	$+1.35(-6)$	$-2.70(-7)$	$-1.08(-6)$
2.00	$+1.10(-6)$	$+6.87(-7)$	$-1.38(-7)$	$-5.50(-7)$
2.40	$+9.52(-7)$	$+5.95(-7)$	$-1.19(-7)$	$-4.76(-7)$
2.80	$+1.10(-6)$	$+6.89(-7)$	$-1.38(-7)$	$-5.52(-7)$
3.20	$+1.44(-6)$	$+8.98(-7)$	$-1.80(-7)$	$-7.18(-7)$
3.60	$+2.06(-6)$	$+1.29(-6)$	$-2.57(-7)$	$-1.04(-6)$
4.00	$+2.97(-6)$	$+1.83(-6)$	$-3.77(-7)$	$-1.45(-6)$
4.40	$+4.46(-6)$	$+2.76(-6)$	$-5.65(-7)$	$-2.19(-6)$
4.80	$+5.88(-6)$	$+3.35(-6)$	$-8.24(-7)$	$-2.46(-6)$
5.00	$+8.41(-6)$	$+5.34(-6)$	$-1.03(-6)$	$-4.32(-6)$
5.40	$+1.31(-5)$	$+8.67(-6)$	$-1.52(-6)$	$-7.23(-6)$
5.45	$+7.07(-6)$	$+2.37(-6)$	$-1.44(-6)$	$-5.61(-7)$
5.50	$-2.97(-6)$	$-7.90(-6)$	$-1.28(-6)$	$+1.03(-5)$
5.55	$-5.44(-6)$	$-1.06(-5)$	$+1.28(-6)$	$+1.32(-5)$
5.60	$-4.63(-6)$	$-1.00(-5)$	$+1.37(-6)$	$+1.27(-5)$
5.80	$-4.39(-5)$	$-5.04(-5)$	$+7.65(-7)$	$+5.53(-5)$
6.00	$-2.16(-4)$	$-2.24(-4)$	$+2.78(-6)$	$+2.37(-4)$
6.20	$-1.11(-3)$	$-1.12(-3)$	$+2.26(-5)$	$+1.18(-3)$
6.40	$-6.88(-3)$	$-7.90(-3)$	$+1.75(-4)$	$+8.27(-3)$
6.60	$-1.07(-1)$	$-1.07(-1)$	$+2.40(-3)$	$+1.12(-1)$
6.80	-7.06	-7.06	$+1.59(-1)$	$+7.39$
6.90	$-1.19(+2)$	$-1.19(+2)$	$+2.68$	$+1.24(+2)$

[†] The radiative zone extends from $r/k = 1.188$ to $r/k = 7.027$ (cf. eq. [25]). The number in parentheses following each entry is the power of ten by which that entry must be multiplied.

SOURCE: Maheswaran, M., *M. N.* **140**, 93, 1968.

generally small, except near the boundary of the convective core ($n = 3/2$) and near the surface of the radiative envelope ($\rho_0 = 0$). Let us examine these two results in succession.

The fact that the circulation pattern consists of two distinct zones separated by a particular level surface is not specific to the Cowling model, and it was independently predicted by Gratton and Öpik. The following analytical proof of this property is due to Mestel. Consider a chemically homogeneous star in uniform rotation. Then, by making use of equations (7)–(12) of Section 7.2, equation (31) reduces to

$$\rho A(\Phi)\mathbf{u}.\text{grad } \Phi = \rho\epsilon_{Nuc} - f(\Phi)(4\pi G\rho - 2\Omega^2) - f'(\Phi)g^2, \qquad (41)$$

where

$$A(\Phi) = c_v\left[\frac{dT}{d\Phi} - (\gamma - 1)\frac{T}{\rho}\frac{d\rho}{d\Phi}\right], \qquad (42)$$

$$f(\Phi) = -\frac{4ac}{3}\frac{T^3}{\kappa\rho}\frac{dT}{d\Phi}, \qquad (43)$$

and $g = d\Phi/dn$ is the magnitude of the effective gravity. (Remember that g varies over a level surface!) Dividing equation (41) by g, and integrating over a level surface, we find

$$[\rho\epsilon_{Nuc} - f(\Phi)(4\pi G\rho - 2\Omega^2)]\langle g^{-1}\rangle - f'(\Phi)\langle g\rangle = 0, \qquad (44)$$

since in a steady state there can be no flux of matter across a closed surface. For convenience, we have defined the mean values

$$\langle g\rangle = \frac{1}{\mathscr{A}_\Phi}\oint_{\mathscr{S}_\Phi} g \, dS \quad \text{and} \quad \langle g^{-1}\rangle = \frac{1}{\mathscr{A}_\Phi}\oint_{\mathscr{S}_\Phi} g^{-1} \, dS, \qquad (45)$$

where $\mathscr{A}_\Phi$ designates the area of the level surface $\mathscr{S}_\Phi$. Combining next equations (41) and (45), we obtain

$$\rho A(\Phi)\mathbf{u}.\text{grad } \Phi = f'(\Phi)\left(\frac{\langle g\rangle}{\langle g^{-1}\rangle} - g^2\right). \qquad (46)$$

In a convectively stable zone, motion against gravity requires an energy supply; the velocity component normal to a level surface is therefore upward-directed if the r.h.s. of equation (46) is positive. In particular, if $f'(\Phi)$ vanishes for a value Φ^*, the circulation does not cross the corresponding level surface. Consider now the realistic case of a star with $\epsilon_{Nuc} = 0$ in its envelope. Then, from equation (44), $f'(\Phi)$ vanishes on that level surface with density $\rho^*(\Phi^*)$ given by

$$\Omega^2 = 2\pi G\rho^*(\Phi^*). \qquad (47)$$

Furthermore, since the effective gravity on a particular level surface attains its maxima at the poles and decreases monotonically towards the equator, the parenthesis in equation (46) is negative at each pole and positive at the equator, with one zero in each hemisphere. Thus, for $\rho > \rho^*$, the meridional circulation is upward at the poles and downward at the equator. The opposite situation takes place in the outer regions where $\rho < \rho^*$, and along the surface $\Phi = \Phi^*$ the circulation is horizontal. Note, however, that Mestel's result is strictly true only for exactly rigid rotation (and for the implausible rotation law $\Omega^2 = c_1 + c_2/\varpi^2$, where c_1 and c_2 are constants). This fact was recently observed by Clement in studying the circulation patterns in differentially rotating stars for which the surface wholly contains the streamlines of the meridional flow. For all but the fastest rotators, Clement found that differential rotation eliminates the circulation reversal that exists in the presence of rigid rotation.

The foregoing results also permit a simple order-of-magnitude estimate of the circulation speed generated over the bulk of the radiative zone in a uniformly rotating star. From equations (37)–(40), we have

$$u_c \approx \frac{LR^2}{GM^2} \alpha. \tag{48}$$

Hence, for stars in general the time taken for a mass element to travel from the core to the surface is, approximately,

$$t_c \approx \frac{R}{u_c} = \frac{GM^2}{LR} \frac{1}{\alpha}. \tag{49}$$

That is to say, the characteristic time of meridional circulation (usually called the Eddington-Sweet time) is equal to the Kelvin-Helmholtz time divided by the ratio α of the perturbing force to gravity. Great care should be exercised when using equation (49), however, for this is a *global* time scale, therefore excluding the outermost surface layers where the term $\bar{\rho}/\rho_0$ becomes very large in the expression of $\mathbf{u}$ in the second-order term (cf. equation [38]). In these regions, for a uniformly rotating star, the appropriate *local* time scale is

$$t_c \approx \frac{GM^2}{LR} \frac{\rho_0}{\bar{\rho}} \frac{1}{\alpha^2}, \tag{50}$$

thus becoming arbitrarily small when a simple, zero density, boundary condition is used. As was shown by Baker and Kippenhahn, the $(\bar{\rho}/\rho_0)$-dependence in the velocity $\mathbf{u}$ also occurs for differentially rotating stars—the unwanted singularity then appearing already in the first-order terms. In any event, whether this $(\bar{\rho}/\rho_0)$-dependence arises in the first- or second-order terms is no object, for in all cases it invalidates one of our basic assumptions, namely, $|\mathbf{u}| \ll |\Omega\varpi|$ at every point of the configuration. Finally, note that a similar difficulty arises near the boundary of the convective core, where a suitable mean value of $(n - 3/2)$ must be incorporated into equation (49) in order to obtain the appropriate *local* time scale.[2]

8.3.   A CRITICAL REVIEW OF THE APPROXIMATIONS

At first glance, we could surmise these singularities to be due to some inadequacy in the expansions (22)–(24) and (29)–(30). That this is not the case has been fully demonstrated by Smith in 1966 by using the Roche potential approximation for a discussion of the rotationally driven circulation in the outer layers of a uniformly rotating star. Indeed, although he used no expansion technique, Smith found that the velocity **u** in this model is also singular at the surface. *It is therefore clear that this weakness must be due to the breakdown of one of the physical assumptions made.* These are:

(i) the radiative flux vector $\mathscr{F}$ is given by the simple diffusion equation (6);
(ii) inertial and viscous effects are negligible;
(iii) magnetic fields play no essential role;
(iv) there exist time-independent, inviscid solutions;
(v) the chemical composition is homogeneous.

It is to these questions that we now turn.

ASSUMPTION (i).   As we know, in stellar interiors the radiative flux vector $\mathscr{F}$ is given to a very good approximation by equation (6). In the surface layers of a star, however, the optical depth is of order unity or smaller, and the flux is no longer determined by this local condition. Instead, we must solve the nonlocal transfer equation normally used in stellar atmospheres. This fact was pointed out, independently, by Osaki and Smith. Detailed calculations by Smith show that the surface singularity does not appear if the transfer of radiation through the outer layers is properly treated. Nevertheless, for small optical depths, the velocities and their gradients are still extremely large, so large in fact that they invalidate the neglect of the inertial and viscous terms in the equations of motion. Finally, note that Smith's treatment remains unable to eliminate the singularity which occurs near the boundary of a deep-seated convective zone.

ASSUMPTION (ii).   Another way to get round the pitfalls of the classical treatment of meridional circulation in stars is to retain viscous dissipation.

---

[2] These local time scales might be largely in error, though, for the circulation patterns derived by the above method do *not* comply with the required boundary conditions, i.e., **n.u** = 0 (with $|\mathbf{u}|$ finite) at the boundaries of the radiative zone.

Such an approach should be most appropriate for discussing currents in the envelope of upper main-sequence stars, since their time scale for (radiative) viscous diffusion is shorter than the main-sequence lifetime of these stars (see Table 7.1). It is thus no longer necessary to prescribe an initial angular velocity distribution, and the transport of angular momentum by the meridional flow can now be adjusted steadily so as to balance the effects of friction. Hence, up to order $|\mathbf{v}|/c$, equation (12) must be generally replaced by

$$\rho \mathbf{u}.\text{grad}\,(\Omega \varpi^2) = \text{div}(\mu \varpi^2 \,\text{grad}\,\Omega), \tag{51}$$

where, for brevity, μ denotes the total coefficient of shear viscosity (cf. §7.5). Further, since the meridional velocity $\mathbf{u}$ and its gradients are expected to be very small in the main parts of a star, we may safely neglect the inertial and viscous terms in the ϖ- and z-components of the equations of motion as well as in the equation expressing conservation of thermal energy. In other words, we must search for well-behaved solutions of equations (3), (5), (6), (10), (11), and (13), *with the function $\Omega(\varpi,z)$ now satisfying equation* (51). In addition to the usual boundary conditions on the pressure and gravitational potential, the tangential viscous stresses across the outer surface $\mathscr{S}$ must vanish. Hence, we must also have

$$\mathbf{n}.\text{grad}\,\Omega = 0, \quad \text{on } \mathscr{S} \tag{52}$$

(cf. §7.5: equation [71]). Finally, because we are considering steady currents, the meridional streamlines must be closed. Thus, the boundary $\mathscr{S}$ must contain its own set of streamlines, and so must the axis of rotation, since no flow can cross it without disturbing the prescribed axial symmetry of the star. The foregoing formulation is originally due to Randers, although he did not investigate any specific model. A detailed discussion of steady meridional currents in a viscous star would be most welcome.[3]

ASSUMPTION (iii). A further difficulty of the classical approach to meridional flows lies in the fact that a rotationally driven circulation inexorably convects angular momentum, thus obliterating the prescribed angular velocity in a global time scale given by equation (49). For the Sun, t_c is much larger than its main-sequence lifetime, and the rotation law will only be slightly disturbed by meridional currents. The situation is quite the

---

[3] Following Mestel, viscosity is small enough to have a negligible effect on the circulation in the main bulk of a slowly rotating star; it is only when the velocity gradients become large that the viscous stresses play a decisive role in preventing the velocity $\mathbf{u}$ from becoming infinite. Since in all cases the layers in which viscous friction acts are presumably very narrow, Mestel attempted to describe the flow in these singular regions by means of a boundary-layer technique. This would be an alternative—and perhaps simpler—approach to the problem, but a satisfactory discussion along these lines is still lacking.

opposite for a typical upper main-sequence star, however, and any initial angular velocity distribution will be rapidly destroyed by the circulation. To partially alleviate this inconsistency of the theory, it has often been assumed that there is a weak poloidal magnetic field that maintains the prescribed rotation law but does not appreciably contribute to the toroidal forces. As a matter of fact, a rigorous approach to this question requires a detailed study of the interaction between rotation, magnetic field, and meridional circulation. Indeed, the convection of angular momentum and magnetic flux by these currents will in general distort the rotational and magnetic fields, and so the circulation itself. Following Chandrasekhar and Mestel, we shall merely assume here that a steady state has been reached, and study those rotations and magnetic fields that are unaffected by the circulation they generate. Such an approach is valid only for stars with circulation time scales short compared to their evolutionary time scales. For the sake of simplicity, viscous and Ohmic dissipations will be neglected.

By virtue of our assumptions, the velocity $\mathbf{v}$ and the magnetic field $\mathbf{H}$ must satisfy the condition

$$\mathrm{curl}(\mathbf{v} \times \mathbf{H}) = \mathrm{curl}[(\mathbf{v}_p + \mathbf{v}_\varphi) \times (\mathbf{H}_p + \mathbf{H}_\varphi)] = 0, \tag{53}$$

where, for reasons of symmetry, we let $\mathbf{v}_p = \mathbf{u}$ and $\mathbf{v}_\varphi = \Omega\varpi\mathbf{1}_\varphi$ (cf. equation [7]); the subscripts "p" and "φ" refer, respectively, to poloidal and toroidal components, i.e., meridional and azimuthal components (cf. §3.6: equations [118]–[120]). If we now separate equation (53) into these two components, we find

$$\mathrm{curl}(\mathbf{v}_p \times \mathbf{H}_p) = 0, \tag{54}$$

and

$$\mathrm{curl}(\mathbf{v}_\varphi \times \mathbf{H}_p + \mathbf{v}_p \times \mathbf{H}_\varphi) = 0. \tag{55}$$

As the system is axially symmetric, the toroidal vector $\mathbf{v}_p \times \mathbf{H}_p$ cannot be the gradient of a single-valued potential; hence, from equation (54), we have

$$\mathbf{v}_p = \kappa_0 \mathbf{H}_p, \tag{56}$$

where κ_0 is a scalar. Similarly, on combining equations (55) and (56), and the condition that the magnetic field is solenoidal, we obtain

$$\mathbf{H}_p \cdot \mathrm{grad}\left(\Omega - \kappa_0 \frac{H_\varphi}{\varpi}\right) = 0, \tag{57}$$

where $H_\varphi = |\mathbf{H}_\varphi|$; we can thus write

$$\Omega - \kappa_0 \frac{H_\varphi}{\varpi} = \alpha_0, \tag{58}$$

where α_0 is constant along a poloidal field line. Equations (56) and (58) imply that the steady motion consists of an arbitrary uniform rotation of each poloidal field line, superimposed on a velocity field $\kappa_0 \mathbf{H}$ which is parallel to the total magnetic field $\mathbf{H}$. Note that if equation (56) is to be satisfied with $\mathbf{v}_p$ non-zero, the field lines of $\mathbf{H}_p$ (which coincide with the streamlines of the circulation) must lie within the star.

We now have

$$\mathrm{div}(\rho \mathbf{v}_p) = \mathrm{div}(\rho \kappa_0 \mathbf{H}_p) = \mathbf{H}_p \cdot \mathrm{grad}(\rho \kappa_0) = 0, \tag{59}$$

or

$$\rho \kappa_0 = \rho \frac{|\mathbf{v}_p|}{|\mathbf{H}_p|} = \rho \frac{v_p}{H_p} = \eta_0, \tag{60}$$

η_0 being constant along a poloidal field line. Let us next consider the φ-component of the equations of motion. By making use of equations (121)–(123) of Section 3.6, we find

$$\rho \mathbf{v}_p \cdot \mathrm{grad}(\Omega \varpi^2) = \varpi \left[\frac{\mathrm{curl}\,\mathbf{H} \times \mathbf{H}}{4\pi} \right]_\varphi = \varpi \mathbf{J}_p \times \mathbf{H}_p. \tag{61}$$

With the help of equations (56) and (58), we may also rewrite this equation in the form

$$\mathbf{H}_p \cdot \mathrm{grad}\left(\varpi \frac{H_\varphi}{4\pi} - \rho \kappa_0 \Omega \varpi^2 \right) = 0; \tag{62}$$

it thus follows that

$$\varpi H_\varphi - 4\pi \rho \kappa_0 \Omega \varpi^2 = \beta_0, \tag{63}$$

where again β_0 is constant along a poloidal field line. The above equation merely says that the transport of angular momentum along a poloidal field line by the circulation is balanced by the transport of angular momentum by the magnetic stresses.

Combining next equations (58), (60), and (63), we find

$$\Omega = \frac{\alpha_0 + \eta_0 \beta_0 / \rho \varpi^2}{1 - 4\pi \eta_0^2 / \rho}, \tag{64}$$

and

$$\varpi H_\varphi = \frac{\beta_0 + 4\pi \eta_0 \alpha_0 \varpi^2}{1 - 4\pi \eta_0^2 / \rho}. \tag{65}$$

In the limit of zero circulation ($\eta_0 \equiv 0$), equation (64) reduces to Ferraro's law of isorotation (cf. §7.5). Also, when $\eta_0 \equiv 0$ and $\beta_0 \neq 0$, equation (65) becomes $\varpi H_\varphi = \beta_0$ and expresses that the poloidal current $\mathbf{J}_p$ maintaining the field H_φ must flow parallel to the vector $\mathbf{H}_p$, so that the magnetic torque $\mathbf{J}_p \times \mathbf{H}_p$ vanishes (cf. equation [61]). With η_0 finite, we have

$$\frac{4\pi\eta_0^2}{\rho} = \frac{4\pi\rho v_p^2}{H_p^2} = \frac{v_p^2}{v_A^2}, \tag{66}$$

where v_A is the local speed of the Alfvén waves along the poloidal field lines.[4] Thus, nonsingular steady-state solutions are possible only if the ratio (66) is greater or less than unity all the way round each poloidal field line. The quantities α_0 and β_0 are free, except that they must be constant along the poloidal field lines.

In principle, self-consistent solutions can be obtained by introducing a stream function $\Psi(\varpi, z)$ for the undetermined circulation, and prescribing the functions $\alpha_0(\Psi)$, $\beta_0(\Psi)$, and $\eta_0(\Psi)$—thus ensuring that they are constant along the field lines of $\mathbf{H}_p$. Application of the perturbation technique given in Section 8.2 leads to a very complicated integro-differential equation for the function Ψ. Self-consistent solutions have been derived by Roxburgh in two extreme cases. In the first one, the energy density of the rotational field is much greater than that of the magnetic field, and nearly uniform rotation is maintained by a weak poloidal field; the second solution has the toroidal component of the magnetic field as the dominant distorting agent. Of these, only the first one is a plausible model for most stars. The possible existence of self-consistent, steady-state structures with slightly non-uniform rotations has been also shown by Maheswaran.

Greater interest is attached to the recent papers of Moss, which describe self-consistent meridional flow patterns in uniformly rotating magnetic stars. As we know, a rotating star with a large-scale magnetic field needs, in general, meridional currents to balance its local energy budget. Because the self-consistent calculation of rotating models containing a magnetic field together with circulation is a difficult one, most studies have so far concentrated on the special case where the circulation produced by a magnetic field is such as to *exactly* cancel that produced by uniform rotation (cf. §15.4). If we let $\alpha = \Omega^2 R^3/GM$ and $\alpha_H = 4\pi\bar{H}^2 R^4/GM^2$, where $\bar{H}$ is a measure of the surface magnetic field, the salient results of these calculations may be summarized as follows: (i) as α/α_H increases, the ratio of interior to surface field strength becomes very large, and (ii) for a given rotation rate and field topology, there is a minimum value of the total internal magnetic flux for which solutions can be found. As was pointed out by Moss, however, there is no obvious reason to expect rotating stars with large-scale magnetic fields to select a zero-circulation configuration, in which the two circulatory motions are exactly equal and opposite. Therefore, Moss has constructed models of uniformly rotating stars with

---

[4] See, e.g., reference 27, pp. 155–157, of Chapter 3.

steady-state meridional currents and large-scale poloidal magnetic fields. As usual, a first-order perturbation technique is used and the velocity- and magnetic-fields are expanded in series of Legendre polynomials truncated after the term in $P_2(\cos \theta)$ or $P_4 (\cos \theta)$, where θ is the colatitude. To allow the field lines to leak out of the surface, *finite* electrical conductivity is also assumed. According to Moss, his models seem to fall into two distinct groups. In the first group, the solutions are essentially small perturbations to the solutions found with zero circulation (see Figures 8.1 and 8.2). In the second group, models are found with relatively weak magnetic fields in which, as the ratio of the centrifugal to magnetic perturbation becomes large, the circulation deep into the star is the unimpeded rotationally driven circulation, whilst near the surface the magnetic forces become

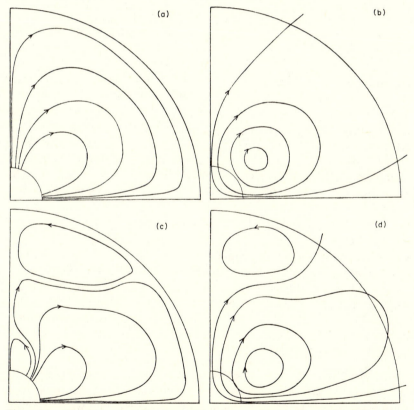

Figure 8.1.   Circulation in a Cowling point-source model. (a): P_2-velocity field ($\alpha/\alpha_H = 10^3$, $\alpha_H = 10^{-6}$); (b): P_2-magnetic field ($\alpha/\alpha_H = 10^3$, $\alpha_H = 10^{-6}$); (c): combined P_2- and P_4-velocity field ($\alpha/\alpha_H = 10^3$, $\alpha_H = 10^{-6}$); (d): combined P_2- and P_4-magnetic field ($\alpha/\alpha_H = 10^3$, $\alpha_H = 10^{-6}$). Source: Moss, D. L., *M.N.* **168**, 61, 1974.

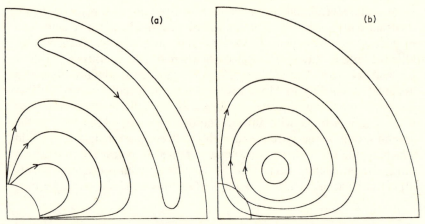

Figure 8.2.   Circulation in a Cowling point-source model. (a): P_2-velocity field ($\alpha/\alpha_H = 10^5$, $\alpha_H = 10^{-6}$); (b): P_2-magnetic field ($\alpha/\alpha_H = 10^5$, $\alpha_H = 10^{-6}$). A few field lines do penetrate the surface, although they are not shown. Source: Moss, D. L., *M.N.* **168**, 61, 1974.

important and the circulation velocity drops rapidly to zero. Because stars with purely poloidal fields might be dynamically unstable in the neighborhood of their neutral lines (cf. §15.4), Moss has also constructed models of uniformly rotating stars containing a poloidal field of even parity, together with a linked toroidal component and the self-consistent thermally-driven circulation. It is found that the range of internal fluxes available to steady-state models is somewhat increased by the inclusion of a toroidal magnetic field, but there remain values of the angular velocity and internal flux for which no solutions can be found. A similar difficulty was encountered by Mestel and Moss for uniformly rotating models with purely poloidal fields, having either basically dipolar or quadrupolar structure. In all cases, however, it is found that *meridional currents tend to concentrate poloidal magnetic fields in the deep interior of a rotating star*. The Mestel-Moss calculations thus offer a possible explanation of why the rapidly rotating A-type stars usually do not possess observable magnetic fields (cf. §15.3), and also of recent observations showing an anticorrelation between the rotation rate and the effective magnetic field among the peculiar A-type (Ap) stars (cf. §15.4).

ASSUMPTION (iv).   If we now neglect viscous friction and magnetic fields, is it possible to have steady meridional currents in a rotating star? *Strictly speaking, the answer is no.* The following two arguments were given by Randers in 1941. *First*, in view of equation (12), the currents must follow the surfaces $\Omega\varpi^2 = constant$. Hence, since a steady circulation has closed steamlines, $\Omega\varpi^2$ must necessarily have a maximum at some interior point.

This, in turn, implies that there exist regions in the star where $\Omega\varpi^2$ decreases outward; and, as we recall from Section 7.3, these regions are most likely to be dynamically unstable! *Second*, because of the assumed equatorial symmetry, the meridional currents cannot cross through the plane $z = 0$. Consequently, the equatorial radius in a meridional plane is a line of constant $\Omega\varpi^2$. Because this radius is linked to the axis of rotation, the constant value of $\Omega\varpi^2$ is, therefore, zero. Now, the continuation of this $\Omega\varpi^2$-line at the outer end is the surface itself, since the currents follow the boundary. We must, therefore, conclude that $\Omega\varpi^2 = 0$ along the stellar surface, which means that the outer boundary does not rotate at all!

Considering these two difficulties, we see that *steady meridional currents cannot exist in a rotationally distorted, nonmagnetic, inviscid star*. For these reasons, unsteady currents in the radiative zones of a rotating star have been discussed in a heuristic way by Osaki, taking into account the effect of the time evolution of the angular velocity (cf. equations [1]–[9]). Two equations play a crucial role in this discussion: equation (8), which now replaces equation (51) or (61), and equation (4), which we rewrite in the form

$$\rho T\left(\frac{\partial S}{\partial t} + \mathbf{u}.\text{grad } S\right) = \rho\epsilon_{Nuc} - \text{div }\mathscr{F}, \qquad (67)$$

where S is the entropy per unit mass (cf. §3.4: equation [65]). Osaki did not actually solve these time-dependent equations for a particular model, but he made a few speculative comments that we will now briefly summarize. *First*, in the standard formulation given in Section 8.2, the convection of energy by the meridional currents is equated to the excess or deficiency of radiative flux (i.e., $\partial S/\partial t \ll |\mathbf{u}.\text{grad } S|$). The validity of this approximation is confirmed for slowly rotating stars ($\alpha \ll 1$) in the main bulk of their radiative zones ($n > 3/2$, $\rho > 0$). *Second*, when the density stratification is nearly neutral ($n \approx 3/2$) and the rotation is moderately rapid, the energy equation (67) reduces to the equation of thermal conduction ($\partial S/\partial t \gg |\mathbf{u}.\text{grad } S|$). That is to say, the excess or deficiency of radiative flux mainly results in local heating or cooling. Also, in this case, the circulation speed u_c never becomes infinite (even when $n = 3/2$), its upper limit being set by the condition $u_c < \text{LR}^2/GM^2$. According to Osaki, the star will then relax to a state of pure thermal equilibrium ($\rho\epsilon_{Nuc} - \text{div }\mathscr{F} = 0$) within its Kelvin-Helmholtz time, redistributing its angular velocity so that there is no meridional circulation in the final state. *Third*, a similar order-of-magnitude argument shows that the surface layers ($\rho \approx 0$) also tend to relax to a state of thermal equilibrium (div $\mathscr{F} = 0$, $\mathbf{u} = 0$) within their thermal time scale. *Fourth*, because thermal imbalance necessarily destroys steady rotation laws of the form $\Omega = \Omega(\varpi,z)$ in a chemically homogeneous star (cf. §7.4), Osaki suggests that a new steady state is likely to be established in which a balance exists between the meridional currents and the irregular motions due to the thermal instability.

A more quantitative discussion of this approach requires the numerical integration of the time-dependent equations (1)–(6), starting from an initial state of rotation, and taking into account the changes in chemical composition through nuclear burning. In this respect, let us mention the work by Sakurai in which he discusses the time-evolution of the inner solar rotation caused by the meridional circulation. In particular, Sakurai was able to show that the singularity in the circulation velocity at the interface between radiative and convective regions can be removed by allowing for a time-dependent angular velocity.

ASSUMPTION (v).   The above results were derived under the explicit assumption of no variations in the chemical composition or the mean molecular weight $\bar{\mu}$. In other words, we have hitherto ignored the possibility of mixing the interior with the outer layers through the agency of rotationally induced meridional currents. From the estimate of the time scale given by equation (49), Sweet pointed out that these currents are far too slow to have any appreciable effects in the radiative zones of late-type stars (such as the Sun), so that rotational mixing can certainly be ignored for these stars. On the basis of the same argument, Sweet was also led to conclude that upper main-sequence stars would be well mixed, since for homogeneous stars the circulation is fast enough to leave and return to the convective core many times during the star's lifetime. As was demonstrated by Mestel, however, *this conclusion needs considerable revision*. This may be seen as follows. Material is transported out of the energy-producing core at different epochs. Hence, the mean molecular weight $\bar{\mu}(\mathbf{x},t)$, at time t and location $\mathbf{x}$ in the radiative zone, is equal to the value of $\bar{\mu}$ in the core at an earlier instant $t-\bar{t}$, where $\bar{t}$ is the time of flow from the core to the point $\mathbf{x}$. Now, because $\bar{t}$ is a function of position, a distribution in mean molecular weight which is itself nonspherical will be set up. This, in turn, will cause "$\bar{\mu}$-currents" in a fashion similar to that in which the rotational disturbance causes the original meridional circulation. These $\bar{\mu}$-currents will strongly perturb the temperature field, and so they will react back on the current velocities. We are thus forced to study these $\bar{\mu}$-currents in order to decide whether they will oppose or enhance the meridional circulation.

This problem has been originally discussed by Mestel, who, for definiteness, considered the radiative envelope of a Cowling point-source model. By means of a linear analysis in the rotational- and $\bar{\mu}$-currents, he was then able to show that the $\bar{\mu}$-current velocities always oppose the circulation velocities near the surface of the convective core. This effect is to restrain the meridional circulation. Actually, rotational current velocities strong enough to overcome this restraint demand that the centrifugal force in a uniformly rotating model should balance gravity at the surface equator. It is concluded that, although rotational currents circulate in the envelope of a uniformly rotating model, they are deflected near the surface of the core, so that no mixing between the core and the envelope can take place. According to Mestel, however, if the angular velocity substantially increases inwards, the meridional circulation would then be faster, and the restraining effect of the $\bar{\mu}$-currents would no longer be sufficient to prevent mixing between the core and the rest of the star. Mestel maintains that the ratio α of the centrifugal and gravitational forces in the equator must be greater than the value $\alpha \approx 0.03$ at the core boundary for the circulation to overcome the restraining effect of the gradient in the chemical composition. In other words, if $\alpha \gtrsim 0.03$ at the core boundary, rotational mixing would keep the star effectively uniform in composition. By coupling this result with observed evolutionary tracks in the Hertzsprung-Russell diagram, we are thus led to believe that for most upper main-sequence stars a strong inward increase in angular velocity can be ruled out.

The role of these $\bar{\mu}$-currents and the ability of a small $\bar{\mu}$-gradient to quench rotationally induced currents has been further investigated by McDonald. For slowly rotating stars with radiative zones of approximate chemical homogeneity, it was found that a small $\bar{\mu}$-gradient can indeed inhibit the circulation. According to McDonald, a large differential rotation in the Sun might have escaped disruption by meridional currents, if a $\bar{\mu}$ stratification of a few parts in one thousand were to exist. A similar quantitative discussion for upper main-sequence stars in rapid rotation is still lacking.

8.4. TURBULENT MOTIONS IN STARS: A BRIEF SURVEY

We have hitherto implicitly assumed that all displacements within a star are characterized by smooth motions of one lamina of fluid past another. As we know from laboratory experiments, however, we may have another type of motion, i.e., a motion that appears to be irregular in nature with transverse eddies covering a wide range of sizes and speeds. This is known as turbulent flow, and it may be regarded as an unsteady fluctuating velocity field superimposed on a laminar mean-flow pattern. As a matter of fact, most flows occurring in nature are predominantly turbulent (e.g., the boundary layer of the Earth's atmosphere, water currents below the surface of the oceans, stellar atmospheres, interstellar gas clouds, . . .). As we shall see, simple theoretical arguments show that turbulent motions must also play an important role in stellar interiors. Since this last problem has been given very little attention so far, a brief survey of turbulence will therefore suffice for our purpose.

Laboratory experiments show that the transition from laminar to turbulent motions in an incompressible fluid only depends on the Reynolds number

$$\mathrm{Re} = \frac{\rho L U}{\mu}, \tag{68}$$

which is a measure of the relative magnitude of the inertial to the viscous forces occurring in the flow. Here U is a characteristic fluid velocity of the mean flow, L is a characteristic dimension for the problem on hand, and μ is the coefficient of shear viscosity. Turbulent flows always occur when the Reynolds number Re exceeds some critical value Re_c(say). This critical number is not a universal constant, but takes different values for each type of flow. (A laminar flow in a pipe normally becomes turbulent when $\mathrm{Re} > \mathrm{Re}_c \simeq 2000$.) This explains why the majority of fluid motions in systems with large dimensions and low viscosity are turbulent. Caution should be exercised when applying the above criterion to stars, however, for the physical conditions in stellar interiors bear no relation to those encountered in a laboratory. For instance, if turbulent motions prevail in the radiative zones of a rotating star, we may then expect the system to exhibit a layered turbulence, with irregular motions mainly along the surfaces of constant density. Although no precise criterion is known for a star, a great deal of insight can be gained from the study of air motions near the ground. In that case, as we know, a convectively stable stratification of the air can inhibit the turbulent motions of an horizontal wind at large Reynolds numbers. The effect of a stable entropy distribution then depends on the Richardson number

$$\mathrm{Ri} = \frac{g}{c_p} \frac{dS}{dx_3} \Big/ \left(\frac{d\bar{v}}{dx_3} \right)^2, \tag{69}$$

which can be identified as the ratio of buoyancy to inertial forces. Here the velocity profile $\bar{v} = \bar{v}(x_3)$ is transverse to the vertical direction x_3, and the positive entropy gradient dS/dx_3 depends on the elevation x_3 only. Measurements in the lower atmosphere at the Brookhaven National Laboratory (Upton, L.I., N.Y.) seem to indicate that no turbulence exists whenever $\mathrm{Ri} > 0.50$, but when $\mathrm{Ri} < 0.15$, the air is always turbulent.[5] (On the theoretical side, a necessary condition for instability of small disturbances in an inviscid shear flow is that $\mathrm{Ri} < 0.25$.) As was shown by

[5] See, e.g., Panofsky, H. A., "The Atmospheric Boundary Layer below 150 Meters" in *Annual Review of Fluid Mechanics* **6**, pp. 147–177, Palo Alto, California: Annual Reviews, Inc., 1974.

Townsend, however, radiative transfer can remove the stabilizing effects of stratification. The reason lies in the fact that radiative diffusion of thermal energy in smoothing temperature differences inhibits density fluctuations, so that the stabilizing buoyancy is reduced. A crude estimate of this effect in rotating stars has been made by Zahn, but no exact criterion is known as yet.

Turbulent motions are always dissipative; that is to say, they cannot maintain themselves, but depend on their surroundings to obtain energy. Turbulence in nature is the result *either* of the growth of small disturbances in a laminar flow *or* of the convective instability of the motion. In the former case, the kinetic energy of turbulence is drawn from the kinetic energy of the mean flow; in the latter case, it derives from the potential energy of the basic state. Both types of turbulence are relevant to stellar interiors. As far as the latter are concerned, the most important feature of turbulence is its ability to cause rapid mixing and increased rates of momentum, heat, and mass transfer. Actually, the rates of turbulent transport and mixing are several orders of magnitude greater than the rates due to molecular (and radiative) diffusion. However, because turbulence is not a feature of fluids but of fluid flows, the momentum exchange mechanism by turbulence only superficially resembles molecular exchange of momentum. Nevertheless, by analogy with molecular diffusion, which gives rise to the concept of viscosity, the turbulent momentum exchange is often represented in astrophysics by an *eddy viscosity*. This analogy will be explored briefly.

Since fully developed turbulence consists of random fluctuations superimposed on a laminar stream, it is therefore convenient to express all physical quantities into mean values and fluctuations with zero mean. Following Reynolds, we shall now derive the equations governing these mean quantities (such as $\bar{\rho}, \bar{p}, \overline{\rho\mathbf{v}}, \ldots$) at each point. By assuming that the mean flow is steady, we can thus write

$$\rho(\mathbf{x},t) = \bar{\rho}(\mathbf{x}) + \rho'(\mathbf{x},t), \tag{70}$$

and similar expressions for the other variables. Here we interpret $\bar{\rho}$ as the time average

$$\bar{\rho}(\mathbf{x}) = \lim_{\tau \to \infty} \frac{1}{\tau} \int_{t_0}^{t_0+\tau} \rho(\mathbf{x},t)\, dt, \tag{71}$$

which, in view of the steadiness of the mean flow, is independent of the instant t_0. By definition, the components of the mean velocity are given by

$$\rho v_k = \overline{\rho v_k}; \tag{72}$$

we thus have

$$\overline{\rho v_k'} = 0 \quad \text{and} \quad \overline{v_k'} \neq 0, \tag{73}$$

so that, on the average, there is no transfer of mass due to turbulence. Accordingly, equation (1) remains valid for the steady mean flow, and we can write

$$\frac{\partial}{\partial x_k}(\bar{\rho}\bar{v}_k) = 0. \tag{74}$$

Consider next the equations of motion for a viscous fluid (cf. §3.3: equations [46]–[47]). By making use of equation (1), we may also recast these equations in the form

$$\frac{\partial}{\partial t}(\rho v_i) + \frac{\partial}{\partial x_k}(\rho v_i v_k) = -\rho\frac{\partial V}{\partial x_i} - \frac{\partial p}{\partial x_i} + \frac{\partial \tau_{ik}}{\partial x_k}. \tag{75}$$

If we suppose the body force to be unaffected by turbulence, and average all terms in these equations, it then follows at once that

$$\frac{\partial}{\partial x_k}(\bar{\rho}\bar{v}_i\bar{v}_k) = -\bar{\rho}\frac{\partial V}{\partial x_i} - \frac{\partial \bar{p}}{\partial x_i} + \frac{\partial}{\partial x_k}(\bar{\tau}_{ik} + \bar{\sigma}_{ik}), \tag{76}$$

since the operations of averaging and differentiation commute. The tensor $\bar{\tau}$ is the average of the viscous stress tensor, and the tensor $\bar{\sigma}$ has the components

$$\bar{\sigma}_{ik} = -\overline{\rho v_i' v_k'}. \tag{77}$$

Equations (76) are identical to the (steady) Navier-Stokes equations with all quantities replaced by their mean values, except for the additional Reynolds stresses (77). This symmetrical tensor actually represents the mean transport of fluctuating momentum by turbulent velocity fluctuations. The term div $\bar{\sigma}$ thus exchanges momentum between the turbulence and the mean flow, even though the three components $\overline{\rho v_k'}$ of the mean momentum of the turbulent velocity fluctuations are zero. Whenever turbulent motions prevail in a star, the average viscous stresses $\bar{\tau}_{ik}$ are usually negligible compared to the Reynolds stresses $\bar{\sigma}_{ik}$.

The central problem in this representation of turbulent motions lies in the fact that the three scalar equations (76) introduce six unknown quantities, namely, the six independent components of the tensor $\bar{\sigma}_{ik}$. Following Boussinesq, it is usually accepted that the turbulent stresses $\bar{\sigma}_{ik}$ within a star should be of the same form as the viscous stresses $\bar{\tau}_{ik}(\bar{\mathbf{v}})$ in an incompressible fluid, with the coefficient μ being replaced by the coefficient of eddy viscosity

$$\mu_t \approx \bar{\rho}\bar{v}_t l, \tag{78}$$

where $\bar{v}_t$ is the r.m.s. speed of turbulence, and l is the so-called Prandtl mixing length. It is assumed, therefore, that discrete eddies can travel over a distance l while keeping their individuality, and then contribute with their average properties to the new location where they are finally absorbed. This heuristic theory of turbulence is the exact analog of the microscopic transfer of momentum between the constitutive particles, when a macroscopic velocity gradient prevails. (In kinetic theory, the coefficient of shear viscosity is of the order of $\mu \approx \rho v_{th} d$, where v_{th} is the mean thermal speed, and d is the mean free path of the particles.) However, even if we are willing to accept the concept of eddy viscosity, the situation is not quite that simple, for $\bar{l}$ and $\bar{v}_t$ are not mere properties of the fluid but properties of the flow. This implies that $\bar{l}$ and $\bar{v}_t$ may vary from one point to another, thus making the coefficient μ_t dependent on the position in the flow. Actually, although turbulence consists of fluctuating motions covering a broad spectrum of sizes, it has been often argued that large eddies contribute more to the momentum transfer than smaller ones. For this reason, $\bar{l}$ is usually taken to be proportional to the size of the larger eddies. In stellar interiors, the mixing length is generally taken to be of the order of the pressure scale height $H_p(=\bar{p}/|\text{grad } \bar{p}|)$. Furthermore, because turbulent motions in a thermally unstable region are generally anisotropic, it has also been suggested that the coefficient of eddy viscosity may be a tensor rather than a scalar (cf. equation [79]). The concept of anisotropic viscosity has been extensively studied by Lebedinski and Wasiutyński in the context of stellar interiors. Various implications of this phenomenological approach to turbulence in stars will be further discussed in Sections 8.5 and 9.2.

8.5.   CIRCULATION IN CONVECTIVE ZONES

Let us now briefly discuss the work of Biermann and Kippenhahn in which they give decisive evidence of a large-scale meridional circulation in the convective regions of a rotating star. For the sake of simplicity, we shall consider a slowly rotating, axially symmetric star that slightly departs from spherical symmetry. Magnetic fields are neglected altogether. Moreover, we shall restrict ourselves to those regions where turbulent convection is so efficient that it maintains the temperature lapse rate closely equal to its adiabatic value.[6] (This is the case in the hydrogen convective layer of the Sun, if a relatively thin region at the outer boundary of the shell is excluded.) However, even if the influence of the slow rotational motion on convection is ignored, the gravitational acceleration

---

[6] We neglect here the influence of rotation upon the onset of convection, i.e., we use the ordinary Schwarzschild criterion for a spherical star (cf. §§6.4 and 7.3). However, because rotation distinguishes a preferred direction in the system, its effects on convective motions must be anisotropic. The influence of rotation upon the onset of convection will be discussed more closely in Section 14.5 from the viewpoint of small oscillations in a rotating body.

determines the main direction of the buoyancy forces. That is to say, the turbulent motions in the convective regions of a star are generally anisotropic, with the radial direction being the preferred one. We thus choose the coefficient of eddy viscosity to be given by a tensor μ_{ik} which, in spherical coordinates (r,θ,φ), has the diagonal form:

$$\mu_{ik} = \mu_t \begin{pmatrix} 1 & 0 & 0 \\ 0 & s & 0 \\ 0 & 0 & s \end{pmatrix} = s\mu_t \begin{pmatrix} 1 & 0 & 0 \\ 0 & 1 & 0 \\ 0 & 0 & 1 \end{pmatrix} + (1-s)\mu_t \begin{pmatrix} 1 & 0 & 0 \\ 0 & 0 & 0 \\ 0 & 0 & 0 \end{pmatrix} \quad (79)$$

where $1 - s$ is a free parameter measuring the departure from isotropy in the turbulent flow. Of course, this can only be a very crude model, but it does allow for a difference in momentum transfer between the radial and lateral directions. As we shall now prove, the presence of this anisotropy in the turbulent velocity distribution necessarily implies the existence of large-scale currents in a convective zone.

Let us first assume that the mean velocity $\overline{\mathbf{v}}$ has the components

$$\overline{v}_r = \overline{v}_r(r,\theta), \quad \overline{v}_\theta = \overline{v}_\theta(r,\theta), \quad \overline{v}_\varphi(r,\theta) = \Omega r \sin\theta, \quad (80)$$

in an inertial frame of reference. By virtue of equation (74), mass conservation is certainly satisfied if we let

$$\overline{v}_r = \frac{1}{\overline{\rho} r^2 \sin\theta} \frac{\partial\Psi}{\partial\theta}, \quad \text{and} \quad \overline{v}_\theta = -\frac{1}{\overline{\rho} r \sin\theta} \frac{\partial\Psi}{\partial r}, \quad (81)$$

where we have introduced the stream function $\Psi(r,\theta)$. Neglecting the viscous stresses $\overline{\tau}$ and making use of equation (74), we may also recast the equations of motion for the mean steady flow in the form

$$\frac{1}{2} \operatorname{grad} |\overline{\mathbf{v}}|^2 - \overline{\mathbf{v}} \times \operatorname{curl} \overline{\mathbf{v}} = -\frac{1}{\rho} \operatorname{grad} \overline{p} - \operatorname{grad} V + \frac{1}{\rho} \operatorname{div} \overline{\boldsymbol{\sigma}} \quad (82)$$

(cf. §3.5: equation [83]). Following Lebedinski and Wasiutyński, we now *choose* the components of the Reynolds stresses $\overline{\boldsymbol{\sigma}}$ as follows:

$$\overline{\sigma}_r{}^r = 2s\mu_t \frac{\partial\overline{v}_r}{\partial r} + 2\mu_t(1-s) \frac{\partial\overline{v}_r}{\partial r}, \quad (83)$$

$$\overline{\sigma}_r{}^\theta = \frac{\overline{\sigma}_\theta{}^r}{r^2} = \frac{s\mu_t}{r^2}\left(\frac{\partial\overline{v}_r}{\partial\theta} + r\frac{\partial\overline{v}_\theta}{\partial r} - \overline{v}_\theta\right) + \frac{\mu_t(1-s)}{r^2}\frac{\partial(r\overline{v}_\theta)}{\partial r}, \quad (84)$$

$$\overline{\sigma}_r{}^\varphi = \frac{\overline{\sigma}_\varphi{}^r}{r^2\sin^2\theta} = s\mu_t\frac{\partial\Omega}{\partial r} + \frac{\mu_t(1-s)}{r^2}\frac{\partial(\Omega r^2)}{\partial r}, \quad (85)$$

$$\overline{\sigma}_\theta{}^\theta = \frac{2s\mu_t}{r}\left(\frac{\partial\overline{v}_\theta}{\partial\theta} + \overline{v}_r\right), \quad (86)$$

$$\bar{\sigma}_\theta{}^\varphi = \frac{\bar{\sigma}_\varphi{}^\theta}{\sin^2\theta} = s\mu_t \frac{\partial\Omega}{\partial\theta}, \tag{87}$$

$$\bar{\sigma}_\varphi{}^\varphi = \frac{2s\mu_t}{r}(\bar{v}_r + \cot\theta\,\bar{v}_\theta). \tag{88}$$

In these equations, the terms proportional to the factor $s\mu_t$ are the components of the Reynolds stresses due to the isotropic part of the viscosity, while the remaining terms are due to the superimposed radial viscosity which produces the anisotropy in the turbulent velocities (cf. equation [79]).

Because the viscosity in a convective zone is much greater than that of the surrounding radiative regions, we can now approximate the condition of continuity of the stress vector by imposing the free-surface condition at both boundaries, i.e.,

$$\bar{\sigma}_r{}^\theta = \bar{\sigma}_r{}^\varphi = 0, \quad \text{at} \quad r = r_i \quad \text{and} \quad r = r_e, \tag{89}$$

where $r_i(\geqslant 0)$ and $r_e(>0)$ designate, respectively, the (mean) inner and outer radii of the convective zone (cf. §3.3: equation [48]). Also, since we assume that the motion is enclosed entirely in the convective region, we must have

$$\frac{\partial\Psi}{\partial\theta} = 0, \quad \text{at} \quad r = r_i \quad \text{and} \quad r = r_e. \tag{90}$$

Finally, besides conditions (89) and (90), the solutions and their derivatives must be regular everywhere in this region.

Now, the φ-component of the equations of motion is

$$\bar{\rho}(\bar{\mathbf{v}} \times \operatorname{curl}\bar{\mathbf{v}})_\varphi + (\operatorname{div}\bar{\boldsymbol{\sigma}})_\varphi = 0, \tag{91}$$

while the azimuthal component of the curl of equation (82) gives

$$[\operatorname{curl}(\bar{\mathbf{v}} \times \operatorname{curl}\bar{\mathbf{v}})]_\varphi + \left[\operatorname{curl}\left(\frac{1}{\bar{\rho}}\operatorname{div}\bar{\boldsymbol{\sigma}}\right)\right]_\varphi = 0. \tag{92}$$

In establishing this last equation, we made use of the fact that the convective zone is an homentropic fluid (cf. §3.5); indeed, the surfaces $\bar{\rho} = constant$ and $\bar{p} = constant$ then coincide, and we can write

$$\operatorname{curl}\left(\frac{1}{\bar{\rho}}\operatorname{grad}\bar{p}\right) = -\frac{1}{\bar{\rho}^2}\operatorname{grad}\bar{\rho} \times \operatorname{grad}\bar{p} = 0. \tag{93}$$

Equations (91) and (92), together with the boundary conditions (89) and (90), determine the functions $\Omega(r,\theta)$ and $\Psi(r,\theta)$ in zones of efficient convection.

Let us now demonstrate that a pure state of permanent rotation (i.e.,
$\Psi \equiv 0$) *is no solution when the turbulent velocities are anisotropic* ($s \neq 1$).
To illustrate the problem, we shall give the proof only in the case when
μ_t is a constant throughout the region. With $\Psi \equiv 0$, equations (91) and
(92) reduce to

$$(\text{div } \bar{\boldsymbol{\sigma}})_\varphi = 0, \tag{94}$$

and

$$[\text{curl}(\bar{\mathbf{v}} \times \text{curl } \bar{\mathbf{v}})]_\varphi = 0. \tag{95}$$

In terms of equations (83)–(88), the first condition provides the following
constraint on the function $\Omega(r,\theta)$:

$$\frac{1}{r^2} \frac{\partial}{\partial r}\left[r^4 \frac{\partial \Omega}{\partial r} + 2(1 - s)r^3\Omega \right] + \frac{1}{\sin^3 \theta} \frac{\partial}{\partial \theta}\left(s \sin^3 \theta \frac{\partial \Omega}{\partial \theta} \right) = 0; \tag{96}$$

this condition is originally due to Lebedinski. Similarly, equation (95)
can be integrated at once to give the second constraint

$$\Omega = \Omega(r \sin \theta), \tag{97}$$

expressing that the angular velocity in an homentropic fluid depends on
the distance $\varpi = r \sin \theta$ from the rotation axis only (cf. §4.3). As was
shown by Biermann and Kippenhahn, equation (96) and the appropriate
boundary conditions can be satisfied only if the angular velocity is constant
on spheres, with the rotation law

$$\Omega = \Omega_0 r^{-2(1-s)}, \tag{98}$$

where Ω_0 is a constant. Thus, if the turbulence is assumed to be isotropic,
the rotation is uniform over the whole zone; and this also satisfies the
constraint (97). On the contrary, if $s \neq 1$, the general solution (98) cannot
fulfill this constraint; it thus follows that $\Psi \equiv 0$ is no acceptable solution.
Therefore, in a zone of efficient convection, large-scale meridional currents
are always present when the eddy viscosity is anisotropic. This result is
very similar to the von Zeipel paradox (cf. §7.2). *However, in contrast with
the case of a radiative zone, it is the necessity to conserve linear momentum,
rather than energy, that drives large-scale currents in a convective zone with
an anisotropic turbulent velocity distribution.* A detailed discussion by
Biermann and Kippenhahn shows that the actual slight departures from
homentropy do not alter this conclusion.

In the present discussion, it has been explicitly assumed that the parameter s is a constant throughout the whole convective zone. This approximation is certainly a reasonable one in the limit of zero rotation; for, then,
the dynamical properties of turbulence are spherically symmetric. In a
rotating star, however, angular momentum reduces the efficiency of con-

vective transfer with increasing latitude. The consequences of a latitude-dependent parameter s has been recently investigated by Roxburgh. In the same phenomenological spirit, he then let

$$s = 1 + \epsilon_0 + \epsilon_2 \sin^2 \theta \tag{99}$$

($\epsilon_0 \ll 1, \epsilon_2 \ll 1$) to observe that the equilibrium of a convective layer with such an anisotropy parameter also drives large-scale currents. Other arguments which favor the existence of a meridional circulation in a convective zone have been proposed by Tayler. Specific solutions for $\Omega(r,\theta)$ and $\Psi(r,\theta)$ will be presented in Section 9.2 with particular reference to the solar differential rotation problem.

BIBLIOGRAPHICAL NOTES

A general account of the subject will be found in:

1. Mestel, L., "Meridian Circulation in Stars" in *Stellar Structure*, (Aller, L. H., and McLaughlin, D. B., eds.), pp. 465–497, Chicago: The Univ. of Chicago Press, 1965.

See also references 1–3 of Chapter 7, and reference 37 below.

Section 8.2. The classical contributions in this context are those of:

2. Vogt, H., *Astron. Nachr.* **223** (Jan. 1925): 229–232.
3. Eddington, A. S., *The Observatory* **48** (March 1925): 73–75.
4. Eddington, A. S., *The Internal Constitution of the Stars*, pp. 282–288, Cambridge: At the Univ. Press, 1926 (New York: Dover Public., Inc., 1959).
5. Eddington, A. S., *Monthly Notices Roy. Astron. Soc. London* **90** (1929): 54–58.

Reference 5 contains the first (but grossly underestimated) determination of the Eddington-Sweet time. Interesting works from the viewpoint of hydrodynamics are those of:

6. Randers, G., *Astroph. Norvegica* **3** (1939): 97–114.
7. Randers, G., *Astroph. J.* **94** (1941): 109–123.
8. Krogdahl, W., *ibid.* **99** (1944): 191–204.
9. Sen, N. R., and Ghosh, N. L., *Bull. Calcutta Math. Soc.* **37** (1945): 141–152.
10. Csada, I. K., *Contributions from the Konkoly Observatory (Budapest)* No 22 (1949): 215–232.

Currently accepted ideas on meridional circulation derive in the main from:

11. Gratton, L., *Mem. Soc. Astron. Italiana* **17** (1945): 5–27.

12. Sweet, P. A., *Monthly Notices Roy. Astron. Soc. London* **110** (1950): 548–558.
13. Öpik, E. J., *ibid.* **111** (1951): 278–288.
14. Mestel, L., *ibid.* **113** (1953): 716–745; *ibid.* **114** (1954): 500.
15. Mestel, L., *Astroph. J.* **126** (1957): 550–558.
16. Kippenhahn, R., *Zeit. f. Astroph.* **46** (1958): 26–65.

See also Schwarzschild's contribution (reference 7, pp. 175–184, of Chapter 3), and

17. Baker, N., und Kippenhahn, R., *Zeit. f. Astroph.* **48** (1959): 140–154.

Equation (18) was first demonstrated in reference 26 of Chapter 5. The method of resolution given in the text largely follows reference 12. A clear derivation of the second-order "corrections" will be found in:

18. Maheswaran, M., *Monthly Notices Roy. Astron. Soc. London* **140** (1968): 93–107.

As was pointed out by Smith (reference 23), Maheswaran's illustrations fail to show that the outer streamlines are in fact not closed! (see reference 13, pp. 284–285). The double circulation pattern was originally discovered in references 11 and 13; see also:

19. Mestel, L., *Zeit. f. Astroph.* **63** (1966): 196–201.
20. Clement, M. J., *Astroph. J.* **175** (1972): 135–145.

The global time scale (49) was first derived in reference 12. The case of differentially rotating stars is generally discussed in reference 17.

Section 8.3. The validity of equation (6) has been reconsidered in:

21. Osaki, Y., *Publ. Astron. Soc. Japan* **18** (1966): 7–22.
22. Smith, R. C., *Zeit. f. Astroph.* **63** (1966): 166–176.
23. Smith, R. C., *Monthly Notices Roy. Astron. Soc. London* **148** (1970): 275–312.
24. Brand, D., and Smith, R. C., *ibid.* **154** (1971): 293–300.

The role of viscosity has been discussed at length in references 6, 7, and 14. A detailed study of magnetic fields in rotating stars will be found in:

25. Chandrasekhar, S., *Astroph. J.* **124** (1956): 232–243.
26. Mestel, L., *Monthly Notices Roy. Astron. Soc. London* **122** (1961): 473–478.
27. Roxburgh, I. W., *ibid.* **126** (1963): 67–76.

See also reference 18, as well as:

28. Maheswaran, M., *ibid.* **145** (1969): 197–216.
29. Maheswaran, M., *Astron. and Astroph.* **37** (1974): 169–178.

But the most interesting results are those of Mestel and Moss:

30. Moss, D. L., *Monthly Notices Roy. Astron. Soc. London* **168** (1974): 61–72; *ibid.* **178** (1977): 51–59.

31. Mestel, L., and Moss, D. L., *ibid.* **178** (1977): 27–49.

The impossibility of having steady currents in an inviscid, nonmagnetic star was first demonstrated by Randers in reference 7. His advice has been systematically ignored. The references to Osaki and Sakurai are to their papers:

32. Osaki, Y., *Publ. Astron. Soc. Japan* **24** (1972): 509–516.

33. Sakurai, T., *ibid.* **24** (1972): 153–176; *ibid.* **25** (1973): 563–566.

34. Sakurai, T., *Monthly Notices Roy. Astron. Soc. London* **171** (1975): 35–52.

See also reference 39 of Chapter 5. Papers of related interest may be traced to:

35. Clark, A., Jr., *Mém. Soc. Roy. Sci. Liège* (6) **8** (1975): 43–46.

A detailed study of the $\bar{\mu}$-currents will be found in references 1, 14, and 15; see also:

36. McDonald, B. E., *Astroph. and Space Sci.* **19** (1972): 309–349.

37. Kippenhahn, R., "Circulation and Mixing" in *Late Stages of Stellar Evolution* (Tayler, R. J., and Hesser, J. E., eds.), pp. 20–40, Dordrecht: D. Reidel Publ. Co., 1974.

Additional contributions to the problem of meridional circulation in radiative zones are those of:

38. Porfirev, V. V., *Astron. Zh.* **46** (1969): 817–823.

39. Brand, D., *Monthly Notices Roy. Astron. Soc. London* **156** (1972): 325–335.

Reference may also be made to:

40. Roxburgh, I. W., *Astroph. and Space Sci.* **27** (1974): 425–435.

Section 8.4. A detailed account of turbulence will be found in:

41. Tennekes, H., and Lumley, J. L., *A First Course in Turbulence*, Cambridge, Mass.: The M.I.T. Press, 1972.

The influence of radiative transfer on the Richardson criterion has been considered in:

42. Townsend, A. A., *J. Fluid Mech.* **4** (1958): 361–375.

A crude application to rotating stars is due to:

43. Zahn, J. P., *Mém. Soc. Roy. Sci. Liège* (6) **8** (1975): 31–34.

The general theory of anisotropic viscosity from the viewpoint of astrophysics is due to:

44. Lebedinski, A. I., *Astron. Zh.* **18** (1941): 10–25.
45. Wasiutyński, J., *Astroph. Norvegica* **4** (1946).

A more recent treatment of the problem will be found in:

46. Elsässer, K., *Zeit. f. Astroph.* **63** (1966): 65–77.

See also:

47. Sakurai, T., *Publ. Astron. Soc. Japan* **18** (1966): 174–200.

Section 8.5. The analysis in this section is taken from:

48. Biermann, L., *Zeit. f. Astroph.* **28** (1951): 304–309.
49. Biermann, L., in *Electromagnetic Phenomena in Cosmical Physics* (Lehnert, B., ed.), pp. 248–257, Cambridge: At the Univ. Press, 1958.
50. Kippenhahn, R., *Zeit. f. Astroph.* **48** (1959): 203–212.
51. Kippenhahn, R., *Mém. Soc. Roy. Sci. Liège* (5) **3** (1960): 249–255.
52. Kippenhahn, R., *Astroph. J.* **137** (1963): 664–678.

The interaction between rotation and convection is further discussed in:

53. Roxburgh, I. W., *Astroph. and Space Sci.* **27** (1974): 419–424.

The reference to Tayler is to his paper:

54. Tayler, R. J., *Monthly Notices Roy. Astron. Soc. London* **165** (1973): 39–52.

This last work also summarizes much of previous research in the field. In particular, it critically discusses the different rotation laws which have been hitherto proposed for a convective zone: $\Omega = constant$, $\Omega \varpi^2 = constant$, and $\Omega \to 0$. Reference may also be made to:

55. Weir, A. D., *Mém. Soc. Roy. Sci. Liège* (6) **8** (1975): 37–42.

9

The Solar Differential Rotation

9.1. INTRODUCTION

The problem presented by the differential rotation of the Sun is one of long standing and many efforts have been made to formulate a plausible flow pattern that reproduces the observed fluid motions on the solar surface. As we repeatedly pointed out in the two previous chapters, very little is known thus far about the angular momentum distribution within a star; and, in spite of the recent work by Dicke and his colleagues, the Sun is no exception (cf. §11.4). Our intent, then, is to devote this brief chapter to those models that attempt to explain the mean solar rotation rate as an effect due to the interaction between rotation and turbulent convection, *and which also allow for a direct comparison with the observational data*. As we recall from Section 2.2, these are:

   (i) the well-documented east-west motion (i.e., the mean rotation rate) and its variation with latitude;
   (ii) the smallness of the mean north-south motion (i.e., meridional currents);
   (iii) the quasi-constancy of the emergent heat flux and temperature on the solar disc.

Many theories have been proposed to explain how the equatorial acceleration originated and is maintained on the photosphere. However, *no commonly accepted model exists at the present time*, though all theories somehow manage to reproduce most of the observational data. The basic difficulty lies in the fact that the observed fluid motions cover wide domains of horizontal sizes, velocities, and time scales (see Figure 9.1). These range from a few minutes for the granules, up to and beyond the solar rotation period for the largest areas. Because we are mainly interested in the largest scales (i.e., the mean rotation rate), it is thus necessary to parametrize the convective motions having much smaller dimensions. This is usually achieved by means of the Reynolds stresses and the mixing-length theory of turbulent convection (cf. §8.4). Magnetic fields will be neglected altogether because they are not likely to be the ultimate cause of the solar differential rotation (though they might be responsible, in part, for the approximate "corotation" of the lower corona). The existing

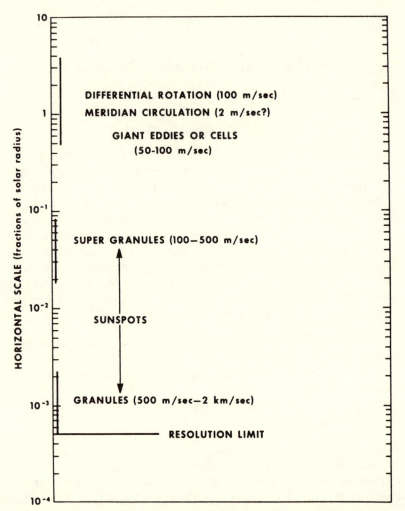

Figure 9.1. Logarithmic plot of the approximate ranges of horizontal scales associated with various photospheric motions, together with their approximate speeds. Source: Gilman, P. A., *Ann. Rev. Astron. Astroph.* **12**, 47, 1974. (By permission. Copyright 1974 by Annual Reviews, Inc.)

theories may be divided into three classes, depending on the mechanism proposed to produce and maintain the equatorial acceleration:

(i) meridional currents caused by anisotropic eddy viscosity in the solar convection zone;

(ii) systematic conversion of the energy of eddy-like motions into mean-flow kinetic energy along the solar parallels;

(iii) direct interaction of rotation with convection in the solar convection zone.

Here we shall briefly comment on these physical mechanisms. The solar-activity cycle, which provides independent information about the large-scale velocity field in the hydrogen convective shell, will be discussed in Section 15.5.

9.2. ANISOTROPIC EDDY VISCOSITY

As we recall from Section 8.5, in a zone of efficient convection with isotropic turbulent motions ($s = 1$) the angular velocity must be uniform over the whole region, and there can be no meridional currents.[1] In contrast, when the turbulent velocities are anisotropic ($s \neq 1$) solid-body rotation is no longer possible, and large-scale meridional currents must necessarily set in. Now, since gravity in the solar convection zone determines the main direction of the buoyancy forces, there is no reason to expect that the coefficient of eddy viscosity in the directions perpendicular and parallel to gravity will be the same. Hence, turbulent mixing in the Sun should be anisotropic, and its outer convective layer cannot rotate uniformly. The concept of anisotropic eddy viscosity was first proposed by Lebedinski in 1941, and again independently by Biermann. Various attempts to apply Lebedinski's original ideas to the solar rotation problem were made by Kippenhahn, Sakurai, Cocke, and Köhler.

To illustrate the main features of this type of solution, we shall briefly summarize the numerical calculations performed by Köhler. The basic equations have been derived in Section 8.5. It is thus assumed that the motions can be described by means of stationary, axially symmetric flow patterns in the hydrogen convective zone, below the solar surface; and turbulent convection is supposed to be so efficient that this layer is essentially a homentropic fluid (cf. §3.5), so that energy transfer need not be taken into account explicitly in the calculations. For the sake of simplicity, Köhler also assumes that the density $\bar{\rho}$, the coefficient μ_t, and the parameter s are all constant in the hydrogen convective zone; this shell covers the outer two-tenths of the solar radius ($0.8R_\odot \leqslant r \leqslant R_\odot$), and the oblateness due to rotation is neglected, as usual. Since the values of μ_t and s are not very well known, it is thus possible to adjust these parameters so as to obtain a good fit to the observed mean rotation rate.

Figures 9.2 and 9.3 illustrate one result from Köhler's calculations, with $s = 1.2$ and $v_t = \mu_t/\bar{\rho} = 4.5 \cdot 10^{12}$ cm²/sec ($= v_\odot$, i.e., the most probable value from mixing-length theory). For this example we observe

[1] In this respect, see also the approaches based on the principle of minimum dissipation (cf. §7.5).

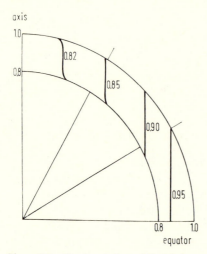

Figure 9.2.   Quadrant of a meridional plane showing curves of constant angular velocity, with $v_t = v_\odot$ and $s = 1.2$ ($\Omega = 1$ at the equator). Source: Köhler, H., *Solar Phys.* **13**, 3, 1970.

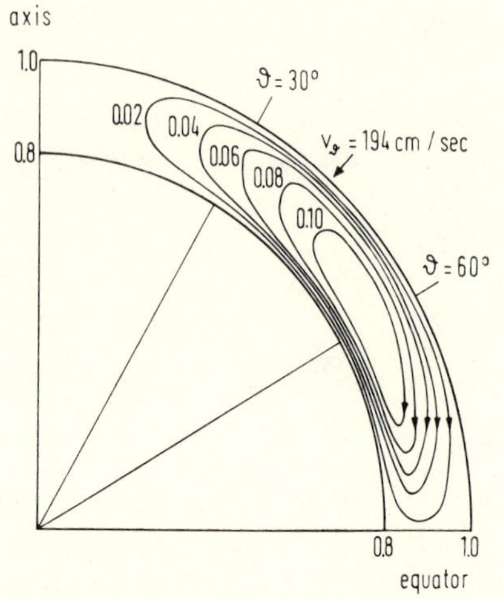

Figure 9.3.   Streamlines of meridional circulation, with $v_t = v_\odot$ and $s = 1.2$. Source: Köhler, H., *Solar Phys.* **13**, 3, 1970.

about a 20 percent drop in the rotation rate from the equator to the poles, and the surface meridional speed is only about a few meters per second. Quite generally, when the eddy viscosity in the radial direction is less than that in the horizontal direction ($s > 1$), the circulation rises at the poles and moves toward the equator on the outer boundary, then streams back to the poles in the inner part of the convective shell. In this case, the outer circulation pattern results in a net transfer of angular momentum per unit mass toward the equator, because in the outer regions along a given radius both the angular momentum $\Omega\varpi^2$ and the angular velocity Ω are larger than in the deep-seated regions of the shell. (When $s < 1$, the direction of meridional circulation would be reversed, and an equatorial deceleration would then result.) Because the actual solar value of the coefficient of eddy viscosity in not very well known, calculations with a fixed value $s = 1.2$ and different values of v_t were also made by Köhler. His numerical results show that the surface rotation rate depends only weakly on v_t if this quantity is sufficiently small ($v_t \lesssim 2v_\odot$). Also, for decreasing values of v_t, the solutions tend asymptotically to a rotation law that shows cylindrical symmetry and a rapidly vanishing circulation (cf. §8.5: equations [97]). On the contrary, for large values of v_t, the rotation laws approach the spherically symmetric form $\Omega = \Omega_0 r^{2(s-1)}$ and the circulation velocities also become small (cf. §8.5: equation [98]). Figure 9.4 depicts the quantity $v_9(\text{max})$, which is a measure of the circulation

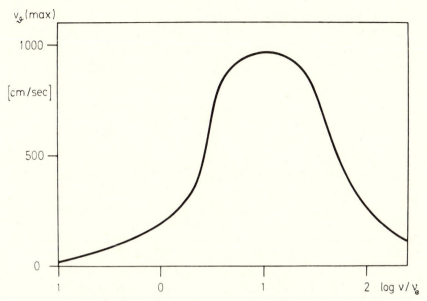

Figure 9.4. The velocity v_9 (max) as a function of the coefficient of eddy viscosity $v = v_t$ (given in units of $v_\odot$). Source: Köhler, H., *Solar Phys.* **13**, 3, 1970.

speed in the whole of the hydrogen convective zone, as a function of the coefficient of eddy viscosity v_t (given in units of $v_\odot$); note that v_3(max) attains its maximum value at about $v_t = 10v_\odot$. Given the uncertainties on the actual value of v_t for the Sun, it is therefore impossible to predict a precise circulation speed from this model.

The main interest of this model is that it allows a detailed comparison with observation. However, it is impossible so far to prove or disprove this type of approach, as this would require a determination of the *fitting* parameter *s* on independent grounds. (According to Durney, if convective motions extend over several scale heights, one would then expect turbulent mixing to be larger in the radial direction, and this would produce an equatorial deceleration!) Also, because *s* is taken to be a constant, anisotropy is assumed to arise only from the preferred direction imposed by gravity. A more realistic approach would be to prescribe latitude-dependent values for *s*, since it is the effective gravity—not the gravitational acceleration alone—which must be considered in determining the degree of anisotropy (cf. §8.5: equation [99]). Finally, with $v_t \approx v_\odot$, the angular velocity is roughly constant over cylinders centered about the rotation axis, and it definitely increases outward (see Figure 9.2). This result disagrees with one very plausible explanation of the difference between tracer and Doppler rotation rates (see Figure 2.2); that is to say, the tracers associated with large surface magnetic fields rotate more rapidly than the photosphere does because these fields might be "anchored" to deep-seated layers whose rotation rate is also fast.[2] If this explanation (due to Foukal) is correct, then it invalidates the model depicted in Figures 9.2 and 9.3.

9.3. BAROCLINIC WAVES AND TWO-DIMENSIONAL TURBULENCE

As we pointed out in Section 2.2, recent observations indicate that large-scale, eddy-like motions are probably present in the photosphere and in the underlying convective zone. Various attempts have been made to ascribe the observed equatorial acceleration to these non-axisymmetric motions, which have much larger horizontal size and larger duration than supergranules. Here we shall consider two types of global flow pattern:

(i) *baroclinic waves* in which nearly horizontal flowing eddies carry angular momentum toward the equator from higher latitudes of each hemisphere. Such a flow pattern is characterized by fluid motions in which the Coriolis force nearly balances the horizontal pressure force (i.e., the so-called heliostrophic balance). The possible existence of these waves on the solar surface was originally proposed by Plaskett and Ward;

(ii) *two-dimensional turbulence*, which maintains a countergradient transport of energy from small eddies to larger eddies. This approach was first suggested by Nickel.

In the Earth's atmosphere, baroclinic waves arise in response to a pole-equator temperature difference caused by non-uniform solar heating. These finite-amplitude disturbances are essentially horizontal (i.e., along spherical surfaces), with streamlines meandering in a wave-like pattern, alternately poleward and equatorward, but in the mean progressively in one direction in longitude. The important feature of baroclinic waves is depicted in Figure 9.5. In the northern hemisphere of the Earth, the rising part of each streamline intersects a latitude circle at a shallow angle, whereas the falling part cuts across virtually at right angles. (The corresponding pattern in the southern hemisphere is obtained by reflecting the figure about the east-west line.) As was first pointed out by Starr, this tilt is such as to imply a net transport of angular momentum *poleward*

---

[2] Another explanation of this phenomenon has been advocated by Yoshimura, namely that the tracers associated with surface magnetic fields might be part of giant convective cells which move prograde relative to the photosphere (cf. §9.4). This alternative theory also runs counter to Köhler's calculations for, then, we must have $s < 1$, and we would observe an equatorial deceleration.

across the latitude circles that intersect the streamline. Since the baroclinic waves occur over a long time and at most longitudes in a belt near the equator, their total effect provides an eddy flux of momentum poleward that sustains the jet streams and the prevailing westerlies in both hemispheres. The kinetic energy thus supplied to the mean zonal flow at mid-latitudes is derived at the expense of the kinetic energy of the large-scale eddies that give rise to the angular momentum transport. This conversion process is, therefore, directly opposed to the effect of viscous friction (i.e., the so-called "negative viscosity" phenomenon).

Now, if such a mechanism is operative in the solar photosphere, one would expect the presence of baroclinic waves that are mirror images of the pattern present in the Earth's atmosphere (see Figure 9.6). As was shown by Starr and Gilman, the streamlines intersect a latitude circle from the poles at smaller angles than they do from the equator, since the net flux of angular momentum must be directed *away from the poles* in both hemispheres. The action of this horizontal flux is to maintain the mean zonal flow (i.e., the observed rotation rate) against frictional dissipation, which would otherwise make the photosphere rotate progressively more nearly as a solid body. Obviously, such a process requires an energy source to sustain the tilted waves, for they continuously give up kinetic energy to the differential rotation. According to Kato and Nakagawa, one possible source of energy could be supplied by the excess of potential energy stored in the hydrogen convective zone, where a radial superadiabatic lapse rate prevails. Another means of excitation of these large-scale eddies is a sizable pole-equator temperature difference caused by the interaction between rotation and convection. According to Gilman,

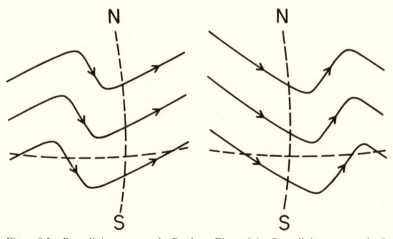

Figure 9.5.   Baroclinic waves on the Earth.   Figure 9.6.   Baroclinic waves on the Sun.

such a latitudinal temperature gradient might exist in the Sun, at least in the deep-seated parts of the convective zone, if not in the photosphere. In any event, we still lack definite observational evidences of baroclinic waves on the Sun, because detailed synoptic maps of large-scale flows on the photosphere are not yet available. (In this respect, the solar physicists are at a disadvantage with respect to the meteorologists.) This is because there are not enough sunspots to trace with precision the actual streamlines of the flow, and because Doppler measurements are usually blurred by the larger velocities of smaller scale turbulence. Further observations are needed to prove or disprove the foregoing approach.

Closely related to the above theory is the two-dimensional turbulence model which favors a countergradient transport of energy from small eddies to the mean zonal flow. Indeed, as we know from the pioneering work of Fjørtoft, turbulent energy preferentially flows from small-scale eddies to larger ones in a two-dimensional motion on a spherical surface, where "two-dimensional" means that the velocity has no component along the vertical direction. (This result should be contrasted with three-dimensional turbulence in which there is a "cascading" of energy from large-scale eddies to smaller ones.) As far as the solar surface is concerned, the restriction of large-scale motions to two dimensions is to be expected, because the horizontal size of the flow greatly exceeds the local density scale height. Detailed calculations by Nickel and Rüdiger show at least a qualitative agreement with the observed rotation in the photosphere.

9.4. INTERACTION OF ROTATION
WITH CONVECTION

The idea that the solar differential rotation could be generated by meridional currents driven by a pole-equator temperature difference was first put forward, independently, by Weiss and by Veronis. This approach is based on the fact that rotation has a small but significant influence upon convection, resulting in a convective heat transport which depends on heliocentric latitude. Thus, if we accept this small deviation from homentropy, the stratification in the solar convective zone must be baroclinic, and an energy balance is possible only with a slow meridional circulation. In a steady state, the angular momentum transported by these currents is then balanced by the viscous transport of momentum, thus leading to a state of differential rotation. Stationary, axially symmetric models that exploit this approach have been constructed by Osaki, and by Durney and Roxburgh. Allowance for the transition from convective transport to radiative transport at the solar surface was also made. Numerical calculations by Durney and Roxburgh exhibit flow patterns which reproduce a surface equatorial acceleration of the right order of magnitude. However, for the equator to rotate faster than the poles, it is necessary to assume that convection is preferentially stabilized at the equator. Moreover, these models predict a surface meridional speed of about 350 cm/sec from the poles to the equator. Finally, an equatorial temperature excess of about 70°K develops, which is not observed.

To alleviate these difficulties, Durney has suggested that rotation affects turbulent convection only deep inside the hydrogen convective zone. The polar regions are then preferentially stabilized, and this results in meridional currents (rising at the equator and sinking at the poles) in the deep-seated parts of the shell. Near the surface, however, the influence of rotation on convection becomes negligible; and a countercell develops, with rising motions at the poles and sinking motions at the equator. According to Durney, this countercell will redistribute the pole-equator differences in flux and temperature that the interaction between rotation and convection generates in the deep-seated parts of the shell. A quantitative model along these lines is still lacking.

Another approach to the solar rotation problem is based on the appealing assumption that the variations in angular velocity arise mainly from the nonlinear interaction of rotation with large-scale convective motions, when a radial superadiabatic gradient of temperature prevails. This theory was first worked out independently by Durney, by Busse, and by Yoshimura and Kato, and it stems from a suggestion by Simon and Weiss that giant convective cells might exist in the hydrogen convective zone.

The principal result of most papers developed along these lines is that giant convective cells can indeed produce and maintain the observed equatorial acceleration. All of the models, however, are still highly simplified when compared to the actual situation which is most likely to prevail in the Sun. In particular, they use the so-called Boussinesq approximation (i.e., variations in density are retained only when coupled with gravity), and they treat small-scale turbulence in a very crude way (i.e., the coefficients of kinematical viscosity and thermal conductivity are merely replaced by constant coefficients of eddy viscosity v_t and eddy conductivity χ_t, respectively). Broadly speaking, these models consider a rotating, spherical shell of a quasi-incompressible fluid subjected to a purely radial gravitational acceleration g. In this case, when the (fixed) temperature difference ΔT across the shell attains a sufficiently high value, the buoyancy forces overcome the stabilizing effects of viscous friction and thermal conductivity. The static state then becomes unstable to small fluctuations, and convective motions set in.[3]

---

[3] See, e.g., reference 27, pp. 9–75, of Chapter 3.

Three dimensionless parameters appear in the problem: the Prandtl number $P = \bar{\rho} c_v v_t / \chi_t$, the Rayleigh number

$$R = \frac{\bar{\rho} c_v g \alpha d^3}{\chi_t v_t} \Delta T, \tag{1}$$

and the Taylor number

$$T = \frac{4 \bar{\Omega}^2 d^4}{v_t^2}, \tag{2}$$

where c_v is the specific heat at constant volume, α is the coefficient of volume expansion, $d (= 0.2 R_\odot = 1.4 \cdot 10^{10}$ cm$)$ is the thickness of the convective zone, and $\bar{\Omega}$ ($\approx 2.6 \cdot 10^{-6}$/sec) is the mean angular velocity of the shell. Hence, with a coefficient of eddy viscosity comparable to Köhler's values ($v_t \approx 10^{14} - 10^{12}$ cm^2/sec), the Taylor number varies by four orders of magnitude, between 10^2 and 10^6. With $P = 1$, the critical Rayleigh number, R_c (say), above which thermal convection sets in depends on the Taylor number only.

Given the above assumptions, various authors (notably Busse, Durney, and Yoshimura and Kato) have shown that, for small rotation rates, the preferred mode of cellular convection should be long rollers that are highly elongated in latitude, and that propagate in a direction opposite to the mean solar rotation. Also, because these rollers continuously transport angular momentum toward the equator, they reproduce in a qualitative way the observed rotation rate at the surface. Unfortunately, they also generate a much stronger heat flux near the equator than at higher latitudes. This is not observed. Let us also note that these calculations do not take into account the feedback effect of the differential rotation on the giant convective cells which produce the equatorial acceleration.

The nonlinear coupling between giant cells and differential rotation has been further investigated by Gilman. For the sake of simplicity, he used simple Cartesian coordinates by restricting the motions to an annular region of fluid, bounded by two latitude circles, which symmetrically straddles the equator. By assumption, the viscous stresses vanish at the top, bottom, and sides; the top and bottom are perfect thermal conductors (temperature fixed), but the sides are thermal insulators. Thus, radial but no latitudinal temperature gradients are imposed. Numerical calculations were carried out in the Boussinesq approximation, with $P = 1$, T in the range 10^2–10^6, and R between R_c and a few times that value. When $R = 1.2 R_c$ and $10^3 < T < 10^6$, the convective rollers do generally make the equatorial belt rotate faster than matter at higher latitudes. The surface layers also rotate faster than the deep-seated parts of the convective shell; and the meridional circulation is in all cases a single cell in each hemisphere, with poleward motions near the outer surface and rising motions at the equator. The equatorial acceleration is maintained primarily through the momentum transfer by giant convective cells (which counterbalances the viscous diffusion of momentum), rather than by meridional currents. Also, the differential rotation tends to stretch out the convective rollers, analogously to what is thought to happen to the solar magnetic regions. As we increase the Rayleigh number above the value $R = 1.2R_c$, the circulation pattern remains essentially unchanged in structure, but the mean rotation rate then shows a marked local *deceleration* near the equator (see Figure 9.7). According to Gilman, solutions in the neighborhood of $T = 3 \cdot 10^4$ seem to compare best with various solar observations. (These include differential rotation amplitude, cell structure, and small meridional velocities.) However, the local deceleration near the equator has not been observed so far. Finally, all flow patterns with $T > 10^3$ generate a vertical heat flux that strongly depends on heliocentric latitude, with a maximum at the equator, no evidence of which is seen on the photosphere.

In summary, although many theories have been proposed to explain the solar rotation rate, no scheme has yet been generally accepted as being basically correct. Most promising are certainly the models based on the

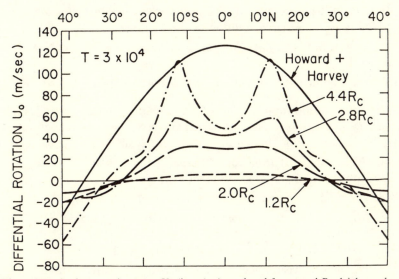

Figure 9.7. Surface rotation rates U_0 (in m/sec) produced for several Rayleigh numbers (in units of R_c), at Taylor number $T = 3 \cdot 10^4$, compared to the mean rotation rate observed by Howard and Harvey (cf. §2.2). Source: Gilman, P. A., *Solar Phys.* **27**, 3, 1972.

nonlinear interaction between rotation and convection. However, such an approach would require a reliable integration of the basic equations for a highly compressible, slowly rotating layer in which turbulent convection prevails. In particular, as was pointed out by various workers in the field, the large vertical gradients of density in the hydrogen convective shell could be essential for the understanding of the velocity fields on the solar surface. Fully nonlinear solutions that include both small-scale turbulence and compressibility are still to be developed.

BIBLIOGRAPHICAL NOTES

A critical survey of the various theories presented in this chapter will be found in:

1. Belvedere, G., Motta, S., and Paternò, L., "Solar Rotation: the Interaction of Rotation and Convection" in *Memorie Soc. Astron. Italiana* **44** (1973): 435–460.

and in reference 2 of Chapter 2. But the most detailed survey is due to Durney:

2. Durney, B. R., "On Theories of Solar Rotation" in *Basic Mechanisms of Solar Activity* (Bumba, V., and Kleczek, J., eds.), pp. 243–295, Dordrecht: D. Reidel Publ. Co., 1976.

Other relevant papers may also be found in this volume. For questions pertaining to the *internal* rotation of the Sun, see:

3. Rasool, S. I., ed., *Physics of the Solar System*, New York: N.A.S.A., 1972.

See especially the papers by Dicke (pp. 23–60) and Spiegel (pp. 61–87).

Section 9.2. Various applications of the concept of anisotropic eddy viscosity to the solar rotation problem have been made in references 44, 47, 49–52, and 53 of Chapter 8. Reference may also be made to:

4. Cocke, W. J., *Astroph. J.* **150** (1967): 1041–1050.
5. Köhler, H., *Solar Phys.* **13** (1970): 3–18.

Papers of related interest are quoted in reference 5. See also:

6. Iroshnikov, R. S., *Astron. Zh.* **46** (1969): 97–112; *ibid.* **52** (1975): 792–803.
7. Rüdiger, G., *Astron. Nachr.* **295** (1974): 229–235.

The difference between tracer and Doppler rotation rates is discussed in:

8. Foukal, P., *Astroph. J.* **173** (1972): 439–444.
9. Durney, B. R., *Solar Phys.* **38** (1974): 301–309.

and in the first reference 44; see also:

10. Foukal, P., and Jokipii, J. R., *Astroph. J. Letters* **199** (1975): L71–L73.
11. Foukal, P., *ibid.* **203** (1976): L145–L148.

Section 9.3. Baroclinic waves were originally discussed by Plaskett (1959) and Ward (1964) in the context of solar physics; see references 17 (and papers quoted therein) and 20 of Chapter 2. Various theoretical aspects of the problem were further investigated by:

12. Starr, V. P., and Gilman, P. A., *Tellus* **17** (1965): 334–340.
13. Starr, V. P., and Gilman, P. A., *Astroph. J.* **141** (1965): 1119–1125.
14. White, M. L., *Icarus* **9** (1968): 364–372.
15. Gilman, P. A., *Solar Phys.* **8** (1969): 316–330; *ibid.* **9** (1969): 3–18.
16. Kato, S., *Astroph. J.* **157** (1969): 827–834.
17. Kato, S., and Nakagawa, Y., *Solar Phys.* **10** (1969): 476–493; *ibid.* **14** (1970): 138–146.
18. Sakurai, T., *Publ. Astron. Soc. Japan* **22** (1970): 177–190.

A paper of related interest is:

19. Wolff, C. L., *Astroph. J.* **194** (1974): 489–498.

Two-dimensional turbulence is discussed in:

20. Nickel, G. H., *Solar Phys.* **10** (1969): 472–475.
21. Rüdiger, G., *ibid.* **51** (1977): 257–269.

Section 9.4. The references to Weiss and Veronis are to their papers:

22. Weiss, N. O., *The Observatory* **85** (1965): 37–39.
23. Veronis, G., *Tellus* **18** (1966): 67–76.

Further discussions of the pole-equator temperature difference are due to:

24. Osaki, Y., *Monthly Notices Roy. Astron. Soc. London* **148** (1970): 391–406.
25. Durney, B. R., and Roxburgh, I. W., *Solar Phys.* **16** (1971): 3–20.
26. Durney, B. R., *ibid.* **26** (1972): 3–7.
27. Durney, B. R., *Astroph. J.* **190** (1974): 211–221; *ibid.* **204** (1976): 589–596.

A closely related approach will be found in:

28. Gierasch, P. J., *ibid.* **190** (1974): 199–210.

Theoretical studies of granules, supergranules, and giant cells are due to:

29. Simon, G. W., and Weiss, N. O., *Zeit. f. Astroph.* **69** (1968): 435–450.
30. Vickers, G. T., *Astroph. J.* **163** (1971): 363–374.

31. Heard, W. B., *ibid.* **186** (1973): 1065–1081.
32. Parker, E. N., *ibid.* **186** (1973): 643–663.

Thermal convection in a rotating spherical shell is treated in:

33. Durney, B. R., *J. Atmospheric Sci.* **25** (1968): 771–778.
34. Busse, F. H., *Astroph. J.* **159** (1970): 629–639.
35. Durney, B. R., *ibid.* **161** (1970): 1115–1127; *ibid.* **163** (1971): 353–361.
36. Yoshimura, H., and Kato, S., *Publ. Astron. Soc. Japan* **23** (1971): 57–73.
37. Busse, F. H., *Astron. and Astroph.* **28** (1973): 27–37.
38. Yoshimura, H., *Publ. Astron. Soc. Japan* **26** (1974): 9–51.
39. Gilman, P. A., *J. Atmospheric Sci.* **32** (1975): 1331–1352.

The problem of a rotating annulus heated from below has been considered by:

40. Davies-Jones, R. P., and Gilman, P. A., *Solar Phys.* **12** (1970): 3–22.
41. Davies-Jones, R. P., and Gilman, P. A., *J. Fluid Mechanics* **46** (1971): 65–81.
42. Gilman, P. A., *Solar Phys.* **27** (1972): 3–26.
43. Gilman, P. A., *J. Fluid Mechanics* **57** (1973): 381–400.

Further relevant papers are those of:

44. Yoshimura, H., *Solar Phys.* **18** (1971): 417–433; *ibid.* **22** (1972): 20–33; *ibid.* **33** (1973): 131–143.
45. Belvedere, G., and Paternò, L., *ibid.* **41** (1975): 289–295.

Departures from the Boussinesq conditions are discussed in:

46. Vandakurov, Yu. V., *ibid.* **40** (1975): 3–21; *ibid.* **45** (1975): 501–520.

10

Solid-Body Rotation vs.
Differential Rotation

10.1. INTRODUCTION

So far, we have described the general principles governing the motion of rotating stars, and we have applied these principles to specific models that have level surfaces that deviate but slightly from spheres. Consider now a single star with fixed angular momentum J and mass M, within which electromagnetic effects may be neglected. To simplify matters even further, let us restrict ourselves to barotropic models and ignore all problems pertaining to energy transport (cf. §7.2). Impose the condition next that the system must rotate with some prescribed rotation law $\Omega = \Omega(\varpi)$, which satisfies the Høiland criterion for stability (cf. §7.3). Under these circumstances, is it always possible to build a model in a state of permanent rotation for all values that we may assign to the total angular momentum? *Even in these simple terms, no general answer can be given at the present time.* Accordingly, once more we must examine, in succession, different idealizations that lend themselves to a tractable formulation.

The question was originally raised by Maclaurin in 1740 in the case of uniformly rotating, homogeneous spheroids. As we recall from Section 4.5, for given mass M, volume $\mathcal{V}$ (or density ρ), and angular momentum J, there exists one and only one spheroid in a state of permanent rotation. However, we also noticed that the angular velocity Ω is *not* a very convenient variable to specify the location of a model along a sequence (see Figure 4.2); instead we should use the parameter

$$\tau = K/|W|, \tag{1}$$

i.e., the ratio of the rotational kinetic energy K to the gravitational potential energy W, which is limited by equilibrium requirements to range from $\tau = 0$ (a spherical body) to $\tau = 0.5$ (an infinitely thin disc that is at rest). Actually, when $\tau \gtrsim 0.14$ there exists another sequence of figures of equilibrium that branches off the Maclaurin sequence: that is, uniformly rotating, homogeneous ellipsoids having a genuine triplanar symmetry, and which are stationary when viewed in a frame of reference rotating with the angular velocity Ω of the ellipsoids.

In 1919, Jeans considered the corresponding problem for uniformly rotating polytropes, i.e., barotropes for which the pressure p and the density ρ are related by the equation

$$p = K\rho^{1 + 1/n}, \tag{2}$$

where K and n are constants ($0 \leqslant n \leqslant 5$, the value $n = 0$ corresponding to a configuration of uniform density). Contrary to the case of homogeneous spheroids, Jeans observed that sequences of uniformly rotating polytropes usually terminated well before the limit $\tau = 0.5$ was reached; also, when $n \gtrsim 0.8$ no uniformly rotating polytrope having a triplanar symmetry could be constructed. Today, with the advent of large computers, much progress has been made in the field. And it is the main purpose of the present chapter to summarize the salient properties of rotating polytropes that greatly depart from spherical symmetry. Solid-body rotation and differential rotation are considered in turn. The final section provides a general comparison between the different results which have some bearing on many astronomical questions. A somewhat more heuristic approach will be followed in the subsequent chapter.

10.2. THE CLASSICAL RESULTS

To clarify some aspects of the general problem, it may not be irrelevant to summarize first the main features of uniformly rotating ellipsoids. For that purpose, consider a system that is at rest with respect to a frame of reference rotating with the constant angular velocity Ω. The equations of relative equilibrium referred to rectangular axes rotating around the x_3-axis are

$$\frac{1}{\rho}\frac{\partial p}{\partial x_i} = -\frac{\partial V}{\partial x_i} + (1 - \delta_{i3})\Omega^2 x_i \tag{3}$$

($i = 1, 2, 3$; $\delta_{13} = \delta_{23} = 0$, $\delta_{33} = 1$; no summation over repeated indices). Now, the components of the gravitational attraction in an homogeneous ellipsoid (with semi-axes a_1, a_2, and a_3) have the form

$$\frac{\partial V}{\partial x_i} = 2\pi G\rho A_i x_i, \tag{4}$$

where

$$A_i = a_1 a_2 a_3 \int_0^\infty \frac{du}{(a_1^2 + u)\Delta}, \tag{5}$$

$$\Delta^2 = (a_1^2 + u)(a_2^2 + u)(a_3^2 + u), \tag{6}$$

and G denotes the constant of gravitation. Hence, by virtue of equation (4), the three equations (3) can be readily integrated to give

$$\frac{p}{\rho} = \tfrac{1}{2}\Omega^2(x_1^2 + x_2^2) - \pi G\rho(A_1 x_1^2 + A_2 x_2^2 + A_3 x_3^2) + constant; \tag{7}$$

and the isobaric surfaces take the form

$$\left(A_1 - \frac{\Omega^2}{2\pi G\rho}\right)x_1^2 + \left(A_2 - \frac{\Omega^2}{2\pi G\rho}\right)x_2^2 + A_3 x_3^2 = constant. \tag{8}$$

In expressing that the boundary of the ellipsoid

$$\frac{x_1^2}{a_1^2} + \frac{x_2^2}{a_2^2} + \frac{x_3^2}{a_3^2} = 1 \tag{9}$$

coincides with one of the surfaces defined in equation (8), we find

$$a_1^2\left(A_1 - \frac{\Omega^2}{2\pi G\rho}\right) = a_2^2\left(A_2 - \frac{\Omega^2}{2\pi G\rho}\right) = a_3^2 A_3. \tag{10}$$

In view of the foregoing equalities, we must have

$$a_1^2 a_2^2 (A_1 - A_2) + (a_1^2 - a_2^2)a_3^2 A_3 = 0, \tag{11}$$

and

$$\frac{\Omega^2}{2\pi G\rho} = \frac{a_1^2 A_1 - a_2^2 A_2}{a_1^2 - a_2^2} = \frac{a_1^2 A_1 - a_3^2 A_3}{a_1^2} = \frac{a_2^2 A_2 - a_3^2 A_3}{a_2^2}. \tag{12}$$

Obviously, the first equality (12) obtains only if $a_1 \neq a_2 \neq a_3$. If we next make use of equation (5), equation (11) becomes

$$(a_1^2 - a_2^2) \int_0^\infty \left[\frac{a_1^2 a_2^2}{(a_1^2 + u)(a_2^2 + u)} - \frac{a_3^2}{a_3^2 + u}\right] \frac{du}{\Delta} = 0. \tag{13}$$

Finally, the three equalities (12) lead to the following relations:

$$\frac{\Omega^2}{2\pi G\rho} = a_1 a_2 a_3 \int_0^\infty \frac{u}{(a_1^2 + u)(a_2^2 + u)} \frac{du}{\Delta}, \tag{14}$$

when $a_1 \neq a_2 \neq a_3$; without any restriction, we also find

$$\frac{\Omega^2}{2\pi G\rho} = \frac{a_2 a_3}{a_1}(a_1^2 - a_3^2) \int_0^\infty \frac{u}{(a_1^2 + u)(a_3^2 + u)} \frac{du}{\Delta}, \tag{15}$$

and a similar expression in which the index 1 replaces the index 2, and conversely.

From equation (15) and its unwritten companion, we first observe that $a_1 \geqslant a_3$ and $a_2 \geqslant a_3$. Thus, the rotation must always take place about the least axis. However, we may have either $a_1 \geqslant a_2$ or $a_1 \leqslant a_2$, since there is no physical difference between any two configurations for which we exchange the indices 1 and 2. Finally, we perceive at once that equation (13) can be satisfied in two different ways. *Either*, we let $a_1 = a_2$; *or*, whenever possible, we let $a_1 > a_2$ (say) and make the integral factor vanish in equation (13). The former solution defines the Maclaurin spheroids (cf. §4.5, where they are viewed from an inertial frame), and the latter corresponds to the Jacobi ellipsoids. (In this case, both equations (14) and (15) define the angular velocity equally well.) A numerical investigation of equation (13) reveals that the Jacobi ellipsoids exist only in the domain $\tau_b \leqslant \tau \leqslant 0.5$, where $\tau_b = 0.1375$; they range from the bifurcation spheroid ($\tau = \tau_b$) to an infinitely long needle that is devoid of rotational motion ($\tau = 0.5$). Numerical data for selected members belonging to the Maclaurin and Jacobi sequences are listed in Appendix D.

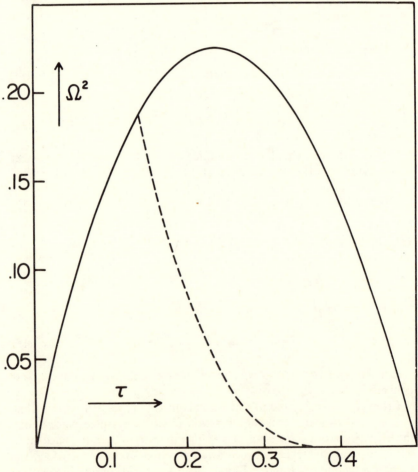

Figure 10.1.   The squared angular velocity Ω^2 along the Maclaurin (*solid line*) and the Jacobi (*dashed line*) sequences, as a function of the ratio $\tau = K/|W|$. The unit of Ω^2 is $2\pi G\rho$. (See also Appendix D.)

Figures 10.1 and 10.2 illustrate the behavior of Ω^2 and J (in normalized units) as a function of τ. Thus, when $0 \leqslant \tau \leqslant \tau_b$, the Maclaurin spheroids are the only possible figures of equilibrium; on the contrary, in the range $\tau_b \leqslant \tau \leqslant 0.5$, to each value of τ correspond two ellipsoidal configurations in relative equilibrium: one Maclaurin spheroid and one Jacobi ellipsoid.

Appendix D shows that for fixed values of J, M, and $\mathscr{V}$, the total mechanical energy $K + W$ is smaller in the body with triplanar symmetry than in the corresponding axisymmetric configuration. Accordingly, *if some dissipative mechanism is operative*, we may expect that beyond the

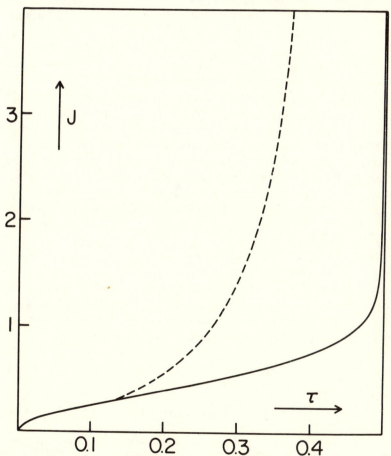

Figure 10.2.   The total angular momentum J along the Maclaurin (*solid line*) and the Jacobi (*dashed line*) sequences, as a function of the ratio $\tau = K/|W|$. The unit of J is $(GM^3\bar{a})^{1/2}$, with $\bar{a} = (a_1 a_2 a_3)^{1/3}$. (See also Appendix D.)

point of bifurcation $\tau = \tau_b$ an incompressible Maclaurin spheroid will evolve gradually to the Jacobi ellipsoid having the same angular momentum (see equation [22] and discussion below). The sectorial oscillations derived in Section 6.7 precisely correspond to such an evolution. A detailed study of these modes is thus in order.

As we recall, in the case of a uniformly rotating, homogeneous spheroid, equation (116) of Section 6.7 is *exact* when dissipative effects may be neglected. In the present instance, we thus obtain

$$\sigma_{-2} = \Omega - (2\omega - \Omega^2)^{1/2}, \quad \text{and} \quad \sigma_{+2} = \Omega + (2\omega - \Omega^2)^{1/2}, \quad (16)$$

and two similar frequencies in which $-\Omega$ replaces Ω; none of them explicitly depends on the compressibility of the system.[1] For Maclaurin spheroids, the parameter ω reduces to

$$\omega = 2\pi G\rho a_1{}^2 a_3 \int_0^\infty \frac{u}{(a_1{}^2 + u)^2} \frac{du}{\Delta}. \quad (17)$$

Following Lebovitz, Figure 10.3 illustrates the behavior of the frequencies σ_{-2} and σ_{+2} along the sequence. We observe that σ_{-2} vanishes when $\Omega^2 = \omega$, i.e., at the point $\tau = \tau_b$ where the Jacobi sequence branches off the Maclaurin sequence. In addition, both frequencies become complex when $\Omega^2 > 2\omega$, i.e., beyond the point $\tau = \tau_i = 0.2738$; clearly, this implies instability by an overstable oscillation of frequency Ω (see, however, Section 11.3). The two critical spheroids corresponding to $\tau = \tau_b$ and $\tau = \tau_i$ are depicted in Figure 10.4. In the case of incompressible spheroids, no further instability arises from the tesseral and zonal modes, which were also discussed in Section 6.7.

Thus, in the absence of dissipation none of the frequencies becomes complex at the point of bifurcation $\tau = \tau_b$, and dynamical instability sets in only past the point $\tau = \tau_i$. As was demonstrated by Roberts and Stewartson, and by Rosenkilde, the situation differs greatly when *viscous dissipation* is taken into account. Indeed, in a first approximation, the secular equation that defines the roots (16) now takes the form

$$\sigma^2 - 2\Omega\sigma - 10i \frac{\langle v \rangle}{a_1{}^2} \sigma + 2(\Omega^2 - \omega) = 0, \quad (18)$$

where

$$\langle v \rangle = \int_V \rho v \, \mathbf{dx} \Big/ \int_V \rho \, \mathbf{dx}, \quad (19)$$

[1] These frequencies were derived in an inertial frame of reference. To obtain their expressions in a frame rotating with the angular velocity Ω we must replace σ by $\sigma \pm 2\Omega$. In this case, the four frequencies simply interchange so that, *formally*, their expressions are similar in both frames of reference. In this and the following section, it is convenient to view σ_{-2} and σ_{+2} in the rotating frame.

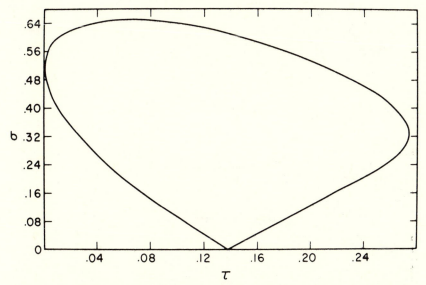

Figure 10.3. Frequencies of the sectorial modes (σ_{+2} and σ_{-2}) along the Maclaurin sequence, as functions of the ratio $\tau = K/|W|$. The frequencies are given in units of $(4\pi G\rho)^{1/2}$; they are not represented beyond the point $\tau = \tau_i$ where they become complex. After Lebovitz (1961). Source: Ostriker, J. P., and Bodenheimer, P., Ap. J. **180**, 171, 1973. (By permission of The University of Chicago Press. Copyright 1973 by the American Astronomical Society.)

and $v(\mathbf{x})$ designates the coefficient of kinematical viscosity.[2] In the limit of small viscosity, the frequency that vanishes at the point $\tau = \tau_b$ becomes

$$\sigma \approx \sigma_{-2} - 5i\frac{\langle v \rangle}{a_1^2}\frac{\sigma_{-2}}{\Omega - \sigma_{-2}}. \tag{20}$$

From equation (20) it follows that the corresponding motion is damped prior to the neutral point $\tau = \tau_b$, and in the range $\tau_b < \tau < \tau_i$ it is amplified with an e-folding time given by

$$t_v = \frac{a_1^2}{5\langle v \rangle}\frac{(2\omega - \Omega^2)^{1/2}}{\Omega - (2\omega - \Omega^2)^{1/2}} \tag{21}$$

(see Table 10.1). Thus, in the region $\tau_b < \tau < \tau_i$, the slightest amount of friction will carry the spheroid into another configuration having a genuine

---

[2] Since we treat the effects arising from viscous dissipation as small perturbations, only the bulk average of $v(\mathbf{x})$ over the volume $\mathscr{V}$ appears in equation (18). However, to comply with the boundary conditions (i.e., the vanishing of the stress vector on the distorted boundary; cf. §3.3: equation [48]) the function $v(\mathbf{x})$ must be chosen so as to vanish on the equilibrium surface $\mathscr{S}$.

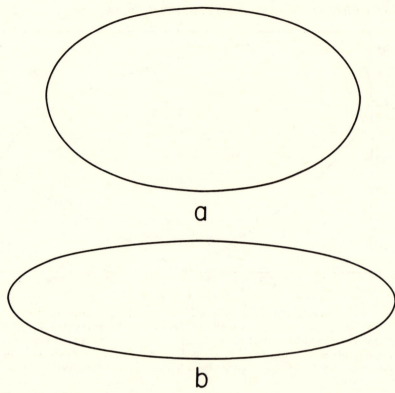

Figure 10.4. Meridional sections of the Maclaurin spheroids corresponding to (a) the point of bifurcation $\tau = \tau_b = 0.1375$ and (b) the limit of dynamical stability $\tau = \tau_i = 0.2738$.

TABLE 10.1

The e-folding times of secular instability[†]

τ	t_v $(a_1^2/\langle v \rangle)$	t_G $(c^5/\pi^2 G^3 I_{11} \rho^2)$
0.1375	∞	∞
0.14	$1.1174(+1)$	$1.0226(+10)$
0.16	1.1306	$1.6922(+5)$
0.18	$5.3851(-1)$	$6.9356(+3)$
0.20	$3.1974(-1)$	$9.4622(+2)$
0.22	$2.0170(-1)$	$2.0534(+2)$
0.24	$1.2366(-1)$	$5.2517(+1)$
0.26	$6.1528(-2)$	$1.1312(+1)$
0.27	$3.1069(-2)$	3.5571
0.2738	$0.$	$0.$

[†] The number in parentheses following each entry is the power of ten by which that entry must be multiplied.

triplanar symmetry. The system is then said to be *secularly unstable.* Beyond the point $\tau = \tau_i$, the Maclaurin spheroid becomes dynamically unstable as in the inviscid case.[3] All remaining modes are damped by viscous dissipation.

The true meaning of secular instability was elucidated by Kelvin and Tait in 1879, and it can be understood as follows.[4] Consider the Navier-Stokes equations in a frame of reference rotating with constant angular velocity Ω (cf. §3.3: equation [52]). Take the scalar product of these equations with the velocity vector that we observe from the rotating frame, and integrate over the total volume $\mathcal{V}$. For an incompressible configuration we find

$$\rho \frac{d}{dt}(T + K + W) = -\int_{\mathcal{V}} \Phi_v \, \mathbf{dx} \leqslant 0, \tag{22}$$

where T designates the relative kinetic energy measured *in the rotating frame,* and $\Phi_v(\mathbf{x},t)$ is the dissipation function (cf. §3.4: equation [62]). Consider now a body which is at rest with respect to our rotating frame ($T \equiv 0$, $\Phi_v \equiv 0$, $K + W = constant$). If the system is given a slight displacement, we can write

$$\rho \frac{d}{dt}(T + \delta K + \delta W) = -\int_{\mathcal{V}} \Phi_v \, \mathbf{dx} \leqslant 0, \tag{23}$$

where δK and δW denote, respectively, small increments in K and W. From equation (23) we perceive at once that the sum $T + \delta K + \delta W$ must continually decrease so long as T differs from zero. Now, if the total mechanical energy $K + W$ is an absolute minimum, $\delta K + \delta W$ is always positive; whence, both T and $\delta K + \delta W$ must go to zero, and the system will eventually return to its equilibrium position. On the contrary, if $K + W$ is not an absolute minimum, $\delta K + \delta W$ decreases; thus, T slowly increases with time until the configuration reaches a neighboring state of relative equilibrium that is secularly stable. This is precisely the fate of the Maclaurin spheroids in the range $\tau \geqslant \tau_b$, since the total mechanical energy $K + W$ is smaller in a Jacobi ellipsoid than in its axisymmetric counterpart having the same angular momentum.

The true role of secular instability has been further explored by Press and Teukolsky, who integrated the Navier-Stokes equations in assuming

---

[3] The singularity occurring in equation (20) when $\Omega^2 = 2\omega$ is only apparent. From equation (18) we obtain $\sigma \approx \Omega \pm (5\langle v \rangle \Omega / a_1{}^2)^{1/2}(1 + i)$, at the point $\tau = \tau_i$.

[4] Thomson, W. (Lord Kelvin), and Tait, P. G., *Treatise on Natural Philosophy* 1, §345 ii, last revised edition, Cambridge: at the University Press, 1912 (New title: *Principles of Mechanics and Dynamics*, New York: Dover Public., Inc., 1962).

ellipsoidal deformations. From their numerical results a consistent picture emerges: beyond the point $\tau = \tau_b$, an incompressible Maclaurin spheroid slowly and monotonically deforms itself into a Jacobi ellipsoid. The intermediate configurations are ellipsoids having internal motions. For all practical purposes, the evolutionary path only depends on the average $\langle v \rangle$ and not on the detailed form of the function $v(\mathbf{x})$. Figures 10.5 and 10.6 illustrate how a viscid Maclaurin spheroid gradually evolves to the Jacobi ellipsoid having the same mass, volume, and *angular momentum*.

Given the above results, we may expect that other dissipative mechanisms will make a Maclaurin spheroid unstable past the point of bifurcation $\tau = \tau_b$. Indeed, Chandrasekhar has shown that the dissipation of energy by *gravitational radiation* also induces a secular instability in one sectorial mode of oscillation. If we assume the slow-motion, weak-field approximation to general relativity, the two sectorial modes (16) are

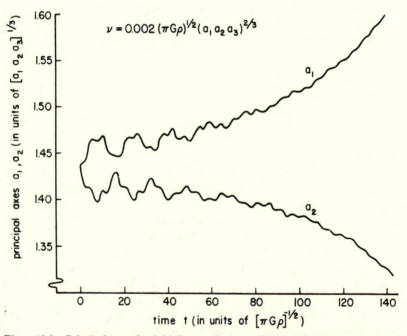

Figure 10.5.   Principal axes of an initially perturbed, secularly unstable Maclaurin spheroid as functions of time. All modes of the perturbation (except one) are seen to damp out in time; the one growing mode corresponds to a monotonic, secular relaxation to a Jacobi ellipsoid. (The behavior of the third principal axis, with initial value $a_3 = 0.488$, is qualitatively similar.) Source: Press, W. H., and Teukolsky, S. A., *Ap. J.* **181**, 513, 1973. (By permission of The University of Chicago Press. Copyright 1973 by the American Astronomical Society.)

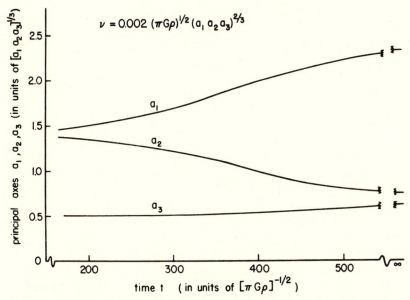

Figure 10.6. Subsequent behavior of the perturbed Maclaurin spheroid depicted on Figure 10.5. Once the damped perturbations have died away, the system follows a unique evolutionary track to the Jacobi sequence. Source: Press, W. H., and Teukolsky, S. A., *Ap. J.* **181**, 513, 1973. (By permission of The University of Chicago Press. Copyright 1973 by the American Astronomical Society.)

approximately given by

$$\sigma^2 - 2\Omega\sigma - \tfrac{4}{5}iD(\sigma - 2\Omega)^5 + 2(\Omega^2 - \omega) = 0, \tag{24}$$

in a frame of reference rotating with angular velocity Ω; we have

$$D = \frac{G}{c^5} \int_{\mathscr{V}} \rho x_1{}^2 \, \mathbf{dx}, \tag{25}$$

where c is the speed of light. From equation (24) we observe that it is now the root σ_{+2} that becomes unstable when gravitational radiation is taken into account. We find

$$\sigma \approx \sigma_{+2} - \tfrac{2}{5}iD \frac{(\sigma_{+2} - 2\Omega)^5}{\Omega - \sigma_{+2}}. \tag{26}$$

While this mode is damped when $0 \leqslant \tau \leqslant \tau_b$, it is amplified in the range $\tau_b < \tau < \tau_i$ (cf. note 3). And the *e*-folding time of this secular instability is given by

$$t_G = \frac{5}{2D} \frac{(2\omega - \Omega^2)^{1/2}}{\left[\Omega - (2\omega - \Omega^2)^{1/2}\right]^5} \tag{27}$$

(see Table 10.1). Thus, gravitational radiation-reaction, like viscosity, makes a Maclaurin spheroid secularly unstable beyond the point $\tau = \tau_b$. But the mode of oscillation that is made unstable by radiation-reaction is not the same one that is made unstable by viscosity.[5]

The response of an inviscid Maclaurin spheroid to the emission of gravitational radiation has been also determined by Miller, for several configurations initially in equilibrium, by a direct integration of the equations of motion. Her analysis indicates that the effects are such as to drive a secularly unstable, inviscid Maclaurin spheroid away from the axisymmetric sequence: beyond the point $\tau = \tau_b$, an inviscid spheroid gradually evolves by radiation-reaction to the Dedekind ellipsoid having the same mass, volume, and *circulation*.

As was recently shown by Lindblom and Detweiler, however, the presence of *both* viscosity and gravitational radiation-reaction moves the point of the onset of secular instability beyond the point $\tau = \tau_b$ to a point determined by the ratio of the strengths of the dissipative forces. In particular, for a spheroid of given mass and volume, one specific value of the coefficient of viscosity will cause the Maclaurin sequence to be secularly stable all the way to the point of the onset of dynamical instability $\tau = \tau_i$! Thus, the presence of both viscosity and gravitational radiation-reaction drastically changes the discussion of the stability of the Maclaurin spheroids from the case where only one or the other of the dissipative forces is acting.

10.3. UNIFORMLY ROTATING POLYTROPES

In the present context, it is convenient to think of the incompressible Maclaurin spheroids as a polytropic sequence of index $n = 0$. Now, to what extent can we extrapolate the foregoing results to centrally condensed polytropes that we force to rotate with constant angular velocity? When $n > 0$, no exact solution can be obtained, and we must resort to numerical

---

[5] According to Chandrasekhar, at the point $\tau = \tau_b$, two sequences of ellipsoids branch off the Maclaurin sequence: (i) the *Jacobi ellipsoids*, and (ii) the *Dedekind ellipsoids*, i.e., homogeneous configurations that are stationary in an inertial frame, and that maintain their ellipsoidal figures by internal motions of uniform vorticity about the least axis (see reference 28 of Chapter 3). Thus, gravitational radiation-reaction makes a Maclaurin spheroid unstable by the Dedekind mode (26), while normal viscous dissipation makes it similarly unstable by the Jacobi mode (20). As was recently shown by Friedman and Schutz, however, *all* inviscid rotating stars are secularly unstable via gravitational radiation to some modes with angular dependence $\exp(im\varphi)$, though only to those with $m \gg 2$ for slowly rotating bodies. However, because the time scale for the instability is likely to be too long for these high modes to be important, the sectorial mode $m = 2$ probably remains the physically interesting phenomenon.

integrations. Broadly speaking, the modern approaches can be divided into the following three classes:

(i) *The numerical computations by James.* To solve the structure problem near the center, the density ρ and the gravitational potential V are expanded in a power series in the radial variable. The coefficients in these expansions are themselves expanded in terms of Legendre polynomials P_l (cos θ). Analytic continuation and finally a step-by-step integration provide the structure in the outer regions. At the present time, this method is regarded as the most rigorous one, and seems to give very accurate results. (The truncation error is about $2 \cdot 10^{-3}$ in the worst case considered.) Physical parameters for uniformly rotating polytropes were obtained in the range $0 < n \leqslant 3$.

(ii) *The Chandrasekhar-Milne expansions* (cf. §5.3). The most detailed results were derived by Aikawa (1971), who built second-order expansions in the parameter

$$v = \frac{\Omega^2}{2\pi G\rho_c}, \tag{28}$$

for a wide range of values for n. To first order in v, the modified Takeda expansion was also used by Monaghan and Roxburgh (cf. §5.4).

(iii) *The variational methods devised by Roberts.* In these approaches, the isopycnic surfaces are chosen to be a family of spheroids of (fixed or variable) eccentricity e. The parameter e and the density ρ are determined by requiring the first-order change in the total energy to be zero for all first-order variations in e and ρ that preserve the mass M and the angular momentum J of the configuration. Many models were constructed along these lines by Hurley and Roberts.

THE AXISYMMETRIC MODELS

In spherical coordinates $(r, \mu = \cos \theta, \varphi)$, the equations governing equilibrium reduce to

$$\frac{1}{\rho} dp = -d[V - \tfrac{1}{2}\Omega^2 r^2(1 - \mu^2)] \tag{29}$$

(cf. §4.3). For a configuration satisfying the pressure-density relation (2), equation (29) integrates to give

$$(n + 1)p = -\rho[V - \tfrac{1}{2}\Omega^2 r^2(1 - \mu^2) - V_0], \tag{30}$$

where V_0 designates the value of V at the poles of the system.

With the usual substitutions

$$\rho = \rho_c \Theta^n, \quad p = K\rho_c^{1 + 1/n}\Theta^{n + 1}, \tag{31}$$

and

$$r = \alpha\xi = \left[\frac{(n+1)K}{4\pi G}\rho_c^{-1+1/n}\right]^{1/2}\xi, \tag{32}$$

equation (30) becomes

$$\Theta = V_0 + \tfrac{1}{6}v\xi^2[1 - P_2(\mu)] - V, \tag{33}$$

where V and V_0 are measured in units of $(n+1)K\rho_c^{1/n}$. By virtue of the Poisson equation, we find

$$\frac{1}{\xi^2}\frac{\partial}{\partial\xi}\left(\xi^2\frac{\partial\Theta}{\partial\xi}\right) + \frac{1}{\xi^2}\frac{\partial}{\partial\mu}\left[(1-\mu^2)\frac{\partial\Theta}{\partial\mu}\right] + \Theta^n = v. \tag{34}$$

TABLE 10.2

Physical properties of uniformly rotating polytropes
$(n = 1.5)$

10^2v	ξ_p	ξ_e	$10g_e$	M	$10^{-2}\mathscr{V}$	$10^{-2}I_{33}$	$10^{-2}I_{11}$
0.00	3.6538	3.6538	2.0330	2.7141	2.0432	0.9316	0.9316
0.20	3.6383	3.6811	1.9829	2.7297	2.0650	0.9391	0.9470
0.40	3.6228	3.7095	1.9315	2.7457	2.0878	0.9469	0.9631
0.60	3.6073	3.7390	1.8790	2.7622	2.1116	0.9551	0.9797
0.80	3.5917	3.7698	1.8251	2.7791	2.1366	0.9636	0.9971
1.00	3.5760	3.8020	1.7697	2.7966	2.1629	0.9725	1.0151
1.20	3.5603	3.8358	1.7128	2.8145	2.1905	0.9819	1.0340
1.40	3.5446	3.8712	1.6542	2.8331	2.2196	0.9917	1.0537
1.60	3.5287	3.9085	1.5936	2.8522	2.2505	1.0020	1.0742
1.80	3.5128	3.9478	1.5309	2.8719	2.2832	1.0128	1.0958
2.00	3.4968	3.9896	1.4659	2.8923	2.3180	1.0242	1.1184
2.20	3.4807	4.0341	1.3983	2.9134	2.3553	1.0362	1.1422
2.40	3.4645	4.0818	1.3277	2.9354	2.3953	1.0489	1.1672
2.60	3.4481	4.1331	1.2537	2.9581	2.4384	1.0623	1.1937
2.80	3.4317	4.1888	1.1756	2.9818	2.4854	1.0766	1.2217
3.00	3.4151	4.2498	1.0928	3.0065	2.5368	1.0918	1.2515
3.20	3.3983	4.3174	1.0040	3.0323	2.5937	1.1080	1.2833
3.40	3.3814	4.3937	0.9080	3.0593	2.6576	1.1255	1.3174
3.60	3.3642	4.4815	0.8021	3.0877	2.7304	1.1444	1.3541
3.80	3.3468	4.5863	0.6825	3.1177	2.8157	1.1649	1.3939
4.00	3.3292	4.7190	0.5410	3.1496	2.9199	1.1874	1.4376
4.16	3.3148	4.8641	0.3983	3.1767	3.0265	1.2072	1.4761
4.24	3.3075	4.9650	0.3063	3.1910	3.0940	1.2179	1.4968
4.32	3.3001	5.1187	0.1769	3.2057	3.1820	1.2292	1.5187
4.36	3.2964	5.3007	0.0400	3.2133	3.2491	1.2351	1.5302

SOURCE: James, R. A., *Ap. J.* **140**, 552, 1964. (By permission of The University of Chicago Press. Copyright 1964 by the University of Chicago.)

Equation (34) must be solved under the conditions

$$\Theta = 1, \quad \text{and} \quad \partial\Theta/\partial\xi = 0, \tag{35}$$

at $\xi = 0$. Moreover, the gravitational force must be continuous across the unknown boundary $\mathscr{S}$ where the pressure (and, hence, Θ) vanishes.

Equation (34) was integrated numerically by James. Characteristics of the sequences $n = 1.5$ and $n = 3$ are given in Tables 10.2 and 10.3, which list the following quantities: the polar and equatorial radii (ξ_p and ξ_e), the effective gravity at the equator g_e, the total mass M, the total volume $\mathscr{V}$, and the moments of inertia:

$$I_{11} = \int_{\mathscr{V}} \rho\xi^2(1 - \mu^2)\,\mathbf{dx}, \quad \text{and} \quad I_{33} = \int_{\mathscr{V}} \rho\xi^2\mu^2\,\mathbf{dx}. \tag{36}$$

In this section only, the units of length, mass, and time are α, $4\pi\alpha^3\rho_c$, and $(4\pi G\rho_c)^{-1/2}$, respectively (cf. equations [31] and [32]); in these units, we

TABLE 10.3

Physical properties of uniformly rotating polytropes
$(n = 3)$

$10^4 v$	ξ_p	ξ_e	$10^2 g_e$	M	$10^{-3}\mathscr{V}$	$10^{-2}I_{33}$	$10^{-2}I_{11}$
0.0	6.89685	6.89685	4.2430	2.01824	1.37417	0.90910	0.90910
2.0	6.88591	6.94462	4.1215	2.02072	1.39101	.91162	0.91352
4.0	6.87504	6.99456	3.9975	2.02323	1.40867	.91418	0.91802
6.0	6.86411	7.04690	3.8708	2.02576	1.42724	.91679	0.92260
8.0	6.85317	7.10189	3.7411	2.02832	1.44681	.91945	0.92726
10.0	6.84227	7.15980	3.6083	2.03090	1.46748	.92216	0.93201
12.0	6.83136	7.22101	3.4719	2.03351	1.48937	.92493	0.93686
14.0	6.82048	7.28587	3.3318	2.03615	1.51263	.92776	0.94180
16.0	6.80949	7.35492	3.1874	2.03882	1.53745	.93064	0.94685
18.0	6.79857	7.42879	3.0380	2.04152	1.56401	.93359	0.95200
20.0	6.78772	7.50803	2.8837	2.04425	1.59258	.93660	0.95726
22.0	6.77682	7.59388	2.7228	2.04701	1.62348	.93969	0.96265
24.0	6.76591	7.68735	2.5550	2.04980	1.65712	.94284	0.96816
26.0	6.75500	7.79007	2.3790	2.05263	1.69401	.94608	0.97381
28.0	6.74406	7.90435	2.1928	2.05550	1.73488	.94940	0.97961
30.0	6.73301	8.03338	1.9943	2.05841	1.78070	.95281	0.98556
32.0	6.72208	8.18197	1.7797	2.06136	1.83289	.95632	0.99168
34.0	6.71111	8.35797	1.5439	2.06435	1.89366	.95994	0.99799
36.0	6.70013	8.57635	1.2763	2.06738	1.96685	.96367	1.00450
36.8	6.69575	8.68182	1.1562	2.06861	2.00111	.96520	1.00716
37.6	6.69138	8.80295	1.0249	2.06985	2.03935	.96675	1.00986
38.0	6.68918	8.87126	0.9540	2.07047	2.06035	.96754	1.01123
38.4	6.68698	8.94619	0.8786	2.07110	2.08289	.96833	1.01261
38.8	6.68477	9.02964	0.7976	2.07172	2.10730	0.96912	1.01400

SOURCE: James, R. A., *Ap. J.* **140**, 552, 1964. (By permission of The University of Chicago Press. Copyright 1964 by the University of Chicago.)

thus compare models having the same central density ρ_c, rather than models having the same mass. We perceive at once that the effective gravity falls to zero at the equator of the more distorted member of a sequence; but this terminal model is far from being an infinitely flat disc. *In general, when $n > 0$, each sequence terminates at a point $\tau = \tau_{max}$ (say) which depends on the polytropic index.* Terminal values of the physical quantities are listed in Table 10.4 for different values of n. When $n \gtrsim 3$, it becomes increasingly difficult to define the outer boundary with any precision; indeed, as n increases, a uniformly rotating polytrope very much resembles a central mass point surrounded by an extended envelope containing a negligible fraction of the total mass. The terminal values τ_{max} are also difficult to obtain. Very roughly, while $\tau_{max} = 0.50$ along the Maclaurin sequence $n = 0$, it is already reduced to the value $\tau_{max} \approx 0.12$ when $n = 1$; for $n = 3$, τ_{max} is a mere few percent, and τ_{max} approaches zero as the polytropic index tends toward its limiting value $n = 5$. *Thus, if we force a centrally condensed polytrope to rotate with constant angular velocity, it simply cannot store much rotational kinetic energy and remain a uniformly rotating figure of equilibrium.*

TABLE 10.4

Terminal values of physical quantities

n	$10^2 v$	ξ_p	ξ_e	M	$10^{-2}\mathscr{V}$	$10^{-2}I_{33}$	$10^{-2}I_{11}$
0.808	10.60296	2.4852	4.7652	5.0248	1.968	2.0890	3.0695
1.0	8.3720	2.6933	4.8265	4.289	2.1825	1.6955	2.3540
1.5	4.3624	3.2962	5.3585	3.2137	3.259	1.2355	1.5309
2.0	2.1604	4.0553	6.307	2.6518	5.527	1.0553	1.209
2.5	0.99300	5.0999	7.7623	2.30563	10.503	0.9818	1.0666
3.0	0.3932	6.58	$\cdots\cdots$	2.089	$\cdots\cdots$	$\cdots\cdots$	$\cdots\cdots$

SOURCE: James, R. A., *Ap. J.* **140**, 552, 1964. (By permission of The University of Chicago Press. Copyright 1964 by the University of Chicago.)

The above remark suggests that the Chandrasekhar-Milne expansion should give fairly good results in the range $n \gtrsim 3$, since this method has been specifically devised to construct configurations which do not greatly depart from spherical symmetry (cf. §5.3). Following Aikawa, to second order in v, we find

$$\Theta(\xi,\mu) = \theta(\xi) + v[\Psi_0(\xi) + A_2\Psi_2(\xi)P_2(\mu)]$$
$$+ v^2\{\chi_0(\xi) + [\chi_2(\xi) + B_2\Psi_2(\xi)]P_2(\mu) + [\chi_4(\xi) + B_4\Psi_4(\xi)]P_4(\mu)\}, \quad (37)$$

where $\theta(\xi)$ defines the spherical solution, i.e., the Lane-Emden function of index n. The functions θ, Ψ_0, Ψ_2, Ψ_4, χ_0, χ_2, and χ_4 are tabulated in Appendix E for different values of n; the constants A_2, B_2, and B_4 are given in Table 10.5. Similarly, the outer boundary can be written in the form

$$\Xi(\mu) = \xi_1 + v[q_{10} + q_{12}P_2(\mu)] + v^2[q_{20} + q_{22}P_2(\mu) + q_{24}P_4(\mu)], \qquad (38)$$

where ξ_1 is the first zero of the function $\theta(\xi)$; the q_{ij}'s are listed in Table 10.5 (compare with §5.3: equation [53]).

TABLE 10.5

Structure parameters for uniformly rotating polytropes

	$n = 1.5$	$n = 2.0$	$n = 2.5$	$n = 3.0$	$n = 3.5$
$-A_2$	0.59794	0.64236	0.68364	0.72330	0.76270
$-B_2$	0.15103(+1)	0.21099(+1)	0.29473(+1)	0.42058(+1)	0.63122(+1)
$+B_4$	0.57638(−1)	0.88927(−1)	0.12619	0.17003	0.22147
$+\xi_1$	0.36538(+1)	0.43529(+1)	0.53553(+1)	0.68969(+1)	0.95358(+1)
$+q_{10}$	0.63669(+1)	0.15052(+2)	0.41519(+2)	0.13749(+3)	0.59611(+3)
$-q_{12}$	0.14080(+2)	0.28487(+2)	0.67043(+2)	0.19224(+3)	0.73611(+3)
$+q_{20}$	0.44884(+2)	0.22932(+3)	0.14490(+4)	0.12014(+5)	0.15466(+6)
$-q_{22}$	0.11114(+3)	0.47343(+3)	0.26210(+4)	0.19881(+5)	0.24074(+6)
$+q_{24}$	0.61333(+2)	0.23863(+3)	0.11749(+4)	0.79194(+4)	0.86343(+5)

SOURCE: Aikawa, T., *Sci. Rep. Tôhoku Univ.* (I) **54**, 13, 1971.

To conclude, it is interesting to compare the equatorial radii obtained by means of the following techniques: (i) the numerical computations by James (J), (ii) the first- and second-order expansions in v (E1 and E2, respectively), and the modified Takeda expansion (MR), and (iii) the variational methods with fixed and variable eccentricities (R and HR, respectively). The equatorial radius ξ_e is by far the most difficult parameter to define with some accuracy. From Figures 10.7 and 10.8 we note that even the second-order expansion (E2) does not closely approximate the numerical results (J), when v approaches its terminal value. The variational methods (R, HR) are somewhat inaccurate since spheroids cannot adequately describe level surfaces in a centrally condensed body (cf. §4.4). When $n \gtrsim 3$, integrated quantities (such as the total mass) are given with better accuracy in all cases, however, for the outer layers then contribute very little to their expressions. In any event, such a comparison provides an estimate of the error that we may expect in the case of uniformly rotating main-sequence stars for which no accuracy test can be made (cf. §12.2).

OSCILLATIONS AND STABILITY

As we know from the previous section, the Jacobi ellipsoids branch off the Maclaurin sequence $n = 0$ at the point $\tau = \tau_b = 0.1375$; and, past this point, the axisymmetric figures become secularly unstable. We may now

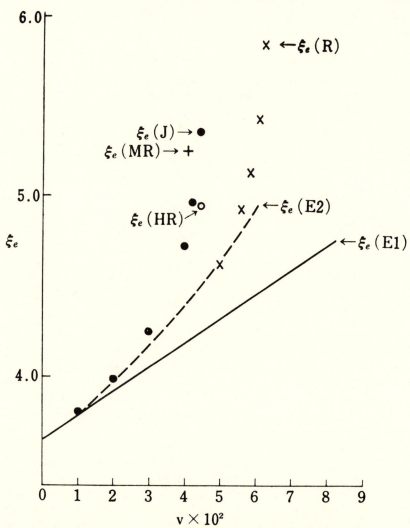

Figure 10.7. Variations of the equatorial radius ξ_e for uniformly rotating polytropes of index $n = 1.5$. The parameter v is defined in equation (28). Solid line and dashed line: first- and second-order expansions in v (Aikawa); ● : the numerical computations by James; ○ : variational method with variable eccentricity (Hurley and Roberts); × : variational method with fixed eccentricity (Roberts); + : the modified Takeda expansion (Monaghan and Roxburgh). The arrows indicate the values for the last member of each sequence of equilibrium. Source: Aikawa, T., *Sci. Rep. Tôhoku Univ.* (I) **54**, 13, 1971.

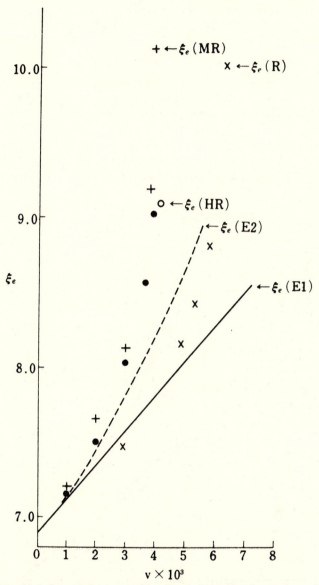

Figure 10.8. Variations of the equatorial radius ξ_e for uniformly rotating polytropes of index $n = 3$. The notations are the same as in Figure 10.7. Source: Aikawa, T., *Sci. Rep. Tôhoku Univ.* (I) **54**, 13, 1971.

ask whether a similar point of bifurcation occurs along a sequence of uniformly rotating polytropes ($n > 0$). As was shown with great accuracy by James, *when*

$$n < 0.808 \qquad (39)$$

each sequence of axially symmetric polytropes in uniform rotation has a point of bifurcation at which non-axisymmetric figures of equilibrium branch off. No bifurcation form exists among uniformly rotating polytropes when condition (39) does not obtain.

To understand this result, let us again consider the sectorial modes (116) derived in Section 6.7. As we know, their *exact* values along the Maclaurin

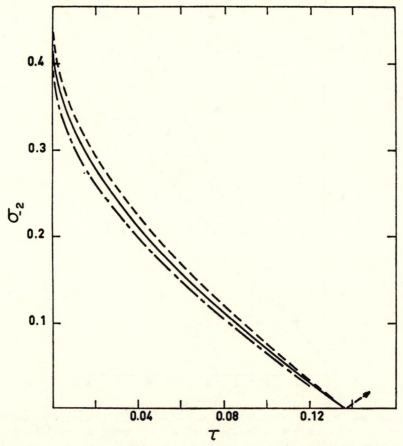

Figure 10.9.   Sectorial frequency σ_{-2} for uniformly rotating polytropes of index $n = 0.6$ (*dashed line*), $n = 0.808$ (*solid line*), and $n = 1.0$ (*dash-dot line*), as a function of the ratio $\tau = K/|W|$. The frequencies are given in units of $(4\pi G\rho_c)^{1/2}$. Source: Tassoul, J. L., and Ostriker, J. P., *A. Ap.* **4**, 423, 1970.

sequence $n = 0$ are given by equations (16) and (17). If we replace equation (17) by

$$\omega = \frac{W_{12,12}}{I_{11}} \tag{40}$$

(cf. §6.7: equation [117]), equation (16) then provides an *approximate* solution for centrally condensed, uniformly rotating bodies. (At this time, we have no way of estimating the errors in σ_{-2} and σ_{+2}; however, in the limiting case $\tau = 0$, it can be shown that the relative error never exceeds five percent when $\rho_c/\bar{\rho} \lesssim 10$.) Figure 10.9 illustrates the variations of σ_{-2} along the polytropic sequences $n = 0.6, n = 0.808$, and $n = 1.0 \, (\tau_{max} \approx 0.16,$ $\tau_{max} \approx 0.138$, and $\tau_{max} \approx 0.12$, respectively). We observe that *whenever a neutral mode exists it always occurs at the point* $\tau \approx 0.138 \, [=\tau_b(n), say]$ *which, to the present accuracy, is independent of the polytropic index.* To be specific, when $\tau_{max} > \tau_b(n)$ (i.e., $n < 0.808$), the root σ_{-2} necessarily vanishes for one member of the sequence. On the contrary, when $\tau_{max} < \tau_b(n)$ (i.e., $n > 0.808$), the root σ_{-2} never goes to zero since no model belonging to the sequence can store enough rotational kinetic energy to reach the fixed (or quasi-fixed) point $\tau = \tau_b(n)$. Finally, when $\tau_{max} = \tau_b(n)$ (i.e., $n = 0.808$), σ_{-2} vanishes exactly for the terminal member of the sequence.

For homogeneous spheroids, the existence of a neutral mode of oscillation implies that secular instability sets in past the point $\tau = \tau_b(0) \equiv \tau_b$, if some dissipative mechanism is operative (cf. equations [20] and [26]); and this implies that secularly stable, non-axisymmetric figures of equilibrium branch off at the point $\tau = \tau_b$. As we shall now see, a similar situation prevails in the case of centrally condensed polytropes. Indeed, when *viscosity* is taken into account, the frequency σ_{-2} then becomes

$$\sigma \approx \sigma_{-2} - 2i \frac{\langle v \rangle}{\langle \varpi^2 \rangle} \frac{\sigma_{-2}}{\Omega - \sigma_{-2}}, \tag{41}$$

where

$$\langle \varpi^2 \rangle = \int_{\mathscr{V}} \rho(x_1^2 + x_2^2) \, dx \bigg/ \int_{\mathscr{V}} \rho \, dx \tag{42}$$

(cf. equations [16], [19], and [40]). Equation (41) thus replaces equation (20) which corresponds to the limiting sequence $n = 0$. (Note that $\langle \varpi^2 \rangle = 2a_1^2/5$, when $n = 0$.) Similarly, as was shown by Miller, *gravitational radiation-reaction* makes centrally condensed, uniformly rotating bodies secularly unstable past the point $\tau = \tau_b(n)$; in this case, however, equation (26) still adequately defines the relevant frequency but σ_{+2} and ω must now be evaluated by means of equations (16) and (40), respectively. All remaining modes discussed in Section 6.7 are stable; they are damped or unaffected by these two dissipative mechanisms.

In summary, whenever $\tau_{\max}$ is larger than $\tau_b(n)$, one of the sectorial frequencies becomes complex beyond the point $\tau = \tau_b(n)$, i.e., the point where σ_{-2} or σ_{+2} vanishes (cf. note 5). That is to say, when $n < 0.808$, secular instability sets in at the point $\tau = \tau_b(n)$; and, above this limit, the axisymmetric models slowly evolve to secularly stable configurations having a genuine triplanar symmetry. In contrast, when $n \geqslant 0.808$, uniformly rotating polytropes are always secularly stable; hence, no bifurcation can possibly occur. This clarifies the results obtained by James on the basis of equilibrium calculations.

10.4. DIFFERENTIALLY ROTATING POLYTROPES

In the two previous sections we have discussed the main properties of uniformly rotating polytropes ($n \geqslant 0$). Obviously, in this case, the viscous stresses rigorously vanish *at equilibrium*, since there are no relative motions in configurations that rotate as solid bodies. On the contrary, for differentially rotating systems, viscosity necessarily implies a continuous change in the angular momentum distribution (cf. §7.5). In this section we shall simply *assume* that viscous effects may be neglected altogether. Accordingly, we are at freedom to prescribe any rotation law, provided it satisfies the Høiland criterion (cf. §7.3).

When departure from sphericity is large, analytical expansions are not very useful, and we prefer to resort to numerical integrations. The problem has been approached in two distinct ways:

(i) *The numerical computations by Stoeckly.* Axisymmetric models were constructed in the case $n = 1.5$ only, with a prescribed angular velocity of the form

$$\Omega(\varpi) = \Omega_c e^{-a\varpi^2/\xi_e^2}, \tag{43}$$

where ϖ is measured from the axis of rotation, Ω_c denotes the angular velocity on this axis, and a is a constant. By virtue of the Høiland criterion, the angular momentum per unit mass $\Omega(\varpi)\varpi^2$ must increase outward; hence, this stability condition demands $a\varpi^2/\xi_e^2 \leqslant 1$, i.e., $0 \leqslant a \leqslant 1$. The numerical solutions were obtained by means of a finite-difference scheme which is very much akin to the Henyey method used for stellar structure problems. No estimate of the errors is available.

(ii) *The self-consistent field method* (cf. §5.5). In this approach, Bodenheimer and Ostriker prescribe the angular momentum distribution $j(m_\varpi)$ rather than the angular velocity $\Omega(\varpi)$. The assumed distributions are obtained from the formula

$$j(m_\varpi) = c_0 + c_1(1 - m_\varpi)^{\alpha_1} + c_2(1 - m_\varpi)^{\alpha_2}, \tag{44}$$

where m_ϖ is the fractional mass interior to the cylinder of radius ϖ. The different constants that appear in equation (44) are chosen to approximate

TABLE 10.6

Coefficients for equation (44)

	$n' = 0$	$n' = 0.5$	$n' = 1$	$n' = 1.5$
c_0	+2.500000	+3.068133	+3.825819	+4.887588
c_1	· · · · · ·	+0.203667	+0.857311	+2.345310
c_2	−2.500000	−3.271800	−4.68313	−7.232898
α_1	· · · · · ·	+0.801297	+0.650981	+0.525816
α_2	+0.666667	+0.500000	+0.400000	+0.333333

SOURCE: Bodenheimer, P., and Ostriker, J. P., *Ap. J.* **180**, 159, 1973. (By permission of The University of Chicago Press. Copyright 1973 by the American Astronomical Society.)

(to within one percent) the angular momentum distribution in a uniformly rotating polytrope of index n' and infinite radius (see Table 10.6). Various axisymmetric models were constructed for different combinations of n and n'. Clearly, the combination $(n,n') = (0,0)$ defines the Maclaurin sequence. (In this case, $c_0 = -c_2 = 5/2$, $\alpha_2 = 2/3$, $c_1 = 0$.) In the series expansions (82) and (84) of Section 5.5, the parameter N is set equal to 15. The overall accuracy of a model is measured by the virial test:

$$VT = \frac{1}{3} \sum_{i=1}^{3} \left| \left(W_{ii} + K_{ii} + \int_{\mathscr{V}} p \, \mathbf{dx} \right) \middle/ W_{ii} \right|, \qquad (45)$$

where all symbols have their standard meanings (cf. §3.7); the quantity VT is thus a measure of the degree of satisfaction of the second-order virial equations.

THE AXISYMMETRIC MODELS

Table 10.7 and Figure 10.10 summarize the physical properties of some of the individual models constructed by Stoeckly (cf. equation [43]); Table

TABLE 10.7

Physical properties of differentially rotating polytropes
($n = 1.5$)

	Sphere	D	E	F	G
a	· · · · · · · ·	0.54223	0.97484	0.77704	0.93890
ξ_e	1.1848(+1)	2.3949(+1)	3.0125(+1)	2.2199(+1)	2.3210(+1)
ξ_p	1.1848(+1)	1.1071(+1)	1.0988(+1)	1.0809(+1)	1.0643(+1)
Ω_c	0.	4.2060(−3)	4.1313(−3)	4.4828(−3)	4.4734(−3)
Ω_c/Ω_e	· · · · · · · ·	1.7198	2.6507	2.1750	2.5572
ρ_c	8.6010(−4)	4.7302(−4)	3.4171(−4)	3.3702(−4)	2.6019(−4)
$\rho_c/\bar{\rho}$	5.9913	10.297	13.198	8.1644	7.3313
$-g_e/(\partial V/\partial \xi)_e$	1.	0.01903	0.21977	0.47404	0.57713
J	0.	0.23106	0.31525	0.29998	0.34463

SOURCE: Stoeckly, R., *Ap. J.* **142**, 208, 1965. (By permission of The University of Chicago Press. Copyright 1965 by the University of Chicago.)

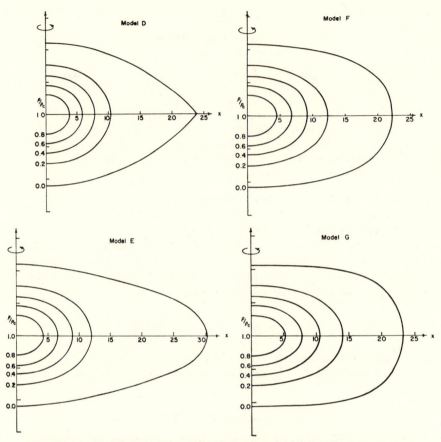

Figure 10.10.   Meridional sections of four differentially rotating polytropes of index $n =$ 1.5 (see Table 10.7). *Model D*: a model with mildly non-uniform rotation marking approximately the end of the sequence of models with zero effective gravity at the equator. *Model E*: a model with *a* just greater than the values for sequences that reach models having zero effective gravity at the equator; this model illustrates the rapid increase of equatorial radius in such sequences. *Model F*: a model in moderately non-uniform rotation with nearly the same total angular momentum as Model E. *Model G*: a model whose isopycnic surfaces are as elliptic throughout as those of the figure of bifurcation along the Maclaurin sequence; the squareness of shape of this model is typical of those with highly non-uniform rotation. Source: Stoeckly, R., *Ap. J.* **142**, 208, 1965. (By permission of The University of Chicago Press. Copyright 1965 by the University of Chicago.)

10.7 includes the spherical polytrope of index $n = 1.5$ and four models chosen as examples. These properties are presented in dimensionless variables based on $4\pi G$, $(n + 1)K$, and M (instead of ρ_c!); for $n = 1.5$, the units of length, mass, and time thus are $(5K/2)(4\pi G)^{-1}M^{-1/3}$, M, and

$(5K/2)^{3/2} (4\pi G)^{-2} M^{-1}$, respectively. In addition to previously defined quantities, Table 10.7 gives the ratio Ω_c/Ω_e of the angular velocities at $\varpi = 0$ and $\varpi = \xi_e$. The ratio of the effective gravity at the equator, g_e, to the magnitude of the gravitational acceleration, $(\partial V/\partial r)_e$, is also listed to indicate how close a configuration is to the terminal model of the sequence (with constant a) to which it belongs.

Stoeckly's sequences of models separate into distinct types by their behavior at high values of the total angular momentum J. For small values of the parameter a (i.e., moderate differential rotations), each sequence terminates at a configuration having zero effective gravity at the equator. On the other hand, when rotation is strongly non-uniform (i.e., when a approaches one), the surface layers become more distorted by comparison with the central regions. In this case, the sequences may even contain rapidly rotating models having cuspate isopycnic surfaces in the interior; in such models, the pressure and density increase outward. Going to the extremes, Stoeckly also observed that sequences at constant $a(> 0.67)$ pass through systems with a detached outer ring of mass, and that the sequences terminate, or are discontinuous, soon after. No stability analysis was performed for the above configurations, and points of bifurcation were not searched for.

Detailed results were obtained more recently by Bodenheimer and Ostriker (cf. equation [44]). Tables 10.8–10.9 and Figures 10.11–10.15 summarize the salient features of their models. This time, the system of units is based on G, M, and ξ_e; and Ω is listed in units of $(4\pi G\rho_c)^{-1/2}$! Along a given sequence (n,n'), a model is specified by the dimensionless angular momentum $J/(GM^3\xi_e)^{1/2}$ or, preferably, by the ratio $\tau = K/|W|$. For fixed values of J, M, n, and n', a sequence thus represents a succession of differentially rotating polytropes with smaller and smaller equatorial radii as the ratio τ increases; and all models belonging to a given sequence have the same angular momentum distribution $j(m_\varpi)$. In this way, the polytropic sequences (n,n') generalize the classical Maclaurin sequence $(0,0)$. A word of caution is necessary. It has never been shown that a contracting, inviscid configuration would in fact follow one of these sequences. For the present, it is thus preferable to view a polytropic sequence (n,n') as a continuous set of unrelated *static* models, rather than as the evolutionary path of a contracting, centrally condensed star.

Figures 10.11–10.14 provide a comparison with the Maclaurin sequence $(0,0)$; they show the range of the ratio ξ_e/ξ_p, the central density ρ_c, the average angular velocity

$$\langle\Omega\rangle = \int_{\mathscr{V}} \rho(\mathbf{x})\Omega(\varpi)\varpi^2 \, \mathbf{dx} \Big/ \int_{\mathscr{V}} \rho(\mathbf{x})\varpi^2 \, \mathbf{dx}, \tag{46}$$

TABLE 10.8

Physical properties of differentially rotating polytropes
$(n = 1.5, n' = 0)$

J	K	W	U_T	I_{11}	Ω_c	ρ_c	v_e	g_p	g_e	ξ_e/ξ_p	$[\xi_e/\xi_p]_c$	VT
0.	0.	−0.857	0.429	0.102	0.	1.43	0.	1.00	1.00	1.00	1.00	1.73(−7)
0.097	0.023	−0.883	0.418	0.103	0.663	1.58	0.242	1.18	0.970	1.13	1.10	3.10(−6)
0.141	0.050	−0.911	0.406	0.103	0.959	1.75	0.351	1.40	0.932	1.28	1.21	7.56(−6)
0.188	0.088	−0.950	0.387	0.104	1.28	2.05	0.470	1.73	0.870	1.53	1.40	9.75(−6)
0.209	0.109	−0.970	0.376	0.104	1.43	2.23	0.524	1.92	0.833	1.68	1.50	1.44(−5)
0.234	0.136	−0.995	0.361	0.104	1.60	2.49	0.584	2.16	0.785	1.89	1.64	1.46(−5)
0.252	0.159	−1.016	0.349	0.103	1.73	2.73	0.630	2.38	0.743	2.08	1.77	3.57(−5)
0.279	0.196	−1.050	0.329	0.103	1.94	3.20	0.697	2.76	0.673	2.43	2.00	9.92(−5)
0.304	0.235	−1.085	0.308	0.102	2.15	3.79	0.759	3.18	0.600	2.85	2.26	1.14(−4)
0.326	0.275	−1.123	0.286	0.100	2.38	4.56	0.815	3.66	0.525	3.36	2.56	5.42(−5)
0.334	0.291	−1.138	0.278	0.100	2.46	4.92	0.833	3.86	0.496	3.59	2.68	1.22(−4)
0.340	0.303	−1.150	0.272	0.099	2.53	5.24	0.848	4.02	0.475	3.78	2.77	2.19(−4)

SOURCE: Bodenheimer, P., and Ostriker, J. P., *Ap. J.* **180**, 159, 1973. (By permission of The University of Chicago Press. Copyright 1973 by the American Astronomical Society.)

TABLE 10.9

Physical properties of differentially rotating polytropes
$(n = 3, n' = 0)$

J	K	W	U_T	I_{11}	Ω_c	ρ_c	v_e	g_p	g_e	ξ_e/ξ_p	$[\xi_e/\xi_p]_c$	VT
0.	0.0	−1.50	1.50	0.0377	0.	12.9	0.	1.00	1.00	1.00	1.00	2.92(−4)
0.055	0.023	−1.52	1.48	0.0380	1.46	13.5	0.137	1.24	0.987	1.12	1.05	3.55(−4)
0.095	0.067	−1.57	1.43	0.0385	2.50	15.0	0.237	1.77	0.959	1.37	1.16	5.23(−4)
0.114	0.097	−1.59	1.40	0.0388	2.99	16.0	0.285	2.15	0.940	1.55	1.23	6.79(−4)
0.130	0.125	−1.62	1.37	0.0391	3.40	17.1	0.326	2.57	0.920	1.72	1.31	8.81(−4)
0.145	0.154	−1.65	1.34	0.0393	3.77	18.2	0.362	3.00	0.900	1.91	1.38	1.14(−3)
0.167	0.204	−1.69	1.28	0.0397	4.34	20.4	0.418	3.82	0.863	2.26	1.52	1.78(−3)
0.179	0.232	−1.71	1.24	0.0398	4.63	21.7	0.447	4.34	0.841	2.48	1.60	2.27(−3)
0.190	0.261	−1.73	1.21	0.0400	4.90	23.1	0.474	4.88	0.819	2.70	1.68	2.87(−3)
0.210	0.316	−1.78	1.13	0.0404	5.40	26.0	0.524	5.90	0.775	3.20	1.86	4.49(−3)
0.228	0.371	−1.81	1.06	0.0408	5.82	29.0	0.569	6.96	0.731	3.76	2.06	6.78(−3)
0.252	0.445	−1.85	0.94	0.0415	6.28	33.1	0.628	8.96	0.667	4.68	2.38	1.14(−2)

SOURCE: Bodenheimer, P., and Ostriker, J. P., *Ap. J.* **180**, 159, 1973. (By permission of The University of Chicago Press. Copyright 1973 by the American Astronomical Society.)

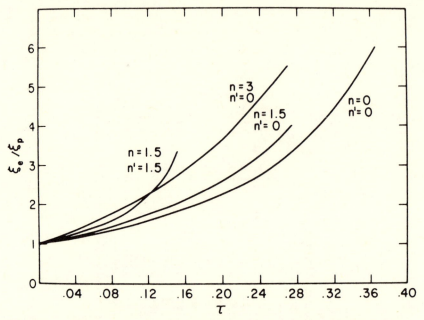

Figure 10.11. The ratio ξ_e/ξ_p for four polytropic sequences as a function of the ratio $\tau = K/|W|$. Source: Bodenheimer, P., and Ostriker, J. P., *Ap. J.* **180**, 159, 1973. (By permission of The University of Chicago Press. Copyright 1973 by the American Astronomical Society.)

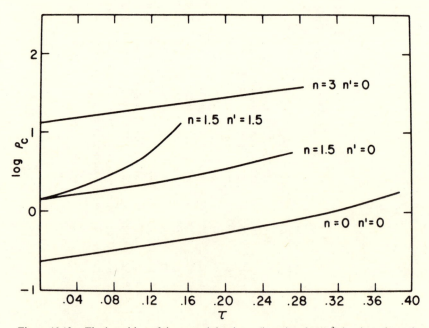

Figure 10.12. The logarithm of the central density ρ_c (in units of M/ξ_e^3) for the polytropic sequences of Figure 10.11, as a function of the ratio $\tau = K/|W|$. Source: Bodenheimer, P., and Ostriker, J. P., *Ap. J.* **180**, 159, 1973. (By permission of The University of Chicago Press. Copyright 1973 by the American Astronomical Society.)

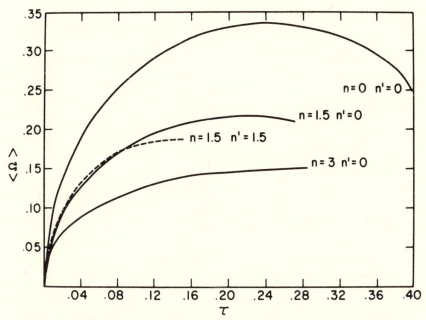

Figure 10.13. The average angular velocity $\langle\Omega\rangle$ for the polytropic sequences of Figure 10.11, as a function of the ratio $\tau = K/|W|$. The unit of $\langle\Omega\rangle$ is $(4\pi G\rho_c)^{1/2}$. Source: Bodenheimer, P., and Ostriker, J. P., *Ap. J.* **180**, 159, 1973. (By permission of The University of Chicago Press. Copyright 1973 by the American Astronomical Society.)

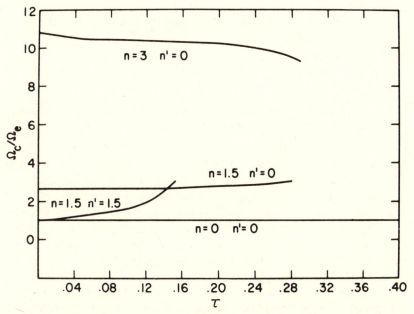

Figure 10.14. The degree of differential rotation (as measured by the ratio Ω_c/Ω_e) for the polytropic sequences of Figures 10.11, as a function of the ratio $\tau = K/|W|$. Source: Bodenheimer, P., and Ostriker, J. P., *Ap. J.* **180**, 159, 1973. (By permission of The University of Chicago Press. Copyright 1973 by the American Astronomical Society.)

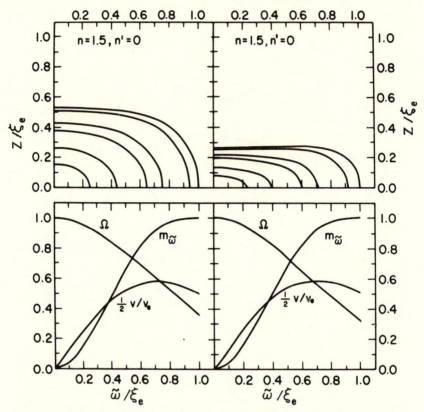

Figure 10.15. Meridional sections of two models belonging to the polytropic sequence (1.5,0). Upper portions depict the isopycnic surfaces of densities $\rho/\rho_c = 0.80$, 0.50, 0.20, 0.10, 0.01, and 0. Lower portions show the ratio Ω/Ω_e, the fraction m_ϖ of the total mass interior to the corresponding cylinder about the rotation axis, and the ratio of the circular velocity v to the surface value v_e. Source: Bodenheimer, P., and Ostriker, J. P., *Ap. J.* **180**, 159, 1973. (By permission of The University of Chicago Press. Copyright 1973 by the American Astronomical Society.)

and, finally, of the ratio Ω_c/Ω_e. Calculation of sequences was terminated when the high degree of flattening led to numerical difficulties and consequent slow convergence. Characteristics of two sequences are given in Tables 10.8 and 10.9, which also list the total internal energy U_T(with $\gamma = 5/3$), the equatorial velocity $v_e = \Omega_e \xi_e$, the effective gravities g_p and g_e at the poles and equator, and the ratio $(\xi_e/\xi_p)_c$ of the axes of the isopycnic surfaces near the center of the polytrope. Finally, detailed views of the meridional plane are displayed for two selected models in Figure 10.15. If we except the Maclaurin sequence for which $v(\varpi) \propto \varpi$, all computed velocity curves $v(\varpi)$ remind us of the galactic rotation law. Let us remember that these curves are *not* assumed *a priori*, but are obtained as a result of the solution of the equations of equilibrium.

In summary, differentially rotating polytropes greatly differ from configurations that we prescribe to rotate as solid bodies. In particular, polytropes in non-uniform rotation can store a large amount of rotational kinetic energy compared to their gravitational potential energy. *Moreover, the polytropic sequences (n,n′) resemble in all essential respects the Maclaurin sequence (0,0), except that they do not maintain uniform rotation.* Of particular importance is the fact that these sequences of differentially rotating polytropes do not terminate. This last result suggests the possible occurrence of a point of bifurcation and, eventually, dynamical instability when departure from sphericity becomes too large.

OSCILLATIONS AND STABILITY

As we recall from Section 6.7, the two sectorial frequencies belonging to second-order harmonics are, approximately,

$$\sigma_{\pm 2} = \langle \Omega \rangle \pm (2\omega + \langle \Omega \rangle^2 - 2\langle \Omega^2 \rangle)^{1/2}, \tag{47}$$

where

$$\langle \Omega^2 \rangle = \int_{\mathscr{V}} \rho(\mathbf{x})\Omega^2(\varpi)\varpi^2 \, \mathbf{dx} \Big/ \int_{\mathscr{V}} \rho(\mathbf{x})\varpi^2 \, \mathbf{dx} \tag{48}$$

(cf. equations [40] and [46]). Note, in particular, that σ_{-2} vanishes when $\langle \Omega^2 \rangle = \omega$. Thus, although the above equation only provides an approximate expression for these modes, we may surmise that the model for which $\sigma_{-2} = 0$ defines a limit of secular stability and, hence, that it corresponds to a figure of bifurcation. As was pointed out by Hunter, however, when viscosity is the dominant dissipative mechanism, a differentially rotating body will change its angular momentum distribution at least as rapidly as any sectorial disturbance is likely to grow; hence, the question of secular instability with respect to second-order sectorial modes is well-posed only if the dissipative mechanism has no effect on the equilibrium configuration, e.g., gravitational radiation-reaction. Furthermore (as was recently shown by Bardeen, Friedman, Schutz, and Sorkin), the *exact* point of bifurcation may be either before or after the point indicated by the tensor virial method along sequences of differentially rotating bodies. (According to Bardeen, numerical calculations for infinitesimally thin, differentially rotating discs indicate that the exact value of τ_b can be as much as 20 percent greater than the tensor virial estimate.) In any event, because the precise location of the point of bifurcation along a sequence of differentially rotating polytropes has yet to be calculated, equation (47) defines: (i) an *approximate* location of the point of bifurcation along a sequence of inviscid, axisymmetric models, i.e.,

$$\langle \Omega^2 \rangle = \omega, \tag{49}$$

and (ii) an *approximate* limit of dynamical stability, i.e.,

$$\langle \Omega^2 \rangle = \omega + \tfrac{1}{2}\langle \Omega \rangle^2. \tag{50}$$

For the Maclaurin sequence (0,0) the method is of course exact, so that one would expect the inaccuracy to be minimal for sequences of low polytropic indices n and n'.

According to Ostriker and Bodenheimer, Figures 10.16 and 10.17 illustrate the two sectorial modes along various polytropic sequences (compare with Figure 10.3). For the sake of completeness, Tables 10.10 and 10.11 also list the six modes that, in the limit of zero rotation, reduce to the five f-modes belonging to the spherical harmonics $l = 2$ and the lowest p-mode (cf. §6.4: equations [37] and [41]). With $\Gamma_1 = \gamma = 5/3$, the tesseral modes (σ_{+1} and σ_{-1}) and the zonal modes (σ_Z and σ_R) are always stable (see, however, Section 14.2). *The essential point of this analysis is that these polytropic sequences mimic the Maclaurin sequence in their stability properties, for the modes considered here, just as they do in their equilibrium properties.*

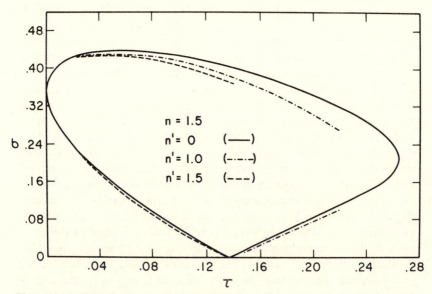

Figure 10.16.   Same as Figure 10.3, for polytropic sequences of index $n = 1.5$ with three different distributions of angular momentum (n'). The frequencies σ are given in units of $(4\pi G\rho_c)^{1/2}$. Source: Ostriker, J. P., and Bodenheimer, P., *Ap. J.* **180**, 171, 1973. (By permission of The University of Chicago Press. Copyright 1973 by the American Astronomical Society.)

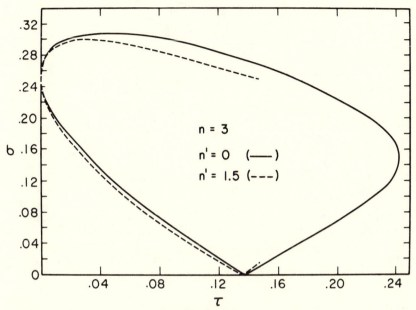

Figure 10.17. Same as Figure 10.3, for polytropic sequences of index $n = 3$ with two different distributions of angular momentum (n'). The frequencies σ are given in units of $(4\pi G\rho_c)^{1/2}$. Source: Ostriker, J. P., and Bodenheimer, P., *Ap. J.* **180**, 171, 1973. (By permission of The University of Chicago Press. Copyright 1973 by the American Astronomical Society.)

TABLE 10.10

Frequencies of oscillation[†]
($n = 1.5$, $n' = 0$)

J	τ	$\rho_c/\bar{\rho}$	σ_{-2}	σ_{+2}	σ_{-1}	σ_{+1}	σ_Z	σ_R	$\langle\Omega\rangle$
0.	0.0	5.99	.3526	.3526	.3526	.3526	.3526	.3942	.0
0.097	0.027	6.60	.2182	.4293	.3069	.4121	.3676	.3972	.1056
0.141	0.054	7.34	.1486	.4387	.2924	.4368	.3668	.4150	.1451
0.188	0.093	8.56	.0728	.4307	.2795	.4574	.3553	.4480	.1788
0.209	0.112	9.36	.0391	.4201	.2735	.4629	.3470	.4643	.1906
0.234	0.137	10.4	.0010	.4049	.2674	.4682	.3372	.4860	.2016
0.252	0.155	11.4	.0272	.3890	.2620	.4694	.3279	.5026	.2078
0.279	0.187	13.5	.0688	.3577	.2521	.4652	.3108	.5257	.2139
0.304	0.217	15.8	.1088	.3241	.2441	.4610	.2956	.5519	.2159
0.326	0.245	19.0	.1512	.2788	.2340	.4503	.2778	.5738	.2144
0.334	0.256	20.6	.1718	.2535	.2286	.4429	.2694	.5783	.2127
0.340	0.264	22.1	.2056	.2158	.2244	.4367	.2628	.5819	.2110

[†] The σ's and $\langle\Omega\rangle$ are given in units of $(4\pi G\rho_c)^{1/2}$.

SOURCE: Ostriker, J. P., and Bodenheimer, P., *Ap. J.* **180**, 171, 1973. (By permission of The University of Chicago Press. Copyright 1973 by the American Astronomical Society.)

TABLE 10.11

Frequencies of oscillation[†]

$(n = 3, n' = 0)$

J	τ	$\rho_c/\bar{\rho}$	σ_{-2}	σ_{+2}	σ_{-1}	σ_{+1}	σ_Z	σ_R	$\langle\Omega\rangle$
0.0	0,0	53.3	.2560	.2560	.2560	.2560	.2560	.2862	.0
0.055	0.015	56.0	.1879	.2984	.2302	.2850	.2618	.2847	.0552
0.095	0.043	61.6	.1292	.3088	.2151	.3033	.2598	.2936	.0898
0.114	0.061	65.6	.1001	.3073	.2086	.3098	.2539	.3027	.1035
0.130	0.078	69.7	.0754	.3031	.2034	.3144	.2474	.3121	.1139
0.145	0.094	73.9	.0534	.2968	.1987	.3172	.2406	.3210	.1217
0.167	0.121	81.5	.0197	.2832	.1916	.3197	.2287	.3358	.1318
0.179	0.136	85.9	.0025	.2741	.1878	.3200	.2219	.3437	.1358
0.190	0.150	90.2	.0137	.2644	.1842	.3197	.2152	.3514	.1390
0.210	0.178	98.5	.0441	.2431	.1779	.3180	.2024	.3665	.1436
0.228	0.204	106.	.0741	.2188	.1727	.3154	.1903	.3821	.1465
0.252	0.241	113.	.1394	.1583	.1676	.3110	.1744	.4064	.1488

[†] The σ's and $\langle\Omega\rangle$ are given in units of $(4\pi G\rho_c)^{1/2}$.

SOURCE: Ostriker, J. P., and Bodenheimer, P., *Ap. J.* **180**, 171, 1973. (By permission of The University of Chicago Press. Copyright 1973 by the American Astronomical Society.)

TABLE 10.12

Properties of models at point where σ_{-2} goes to zero

n	0	1.5	1.5	1.5	3	3
n'	0	0	1	1.5	0	1.5
$\tau_b(n,n')$	0.1376	0.1377	0.1375	0.1375	0.1381	0.1348
VT	0	0.00002	0.00001	0.00030	0.00230	0.04800
$\tau_b(n,n') - \tau_b(0,0)$	0	+0.0001	−0.0001	−0.0001	+0.0005	−0.0028

SOURCE: Ostriker, J. P., and Bodenheimer, P., *Ap. J.* **180**, 171, 1973. (By permission of The University of Chicago Press. Copyright 1973 by the American Astronomical Society.)

Particularly striking is the invariance of the point $\tau_b(n,n')$, i.e., the point where σ_{-2} goes to zero. Table 10.12 shows that the variation of $\tau_b(n,n')$ is never significantly greater than the estimated numerical inaccuracy of the models. Let us recall that a similar puzzling behavior was already observed for uniformly rotating polytropes[6] (cf. §10.3). As was shown by Roberts, when the level surfaces of a self-gravitating configuration are homothetic spheroids, the point $\tau_b = W_{12,12}/|W|$ along a given sequence is *rigorously* independent of the mass distribution; however, as we recall from Section

[6] In this respect, it is also worth noting that $\tau \approx 0.14$ represents approximately the maximum rotational energy an axisymmetric *stellar* system can contain and remain stable to the formation of a bar. See, e.g., Ostriker, J. P., and Peebles, P. J. E., *Ap. J.* **186**, 467, 1973 (see especially their Table 1, which summarizes many previous studies in this field).

4.4, such models are definitely *not* figures of equilibrium in the case of barotropes. Nevertheless, since the inner level surfaces of all models discussed by Bodenheimer and Ostriker never greatly deviate from spheroids, the above result might provide the clue as to why the values of $\tau_b(n,n')$ are nearly independent of n and n'. Clearly, this problem requires further study.

Finally, in all cases for which very flattened configurations could be constructed numerically, the point of dynamical instability (50) is reached when $\tau = \tau_i(n,n') = 0.26 \pm 0.02$. Again, the departure from the value $\tau_i = 0.2738$ for the Maclaurin sequence is of the same order as the integral error VT on the models at that point (cf. equation [45]). In this case, however, the significance of the numerical results is not obvious, since the errors on the equilibrium computations of the flattened models are relatively large.

10.5. A GENERAL SUMMARY

This chapter has been concerned with an investigation of self-gravitating barotropes rotating freely in space. Basically, we raised three questions:

(i) Given a total mass M, a total angular momentum J, and a prescribed rotation law, is it always possible to construct a sequence of axisymmetric, inviscid models along which the ratio $\tau = K/|W|$ steadily increases from $\tau = 0$ to $\tau = 0.5$ (the limit allowed by the virial theorem)?

(ii) As we proceed along such a sequence of models in a state of permanent rotation, can we eventually reach a point of bifurcation, i.e., a point at which non-axisymmetric figures of equilibrium branch off?

(iii) Are the axisymmetric configurations stable with respect to infinitesimal disturbances?

In the present context, the Maclaurin spheroids are particularly worth discussing for they range from a sphere to an infinitely thin disc, when the ratio τ steadily increases from $\tau = 0$ to $\tau = 0.5$. Thus, this sequence of uniformly rotating, homogeneous spheroids does not terminate. However, the point $\tau = \tau_b = 0.1375$ corresponds to a figure of bifurcation at which uniformly rotating ellipsoids with genuine triplanar symmetry branch off, i.e., the Jacobi ellipsoids. Actually, the total mechanical energy $K + W$ is smaller in the triaxial figure than in the corresponding spheroid having the same values of J, M, and $\mathscr{V}$. Hence, past the point $\tau = \tau_b$, all Maclaurin spheroids are secularly unstable. That is, they slowly evolve to the Jacobi sequence whenever viscous friction is operative. (Gravitational radiation-reaction, like viscosity, also makes an inviscid Maclaurin spheroid secularly unstable past the point $\tau = \tau_b$, but interference between viscous friction and gravitational radiation-reaction can at least partially suppress secular instability.) Finally, further along the Maclaurin sequence, when

$\tau > \tau_i = 0.2738$, the spheroids become dynamically unstable with respect to disturbances that transform them into barlike configurations; the time scale of this instability is of the order of $(\pi G \rho)^{-1/2}$.

Uniformly rotating polytropes of index n were considered next. They differ in one important respect from the Maclaurin sequence. That is, when $n > 0$ each sequence does terminate at a point $\tau = \tau_{max}$ (say), and the values of τ_{max} decrease sharply with polytropic index. (While $\tau_{max} = 0.50$ along the Maclaurin sequence, $\tau_{max} \approx 0.12$ when $n = 1$, and τ_{max} is already reduced to a mere few percent when $n = 3$.) In other words, the amount of rotational kinetic energy that a uniformly rotating, centrally condensed polytrope is able to sustain (when compared to its gravitational potential energy) is not very large. This is not surprising. In the regions of a centrally condensed body within which the pressure forces are small compared to the inertial forces, the rotation law must approach Kepler's third law, i.e., $\Omega^2 \propto \varpi^{-3}$. Accordingly, if we restrict ourselves to uniformly rotating bodies, we cannot construct centrally condensed figures of equilibrium beyond the small range $0 \leqslant \tau \ll 0.5$. As a consequence, bifurcation—and the ensuing secular instability—does not occur along sequences of axisymmetric polytropes for which $n \geqslant 0.808(\tau_{max} \leqslant 0.138)$. Indeed, configurations having genuine triplanar symmetry always branch off at the point $\tau \approx 0.138$, which is quite independent of n. Since $\tau_{max} \leqslant 0.138$ when $n \geqslant 0.808$, no figure of equilibrium belonging to these sequences can store enough rotational kinetic energy to reach the point $\tau \approx 0.138$. In summary, uniformly rotating, centrally condensed polytropes are always secularly and dynamically stable because they cannot rotate very rapidly!

A completely different picture emerges from the study of *inviscid*, differentially rotating polytropes for which we prescribe a given angular momentum distribution (cf. equation [44]). In this case, the models closely simulate, in all essential respects, Maclaurin spheroids, except that they do not maintain uniform rotation. In particular, the sequences do not terminate; and, once more, bifurcation might occur when τ is very closely equal to the value $\tau = \tau_b$ that was obtained for the Maclaurin spheroids. Beyond this point, the models are expected to be secularly unstable if gravitational radiation-reaction is operative. *None of these results has been properly demonstrated.* In addition, dynamical instability always sets in beyond the point $\tau \approx \tau_i$, which again does not greatly depend on the particular sequence. As a matter of fact, changes in the polytropic index or in the angular momentum distribution have little effect upon these results. And, as we shall see in Section 13.2, similar results were also obtained for zero-temperature white dwarfs in non-uniform rotation. Thus, the classical work on uniformly rotating, homogeneous spheroids has a range of validity much greater than was usually anticipated. This is fortunate, since it will now enable us to reconsider, among other questions, the classical hypothesis of fission and equatorial break-up.

BIBLIOGRAPHICAL NOTES

The classical references in this subject are those of Jeans; see:

1. Jeans, J. H., *Problems of Cosmogony and Stellar Dynamics*, Cambridge: At the Univ. Press, 1919.

and reference 44 of Chapter 7. Despite some minor flaws, both books still make interesting reading today. A more rigorous treatment of stability is contained in:

2. Lyttleton, R. A., *The Stability of Rotating Liquid Masses*, Cambridge: At the Univ. Press, 1953.

Section 10.2. A modern account of the Maclaurin and Jacobi ellipsoids is contained in reference 2. In this book, the necessary and sufficient conditions for equilibrium are derived from the equations of motion. Like Poincaré, Lyttleton discusses the stability problem by means of the very cumbersome Lamé functions. In reference 28 of Chapter 3, Chandrasekhar derives all the necessary conditions for equilibrium from the second-order virial equations. This approach is thus less general (as far as equilibrium is concerned), but it greatly simplifies the stability analysis. Equations (16) and (17) were first derived and solved by Lebovitz (reference 46 of Chapter 6). The role of viscosity has been investigated by:

3. Roberts, P. H., and Stewartson, K., *Astroph. J.* **137** (1963): 777–790.
4. Rosenkilde, C. E., *ibid.* **148** (1967): 825–832.

A comprehensive study of secular stability is due to:

5. Robe, H., *Mém. Soc. Roy. Sci. Liège* (5) **18** (1969): 1–69.

This paper contains, in particular, a detailed derivation of equation (22) and a rigorous treatment of the boundary conditions. Other approaches will be found in:

6. Kopal, Z., *Astroph. and Space Sci.* **24** (1973): 145–174.
7. Bertin, G., and Radicati, L. A., *Astroph. J.* **206** (1976): 815–821.

The following paper is further quoted in the text:

8. Press, W. H., and Teukolsky, S. A., *ibid.* **181** (1973): 513–517.

Detailed studies of the influence of gravitational radiation on homogeneous ellipsoids will be found in:

9. Chandrasekhar, S., *ibid.* **161** (1970): 561–569.
10. Miller, B. D., *ibid.* **187** (1974): 609–620.

The result of Friedman and Schutz may be traced to reference 36. The combined effects of viscosity and gravitational radiation-reaction are discussed in:

11. Lindblom, L., and Detweiler, S. L., *ibid.* **211** (1977): 565–567.
12. Detweiler, S. L., and Lindblom, L., *ibid.*, **213** (1977): 193–199.

Section 10.3. The most satisfactory results will be found in:

13. James, R. A., *ibid.* **140** (1964): 552–582.

Expansions in the parameter v are discussed in references 14–19 of Chapter 5. For uniformly rotating polytropes, the following papers may also be noted:

14. Thüring, B., *Astron. Nachr.* **253** (1934): 73–90.
15. Steensholt, G., *Zeit. f. Astroph.* **10** (1935): 391.
16. Kopal, Z., *Monthly Notices Roy. Astron. Soc. London* **99** (1939): 266–271.
17. Papoian, V. V., Sedrakian, D. M., and Chubarian, E. V., *Astrofizika* **3** (1967): 41–54.
18. Williams, P. S., *Astroph. and Space Sci.* **34** (1975): 425–430.

See also reference 36 of Chapter 5. Second-order expansions in v were derived with various measures of success by:

19. Occhionero, F., *Memorie Soc. Astron. Italiana* **38** (1967): 331–348.
20. Aikawa, T., *Sci. Reports Tôhoku Univ.* (I) **51** (1968): 61–66.
21. Anand, S. P. S., *Astroph. J.* **153** (1968): 135–145.

Further improvements are due to:

22. Linnell, A. P., *Astroph. and Space Sci.* **37** (1975): 73–86.
23. Smith, B. L., *ibid.* **35** (1975): 223–240; *ibid.* **43** (1976): 411–424.

See also reference 41 of Chapter 16. The variational methods are discussed in the following papers:

24. Roberts, P. H., *Astroph. J.* **137** (1963): 1129–1141; *ibid.* **138** (1963): 809–819.
25. Hurley, M., and Roberts, P. H., *ibid.* **140** (1964): 583–598.
26. Hurley, M., and Roberts, P. H., *Astroph. J. Supplements* **11** (1965): 95–119.

See also:

27. Auchmuty, J. F. G., and Beals, R., *Astroph. J. Letters* **165** (1971): L79–L82.
28. Auchmuty, J. F. G., and Beals, R., *Archive Rat. Mechanics and Analysis* **43** (1971): 255–271.
29. Auchmuty, J. F. G., *Lecture Notes in Mathematics* **430**, pp. 28–60, Berlin: Springer-Verlag, 1974.

A critical comparison between the different results is due to:

30. Aikawa, T., *Sci. Reports Tôhoku Univ.* (I) **54** (1971): 13–27.

This paper contains the most detailed numerical results (see Appendix E). The secular instability of uniformly rotating polytropes is discussed in:

31. Tassoul, J. L., and Ostriker, J. P., *Astron. and Astroph.* **4** (1970): 423–427.

The reference to Miller is to her paper:

32. Miller, B. D., *Astroph. J.* **181** (1973): 497–512.

A criterion for the occurrence of a point of bifurcation along sequences of uniformly rotating bodies was originally discussed by:

33. Chandrasekhar, S., and Lebovitz, N. R., *ibid.* **135** (1962): 238–247.

In this paper, it is claimed that bifurcation occurs when $\Omega^2 = \omega$ (cf. equation [40]). Although this is what happens at the point where the Jacobi sequence branches off the Maclaurin sequence, it can be shown that the condition $\Omega^2 = \omega$ is in fact approximate in the case of centrally condensed bodies, and that bifurcation might well occur when Ω^2 is less than ω. In other words, $\Omega^2 = \omega$ is a *sufficient condition* for the existence of a point of bifurcation for centrally condensed bodies, *but it has never been proved rigorously that this condition is also necessary in the general case of nonhomogeneous configurations.* (The same misapprehension will be noticed also in reference 31.) Hence, the limit $\Omega^2 = \omega$, which was originally thought to be exact, is in fact approximate; that is, bifurcation might well occur when Ω^2 is less than ω. (This excludes of course the Maclaurin spheroids for which the limit $\Omega^2 = \omega$ *is* exact.) Clarification of this point has been made by:

34. Friedman, J. L., and Schutz, B. F., *Astroph. J. Letters* **199** (1975): L157–L159.
35. Hunter, C., *Astroph. J.* **213** (1977): 497–517.
36. Bardeen, J. M., Friedman, J. L., Schutz, B. F., and Sorkin, R., *Astroph. J. Letters* **217** (1977): L49–L53.

See especially reference 36 (and further work quoted therein).

Section 10.4. Models in slow differential rotation will be found in:

37. Danoz, N., *C.R. Soc. Phys. Genève* **47** (1930): 88–92.
38. Krat, W., *Zeit. f. Astroph.* **12** (1936): 192–202.
39. Bhatnagar, P. L., *Bull. Calcutta Math. Soc.* **32** (1940): 21–41.
40. Roy, S. K., *ibid.* **35** (1943): 85–98.

See also:

41. Blinnikov, S. I., *Astron. Zh.* **49** (1972): 654–658.

The results quoted in this section are those of:

42. Stoeckly, R., *Astroph. J.* **142** (1965): 208–228.
43. Bodenheimer, P., and Ostriker, J. P., *ibid.* **180** (1973): 159–169.
44. Ostriker, J. P., and Bodenheimer, P., *ibid.* **180** (1973): 171–180.

See, however, paper 36. The reference to Roberts is to his paper:

45. Roberts, P. H., *ibid.* **136** (1962): 1108–1114.

Centrally condensed models with triplanar symmetry were actually constructed by:

46. Clement, M. J., *ibid.* **148** (1967): 159–174.

A preliminary discussion of polytropic discs in the limit $\tau \to 0.5$ will be found in:

47. Spiegel, E. A., *Symposium on the Origin of the Solar System*, pp. 165–178, Nice, 1972.

11

Collapse and Fission

11.1. INTRODUCTION

Although there is as yet no general agreement about the process of star formation, it is commonly accepted that stars are born out of diffuse gas and dust found in interstellar space. The most detailed theories postulate conditions in the interstellar medium such that large-scale hydrodynamical or thermal instabilities will occur over extended regions of the Galaxy. For instance, according to the classical picture first put on a quantitative basis by Jeans in 1902, the initial step in star formation is the pulling together, by gravitational forces, of some large mass of interstellar matter. This requires, in essence, that the gravitational binding energy of the cloud must exceed the thermal kinetic energy, plus the energy of rotation and magnetic fields. Given the typical conditions likely to prevail in the interstellar medium, collapse occurs only for massive clouds ($M \gtrsim 10^3 - 10^4 M_\odot$); hence, in this approach, star formation must generally proceed via the fragmentation of a massive collapsing cloud, the ultimate fragments being protostars with a wide range of masses. At present, however, it is by no means clear what the detailed mechanism of fragmentation of a protostellar cloud is. In any event, observation strongly suggests that new stars are formed in clusters, although the creation of a single star out of an individual cloud cannot be excluded on observational grounds.

A number of basic questions arise at once when rotation is considered in connection with the pre-main-sequence phases of stellar evolution:

(i) How does the angular momentum in single stars originate?

(ii) How can we explain the existence of so many double and multiple stars? And how do we account for the existence of planets around the Sun (and probably around other stars)?

(iii) To what extent does rotation modify the hydrodynamical collapse to stellar densities and the subsequent quasi-static contraction to the main sequence?

(iv) Why are the observed rotational velocities in upper main-sequence stars considerably higher than those in main-sequence stars of spectral type F5 or later?

This chapter will discuss a few specific aspects of these questions. Unfortunately, as we shall see, most available theories are still conjectural, and all of them require further quantitative studies. Section 11.2 is devoted to the initial stages of stellar formation. The various fission theories of close binaries are expounded in Section 11.3. We conclude the chapter with the quasi-static contraction phase toward the main sequence, together with a brief comparison with the meagre observational data. A basic knowledge of the pre-main-sequence evolution of spherically symmetric stars (i.e., models that are devoid of rotation and magnetic fields) is expected of the reader. The role of magnetic fields will be further discussed in Chapter 15.

11.2. THE EARLY PHASES

The evolution of an idealized spherical star—from "birth" to the main sequence—has an extensive literature. Not surprisingly, the gravitational collapse of a rotating gas cloud has received, comparatively, much less attention. Indeed, when considering that problem, we are faced at once with a severe difficulty: the initial angular momentum of a typical collapsing cloud must apparently be reduced by many orders of magnitude to produce the stars that we observe. For example, if a protostellar cloud acquires its angular momentum from the galactic rotation, it is then much larger than that inferred for single or even double stars on the main sequence. (To be specific, with an original density of about 10^{-24} g/cm^3, we would obtain $J \approx 10^{56}(M/M_\odot)^{5/3}$ g cm^2/sec; in comparison, the most rapidly rotating main-sequence stars have at most $J \approx 10^{51} - 10^{52}$ g cm^2/ sec!) This assumption may be questioned, though, since neither the stellar rotation axes nor the axes of binaries exhibit any preferential orientation with respect to the galactic plane. The observed randomness of these axes suggests, rather, that the angular momentum may have originated in the turbulent motions that are present in the interstellar medium. In this case, the rotational kinetic energy of a protostellar cloud would be expected to be comparable with its translational kinetic energy, which would then lead to an angular momentum in equilibrium stars that is even *higher* than that derived from the galactic rotation. Thus, unless stars are formed in very quiet regions of the Galaxy (with angular momenta far below the average), the effects of rotation in a collapsing cloud must become of paramount importance well before stellar conditions are reached. In all likelihood, both the galactic rotation and random motions of subregions will contribute to the final angular momentum of a star. We are led, therefore, to seek some mechanism to dispose of this excess angular momentum in a collapsing cloud. This is the so-called "angular momentum problem."

In order to overcome the above difficulty, it has often been suggested that magnetic braking torques may be able to transfer most of the angular momentum of a collapsing cloud to the galactic background. As was shown by Gillis, Mestel, and Paris, however, the time scale for magnetic transport of angular momentum is probably larger than the free-fall time of a protostellar cloud, so that the total angular momentum of this large mass should be approximately conserved during the early stages of its collapse (see, however, Section 11.4). Another tentative approach to the angular momentum problem is the formation of multiple systems during the accelerating collapse of a protostellar cloud. Indeed, multiple stars can store much more angular momentum in the form of orbital motions than single stars can contain in their rotational motions. Accordingly, our problem would be greatly reduced (if not solved) were it possible to transport the *spin* angular momentum of a massive collapsing cloud into *orbital* angular momenta of protostars in a cluster. At this time, the detailed mechanism of fragmentation of a rotating gas cloud undergoing collapse is still vague and unsatisfactory. Nevertheless, if we assume that such a process does take place during the initial contraction of a protostellar cloud, the subsequent evolution of the cluster of protostars may be described, to first approximation, by the gravitational interaction of an assembly of point masses. Numerical calculations by van Albada indicate that groups of stars with some tens of members, and radii ranging from 10^2 to 10^5 AU, disintegrate in a time interval of 10^3 to 10^8 years, yielding wide (visual) binaries with major axes one or two orders of magnitude smaller than the original radius of the group. These double stars are in general surrounded by a few companions of smaller mass moving in extended orbits about the central binary. Close binaries cannot be formed in this manner, though, for the disintegrating cluster must then be so small that collisions between protostars will occur during the dynamical evolution of the cluster. Thus, even if the above suggestions minimize the angular momentum problem, they still leave many fundamental questions unanswered: how does rotation affect the collapse of a protostar? How can we explain the formation of wide *and* close binaries? And what role did rotation play in the formation of the solar system? It is to these questions that we now turn.

DYNAMICAL COLLAPSE. At present, *reliable* hydrodynamical calculations have been completed only for spherically symmetric protostars, neglecting rotation and magnetic fields. In this ideal case, the following sequence of events is thought to take place. Because the spherical protostar initially has a very low pressure, the system undergoes gravitational collapse on the free-fall time scale of the order of 10^5 to 10^6 years. This initial phase is very nearly isothermal, since the protostar is still transparent to its own radiation. As the collapse progresses, the central density increases more rapidly than in the outer regions. The inner core thus becomes more opaque and, as more energy becomes trapped inside the body, the central temperature begins to rise substantially. The concomitant increase in pressure begins to decelerate and eventually halt the collapse of the central core. However, this region is still compressed and heated, because material from the outer layers continues to move inward almost in free fall. When the central temperature reaches 1,800°K, hydrogen molecules begin to dissociate. This effect reduces the adiabatic exponent below the critical value 4/3 and causes a second phase of dynamical collapse (cf. §6.4). The bulk of the material eventually comes into hydrostatic equilibrium, but only after the dissociation and ionization of the dominant chemical constituents have been achieved. This original phase has a total duration of about 10^6 years. Thence, as we shall see in Section 11.4, the evolution proceeds through a slow contraction phase that may be viewed as a sequence of quasi-equilibrium states.

The influence of rotation upon the accelerating collapse of a protostar is still a very controversial issue. Numerical calculations which include radiative energy transport have been made independently by Larson, by Tscharnuter, and by Black and Bodenheimer to describe the early stages of collapse of axially symmetric protostars of 1, 2, 5, and $60M_\odot$. For the sake of simplicity, it is assumed that the fragmentation of the primeval cloud has already taken place during the initial step of its collapse, and that each fragment (i.e., protostar) is endowed with an angular momentum below the limiting value above which the collapse of a given protostar cannot occur. Most of these calculations consider initial configurations with constant density and constant angular velocity. Also, although a net transfer of angular momentum may take place in nature, all mechanisms that could produce such an effect are neglected, i.e., magnetic fields and frictional forces are not taken into account.

In Larson's computations (with $M = M_\odot$), the collapse proceeds initially in the same nonhomologous fashion as for a nonrotating model, and the density distribution again becomes centrally peaked. Infall of material takes place preferentially along the rotation axis, where the collapse is least inhibited by centrifugal forces. This causes the central density to rise by a substantial factor before centrifugal forces halt further collapse. Because of the rapid nonhomologous increase in central concentration, centrifugal forces become sufficient to halt the collapse first at the center of the protostar; meanwhile, material in the outer regions continues to fall inward and accumulate in a ring-shaped domain around the periphery of the central region which has stopped collapsing. In Larson's opinion, the ensuing instabilities in this rotating ring would result in fragmentation and the formation of two or more stars in orbit around each other. (In this picture, the single stars would be accounted for as escapers from unstable multiple systems.) The formation of a ring structure in the interior of a collapsing protostar has been confirmed by the independent numerical calculations of Black and Bodenheimer (with $M = 1$, 2, and $5M_\odot$, and various total angular momenta).

Unfortunately, the appearance of a ring-shaped structure has *not* been confirmed by the independent numerical work of Tscharnuter. Using the same initial conditions as Larson's (but another numerical program, in which the angular dependence of every function is expanded into Legendre polynomials), he found that the whole mass reexpands after the initial collapse, showing nowhere a tendency to form a rotating "doughnut" with a density minimum in its center. In all other model calculations carried out by Tscharnuter (with $M = M_\odot$ and $M = 60M_\odot$, and various total angular momenta) the collapse always proceeds with a density distribution that is highly peaked at the center. His calculations indicate, therefore, that a "stellar nebula" with a protostar in its center would evolve from the very beginning of the isothermal collapse. If correct, this could well lead to the formation of a planetary system.

At this moment, it is not clear whether the foregoing discrepancy is due to some inappropriate formulation of the boundary conditions, whether it is due to rounding-off errors which grow in an uncontrolled way, or whether a larger number of terms should have been retained in Tscharnuter's truncated expansions.[1] Moreover, it is not at all obvious that the accelerating collapse of a rotating protostar will necessarily proceed along a sequence of axisymmetric models, were they ring-shaped structures or "stellar nebulae." (Non-axisymmetric motions may also occur at the very early stages of the collapse, thus modifying the whole subsequent picture!) Finally, nobody knows so far to what extent the above models are sensitive to the chosen initial conditions and to the various physical processes that have been hitherto neglected (such as the transport of angular momentum by turbulence or magnetic torques) and that may become important during some phases of the isothermal collapse of a protostar.

DOUBLE STARS.   On the evidence we have, more than half of the stars are members of gravitationally bound double (or multiple) systems. The binaries range from close pairs in which the components are almost in contact (with centers 10^{-2} AU apart) to wide pairs with major semi-axes (a, say) up to 10^4 AU. Systems with $a < 10^0$ AU are mainly spectroscopic or eclipsing binaries, whereas those with $a > 10^1$ AU are almost entirely

---

[1] After this text was written, however, it was reported at the I.A.U. Symposium No 75 (Geneva, September 1976) that Tscharnuter, who did not observe ring-shaped structures in his first calculations, now obtains rings.

visual binaries. Discoveries of stars in the intermediate range are few to date. Moreover, in spite of many statistical studies of the frequency distribution of the various physical parameters of binaries, we do not know yet whether we may regard all double stars as objects of a single genus, or whether we must look for a separate mode of formation for the wide and close binaries. For example, as was shown by Blaauw and van Albada, the frequency distribution of $\log_{10} a$ for O9 to B5 stars in the nearest associations lends support to the existence of two groups of early-type binaries, close and wide pairs, with a deficiency of double stars having separations in the range 10^0 AU $< a < 10^2$ AU. However, because this break is located near the point where the incidence of selection is the largest, it is not known so far to what extent their result is real and not due to some selection effect. By contrast, more detailed studies covering a wide range of spectral types and luminosity classes (notably those of Dommanget and Heintz) do not lend support to the existence of any clear-cut cleavage in the distribution of $\log_{10} a$, thus favoring the common-origin argument. As was recently shown by Abt and Levy, however, a quite different picture emerges from their statistical analysis of multiplicity among 135 F3–G2 (IV or V) bright field stars. Indeed, it is found that binaries with periods less than 100 days have a frequency distribution proportional to the cube root of the secondary mass, whereas the longer-period systems fit the van Rhijn function. On the basis of this result, it would thus appear that there are two distinct types of binaries involved: (i) short-period binaries that *might* be fission systems in which a single protostar subdivided because of excessive initial angular momentum, and (ii) long-period binaries that *might* represent pairs of protostars that contracted separately but are gravitationally held to each other. Obviously, this is a purely observational problem that will be resolved only by adding new material to the existing data.

Three principal theories have been proposed to explain the formation of double stars: (i) simple capture (Stoney, 1867), (ii) formation by separate nuclei (Laplace, 1796), and (iii) fission and subsequent separation (Kelvin and Tait, 1883). Let us examine these working hypotheses in sequence.

At present, very few people still accept the simple *capture theory*, since it requires a third body (i.e., a third star or a resisting medium) to absorb the excess energy liberated during the capture of one star by another to form a gravitationally bound double star. According to the recent calculations by Mansbach, the rate of binary formation due to three-body collisions is too small to account for any binaries in the solar neighborhood. The motion of two point masses in a resisting medium has been also considered by van Albada, the resisting force arising from gravitational interaction between interstellar gas and stars. Again, only small amounts of energy could be absorbed in an encounter. This also makes the second version of the capture theory a very inefficient process for double-star formation.

The *separate nuclei theory* apparently affords sufficient latitude for the explanation of any binary (or multiple) system, except perhaps the very close double stars. This theory supposes that, in some manner, nuclei of condensation developed that began to go round each other under conditions not much different from those observed in double stars. It has often been criticized as not constituting an explanation but as explaining the little known by the less known. In any case, because this approach does not account for spectroscopic binaries, it should be supplemented by some other mechanism that can produce very close systems. The process of fission might be regarded as such.

The *fission theory* is perhaps the most appealing mechanism, but (as we shall see below) it also meets with some difficulties. This theory postulates conditions such that a contracting protostar in its primal nebulous stage (or perhaps at a later one) divides into two masses which would at first revolve into surface contact. Obviously, although a double star produced in this manner may appear as a close binary while the components still have protostellar dimensions, the system will inevitably develop into a wide binary, if both components contract to the main sequence without any marked decrease in their orbital angular momentum. Thus, the fission theory also has difficulties explaining spectroscopic binaries. Two possibilities have been suggested to explain the formation of close binaries: *first*, retarded fission of a slowly contracting, pre-main-sequence star into two components during its quasi-static phase; and, *second*, gradual loss of orbital angular momentum during the contraction of both components toward the main sequence.

To be specific, the first alternative supposes that close and wide binaries do have a common origin, i.e., fission of a gaseous mass that is endowed with a too large amount of angular momentum: *either* fission occurs during the accelerating collapse of a rotating protostar, the two collapsing fragments eventually forming a wide (visual) binary by the time they reach the main sequence, *or* fission is somehow delayed and occurs when a single collapsing mass has reached stellar proportions only, thus leading to a close (spectroscopic) binary when both components emerge on the main sequence. This approach will be further discussed in the next section.

The second alternative (mainly propounded by Huang and Mestel) postulates that fission always occurs during the accelerating phase of the collapse. In other words, all binaries consist initially of a close pair of protostars. Thence, their subsequent evolution depends on the relative importance of magnetic activity and gas ejection during the collapse of each protostellar fragment. Indeed, if there is a strong coupling between the spin and orbital motions, any kind of magnetic braking of the individual fragments (such as those discussed at the end of Section 11.4) will then be offset at the expense of the orbital angular momentum. Thus, whenever this mechanism is operative, it implies a steady reduction in the mutual distance a, so that the system always remains a close binary in spite of the contraction of the individual components. Because a large number of unknown parameters appears in this theory, it still remains to prove that the losses will occur at the exact rate required to maintain both stars in close proximity.

11.3. THE FISSION PROBLEM

Broadly speaking, the fission problem may be stated as follows. Will a single contracting body—initially endowed with some angular momentum and free from any appreciable external disturbance—eventually separate into a pair of detached fragments; and, if so, how will the separation take place? Since the basic idea is that of a single body first becoming elongated and then breaking up into two pieces, any approach to this question is necessarily an "exercise" in three-dimensional hydrodynamics. Hence, in spite of many attempts, no conclusive solution to this problem has yet been developed. As a matter of fact, the only models that have been treated with some rigor consider the small oscillations of various homogeneous ellipsoids that pass through series of equilibria as they slowly contract. (The accelerating collapse has received, comparatively, very little attention.) Thus, the three evolutionary sequences that we will now briefly describe in turn refer, at best, to the formation of close binaries during the quasi-static contraction phase of a star toward the main sequence (see also Section 11.4). We will conclude this section with a critical assessment of these idealized models.

THE CLASSICAL APPROACH.   As we recall from Section 10.2, the Maclaurin and Jacobi ellipsoids represent the simplest figures of equilibrium for self-gravitating bodies in uniform rotation. The classical formulation of Poincaré considers the slow evolution of a uniformly rotating, viscous spheroid as it moves along the Maclaurin sequence and, thence, bifurcates to the Jacobi sequence. The latter circumstance was thought of as arising from the gravitational contraction of the original mass, and friction was supposed to be large enough to keep the system in uniform rotation. Following Jeans (the main proponent of this model), we shall assume a secular increase in angular momentum, with the density remaining uniform, since this is formally equivalent to constant angular momentum with density increasing gradually. Alternately, we may also describe the evolution of the slowly contracting body in terms of the ratio $\tau = K/|W|$, where K and W are, respectively, the rotational kinetic energy and the gravitational potential energy $(0.0 \leqslant \tau \leqslant 0.5)$. Then, for small values of τ (i.e., $\tau < \tau_b = 0.1375$), the contracting mass takes the form of an oblate spheroid. At the point of bifurcation $\tau = \tau_b$, the viscous body becomes secularly unstable and an ellipsoid with three unequal axes develops. Obviously, for this evolutionary scheme to occur it is necessary for the Kelvin-Helmholtz contraction time t_K to be much larger than the viscous time scale t_v; that is to say,

$$t_v \ll t_K. \tag{1}$$

Now, if we assume that inequality (1) holds true, how does the subsequent evolution of the configuration take place? At first, the triaxial body slowly contracts and elongates, while retaining its ellipsoidal shape. Later on, as the largest diameter becomes about three times the length of the shortest one, a point is reached where the ellipsoid becomes secularly unstable with respect to some third-order harmonics; and, at this point, a series of pear-shaped figures branches off the Jacobi sequence.[2] It was originally believed that these new figures would prove secularly and dynamically stable, so that the contracting ellipsoid would bifurcate and then evolve by means of equilibrium forms along this series. This would have implied a transfer (on the time scale t_v) to the pear-shaped figures, followed by a quasi-static contraction along that sequence (on the time scale t_K), with the furrow continually deepening until a stage of complete separation occurred. This idealized scenario became much less credible, however, when Liapunov and Jeans proved that the pear-shaped figures are secularly unstable, and Cartan showed that the Jacobi ellipsoids become dynamically unstable precisely at the point where these pear-shaped figures branch off the Jacobi sequence. Therefore, even if the above sequence of events adequately represents the formation of a close binary, the final break-up of the initial mass into two separate pieces must occur in a cataclysmic way (i.e., on a dynamical time scale) after the point of bifurcation along the Jacobi sequence is passed. So far, no one has yet been able to prove or disprove that an elongated pear-shaped figure with one end larger than the other will eventually split into two detached masses in mutual orbital motion. This would require a *fully* nonlinear, time-dependent study of the evolution of a pear-shaped figure, the actual boundary of the configuration being itself an unknown at every instant.

[2] Bifurcation occurs when the semi-axes of the Jacobi ellipsoid are in the ratios of $a_3/a_1 = 0.345026$ and $a_2/a_1 = 0.432159$. This particular figure corresponds to $\tau = 0.1628$. Note that the qualifier *pear-shaped* is actually a misnomer since the bifurcation figures very much resemble *egg-shaped* bodies, with no change in curvature in the equatorial planes (see, e.g., Figure 11.2).

THE OSTRIKER APPROACH.   The above formulation crucially depends on the action of viscosity to make a contracting Maclaurin spheroid transform itself into a Jacobi ellipsoid at the point of bifurcation $\tau = \tau_b$. Assume now that viscous friction does *not* play an important role during the quasi-static contraction of a homogeneous body which, somehow, maintains uniform rotation. Putting it another way, consider the opposite situation in which

$$t_K \ll t_v. \tag{2}$$

Then, as we know, the point of bifurcation $\tau = \tau_b$ along the Maclaurin sequence has no significance in the present discussion, because no secular instability occurs at this point so long as viscosity can be neglected. Thus, if inequality (2) holds true, a slowly contracting Maclaurin spheroid will not enter the Jacobi sequence at all, but will steadily evolve along the Maclaurin sequence until the point $\tau = \tau_i = 0.2738$ is encountered. There, as we recall from Section 10.2, the configuration becomes dynamically unstable with respect to second-order infinitesimal disturbances. What happens next? A partial answer to this question is provided by the work of Rossner and Fujimoto, who considered the finite-amplitude oscillations of a homogeneous spheroid beyond the point $\tau = \tau_i$. In both works, the problem is made tractable by considering oscillations that preserve the ellipsoidal nature of the surface and generate internal motions having a uniform vorticity. (Rossner specifically deals with the finite-amplitude motions of an incompressible body, whereas Fujimoto considers the non-axisymmetric collapse of a compressible spheroid having a spatially uniform but time-varying density.) The motion of these configurations is quite complex but might very roughly be described as the end-over-end tumbling of a spheroid that is highly prolate about one of its principal axes transverse to the rotation axis. According to Ostriker, there are strong arguments for believing that, after the point $\tau = \tau_i$ is reached, the needle-like configuration will break up into two or more fragments orbiting about their common center of mass.[3] Again, we cannot prove or disprove this conjecture, since it would require a careful study of the nonlinear growth of third- and higher-order disturbances during the dynamical evolution of a Maclaurin spheroid past the point $\tau = \tau_i$. Thus, dynamically induced fission is plausible, but remains unproven.

[3] This conjecture derives from the similarity between our needle-like configuration and a self-gravitating cylinder of finite radius. Indeed, as we know from the work of Chandrasekhar and Fermi, an infinitely long column breaks into pieces when the wavelengths of axisymmetric disturbances exceed, approximately, the circumference of the cylinder (see, e.g., reference 27, pp. 516–523, of Chapter 3). See also: Ostriker, J. P., *Ap. J.* **140**, 1529, 1964; Tassoul, J. L., and Aubin, G., *J. Math. Anal. Applic.* **45**, 116, 1974.

THE LEBOVITZ APPROACH.   This third formulation of the fission problem considers inviscid, homogeneous ellipsoids that slowly contract by virtue of their radiating energy at the rate L; that is to say, the density—while spatially uniform—is explicitly allowed to be a function of time, so that the schematic configurations are not assumed incompressible. (In the two previous approaches, radiation was invoked only to justify considering series of equilibria of progressively greater density.) Furthermore, instead of assuming exact axial symmetry, we now describe the evolution of a contracting mass along a sequence of Riemann ellipsoids (with semi-axes, a_1, a_2, and a_3) rotating with the angular velocity Ω, and characterized by internal motions of uniform vorticity ω. (By assumption, Ω and ω are both directed along the x_3-axis.) For present purposes, let us call attention to some of their properties.[4] Defining, in place of ω, the parameter Λ by

$$\Lambda = -\frac{a_1 a_2}{a_1{}^2 + a_2{}^2}\,\omega, \tag{3}$$

we can easily prove that the quantities

$$2K_1 = (a_1 - a_2)^2(\Omega + \Lambda) = \frac{5}{M}\,J - \frac{1}{\pi}\,C \tag{4}$$

and

$$2K_2 = (a_1 + a_2)^2(\Omega - \Lambda) = \frac{5}{M}\,J + \frac{1}{\pi}\,C \tag{5}$$

are constants along a trajectory. (In fact, they are linear combinations of the angular momentum J and circulation C.) Figure 11.1 depicts the domain of existence of these configurations in the $(a_2/a_1,\ a_3/a_1)$-plane. The Riemann ellipsoids of the considered type all lie in the horn-shaped region bounded by the curves marked $K_1 = 0$ and $K_2 = 0$, and by the Maclaurin line $a_2/a_1 = 1$, where $K_1 = 0$ also. (The curve labeled $\Lambda = 0$ represents the Jacobi sequence.)

Consider now the evolution of a compressible mass that starts out as a Riemann ellipsoid with slightly unequal axes. By taking a_2/a_1 very closely equal to unit, we make the departure from axial symmetry very small, but not zero. Figure 11.1 illustrates how such a contracting body evolves through a series of equilibria that not only are rotating but also have internal motions ($\Lambda \neq 0$). Its evolution parallels the Maclaurin series (i.e., a_2/a_1 remains nearly unity) until a_3/a_1 approaches the value 0.3033; thence, the ratio a_2/a_1 changes rapidly, and its path begins to parallel the lowest Riemann sequence with $K_1 = 0$. (Note that this series plays the role played by the Jacobi series in the classical approach.) Now,

---

[4] See, e.g., reference 28, pp. 129–156, of Chapter 3.

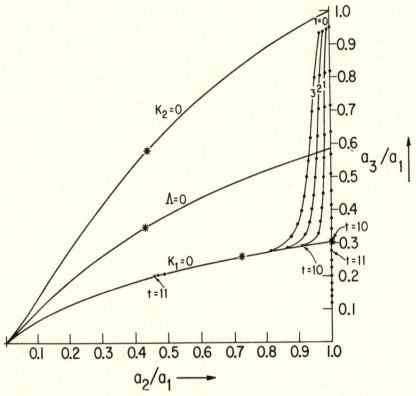

Figure 11.1. The Maclaurin trajectory (with $K_1 = 0$) as compared with Riemann trajectories (marked 1, 2, and 3) having $K_1 = 0.54 \cdot 10^{-4}, 0.23 \cdot 10^{-3}$, and $0.57 \cdot 10^{-3}$, respectively. The three ellipsoids have the same mass, angular momentum, and initial temperature as the Maclaurin spheroid that starts at $a_3/a_1 = 0.95$. The Jacobi series is marked $\Lambda = 0$. Asterisks indicate points at which dynamical instability sets in (under the assumption of incompressibility). The intervals between dots are intervals of equal time. The luminosity is assumed constant and is the same for each configuration, and $\gamma = 5/3$. Source: Lebovitz, N. R., *Ap. J.* **175**, 171, 1972. (By permission of The University of Chicago Press. Copyright 1972 by the American Astronomical Society.)

since these equilibrium forms do not lie exactly on the Maclaurin line, they never encounter the point of dynamical instability $\tau = \tau_i$ along the Maclaurin sequence. Accordingly, their motions remain at all times on the contraction time t_K, and evolution always proceeds on or near the lowest Riemann sequence. What can we say about the stability of these slowly contracting equilibria? As was shown by Lebovitz, ellipsoids starting nearly axisymmetric do become dynamically unstable to infinitesimal disturbances associated with third-order harmonics. An extraordinary narrow band of instability exists just above the $K_1 = 0$-line, and it is necessarily traversed by all evolutionary paths with sufficiently small initial departures from axial symmetry. (When $\alpha = 5/3$, the trajectory marked 1 on Figure 11.1 does not become unstable until $a_2/a_1 = 0.640$, not very far away from the point marked by an asterisk on the lowest Riemann sequence.) The basic conjecture of the present theory is that evolution continues along a new series of steady-state configurations (which take over after further evolution along a Riemann sequence is no longer possible), and that fission may conceivably occur as the result of evolution along this new bifurcating series (see Figure 11.2). However, the evolutionary paths encounter the narrow band of instability described above only if the initial departure from axial symmetry is smaller than a

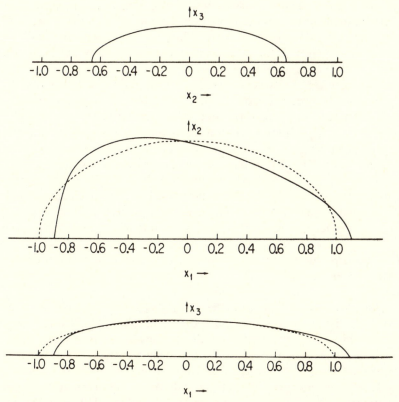

Figure 11.2. The shape of the bifurcating series. The shape of the ellipsoid (with $a_2/a_1 = 0.656$ and $a_3/a_1 = 0.243$) is given by the dotted lines, that of the bifurcation figure by the solid lines. From top to bottom, the figures are the cross-sections in the $x_2 x_3$-, $x_1 x_2$-, and $x_1 x_3$-planes. Only half the cross-section is shown; the other half is obtained by reflection in the horizontal plane. Source: Lebovitz, N. R., *Ap. J.* **190**, 121, 1974. (By permission of The University of Chicago Press. Copyright 1974 by the American Astronomical Society.)

certain critical value; if it exceeds this critical value, the contracting body escapes instability to third-order harmonics, and probably becomes unstable to disturbances associated with fourth- or higher-order harmonics. Again, the foregoing discussion is based upon a linear stability analysis only, and nothing is known so far about the actual motion of a perturbed Riemann ellipsoid in the nonlinear regime.

CONCLUDING REMARKS.   Among the many objections to the various fission theories for the origin of binary stars, the use of models of uniform density may seem the most serious. Indeed, as was demonstrated in Section 10.3, sequences of centrally condensed, uniformly rotating polytropes never reach the points of bifurcation or dynamical instability. That is to say, when $n > 0.808$, each sequence of axially symmetric polytropes in uniform rotation terminates at a point of so-called "equatorial break-up" (i.e., a terminal model in which the effective gravity falls to zero at the equator) before a point of bifurcation may occur. However, such series are not evolutionary series for an inviscid, centrally condensed mass, since the latter must necessarily develop *differential* rotation by virtue of local conservation of angular momentum. In the present context, it thus appears more realistic to consider evolutionary sequences of polytropes in non-uniform rotation, the assumption of constant angular velocity being replaced by the assumption of assigned angular momentum per unit mass. This has been done by Bodenheimer and Ostriker (cf. §§10.4 and 11.4). In brief, such sequences of differentially rotating polytropes resemble in all essential respects the Maclaurin sequence; and, in particular, equatorial break-up never occurs. Therefore, although the context of models of uniform density is certainly remote from the actual conditions in pre-main-sequence stars, the pattern of behavior of homogeneous ellipsoids appears to be qualitatively (and, to some extent, semi-quantitatively) the same as that of slowly contracting, centrally condensed models.

A second objection (due to Lyttleton) refers to the crucial phase during which the fission process is supposed to take place. In all three cases, the final instability of the ellipsoidal figures is always dynamical; hence, it is independent of friction, and any ensuing motion must be time-reversible. Now, if we reverse the arrow of time in an existing binary, the latter does not revert to a single body but simply remains the same binary rotating in the opposite sense. It thus follows that a double star could not have evolved without dissipation from a single body that has become dynamically unstable. Lyttleton asserts that the only way an unstable ellipsoid can resolve its dynamical instability is by breaking up into pieces whose kinetic energy is greater than their mutual gravitational energy, so that the two fragments separate indefinitely. Obviously, dynamical instability implies a release of energy, and any new steady state (such as a binary star) must have less energy than the unstable ellipsoid having the same values for M and J. However, because stellar evolution always proceeds in a *nonconservative* way, it seems unreasonable to assume that the fission process alone should be energy-conservative. Accordingly, there is no physical reason why gradual dissipation should not allow a massive body to transform itself into a steady state consisting of a binary star.

Let us now briefly discuss the time scales t_K and t_v, which occur in inequalities (1) and (2). On the basis of present evidence, the Kelvin-Helmholtz contraction time t_K lies in the range 10^4 to 10^9 years, as M ranges from $10^2 M_\odot$ to $10^{-1} M_\odot$. Hence, if t_v is taken as the viscous diffusion time (see, e.g., Table 7.1), we generally have $t_K \ll t_v$. This fact apparently rules out the classical theory in favor of the most recent approaches. However, it is not decisively established that t_v should be taken to be the viscous diffusion time. As Lebovitz pointed out, it is possible that, for some if not all pre-main-sequence stars, an appropriate coefficient of eddy viscosity should replace the usual coefficient of kinematical viscosity, thus requiring a downward revision of the viscous time scale t_v. Alternatively, it may turn out that the spin-up time (i.e., the geometrical mean of the viscous time scale and the rotation period) is the correct choice for t_v. These and other poorly understood phenomena would thus bring t_K and t_v to roughly comparable values, with the possibility of either inequality (1) or (2) occurring under certain circumstances. Therefore, even if the modern approaches look promising for the future, it is not at all obvious at this stage that the classical theory should be disregarded altogether.

Also, the occurrence of a dynamical instability along an evolutionary trajectory need *not* imply the onset of motions on the dynamical time scale t_D, which is by far the shortest one. Indeed, related calculations by Lebovitz and Schaar suggest that motions might take place on the time scale t_K before *and* after the point of bifurcation (where dynamical instability sets in) is reached, except for a time interval of order $(t_D t_K)^{1/2}$ during which motions occur on the same time scale $(t_D t_K)^{1/2}$. If this result holds true for the fission problem, the behavior of a pear-shaped figure would not require solving the full dynamical equations; only the equilibrium solutions would then be needed to infer the subsequent behavior of a pear-shaped figure.

To summarize, the basic conjecture common to all three approaches is that a contracting mass will develop a constriction that becomes progressively narrower as the evolution proceeds, resulting in a pair of detached fragments. The main *technical* problem is, therefore, how to describe the finite-amplitude motions of a body past the point where it becomes unstable to infinitesimal disturbances? An attempt to follow the dynamical evolution of a pear-shaped figure has been made by Aubin. Unfortunately, it has not been possible so far to confirm the basic conjecture. In all likelihood, this is due to the fact that Aubin's original analysis has been carried out in the quasi-linear approximation only. (That is to say, he used expansions that retain terms of order ϵ and ϵ^2 only, where ϵ is the initial amplitude of the third-order disturbances.) It thus appears that no solution short of a fully nonlinear treatment will definitively answer to the basic conjecture of the fission theories. Of course, the appropriate extension of Larson's original approach to axisymmetric collapse would be most interesting in the present context (cf. §11.2); but, then, one would be faced again with the difficult task of estimating the limitations of the numerical work!

11.4.   THE QUASI-STATIC CONTRACTION

The behavior of quasi-static evolutionary sequences of spherical stars toward the main sequence is now fairly well established over the mass range 10^{-1} to $10^2 M_\odot$. As we know, during this slow contraction phase, a nonrotating star may be thought of as evolving along a sequence of quasi-hydrostatic equilibria. Actually, it is the small departure from strict hydrostatic equilibrium that allows a spherical star to evolve. The configuration then radiates at the expense of its gravitational potential energy. However, when central temperatures become high enough, nuclear reactions also come into play; and the contraction eventually ends as the star makes its transition to the main sequence. At present, very few papers have appeared that are devoted to the effects of rotation on the structure and evolution of stars during their quasi-static contraction. Moreover, since the lifetime of a (spherical) star during this phase is only of the order of 1 percent of its main-sequence lifetime, observational material regarding rotational velocities and related parameters is also very sparse. As we shall now see, the slow contraction phases of high-mass and low-mass stars differ in many essential respects; hence, it is convenient to discuss each group of stars in turn. We conclude the chapter with a brief discussion of the sharp difference in rotational velocities between high-mass and low-mass stars on the main sequence.

HIGH-MASS STARS.   A promising line of research has been followed by Bodenheimer and Ostriker, who have examined, in an approximate way, the effects of rapid, differential rotation during the quasi-static contraction of massive stars. The results that will be described are restricted to the radiative portion of the evolution of stars in the range $3-12M_\odot$. For these masses, convection is important only during a negligible fraction of the contraction time; and, to a fair degree of approximation, the pressure and density can be related by the polytropic relation of index $n = 3$ (cf. §10.1). Furthermore, since ordinary viscosity and circulation currents that transport angular momentum have characteristic times much larger than the radiative contraction time (cf. §§7.4 and 8.2), one can reasonably assume that the angular momentum is conserved on each annular element of mass coaxial with the axis of rotation. It is also assumed that magnetic fields are absent, and that stellar winds or thin surface convective zones transport a negligible amount of angular momentum. Finally, the configurations are assumed to be axisymmetric, and circulation currents are neglected altogether. Under these circumstances, the total mass M and angular momentum J are conserved, and the angular velocity Ω is a function only of the distance ϖ from the axis of rotation (cf. §4.3).

An evolutionary sequence is determined once the following parameters are specified: M, J, chemical composition (X,Y,Z), and the distribution of angular momentum per unit mass $j(m_\varpi)$, where m_ϖ is the mass fraction interior to the cylinder of radius ϖ about the axis of rotation. The function $j(m_\varpi)$, which remains invariant during evolution and satisfies the Høiland stability criterion (cf. §7.3), is taken to be that of a star initially formed as a uniformly rotating sphere of uniform density (cf. §10.4: equation [44]). At the actual starting point of the numerical calculations, with an equatorial radius corresponding to $T_{\text{eff}} = 5,000°K$, a differentially rotating polytrope of index $n = 3$ is constructed by means of the self-consistent field method (cf. §5.5). The temperature on each level surface is obtained from the equation of state for an ideal gas. The luminosity crossing each level surface is then calculated by integrating, over this surface, the flux obtained from the equation of radiative transfer. In the outer parts of the model, the luminosity levels off, so that the total luminosity L of the star is well determined in most cases. An evolutionary sequence is obtained by constructing a set of such differentially rotating polytropes for decreasing values of the equatorial radius, or, equivalently, for increasing values of the ratio $\tau = K/|W|$. The time interval between two models in a sequence is obtained from conservation of energy: it is equal to the difference in total energy of the two configurations divided by the average luminosity. The final point of an evolutionary sequence is taken to occur when nuclear energy sources produce roughly 25 percent of the total luminosity of the star; beyond this point, a convective core develops in high-mass models, the structure changes, and the approximations are no longer valid.

Several principal conclusions emerge from the model calculations of Bodenheimer and Ostriker:

First, an increase in the total angular momentum J, for a given mass M, has a substantial effect on the evolutionary track (see Figure 11.3).

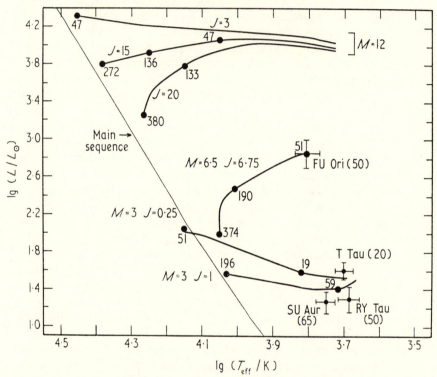

Figure 11.3. Theoretical pre-main-sequence evolutionary tracks of differentially rotating stars. Masses are given in solar units, and angular momenta in units of $(GM_\odot{}^3R_\odot)^{1/2} = 6.05 \cdot 10^{51}$ g cm²/sec. Numerals on curves give calculated rotational velocities at the equator (in km/sec). The observed positions of four objects are given together with their observed values for $v_e \sin i$ (in parentheses, and in the same units). These observations are taken from: Herbig, G. H., *Ap. J.* **125**, 612, 1957; Herbig, G. H., *Vistas in Astronomy* (Beer, A., ed.) **8**, 109, 1966; Mendoza, E. E., *Ap. J.* **151**, 977, 1968. Source: Bodenheimer, P., *Rep. Progress Phys.* **35**, 1, 1972. (By permission. Copyright 1972 by The Institute of Physics.)

If J is high enough, differential rotation causes the luminosity to decrease as the star evolves, in contrast to the usual slow increase characteristic of the radiative phase of nonrotating models.

Second, although the transition to the main sequence has not been calculated, Figure 11.3 indicates that the contracting models arrive at a position close to the main sequence with equatorial velocities v_e in the observed range (up to 400 km/sec) and with luminosities substantially reduced from those of nonrotating stars of the same mass.

Third, the (M–J/M)-relation inferred from these calculations for main-sequence stars with v_e near the average of the observed range is significantly different from that derived under the assumption of constant angular velocity. The two relations are depicted approximately as curves (3) and (1), respectively, in Figure 11.4.

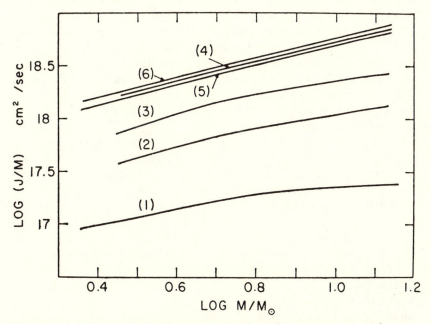

Figure 11.4. Estimated relations between mass and angular momentum per unit mass: (1) for main-sequence stars, based on average observed velocities and the assumption of uniform rotation; (2) and (3) for main-sequence stars, based on the observations but for stars initially formed as uniformly rotating polytropes of indices $n = 1.5$ and $n = 0$, respectively; (4) for stellar models (initially formed as uniformly rotating spheres with constant density) that are suspected of having undergone fission just prior to the onset of nuclear reactions; (5) and (6) for main-sequence contact binaries with the given total mass and mass ratios 2 and 1, respectively. Curves (1), (5), and (6) are adapted from: Kraft, R. P., *Spectroscopic Astrophysics* (Herbig, G. H., ed.), p. 385, Berkeley: Univ. of California Press, 1970. Source: Bodenheimer, P., and Ostriker, J. P., *Ap. J.* **161**, 1101, 1970. (By permission of The University of Chicago Press. Copyright 1970 by the University of Chicago.)

Fourth, for given M, the contraction time to the main sequence increases with J because of the reduction in luminosity. However, sequences (with different M and J) that arrive near the main sequence at the same luminosity have nearly identical contraction times. Thus, the contraction time to a given luminosity on the main sequence seems to be little affected by the rotation law chosen.

Fifth, the rotational velocities of T Tauri stars, when analyzed in connection with evolutionary tracks, support the idea that angular momentum is conserved, to a good degree of approximation, during the radiative portion of the pre-main-sequence contraction. Indeed, the three T Tauri stars whose rotational velocities have been observed by Herbig (T Tau: $v_e \sin i = 20$ km/sec; RY Tau: $v_e \sin i = 45$–50 km/sec; SU Aur: $v_e \sin i = 65$ km/sec) fit approximately to evolutionary tracks with M = $3M_\odot$ and constant J. In all cases, the models reach the main sequence with spectral types between A0 and A5, T Tauri with $v_e \approx 100$ km/sec, and the other two stars with $v_e \approx 200$ km/sec. These estimated velocities are consistent with the observed range for main-sequence A stars. A similar statement can be made for the star FU Orionis ($v_e \sin i = 50$ km/sec), except that we must now consider an evolutionary track with M = $6.5M_\odot$ and constant J (see Figure 11.3).

To conclude, let us briefly comment on the possible occurrence of fission along the evolutionary tracks as derived by Bodenheimer and Ostriker. If we accept Ostriker's conjecture that fission occurs if a point of dynamical instability is reached before nuclear reactions can stop the slow contraction (cf. §11.3), high values of J for a given mass might well result in the formation of a binary during the quasi-static phase. In analogy to the sequence of Maclaurin spheroids, we thus *assume* that binary formation will occur when a star contracts to the point at which the ratio $\tau_i \approx 0.27$ is attained. The shape of a model at this value of τ is depicted in Figure 11.5. Numerical calculations indicate that associated with each mass is a critical value of the total angular momentum, J_c (say), above which dynamical instability occurs before the star has contracted to the main sequence. This result is shown on Figure 11.4, where it is compared with the (M–J/M)-relation for contact main-sequence binaries with mass ratios of 1 and 2, as given by Kraft. Note the good agreement between the *threshold* value J_c required for fission (under the criterion used by Bodenheimer and Ostriker) and the *minimum* angular momentum of binaries with the most commonly observed mass ratios. (As we shall see below, a similar agreement was found by Roxburgh for low-mass stars.) Independent support of the fission hypothesis is provided by the observed rotational correlation in close binaries (cf. §2.4). In brief, although much theoretical work remains to be done to demonstrate that the dynamical instability ($\tau \gtrsim \tau_i$) in a centrally condensed body necessarily leads to fission, there is now ample evidence (both theoretical and observational) in favor of this hypothesis for the formation of close binary

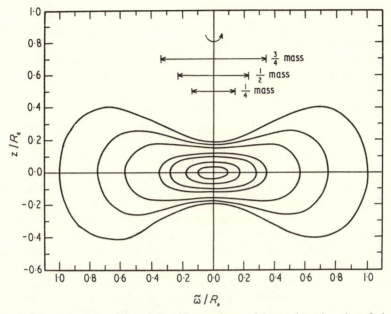

Figure 11.5. Isopycnic surfaces of a rapidly rotating model near the point where fission is expected to occur ($M = 12M_\odot$ and $J = 30[GM_\odot^3 R_\odot]^{1/2}$). R_e is the total equatorial radius. Diameters of cylinders that contain 1/4, 1/2, and 3/4 of the total mass are indicated. A toroid-like configuration is obtained when the diagram is rotated about the z-axis. Source: Bodenheimer, P., *Rep. Progress Phys.* **35**, 1, 1972. (By permission. Copyright 1972 by The Institute of Physics.)

systems. And, as we have said in Section 11.2, this would also resolve the angular momentum problem for massive stars.

LOW-MASS STARS.   The influence of rotation during the pre-main-sequence evolution of largely convective, low-mass stars is still poorly understood. Because turbulent convection produces a large eddy viscosity that results in rapid transport of angular momentum, it is generally assumed that convection forces the whole star to rotate uniformly—presumably with the critical angular velocity at which the centrifugal force balances gravity at the equator. As we recall from Section 8.5, however, the rotation law $\Omega = constant$ holds true only if the turbulent motions are strictly isotropic. Furthermore, even if the coefficient of eddy viscosity is isotropic, there is now strong evidence that viscous friction does lead to solid-body rotation only if the rotational forces are small compared to the pressure forces. That is to say, if the initial motion is predominantly Keplerian below the surface of a rapidly rotating star (i.e., $\Omega \propto \varpi^{-3/2}$), the presence of an isotropic eddy viscosity will cause an outward transport of angular momentum, but differential rotation will probably be maintained in the outermost layers of the star, and uniform rotation will *not* occur.[5]

Qualitative discussions regarding the effects of uniform rotation during the Hayashi phase were originally presented by Schatzman and Roxburgh. In these crude approaches, a low-mass star is assumed to evolve at first with constant J, but angular momentum is transported within the body so as to maintain uniform rotation. Thence, as the star slowly contracts, its constant angular velocity increases until the centrifugal force balances gravity at the equator. Beyond this point, further contraction can occur only if some angular momentum is lost. A possible mechanism is equatorial shedding, which leaves the material in Keplerian orbits around the star, thus resulting in the formation of an equatorial disc. Roxburgh's estimates show that relatively little mass is lost in the process, but that the total angular momentum $J(t)$ decreases as $R^{1/2}(t)$. As was pointed out by Nobili and Secco, however, the total amount of mass loss during the contraction strongly depends on the duration of the fully convective phase. By using the simplifying assumptions of rigid rotation and spherical symmetry, they found an appreciable mass loss for a $2M_\odot$ star when starting with the critical angular velocity at a radius of $50R_\odot$. Numerical calculations for nonspherical stars contracting homologously to the main sequence and rotating with the critical angular velocity have also been performed by Moss. Broadly speaking, the effect of uniform rotation is to move the models built on polytropes of index $n = 3/2$ to lower effective temperatures and luminosities at the same volume. Thus, rotation causes a star to resemble one of lower mass, as is found for some properties of rotating main-sequence stars (cf. §12.2). As far as the question of contraction time scales is concerned, the results are still very uncertain. It does seem that the contraction time for a critically rotating, low-mass star is considerably greater than for a nonrotating model that arrives at the same place on the main sequence. As a matter of fact, definitive evolutionary tracks for rapidly rotating, largely convective stars contracting to the main sequence are not yet available.

Let us now briefly report on the fission problem as it was formulated by Roxburgh. With continued contraction, a low-mass star eventually develops a radiative core in which eddy viscosity is no longer operative. Hence, each mass element of this growing core conserves its angular

---

[5] This property was actually derived by Durisen in the case of massive white-dwarf models (cf. §13.3). The relevance of his result for pre-main-sequence stars is thus only tentative, and should be confirmed by a quantitative study of slowly contracting, nondegenerate models.

momentum, thus causing an inward increase in angular velocity. Roxburgh then *assumes* that the nonhomogeneous, differentially rotating core behaves like a contracting Maclaurin spheroid, and that fission will occur when the growing ratio τ reaches the critical value $\tau_b \approx 0.14$ (cf. §11.3). Two interesting results emerge from his heuristic analysis. *First*, stars with $M < 0.8M_\odot$ reach the main sequence and stop contracting before rotationally induced fission can occur in the core. The theoretical limit $M = 0.8M_\odot$ is in agreement with the lower limit observed on the total masses of W Ursae Majoris systems. *Second*, the (M–J/M)-relation of the models at the time fission supposedly occurs is in good agreement also with the observed angular momenta of all close binaries. However, Roxburgh's conjecture that stars with $M < 4M_\odot$ can form a contact binary whereas more massive stars will produce detached binaries has not been confirmed by the recent observational discussion of Mammano and his collaborators. Given the many uncertainties and approximations which are entailed in this very crude approach to the fission problem, one may wonder, however, whether any *accurate* theoretical prediction (such as a clear-cut distinction between close and detached systems) can be properly made at this time. In our opinion, rotationally induced fission is still attractive as a way of producing close binaries among the low-mass stars, but it also remains unproven.

THE BREAK IN ROTATIONAL VELOCITIES NEAR SPECTRAL TYPE F5V. It has long been known that appreciable rotational velocities are common among the *normal* O-, B-, and A-type stars along the main sequence, and that they virtually disappear somewhere near spectral type F5 (cf. §2.3). As Struve suggested in 1930, the slow rotation observed for main-sequence stars later than type F5 ($M \lesssim 1.25M_\odot$) is possibly due to the systematic occurrence of planets (or black dwarfs?) around these stars. Although this explanation retains its attractiveness to some, there is now ample evidence that this is not the most likely cause of the remarkable decline in rotational velocities in the middle F's along the main sequence.

The relevance of magnetic braking for stars having outer convective layers (such as low-mass stars on their Hayashi tracks and main-sequence stars later than type F5) was first recognized by Schatzman. In this picture, it is assumed that stars with convective envelopes generate solar-like flares that act as an expanding plasma in a magnetic field. As material is ejected from the flare zones, the magnetic field can enforce approximate co-rotation until the gas has moved out to distances much greater than the stellar radius. Beyond this region, because the magnetic stresses become less and less important, the outflowing material can thus leave the star, with each mass element carrying away its angular momentum. As was shown by Schatzman and Okamoto, a quite small amount of mass loss that results from jets yields proportionally a much greater loss of angular momentum. With $M < 2M_\odot$ and magnetic fields in the range 10^2–10^3 gauss, the total fractional rate of mass loss $\Delta M/M$ necessary to throw away all of the angular momentum of a star during its whole quasi-static contraction (at the critical angular velocity) is smaller than, or at most equal to, $2 \cdot 10^{-4}$. This amount of mass loss lies well within the range deduced by Kuhi from observations of T Tauri stars.

Given the efficiency of this mechanism for extracting angular momentum from rotating stars with convective envelopes, the break in main-sequence rotational velocities can now be explained as follows. Since high-mass stars spend relatively little time in the convective phase, magnetic braking is therefore virtually inoperative for these stars. It thus follows that they suffer very little loss of angular momentum during their quasi-static contraction. On the contrary, low-mass stars have a more important convective phase, for they always retain an outer convective zone all the way to the main sequence. Magnetic braking can thus operate during their entire quasi-static contraction and/or during their much longer stay on the main sequence. Spherical stellar models published by Demarque and Roeder lead theoretical support to these views. Moving from low-mass stars to high-mass stars along the main sequence, they found that the thickness of the hydrogen convective envelope of their models decreases rapidly and goes to zero very sharply among the middle F's. In other words, *the spectral type that separates the rapid rotators from the slow ones also separates the stars that are in radiative equilibrium near the surface from those that have subphotospheric convective zones.* This is, in essence, the basic argument in favor of the Schatzman mechanism: turbulent convection below the surface of a star is responsible for flare activity, with concomitant corpuscular emission and magnetically enhanced loss of angular momentum. As we shall see in Section 15.3, the magnetic field itself is either generated by a dynamo process driven by rotational and convective motions below the stellar surface or could exist as remnant of the primeval field in the protostellar cloud.

Another mechanism by which stars with convective envelopes can dispose of a considerable fraction of their initial angular momenta is provided by stellar winds, the generalized concept of Parker's solar wind. Indeed, from a comparison of the directions of the tails of direct and retrograde comets, Brandt has amply confirmed the idea that the solar wind carries away a net flux of angular momentum, and that it produces a torque that is sufficient to halve the solar rotation rate in about $5 \cdot 10^9$ years (assuming the Sun to be rotating as a solid body). It therefore appears that we can confidently apply the same idea to all stars with surface convective zones.

Following Mestel, it is subphotospheric convection that is again the essential feature of the mechanism. Waves generated in the convective zone are dissipated above the photosphere, thus supplying the heat responsible for the formation of a chromosphere and a corona. In the present instances, however, when the coronal temperatures are too low to generate a *thermal wind*, large centrifugal forces acting on the co-rotating material can generate an outwardly moving flow, i.e., a *centrifugal wind*. In both cases, the wind motion accelerates outward from very low values at the bottom of the corona to supersonic values far away from the stellar surface. Various authors have independently shown that within the so-called Alfvén surface (i.e., the surface where the gas speed matches the Alfvén speed) the magnetic field of the star is strong enough to force the wind to follow the field and to keep its motion in approximate co-rotation with the star. Beyond that surface, the gas flow drags the magnetic field to follow the motion, and so approaches a state of angular momentum conservation. Moreover, if the magnetic field has an energy density comparable with the thermal energy at the basis of the corona, there is then no difficulty in ensuring that the "radius" of the Alfvén surface (R_a, say) is much larger than the radius of the star. As derived from solar wind data, we have $R_a \approx 25 R_\odot$ for the Sun; and even higher values are obtained for stars on the lower main sequence. Thus, although there is as yet no definitive model in the theory, these figures clearly demonstrate the efficiency of magnetic braking by a stellar wind as an alternative means of extracting angular momentum from rotating stars with subphotospheric convective zones.

An important piece of evidence in support of the above picture comes from the observations of Wilson, who suggested that stars with active chromospheres (and presumably torque-exerting winds) may be defined as those showing Ca II (K2 and H2) emission in their spectra. Indeed, as one moves down the main sequence, there is a sudden appearance of Ca II emission in the middle F's, whereas it is never observed among stars earlier than type F5V. Moreover, *slow rotation sets in rather abruptly just where Ca II emission begins.* (As we recall, this is also the place where outer convective layers appear!) Finally, among the main-sequence stars with spectral types F and G, those with Ca II emission (less than 10 percent of Wilson's sample at 10 Å/mm dispersion) are nearer the zero-age main sequence than those without it. In view of these results, the presumption is therefore strong that the normal main-sequence stars above the middle F's are, on the average, rapid rotators because they exhibit no chromospheric activity (i.e., Ca II emission) due to subphotospheric convection that generates stellar winds. However, if F- and G-type stars with Ca II emission are indeed nearer the zero-age main sequence than those without it, this would also mean that chromospheric activity and J-cutting winds must necessarily decay as the stars leisurely evolve on the main sequence.

Another property of the F- and G-type main-sequence stars related to the presence of Ca II emission was also found by Wilson: for a fixed spectral type, Ca II emission becomes progressively more important among the younger stars. For example, G0V stars in the Pleiades (age $\approx$ $4-5 \cdot 10^7$ years) have stronger Ca II emission than G0V stars in the Hyades (age $\approx 4-9 \cdot 10^8$ years), which in turn have stronger Ca II emission than field stars with spectral type G0V (cf. §12.5). In subsequent work, Kraft also found that, among the field stars later than type F5V, those with Ca II emission rotate, on the average, faster than those without it, and that stars in galactic clusters rotate faster than field stars without Ca II emission. According to Kraft, stars with M = 1.2 M$_\odot$ and ages near $4 \cdot 10^7$ years have a mean equatorial velocity of about 40 km/sec; this rotation rate then decays by a factor of two in a time equal to the age of the Hyades, and by a factor of three up to the age of old field stars. A more recent discussion by Skumanich indicates that *Ca II emission and surface rotation rates both decay as the inverse square root of the age.*

We may, thus, plausibly assume that, whenever a star possesses an outer convective shell, this layer is decelerated within the observed time scale, presumably in response to the torque exerted by a stellar wind in the presence of a magnetic field. It is not yet clear, though, whether the pre-main-sequence contraction phase accounts for a significant loss of angular momentum, or whether late-type stars begin their main-sequence evolution rotating rapidly, and subsequently spin down (at least in their observable layers) during the much longer main-sequence phase. Moreover, because solar-type stars eventually develop a radiative core, it is also possible that these stars approach the main sequence with a core that rotates much more rapidly than their outer layers. Finally, the question of whether and how the deceleration of the surface layers makes itself felt inside a solar-type star is still a matter of considerable debate, and has so far resisted a definitive solution. The current literature (and especially the comprehensive review papers by Dicke and Spiegel) presents a detailed discussion of this problem.

BIBLIOGRAPHICAL NOTES

Among the many survey papers relating to the subject matter of this chapter, reference may be made to:

1. Spitzer, L., Jr., *Diffuse Matter in Space*, pp. 214–246, New York: Wiley and Sons, 1968.

2. Mestel, L., "Effects of Rotation, Magnetic Fields, Planetary Systems, Angular Momentum in Stars, Spin-Down, Binary and Multiple Stars" in *Mém. Soc. Roy. Sci. Liège* (5) **19** (1970): 167–194.
3. McNally, D., "Theories of Star Formation" in *Rep. Progress Phys.* **34** (1971): 71–108.
4. Bodenheimer, P., "Stellar Evolution toward the Main Sequence" in *Rep. Progress Phys.* **35** (1972): 1–54.
5. Larson, R. B., "Dynamical Problems of Pre-Main-Sequence Evolution" in *Mém. Soc. Roy. Sci. Liège* (6) **8** (1975): 451–464.

See also reference 2 of Chapter 7. Most of the early contributions may be traced to these papers.

Section 11.2. Much of our knowledge about the angular momentum problem derives from the pioneering work of Hoyle, Mestel, Spitzer, and von Weizsacker. Various solutions to this problem will be found in:

6. Mestel, L., "Problems of Star Formation" in *Quart. J. Roy. Astron. Soc. London* **6** (1965): 161–198, 265–298.
7. Spitzer, L., Jr., "Dynamics of Interstellar Matter and the Formation of Stars" in *Nebulae and Interstellar Matter*, (Middlehurst, B. M., and Aller, L. H., eds.), pp. 1–63, Chicago: The Univ. of Chicago Press, 1968.

Further contributions are due to:

8. Prentice, A. J. R., and ter Haar, D., *Monthly Notices Roy. Astron. Soc. London* **151** (1971): 177–184.
9. Nakano, T., and Tademaru, E., *Astroph. J.* **173** (1972): 87–101.
10. Nakano, T., *Publ. Astron. Soc. Japan* **25** (1973): 91–100.

The difficulties connected with magnetic braking during the accelerating collapse have been studied by:

11. Gillis, J., Mestel, L., and Paris, R. B., *Astroph. and Space Sci.* **27** (1974): 167–194.

See also:

12. Fleck, R. C., and Hunter, J. H., *Monthly Notices Roy. Astron. Soc. London* **175** (1976): 335–343.

The evolution of a small cluster is discussed in:

13. van Albada, T. S., *Bull. Astron. Inst. Netherlands* **20** (1968): 57–68.

See also:

14. Kumar, S. S., *Astroph. and Space Sci.* **17** (1972): 453–458.
15. Arny, T., and Weissman, P., *Astron. J.* **78** (1973): 309–315.
16. Harrington, R. S., *ibid.* **80** (1975): 1081–1086.

Various approaches to the nonspherical collapse and/or fragmentation problems will be found in:

17. Lynden-Bell, D., *Proc. Cambridge Phil. Soc.* **58** (1962): 709–711.
18. Lynden-Bell, D., *Astroph. J.* **139** (1964): 1195–1216; *ibid.* **142** (1965): 1648–1649.
19. Lin, C. C., Mestel, L., and Shu, F. H., *ibid.* **142** (1965): 1431–1446.
20. Arny, T., *Ann. Astroph.* **30** (1967): 1–11.
21. Fujimoto, M., *Astroph. J.* **152** (1968): 523–536; *ibid.* **170** (1971): 143–152.
22. Mizuno, T., and Fujimoto, M., *Publ. Astron. Soc. Japan* **22** (1970): 413–422.
23. Falle, S. A. E. G., *Monthly Notices Roy. Astron. Soc. London* **156** (1972): 265–273.
24. Ferraioli, F., and Virgopia, N., *Mem. Soc. Astron. Italiana* **44** (1973): 181–191.
25. Hara, T., Matsuda, T., and Nakazawa, K., *Progress Theor. Phys.* **49** (1973): 460–478.
26. McNally, D., *Mém. Soc. Roy. Sci. Liège* (6) **8** (1975): 479–486.
27. Weber, S. V., *Astroph. J.* **208** (1976): 113–126.

The following papers are discussed in the text:

28. Larson, R. B., *Monthly Notices Roy. Astron. Soc. London* **156** (1972): 437–458.
29. Black, D. C., and Bodenheimer, P., *Astroph. J.* **199** (1975): 619–632; *ibid.* **206** (1976): 138–149.
30. Tscharnuter, W., *Astron. and Astroph.* **39** (1975): 207–212.
31. Fricke, K., Möllenhoff, C., and Tscharnuter, W., *ibid.* **47** (1976): 407–412.

Related studies of discs and rings are those of:

32. Ostriker, J. P., *Astroph. J.* **140** (1964): 1067–1087.
33. Cameron, A. G. W., and Pine, M. R., *Icarus* **18** (1973): 377–406.
34. Drobyshevski, E. M., *Astron. and Astroph.* **36** (1974): 409–413.
35. Lynden-Bell, D., and Pringle, J. E., *Monthly Notices Roy. Astron. Soc. London* **168** (1974): 603–637.
36. Wong, C.-Y., *Astroph. J.* **190** (1974): 675–694.
37. Stewart, J. M., *Astron. and Astroph.* **42** (1975): 95–101.

A well-balanced survey of binary formation is due to:

38. Batten, A. H., "On the Interpretation of Statistics of Double Stars" in *Annual Review of Astronomy and Astrophysics* **5**, pp. 25–44, Palo Alto, California: Annual Reviews, Inc., 1967.

Detailed discussions of the observational data will be found in:

39. van Albada, T. S., *Bull. Astron. Inst. Netherlands* **20** (1968): 47–56.
40. Heintz, W. D., *J. Roy. Astron. Soc. Canada* **63** (1969): 275–298.
41. Dommanget, J., *Ciel et Terre* **86** (1970): 463–485.
42. Abt, H. A., and Levy, S. G., *Astroph. J. Supplements* **30** (1976): 273–306.

Among the recent conflicting reports on the capture mechanism, let us mention:

43. van Albada, T. S., *Bull. Astron. Inst. Netherlands* **20** (1968): 40–46.
44. Mansbach, P., *Astroph. J.* **160** (1970): 135–145.
45. Szebehely, V., *Astron. J.* **79** (1974): 981–983, 1449–1454.

The evolution of a wide binary into a close binary is discussed in:

46. Huang, S.S., *Ann. Astroph.* **29** (1966): 331–338.
47. Mestel, L., *Monthly Notices Roy. Astron. Soc. London* **138** (1968): 359–391.

See also:

48. Okamoto, I., and Sato, K., *Publ. Astron. Soc. Japan* **22** (1970): 317–333.

Section 11.3. The classical approach is expounded at length in reference 2 of Chapter 10. (All standard references on the subject may be found there.) The Ostriker approach is contained in:

49. Ostriker, J. P., in *Stellar Rotation* (Slettebak, A., ed.), pp. 147–156, New York: Gordon and Breach, 1970.

The references to Rossner and Fujimoto are to their papers; see:

50. Rossner, L. F., *Astroph. J.* **149** (1967): 145–168.

and references 21. The Lebovitz approach will be found in:

51. Lebovitz, N. R., *ibid.* **175** (1972): 171–183; *ibid.* **190** (1974): 121–130.

See also:

52. Lebovitz, N. R., *Mém. Soc. Roy. Sci. Liège* (6) **8** (1975): 47–53.
53. Lebovitz, N. R., and Schaar, R. J., *Studies in Applied Mathematics* **54** (1975): 229–260.

The reference to Aubin is to his thesis:

54. Aubin, G., Ph.D. Thesis (unpublished), Univ. of Montreal, 1973.

Section 11.4. The presentation in the text largely follows reference 4 above, reference 2 of Chapter 7, and references 1–2 of Chapter 12. High-mass stars are discussed in:

55. Bodenheimer, P., *Mém. Soc. Roy. Sci. Liège* (5) **19** (1970): 199–205.

56. Bodenheimer, P., and Ostriker, J. P., *Astroph. J.* **161** (1970): 1101–1113.

Low-mass stars have been studied by:

57. Schatzman, E., in *Star Evolution* (Gratton, L., ed.), pp. 177–242, New York: Academic Press, 1963.
58. Roxburgh, I. W., *Astroph. J.* **143** (1966): 111–120.
59. Nobili, L., and Secco, L., *Mém. Soc. Roy. Sci. Liège* (5) **19** (1970): 207–211.
60. Moss, D. L., *Monthly Notices Roy. Astron. Soc. London* **161** (1973): 225–237.
61. Roxburgh, I. W., and Williams, P. S., *Mem. Soc. Astron. Italiana* **45** (1974): 477–483.

The Roxburgh approach to the fission problem will be found in reference 58; see also his earlier paper:

62. Roxburgh, I. W., *Nature* **208** (1965): 65–66.

Criticisms have been made by:

63. Mammano, A., *Mem. Soc. Astron. Italiana* **38** (1967): 425–432.
64. Mammano, A., Margoni, R., and Stagni, R., *Astron. and Astroph.* **35** (1974): 143–147.

A paper of related interest is:

65. Sakurai, T., *Astroph. and Space Sci.* **41** (1976): 15–25.

Related matters are treated in:

66. Kopal, Z., *Publ. Astron. Soc. Pacific* **83** (1971): 521–538.
67. Van't Veer, F., *Astron. and Astroph.* **40** (1975): 167–174.

The Schatzman mechanism is considered in:

68. Schatzman, E., *Ann. Astroph.* **25** (1962): 18–29.

Further developments along similar lines are due to:

69. Okamoto, I., *Publ. Astron. Soc. Japan* **21** (1969): 25–53, 350–366.
70. Okamoto, I., in *Stellar Rotation* (Slettebak, A., ed.), pp. 73–81, New York: Gordon and Breach, 1970.

Among the many theoretical papers on stellar winds and losses of angular momentum, reference may be made to:

71. Brandt, J. C., *Astroph. J.* **144** (1966): 1221–1222.
72. Ferraro, V. C. A., and Bhatia, V. B., *ibid.* **147** (1967): 220–229.
73. Mestel, L., in *Plasma Astrophysics* (Sturrock, P. A., ed.), pp. 185–228, New York: Academic Press, 1967.
74. Modisette, J. L., *J. Geophys. Research* **72** (1967): 1521–1526.
75. Weber, E. J., and Davis, L., Jr., *Astroph. J.* **148** (1967): 217–227.

See especially reference 47, as well as the following papers:

76. Mestel, L., *Monthly Notices Roy. Astron. Soc. London* **140** (1968): 177–196.
77. Nariai, K., *Astroph. and Space Sci.* **3** (1969): 150–159.
78. Mestel, L., and Selley, C. S., *Monthly Notices Roy. Astron. Soc. London* **149** (1970): 197–220.
79. Grzedzielski, S., *Acta Astron.* **21** (1971): 199–219.
80. Okamoto, I., *Monthly Notices Roy. Astron. Soc. London* **166** (1974): 683–701; *ibid.* **173** (1975): 357–379.
81. Yeh, T., *Astroph. J.* **206** (1976): 768–776.

Support to the Schatzman mechanism is given in:

82. Wilson, O. C., *ibid.* **144** (1966): 695–708.
83. Demarque, P., and Roeder, R. C., *ibid.* **147** (1967): 1188–1191.
84. Kraft, R. P., *ibid.* **150** (1967): 551–570.
85. Skumanich, A., *ibid.* **171** (1972): 565–567.

See also:

86. Boesgaard, A. M., *ibid.* **188** (1974): 567–569.

The rotational behavior of the main-sequence stars and its consequences concerning the formation of planetary systems is discussed in:

87. Huang, S.S., *ibid.* **141** (1965): 985–992; *ibid.* **150** (1967): 229–238.

See also McNally's paper (reference 44 of Chapter 2), and:

88. van den Heuvel, E. P. J., *The Observatory* **86** (1966): 113–115.
89. Kumar, S. S., *Zeit. f. Astroph.* **66** (1967): 264–267.
90. Fahlman, G. G., and Anand, S. P. S., *J. Roy. Astron. Soc. Canada* **63** (1969): 36–41.
91. Brosche, P., *ibid.* **63** (1969): 150.
92. Virgopia, N., *Mem. Soc. Astron. Italiana* **43** (1972): 813–821.

Mass loss from T Tauri stars is discussed in:

93. Kuhi, L. V., *Astroph. J.* **140** (1964): 1409–1433; *ibid.* **143** (1966): 991–993.

Various theoretical estimates of angular momentum loss from rotating stars (in this and related contexts) are those of:

94. Williams, I. P., in *Mass Loss from Stars* (Hack, M., ed.), pp. 139–145, New York: Springer-Verlag New York, Inc., 1969.
95. Guseinov, O. Kh., and Kasumov, F. K., *Astron. Zh.* **48** (1971): 722–725.
96. De Grève, J. P., De Loore, C., and De Jager, C., *Astroph. and Space Sci.* **18** (1972): 128–134.

97. Weidelt, R. D., *Astron. and Astroph.* **27** (1973): 389–394; *ibid.* **46** (1976): 213–218.

The spin-down problem for solar-type stars is discussed at length by Dicke and by Spiegel in reference 3 of Chapter 9; other papers may be traced to reference 35 of Chapter 8.

12

Stellar Models:
Structure and Evolution

12.1. INTRODUCTION

With the advent of high-speed computers, significant advances have been made in our understanding of the structure and evolution of rotating stars. However, as we indicated in Chapter 7, the actual distribution of angular momentum within a star is still largely unknown; hence, in all models proposed to date, the rotation law is always specified in an *ad hoc* manner. In other words, even the most detailed calculations are still very preliminary in character, their aim being essentially to evaluate the gross changes caused by rotation on stellar models. The modifications brought by rotation on the interiors of chemically homogeneous, main-sequence stars are discussed in Section 12.2. The subsequent section is devoted to the effects of rotation on the observable parameters that now depend on the angle between the rotation axis and the line of sight. The abnormally low rotation rates of the Am- and Ap-stars are considered in Section 12.4. Rotation effects in stars belonging to open clusters are discussed in Section 12.5. We conclude the chapter with a general discussion of the role of rotation during the post-main-sequence phases of stellar evolution. It is assumed that the reader is already familiar with the theory of spherical stellar models.

12.2. MAIN-SEQUENCE MODELS

Broadly speaking, rotation has two general effects on the structure of a chemically homogeneous, main-sequence star: (i) a global expansion of the star due to the local centrifugal force, and (ii) a departure from sphericity due to the nonspherical part of the effective gravity. The first effect causes a reduction in the total luminosity of a star when it is compared to its nonrotating counterpart having the same mass. The second effect induces a dependence of effective temperature and magnitude on the inclination of the rotation axis to the line of sight (cf. §12.3). At present, two groups of calculations have been aimed primarily at evaluating these changes: (i) uniformly rotating models, and (ii) differentially rotating models in which the angular velocity depends on the distance ϖ from the rotation axis only. In both cases, viscosity, circulatory currents, and magnetic fields are neglected altogether. By virtue of the Poincaré-Wavre theorem (cf. §4.3), both sets of models thus share all the mechanical properties of a barotrope, and this in spite of the fact that the total pressure p is related to the density ρ, temperature T, and chemical composition (X,Y,Z) through an assigned equation of state! We may therefore surmise that the clear-cut distinction that was found to exist between polytropes in uniform and non-uniform rotation will occur also for the present sequences of pseudo-barotropic, main-sequence models (cf. §10.5). That is to say, sequences of uniformly rotating stars should terminate with a model in which the effective gravity vanishes at the equator (i.e., the point of so-called "equatorial break-up"), whereas sequences of differentially rotating stars should not terminate, therefore allowing for much larger observable effects than in the former sequences. That this is indeed the case will become apparent from the subsequent discussion.

UNIFORMLY ROTATING MODELS

Sweet and Roy were the first to consider the effects of uniform rotation on the luminosity and effective temperature of a main-sequence star. Their model was a slowly rotating, slightly distorted Cowling model star. However, the first detailed structures for main-sequence stars in uniform rotation were constructed by Faulkner, Roxburgh, and Strittmatter in 1968. By making use of Takeda's perturbation technique, they considered models for intermediate-type dwarfs of $1M_\odot$, $2M_\odot$, and $5M_\odot$. (These calculations include detailed opacity and nuclear-energy-generation data, as well as a careful treatment for determining the outer surface boundary conditions; $X = 0.710, Z = 0.020$.) Takeda's method was subsequently applied to main-sequence stars in the mass range $5-15M_\odot$ by Sackmann and Anand. (The radiation pressure is now included, but the simple condition $p = T = 0$ is used on the surface; $X = 0.670, Z = 0.030$.) This type of analysis was further used by Sackmann to construct models in the mass range $0.8-20.0M_\odot$, with radiation pressure included and with a careful treatment of the outer layers. (Her spherical solutions are obtained by means of the Kippenhahn-Weigert-Hofmeister program; $X = 0.739$, $Z = 0.021$.) The perturbation technique described in Section 5.4 was also applied by Kippenhahn and Thomas to main-sequence stars of $1M_\odot$, $5M_\odot$, and $10M_\odot$ $(X = 0.739, Z = 0.021)$. Finally, by making use of a technique which is quite similar to the foregoing method, Papaloizou and Whelan recently presented detailed calculations of uniformly rotating stars in the mass range $0.6-62.7M_\odot$ (M $\geqslant 10M_\odot$: $X = 0.739$, $Z = 0.021$; M $\leqslant 2M_\odot$: $X = 0.700, Z = 0.050$).

First of all, because no estimate of the errors is available, a general comparison between these uniformly rotating models is in order. (In this connection, see especially Figures 10.7 and 10.8!) Table 12.1 assembles the changes in luminosity ($\Delta L/L_0$) and polar radius ($\Delta R_p/R$) for all models on the verge of equatorial break-up, i.e., the most rapidly rotating models. (L_0 and R refer, respectively, to the luminosity and radius of the spherical model having the same mass.) Obviously, despite the many differences in method and physical input, there is no serious discrepancy between these various results. In particular, the maximum luminosity change caused by uniform rotation is between about 7 and 10 percent for high-mass models, somewhat smaller for low-mass models with a radiative envelope, and between about 15 to 35 percent for low-mass stars with a convective envelope. However, as was pointed out by Papaloizou and Whelan, this last result is certainly the most inaccurate one, as it is most seriously

TABLE 12.1

A summary of the luminosity and polar radius changes at break-up rotation, the results of several authors

Authors	FRS		SA		S		KT		PW	
$M/M_\odot$	$\dfrac{-\Delta L}{L_0}$	$\dfrac{-\Delta R_p}{R}$	$\dfrac{-\Delta L}{L_0}$	$\dfrac{-\Delta R_p}{R}$	$\dfrac{-\Delta L}{L_0}$	$\dfrac{-\Delta R_p}{R}$	$\dfrac{-\Delta L}{L_0}$	$\dfrac{-\Delta R_p}{R}$	$\dfrac{-\Delta L}{L_0}$	$\dfrac{-\Delta R_p}{R}$
	(in %)		(in %)		(in %)		(in %)		(in %)	
62.7	....	....	....	....	....	....	....	....	7.3	1.1
40.0	....	....	....	....	....	....	....	....	7.7	1.9
28.25	....	....	....	....	....	....	....	....	8.0	2.3
20.0	....	....	....	....	7.5	2.1	....	....	8.1	2.5
15.0	....	....	6.5	1.2	7.7	2.5	....	....	....	....
12.0	....	....	6.7	1.5	....	....	....	....	....	....
10.0	....	....	6.7	1.7	7.2	2.5	6.2	2.5	8.0	2.9
9.0	....	....	6.7	1.8	7.2	2.4	....	....	....	....
7.0	....	....	6.7	1.9	7.1	2.3	....	....	....	....
5.0	8.6	2.9	6.6	2.0	6.6	2.1	5.4	1.7	....	....
3.0	....	....	....	....	5.9	1.7	....	....	....	....
2.0	8.8	2.3	....	....	5.5	1.3	....	....	4.6	1.0
1.8	....	....	....	....	5.0	1.2	....	....	....	....
1.5	....	....	....	....	5.6	1.2	....	....	5.9	2.2
1.4	....	....	....	....	5.5	2.6	....	....	....	....
1.0	16.1	4.2	....	....	15	6.4	17.7	19.0	27.0	11.1
0.8	....	....	....	....	23	8.2	....	....	32.8	7.1
0.6	....	....	....	....	....	....	....	....	26.0	14.5

SOURCES: Faulkner, J., Roxburgh, I. W., and Strittmatter, P., *Ap. J.* **151**, 203, 1968 (FRS); Sackmann, I-J., and Anand, S. P. S. *Ap. J.* **162**, 105, 1970 (SA); Sackmann, I-J., *A. Ap.* **8**, 76, 1970 (S); Kippenhahn, R., and Thomas, H. C., in *Stellar Rotation* (Slettebak, A., ed.), p. 20, New York: Gordon and Breach, 1970 (KT); Papaloizou, J. C. B., and Whelan, J. A. J., *M. N.* **164**, 1, 1973 (PW).

affected by chemical composition, mixing length, and other poorly understood physical phenomena (cf. §8.4). Given the good agreement between
the above calculations, we shall now closely follow the work of Sackmann.

As early as 1923, Milne observed that solid-body rotation will decrease
the total luminosity of a star; that is to say, a uniformly rotating star of
mass M should have similar central properties as a nonrotating star with
mass $M + \Delta M$. Figure 12.1 presents a detailed numerical proof of this
statement. We observe that the values of T_c and ρ_c for rotating stars on
the verge of equatorial break-up fall exactly along the curve for nonrotating stars, their positions being somewhat shifted in the direction of
the lower masses. (Note that the largest deviation between the values for
critically rotating stars and nonrotating stars is as small as 0.001 in
$\log_{10} T_c$ and 0.004 in $\log_{10} \rho_c$!) As was shown by Sackmann, the ratio
$\Delta M/M$ is given, to a good degree of approximation, by

$$\frac{\Delta M}{M} = -\tfrac{3}{2}\bar{\chi}, \tag{1}$$

where the positive quantity $\bar{\chi}$ is the pressure-weighted average of the
ratio of the centrifugal force to the gravitational attraction over the whole

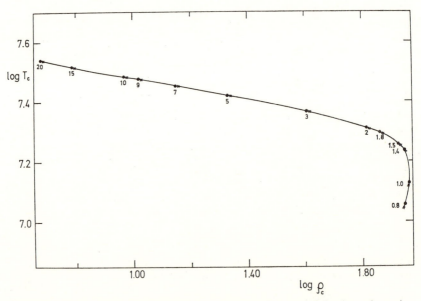

Figure 12.1. The central temperature T_c as a function of the central density ρ_c for main-
sequence stars. The curve is drawn through the data referring to nonrotating stars (*dots*).
The crosses refer to critically rotating stars. The numerals along the curve define the mass
of the models. Source: Sackmann, I-J., *A. Ap.* **8**, 76, 1970.

TABLE 12.2

The percentage decrease in mass necessary to make the central pressure of critically rotating models equal to that of a nonrotating model

$M/M_\odot$	$-\Delta M/M$	$M/M_\odot$	$-\Delta M/M$
0.8	3.0	3	2.2
1.0	4.1	5	2.0
1.4	0.7	7	2.7
1.5	0.0	9	2.3
1.8	1.2	10	2.0
2.0	1.4	20	2.8

SOURCE: Sackmann, I-J., A. *Ap.* **8**, 76, 1970.

star. Table 12.2 illustrates this mass-lowering effect at break-up rotation along the main sequence.

How are the central pressure and temperature affected by rotation along the main sequence? Figure 12.2 illustrates their changes for uniformly rotating models on the verge of equatorial break-up. We observe that there is a critical point along the main sequence where the effects of rotation change in a drastic manner. When $M > 1.5M_\odot$, p_c is increased by rotation, whereas it is reduced for masses below $1.5M_\odot$; on the other hand, T_c is always decreased by rotation, although its changes become much steeper for masses below $1.5M_\odot$. In order to understand these features, it is necessary to look at the variations of p_c and T_c for nonrotating stars along the main sequence (see Figures 12.3 and 12.4). In this case, as we proceed down the main sequence, the central pressure does not steadily increase but reaches a maximum at $1.5M_\odot$, while there is no such peak in the curve for central temperature.[1] Now, because rotation has a mass-lowering effect, the changes depicted on Figure 12.2 can be obtained, therefore, by shifting the points on Figures 12.3 and 12.4 in the direction of the lower masses. This explains at once the modifications in central pressure and temperature caused by solid-body rotation. The steep slope observed on Figure 12.4 below $1.5M_\odot$ also explains why the magnitude of the central-temperature changes produced by rotation becomes so large for late-type stars.

[1] A phenomenological explanation of these effects in nonrotating stars has been presented by Sackmann. By making use of homological transformations and a mass-radius relation of the form $R \propto M^a$ along the main sequence, she obtains $p_c \propto M^{2-4a}$ and $T_c \propto M^{1-a}$. Under reasonable assumptions, it can then be shown that $a \approx 0.7$ for upper main-sequence stars having carbon-nitrogen burning ($\epsilon_{Nuc} \propto \rho T^{16}$), and $a \approx 0.1$ for lower main-sequence stars with the proton-proton chain ($\epsilon_{Nuc} \propto \rho T^4$). Inserting the above values for a in the crude expressions for p_c and T_c, one reproduces at once the qualitative features found in Figures 12.3 and 12.4.

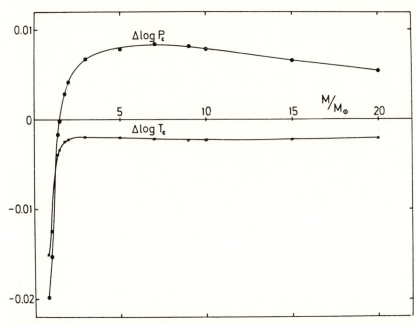

Figure 12.2. The change of the central pressure p_c and of the central temperature T_c for a critically rotating star from that of a nonrotating star of the same mass, as a function of the mass of the models. Source: Sackmann, I-J., *A. Ap.* **8**, 76, 1970.

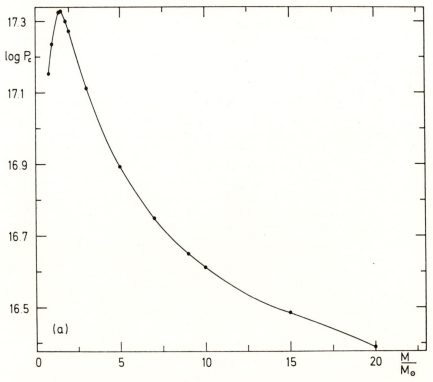

Figure 12.3. For nonrotating stars, the central pressure p_c as a function of mass along the main sequence. Source: Sackmann, I-J., *A. Ap.* **8**, 76, 1970.

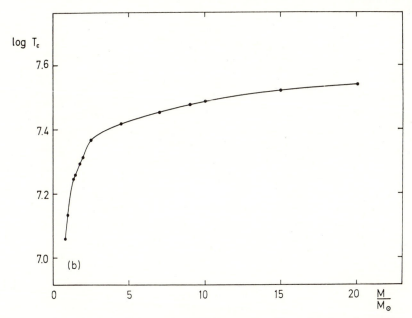

Figure 12.4.   For nonrotating stars, the central temperature T_c as a function of mass along the main sequence. Source: Sackmann, I-J., *A. Ap.* **8**, 76, 1970.

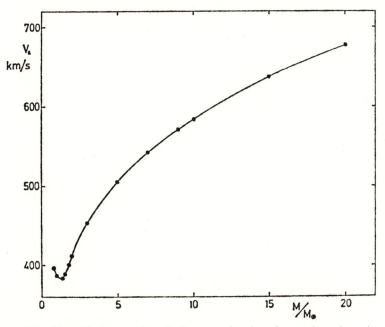

Figure 12.5.   The critical equatorial velocity v_c as a function of mass along the main sequence. Source: Sackmann, I-J., *A. Ap.* **8**, 76, 1970.

The fractional changes in luminosity and polar radius at maximum rotation are well illustrated in Table 12.1. For example, in Sackmann's results, $\Delta L/L_0$ is practically constant at 7 percent for masses ranging from $20M_\odot$ to $5M_\odot$, then dropping to a minimum of 5 percent at $1.8M_\odot$ and thereafter rising so sharply that the maximum percentage in luminosity change for a $1M_\odot$ star is 2.5 times as large as for a $10M_\odot$ star. The near constancy of $\Delta R_p/R$ for early-type stars, the drop to about 1 percent at $1.8M_\odot$, and the steep rise to 6.4 percent at $1M_\odot$ have the same features as the luminosity changes. Again, it is the cross-over from the carbon-nitrogen cycle to the proton-proton chain that produces these large modifications in luminosity and polar radius for the late-type stars (cf. note 1).

For the sake of completeness, let us also examine the variations of the critical equatorial velocity v_c at the point of break-up along the main sequence (see Figure 12.5). For a $10M_\odot$ star (corresponding to a spectral type B2.5, where the most rapidly rotating Be stars are observed), v_c is computed as 584 km/sec. This is in fair agreement with the most rapidly rotating star observed, namely the star ϕ Persei with $v_e \sin i \approx 550$ km/sec

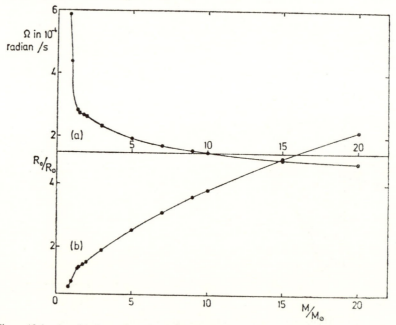

Figure 12.6.   At critical rotation, the variation of (a) the angular velocity and (b) the mean radius, as functions of mass along the main sequence. Source: Sackmann, I-J., *A. Ap.* **8**, 76, 1970.

(as derived by Slettebak). The velocity v_e steadily decreases as one passes down the main sequence from $20M_\odot$ to $1.4M_\odot$, and it rises again as the mass is decreased below $1.4M_\odot$. An explanation of this turn-over is readily found in Figure 12.6, which displays, at critical rotation, the angular velocity Ω_c and the mean radius R_0, as a function of the mass. We observe that Ω_c does not vary much with mass on the upper main sequence, so that the radius dominates the behavior of v_c there. However, for masses below $1.4M_\odot$, Ω_c rises more sharply than R_0 decreases, therefore causing a net increase in v_c as the mass is reduced.

In summary, for each mass along the main sequence it is possible to construct a series of uniformly rotating models with the quantities $\Delta L/L_0$ and $\Delta R_p/R$ steadily decreasing. Moreover, each series terminates with a model for which the effective gravity vanishes at the equator. The above result strongly suggests that uniform rotation may indeed be considered as a small perturbation superimposed on the structure of a nonrotating star. It should be stressed, however, that the numerical values listed in Table 12.1 provide a general trend of the effects of uniform rotation only, for meridional currents and other mechanisms that continuously transport angular momentum have not been considered at all. For the very same reason, these models cannot explain the break in rotational velocities at about spectral type F5V; this problem has been discussed at length in Section 11.4.

DIFFERENTIALLY ROTATING MODELS

The effects of differential rotation on the structure of upper main-sequence stars were originally considered by Mark in 1968. His calculations, based on the self-consistent field method and a polytropic approximation (cf. §5.5), gave the important result that rapid, differential rotation has a significant effect on upper main-sequence stars and, in particular, on their mass-luminosity relation. Independent calculations that actually satisfy the four equations of stellar structure—including those of energy balance and energy transport—have subsequently been carried out by Bodenheimer, and by Monaghan and Smart. Since both sets of models are in qualitative agreement with those obtained by Mark, we shall now illustrate the gross features of differentially rotating models by means of the most detailed and numerous calculations of Bodenheimer.

As we recall from Section 5.5, an iterative scheme has been developed by Jackson to describe the structure and evolution of pseudo-barotropic models. This technique (which combines the Henyey method for solving the equations of stellar structure with the self-consistent field method devised by Ostriker and Mark) has been successfully applied by Boden-heimer to calculate the structure of chemically homogeneous, main-sequence stars of $15M_\odot$, $30M_\odot$, and $60M_\odot$ ($X = 0.70$, $Z = 0.03$). As is usual in this type of analysis, the models were computed under the assump-tion of axial symmetry—with magnetic fields, viscous friction, and merid-ional currents neglected. Parameters that must be specified in addition to the total mass and chemical composition are: (i) the total angular momentum J, and (ii) a rotation law that defines the angular momentum distribution $j(m_\varpi)$ as a function of m_ϖ, the mass fraction interior to the cylinder of radius ϖ about the rotation axis. Figure 12.7 illustrates three

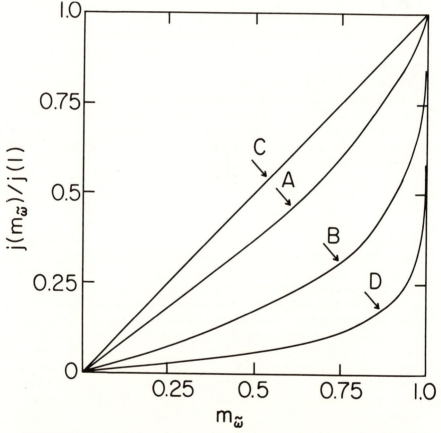

Figure 12.7. The angular momentum per unit mass $j(m_\varpi)$ for three assumed laws of differ-ential rotation (cases A, B, and C) and for a uniformly rotating model (case D). Source: Bodenheimer, P., *Ap. J.* **167**, 153, 1971. (By permission of The University of Chicago Press. Copyright 1971 by the University of Chicago.)

assumed laws of differential rotation (cases A, B, and C) and the function $j(m_\varpi)$, which is appropriate to a model in solid-body rotation (case D). In the latter case, most of the angular momentum is concentrated in the outer layers, whereas cases B, A, and C represent progressively increasing degrees of concentration of angular momentum toward the center. Also, since $j(m_\varpi)$ is an increasing function of m_ϖ in all cases, the Høiland criterion for dynamical stability is always satisfied (cf. §7.3). No special treatment is given to the distribution of angular momentum in the cylinder containing the convective core, and the influence of rotation on the onset of convection is not taken into account (cf. §14.5).

Seven series of chemically homogeneous, main-sequence stars have been constructed by Bodenheimer, each with fixed mass, composition, and $j(m_\varpi)$, but with increasing values for J. Figures 12.8 shows, *first*, that the mass-luminosity relation is significantly affected by rapid, differential rotation. For example, the luminosity of a model with $60M_\odot$ (case A) is reduced by factors up to 5.5 from that of the corresponding nonrotating model. Actually, since the last models along the various sequences depicted on Figure 12.8 are not limiting models in any physical sense, configurations with even larger reduction in luminosity are possible. *Second*, the displacement of differentially rotating models is to the right of the zero-age, zero-rotation main sequence. The calculated equatorial velocities v_e fall well within the range of observed velocities on the upper main sequence, with a mean v_e of about 180–200 km/sec for late O- and early B-stars and a maximum velocity about twice as large. Finally, we see that differentially rotating models populate the same region of the Hertzsprung-Russell diagram as uniformly rotating models and have comparable equatorial velocities.

The rotational characteristics of three models of $30M_\odot$ (case A) are illustrated in Figure 12.9. Note that considerable polar flattening occurs, with the ratio of equatorial to polar radii ranging up to about 4. The accompanying distortion of the outer layers results in the formation of a toroid-like structure. However, the inner regions are less distorted, with the ratio of radii on the level surface enclosing 20 percent of the mass ranging up to 2.5 only. Also, in the most distorted models, the surface gravity at the poles becomes larger than that at the equator by factors up to 11. Therefore, large variations in visual magnitude as a function of aspect angle would be expected for such configurations (cf. §12.3). However, none of these models—despite the high values for J in some cases—approaches the limit of zero effective gravity at the equator. (For example, the ratio of centrifugal force to gravitational attraction at the equator is equal to 0.2 for the most distorted model depicted on Figure 12.9.) Finally, in contrast to the case of solid-body rotation, conditions in the central

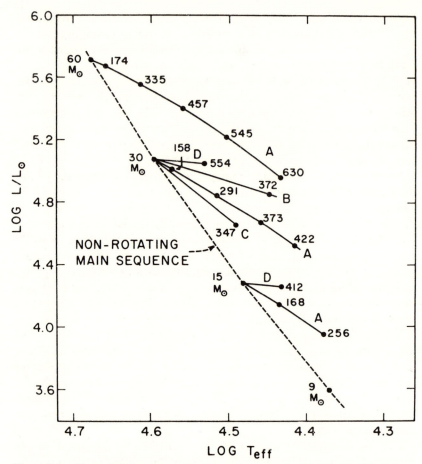

Figure 12.8.　Theoretical Hertzsprung-Russell diagram showing model sequences of increasing angular momentum (*solid curves*). Numbers on curves give calculated velocities at the equator (in km/sec). The distribution of angular momentum for each sequence is indicated by the letter A, B, C, or D. Source: Bodenheimer, P., *Ap. J.* **167**, 153, 1971. (By permission of The University of Chicago Press. Copyright 1971 by the University of Chicago.)

regions now show large changes caused by differential rotation. Indeed, as we observe on Figure 12.10, the effect of an increase in J is to shift the configurations closely parallel to and downward along the zero-rotation main sequence. (For example, with rotation law A, the last model of 60M$_\odot$ has central conditions approximating those of a nonrotating model of one-third its mass.) Similarly, the central pressure increases slowly with J, thus paralleling the similar behavior of nonrotating models as the mass is reduced (see Figure 12.3).

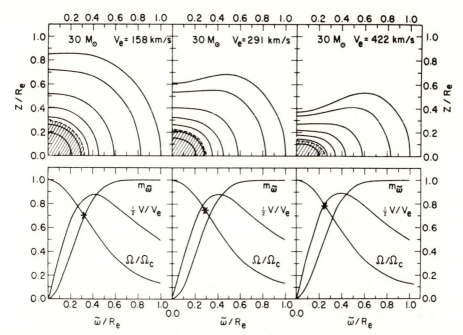

Figure 12.9. Detailed structure of three models for $30 M_\odot$ (case A). R_e is the total equatorial radius. The shaded area indicates the convective core. The upper portions show isopycnic contours enclosing mass fractions 0.2, 0.4, 0.6, 0.8, 0.95, 0.999, and 1.0. The lower portions give the ratio of the angular velocity Ω to the central value Ω_c, the fraction $m_{\tilde\omega}$ of the total mass interior to the corresponding cylindrical surface about the rotation axis, and the ratio of the circular velocity v to the surface value v_e. The boundary of the convective core is indicated by an asterisk. Source: Bodenheimer, P., *Ap. J.* **167**, 153, 1971. (By permission of The University of Chicago Press. Copyright 1971 by the University of Chicago.)

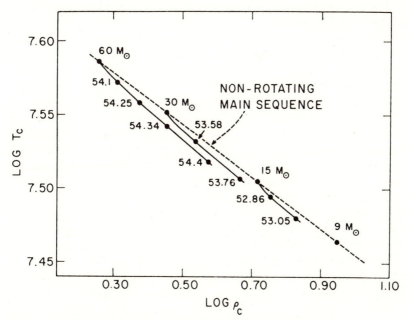

Figure 12.10. Sequences of rotating models with increasing angular momentum J (*solid curves*) in the ($\log \rho_c$–$\log T_c$)-plane. Numbers on curves give the logarithm of J in c.g.s. units. The rotation law A is used in all cases. Source: Bodenheimer, P., *Ap. J.* **167**, 153, 1971. (By permission of The University of Chicago Press. Copyright 1971 by the University of Chicago.)

Now, how are these results dependent on the assumed distribution of angular momentum within a model? From Figure 12.8 it is evident that models in solid-body rotation are displaced nearly horizontally, whereas models belonging to case C are displaced downward and to the right, nearly parallel to the zero-rotation main sequence. Thus, the greater the degree of central concentration of angular momentum, the steeper the slope of the displacement due to an increase in J. This behavior clearly reflects the effect of rotation on the *radius* of a star, as a function of $j(m_\varpi)$. Indeed, for given J, a higher concentration of angular momentum toward the surface results in a larger radius, due to the increased effect of centrifugal force in the outer layers. In other words, *given the mass M and the surface quantity v_e, the change in bolometric magnitude depends very definitely on the distribution of angular momentum within the star.* In the limit of uniform rotation, M_{bol} is practically unchanged from its nonrotating value even at $v_e \approx 500$ km/sec; in the opposite limit of substantial central concentration of angular momentum, the change in M_{bol} may be considerably greater than 1^m even at $v_e \approx 300$ km/sec. However, *given the total mass M and angular momentum J, the bolometric magnitude of a rotating star is practically independent of its distribution of angular momentum.* The degree to which this statement holds true is illustrated in Figure 12.11 for $30M_\odot$ stars. (Further calculations are required to check its validity for masses lower than those considered here.) The other portions of Figure 12.11 indicate that the values of ρ_c and T_c, for given J, are fairly insensitive to $j(m_\varpi)$, but not nearly so independent of it as in the case of the luminosity.

Let us next discuss the effect of rotation on the main-sequence lifetime of a star. For this purpose, Bodenheimer defines the quantity $t_{ms} = 1.4 \cdot 10^{11} M_{core}/L$, which is an estimate of the main-sequence lifetime in years based on the nuclear energy available in the entire zero-age convective core. (t_{ms} is about a factor of 2 higher than the actual main-sequence lifetime, but is nevertheless a useful quantity for comparing rotating and nonrotating models.) According to Bodenheimer, the lifetime t_{ms} can be increased by more than a factor of 2 along the sequence of models with $M = 30M_\odot$, and by about a factor 3 for $60M_\odot$. However, when compared with nonrotating models at the same *luminosity*, a model with $v_e \approx 300$ km/ sec has a lifetime typically only 25 percent longer, and one with $v_e \approx 400$ km/sec only 50 percent longer. Given M and J, t_{ms} is nearly independent of $j(m_\varpi)$; but again, given M and v_e, t_{ms} does depend on $j(m_\varpi)$, being almost unchanged from its nonrotating value in the limit of uniform rotation.

The above results can be summarized as follows. The location of a main-sequence model (with given M and J) in the Hertzsprung-Russell diagram crucially depends on the distribution of angular momentum

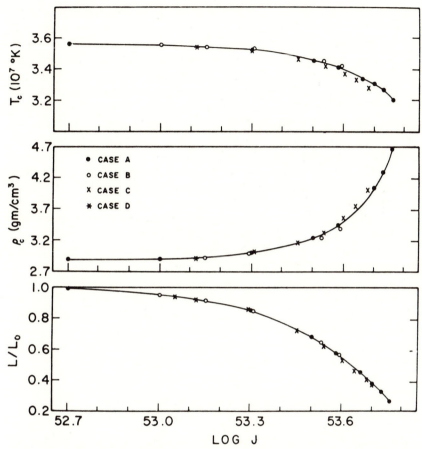

Figure 12.11. The central temperature, central density, and luminosity, as functions of the angular momentum J for rotating models of $30M_\odot$, showing the effects of changing the angular momentum distribution $j(m_\varpi)$. The solid lines connect points corresponding to the rotation law A. L_0 is the luminosity of the nonrotating model. Source: Bodenheimer, P., *Ap. J.* **167**, 153, 1971. (By permission of The University of Chicago Press. Copyright 1971 by the University of Chicago.)

because of its effect on the equatorial radius. A sequence of models with increasing J falls nearly along the zero-rotation main sequence only if the distribution of angular momentum is quite strongly concentrated toward the center; otherwise the configurations fall to the right of the zero-rotation main sequence. The fact that we have no direct knowledge of the function $j(m_\varpi)$ in stars, together with the uncertainty in the inclination of the rotation axis on the line of sight, will make comparison of these models with observations rather difficult. Moreover, although a definitive estimate of the Eddington-Sweet time for differentially rotating stars is not available (cf. §8.3), a crude calculation by Monaghan and Smart indicates that the circulation time scale is perhaps about 10^5 years for a $60M_\odot$ star with the largest (assumed) angular momentum. Since the typical evolutionary time scale for these massive stars is only about 10^6 years, one may wonder whether the great accuracy afforded by highly involved iterative schemes is justified at this stage. Obviously, a consistent theory of meridional currents in stars that greatly deviate from spherical symmetry is needed to make any further progress in our understanding of main-sequence stars (cf. §8.3).

To conclude this brief survey of differentially rotating, main-sequence models, let us remark on the relation between M and J for stars with v_e near the average of the observed range. For given v_e, an increase in the degree of concentration of angular momentum toward the center corresponds to an increase in J for a given M. Bodenheimer's results strongly suggest, therefore, that a star of given mass and v_e may have a factor of 5–8 more angular momentum than that derived by Kraft under the assumption of solid-body rotation. As was pointed out by Bodenheimer and Ostriker (cf. §11.4), the substantial gap in total angular momentum per mass inferred for single stars, assumed uniformly rotating, and double stars may thus be filled in to a considerable extent if it can be shown that at least some of the single main-sequence stars are in a state of differential rotation.

12.3. EFFECTS OF ROTATION ON THE OBSERVABLE PARAMETERS

As we recall from Section 7.2, a pseudo-barotropic model with a radiative envelope has an emergent flux $\mathscr{F}$, which varies in proportion to the surface effective gravity g. Since the latter quantity is smaller at the equator than at the poles, both the local effective temperature and surface brightness of a rotating star are, therefore, lower at the equator than at the poles. This implies in turn that the various magnitudes and color indices of a rotating star will be functions of the aspect angle i between the line of sight and the rotation axis (see Figure 2.1). The theoretical problem divides naturally into three parts: (i) to build an interior model so that the effective temperature and gravity become known as functions of latitude on its surface; (ii) to compute the energy spectrum of radiation as a function of aspect angle when a suitably realistic model atmosphere is fitted at each point of the surface; and (iii) to integrate the emergent flux in order to obtain the usual photometric parameters for each aspect angle. Detailed interior structures are now available for different rotation laws; they have been discussed in Section 12.2. The exact treatment of the atmosphere surrounding a rotating star is made difficult by the presence of meridional currents that convect energy in and out of a given region. However, order-of-magnitude calculations by Hardorp and Strittmatter indicate that for upper main-sequence stars the circulation speed is substantially lower than both the escape and sound speeds, so that the energy flux that might be expected from meridional currents is small compared to the radiative flux. Moreover, as was shown by Collins, the "horizontal" flux caused by a pole-equator temperature difference in the atmosphere of a rotating star is also small compared to the "radial" radiative flux. Given these results, it would appear that the approximate properties of the emitted radiation from a rotating star can be computed adequately on the basis of hydrostatic equilibrium, the emergent flux being determined by the von Zeipel relation $\mathscr{F} \propto g$ at each point of the surface.[2] We now turn to the changes brought by rotation on the continuum and on the spectral lines.

The earliest effort to describe the effects of varying the aspect of an axisymmetric star appears to be due to Sweet and Roy.[3] However, their calculations refer to the total luminosity of a star only, and not to that quantity that would be measured by an observer. Following the pioneering papers of Collins and Zhu in 1963, many authors have calculated the variations in the spectra, magnitudes, and color indices due to aspect effects for various rotating models. Unfortunately, most of these studies are based either on the Sweet-Roy interiors or on the erroneous models of Roxburgh, Griffith, and Sweet. (Both sets of interiors predict changes $|\Delta L/L_0|$ and $|\Delta R_p/R|$, which are too large by at least a factor of 3 when compared to the values listed in Table 12.1.) The most consistent calculations are those of Maeder and Peytremann. By making use of the Faulkner-Roxburgh-Strittmatter interiors, they have computed the energy spectrum of radiation for uniformly rotating stars of $5M_\odot$, $2M_\odot$, and $1.4M_\odot$, including not only continuous opacities but also hydrogen-line opacities for the Balmer and Lyman series. Calculations were also carried out for a $1.4M_\odot$ star in uniform rotation, then including continuous and

[2] This law holds strictly only for pseudo-barotropes with radiative atmospheres in which the transfer of energy is approximated by a diffusion equation (cf. §8.3). As was shown by Smith and Worley, the emergent flux of a pseudo-barotrope satisfies approximately the relation $\mathscr{F} \propto g^{1/2}$ if the transfer equation is used. The inclusion of circulation, however, seems to restore the von Zeipel law $\mathscr{F} \propto g$. Finally, for genuine baroclines with $\partial\Omega/\partial z \neq 0$, $\mathscr{F}$ is no longer a simple power of g, even assuming the diffusion equation.

[3] The (variable) appearance of an isolated stellar model having genuine triplanar symmetry has been also described for different values of the aspect angle i (Sčigolev, B., *Astron. Zh.* **14**, 447, 1937). His theoretical light curves might be relevant to an observational test of the various fission theories (cf. §11.3).

hydrogen-line opacities, as well as metallic-line opacities. Such refinements and accurate interiors are essential when comparing theory with observation.

Figure 12.12 shows a color-magnitude array for the Maeder-Peytremann models without metallic-line opacities. Each rotational track represents configurations ranging from the nonrotating model to the uniformly rotating model for which $\Omega/\Omega_c = 0.99$, where Ω_c is the angular velocity at break-up rotation. For each mass, different values of the inclination i

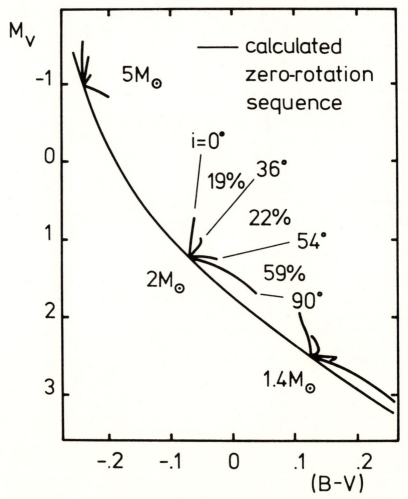

Figure 12.12. Color–magnitude diagram with rotational tracks for $5M_\odot$, $2M_\odot$, and $1.4M_\odot$, and various angles i. The termination points are for $\Omega/\Omega_c = 0.99$. Source (revised): Maeder, A., and Peytremann, E., *A. Ap.* 7, 120, 1970. (Courtesy Dr. A. Maeder.)

have been considered, the aspect angle increasing from $i = 0°$ ("pole-on" stars) to $i = 90°$ ("equator-on" stars). For the models of $2M_\odot$, the percentage of stars under the random-orientation hypothesis is also indicated. (This is of course valid for all masses.) Similarly, Figure 12.13 illustrates some rotational tracks at various angles i, for a $1.4M_\odot$ model in solid-body rotation, with or without metallic-line opacities. We observe that a "pole-on" model appears brighter with almost the same color, while an "equator-on" model appears fainter and considerably redder than a nonrotating model of the same mass. However, a rotating configuration seems always brighter than the star of the same color on the zero-rotation main sequence, the change in luminosity being dependent mainly on v_e and very little on the aspect angle i. Also, for M $= 1.4M_\odot$, the vertical shifts ΔM_V due to uniform rotation are 10–25 percent larger when metallic lines are included. In all cases, ΔM_V is at most equal to $0.^m5$ in the color-magnitude diagrams. Figure 12.14 shows rotational tracks in the two-color diagram, with and without metallic-line opacities. One can observe the large displacement of the zero-rotation main sequence due to the inclusion of metallic lines for stars later than spectral type A0. For these stars, the maximum change in $(U - B)$ is equal to $0.^m075$, compared to $0.^m135$ for a $5M_\odot$ star in uniform rotation.

How do these results compare with the available observational data for *normal* main-sequence stars? As was originally pointed out by Strittmatter, the deviation ΔM_V from the zero-rotation main sequence can be approximated by

$$\Delta M_V = 10^{-5}hv_e^2, \tag{2}$$

the quantity ΔM_V being thus weakly dependent on sin i (see Table 12.3). (As usual, v_e is expressed in km/sec.) Following Golay and Maeder, however, the rotational spread varies along the main sequence, becoming greater for late-type stars. (For A-type stars, observation reveals a quadratic dependence of ΔM_V on v_e; for late-type stars, even if this dependence is stronger than a power of 2, the above formula is nevertheless very useful when confronting theory to observation.) Also, the change $\Delta(B - V)$ caused by rotation in the color index $(B - V)$ may be represented by

$$\Delta(B - V) = 10^{-6}k(v_e \sin i)^2, \tag{3}$$

the models now showing a dependence of $\Delta(B - V)$ on $v_e \sin i$, and not only on v_e. (As usual, v_e is expressed in km/sec.) Table 12.3 compares the observed values with their theoretical counterparts derived from the most extreme models of Maeder and Peytremann. For main-sequence stars earlier than spectral type A7, the observed quantities are in good agreement with the largest values obtained under the assumption of solid-body rotation. Later than type A7V, however, there is no agreement

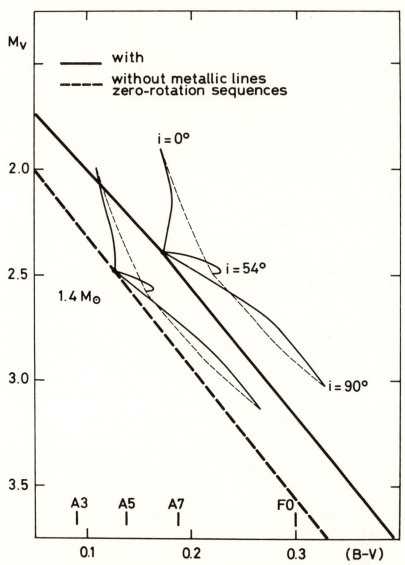

Figure 12.13.   Color–magnitude diagram with rotational tracks for various angles i, for uniformly rotating models with and without metallic-line opacities. The small dashed lines indicate the loci of maximum equatorial velocities. Source: Maeder, A., and Peytremann, E., *A. Ap.* **21**, 279, 1972.

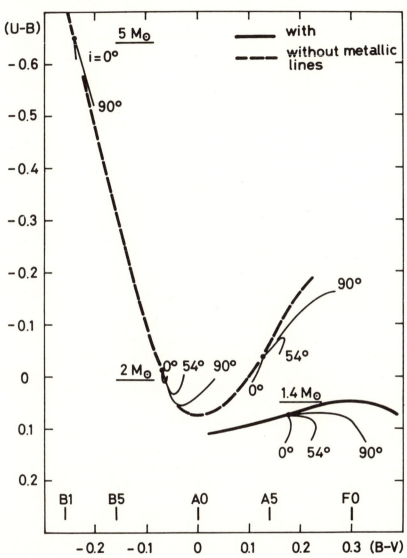

Figure 12.14. Rotational tracks in the (U − B) vs. (B − V) diagram for uniformly rotating models with and without metallic-line opacities. Source: Maeder, A., and Peytremann, E., *A. Ap.* **21**, 279, 1972.

TABLE 12.3

Theoretical and observed values of h and k

Type	Theory	Observation	Observers
A0V-A7V	$h \approx 0.4$	$h \approx 0.5$	G
	$0.5 \lesssim k \lesssim 0.7$	$k \approx 0.6$	G
After A7V	$h < 0.45$	$1.0 \lesssim h \lesssim 1.7$	G
		$h \approx 1.1$	SS
		$h \approx 1.3$	M
		$h = 0.8\,(\pm 0.3)$	S1
		$h = 2.2\,(+1.8,\,-0.8)$	S2
	$k < 0.14$	$k \approx 2.0$	G

SOURCES: Maeder, A., and Peytremann, E., *A. Ap.* **7**, 120, 1970; *ibid.* **21**, 279, 1972; Golay, M., *Arch. Sci. Genève* **21**, 105, 1968 (G); Strittmatter, P. A., and Sargent, W. L. W., *Ap. J.* **145**, 130, 1966, for Praesepe (SS); Maeder, A., *Thesis* (unpublished), Univ. of Geneva, 1971, for Praesepe (M); Smith, R. C., *M. N.* **151**, 463, 1971, for Praesepe (S1) and Hyades (S2).

at all, the observed values of h and k being larger by at least a factor of 2 than the corresponding quantities derived for uniformly rotating models on the verge of equatorial break-up. A similar conclusion was reached by Maeder and Peytremann by comparing the observed and theoretical changes in the Balmer discontinuity and in the Strömgren c_1-index. *Thus, for main-sequence stars later than spectral type A7 (up to about type F5 where rotational velocities abruptly decrease), the maximum effects given by uniformly rotating models are always smaller, at least by a factor of 2, than the observed rotational spread of the main sequence.* In other words, whereas the small changes produced by solid-body rotation adequately reproduce the observed effects for stars earlier than type A7V, the photometric deviations due to rotation in stars later than type A7V are too large to be explained by uniformly rotating models, even allowing for metallic-line-blanketed atmospheres.

This important conclusion calls for a few remarks. *First*, although the above results are explicitly based on the von Zeipel law (cf. note 2), it has been shown by Smith and Worley that the coefficient h is very insensitive to the assumed gravity-darkening law. Since the Maeder-Peytremann spectra have been computed with fairly realistic opacities and consistent interiors, one may therefore conclude that uniform rotation appears to be excluded among the main-sequence stars of spectral type later than A7. *Second*, the fact that uniformly rotating models adequately represent the observations of A-type stars earlier than A7V is no complete proof of strictly uniform rotation among these stars. Indeed, by making use of the Bodenheimer's models (cf. §12.2), Smith and Worley have shown that a large concentration of angular momentum toward the center can also lead to the value $h \approx 0.4$. The small observed spread of stars earlier than type A7V may therefore mean either uniform rotation or strong differential rotation. *If so, then, what rotation law do main-sequence stars actually follow?* This purely observational problem has been recently considered by Smith, who made a statistical analysis of the data available for rotating stars in the Praesepe and Hyades clusters. It is found that the observed spread of the main sequence can only be used to eliminate possibilities (e.g., solid-body rotation). Unfortunately, a detailed study of the errors involved shows the uncertainties to be such that the observations cannot be said to support any particular law of non-uniform rotation.

Let us now briefly discuss the effects of rotation upon the spectral lines of a star. Since the most obvious observational aspect of stellar rotation is the Doppler broadening of spectral lines, it is therefore not very surprising that all the initial efforts focused on this specific aspect of the problem. As we recall from Section 2.3, a careful analysis of rotationally broadened lines allows a simple way to determine the *projected* equatorial velocities for many rotating stars. Broadly speaking, the standard method may be described as follows. One assumes first a form for the local line profile on the apparent disc of a star and then, correcting for the rotationally induced Doppler shift, convolutes all these profiles over the disc (cf. §2.3: equation [6]). The value of $v_e \sin i$ is finally obtained by comparing the resulting profile to the observed profile of a suitably chosen line in the spectrum of the star. Following the pioneering work of Slettebak, the local line profile is usually determined by using carefully measured profiles for known sharp-line stars, and it is assumed to be the same at all points on the visible disc of the star. Also, because most of the current investigation is directed toward estimating the $v_e \sin i$ values only, second-order effects (such as departure from spherical symmetry, non-uniform rotation, gravity darkening, and the aspect that stars present to the observer) are not customarily taken into account. With the revival of interest in stellar rotation in the 1960s, the importance of these effects on the profiles of absorption lines has now been carefully examined. As was originally shown by Collins and Harrington, this can be done by fitting a constant-flux, plane-parallel atmosphere at a number of latitudes to interior models and then, in accordance with the theory of radiative transfer, calculating a local line profile at each latitude on the disc. Thence, provided with the local profile shape, the results can be integrated over the apparent disc for different values of the aspect angle so as to obtain the total flux profile. In this connection, let us mention that Slettebak, Collins, Boyce, White, and Parkinson have recently measured profiles for the He I 4471-, Mg II 4481-, and Fe I 4476-lines in the spectra of 217 bright northern and southern stars of types O9-F9, using photoelectric scans and photographic coudé spectrograms. Half-intensity widths of these observed line profiles were compared with the corresponding quantities in a set of theoretical rotationally-broadened line profiles computed using the model-atmosphere approach. The resulting $v_e \sin i$'s were then used to establish a system of standard rotational velocity stars, for use in estimating $v_e \sin i$ directly from visual inspection of spectrograms. This is the most detailed atlas of rotationally broadened spectra.

The effects of solid-body rotation on the equivalent widths of some selected lines have been investigated by many authors. In view of the many uncertainties present in the calculations and the difficulties encountered in observing second-order effects, we shall tentatively summarize some of their results. *First,* by making detailed computations for the line He I 4471, Hardorp and Scholz have concluded that the weakness of the helium lines in certain sharp-lined B-type stars cannot be explained by means of uniformly rotating stars observed pole-on. Indeed, Figure 12.15 indicates that, although the "pole-on" models lie below the sequence of nonrotating stars, the maximum displacement at a given color is only a matter of 20 percent. For the very same reason, the He I strengths do not seem to provide an effective means of differentiating between rapidly rotating stars seen pole-on and intrinsically slow rotators. *Second,* as was shown by Hardorp and Strittmatter, solid-body rotation introduces a considerable scatter in the calibration of luminosity against Balmer-line index for early-type stars. In fact, for models rotating near their break-up velocities, the maximum displacement in M_V at a given Balmer-line intensity may be as much as $\pm 0\overset{m}{.}5$ depending on aspect. *Third,* independent calculations by Collins and Harrington, and by Hardorp and Strittmatter unambiguously indicate that the equivalent widths of the lines $H\beta$ and $H\gamma$ should decrease roughly as $(v_e \sin i)^2$. The observed correlation between Balmer-line index and projected rotational velocity (as discussed by Guthrie for stars in young clusters) was found to be in reasonable agreement with this theoretical result; a similar agreement was also obtained by Balona. Very much in the same vein, Hardorp and Scholz have predicted, for stars around spectral type B2, a correlation between $v_e \sin i$ and the equivalent width of the line He I 4471 in the sense that the latter should be smaller for larger velocities. According to Norris and Scholz, in early B-type stars of a given Balmer discontinuity, the equivalent width of this line is indeed systematically larger in sharp-lined than in broad-lined stars (when corrections are applied for evolutionary effects). *Fourth,* for stars seen pole-on, Hardorp and Strittmatter have concluded that uniform rotation alone has a negligible effect on abundance determinations based on lines with equivalent widths in the range 25–150 mÅ, and

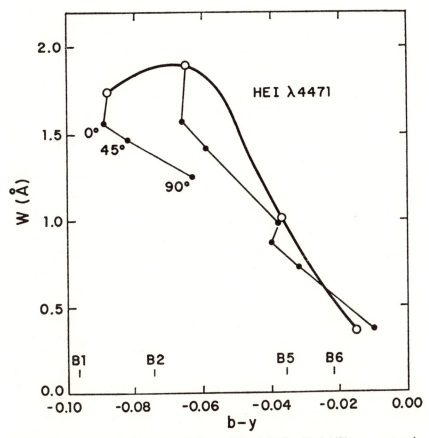

Figure 12.15.   Dependence of the equivalent width of the line He I 4471 on aspect and color. *Heavy line*: nonrotating models. *Dots*: the most rapidly rotating models viewed under the angle $i = 0°$, $45°$, and $90°$. Thin lines connect models of the same mass. For sharp-lined stars, the weakening of the line He I 4471 due to uniform rotation is less than 20 percent. Source: Hardorp, J., and Scholz, M., *A. Ap.* **13**, 353, 1971.

that it does not provide an explanation for the excessive strength of Si III lines in early B-type stars. (As was pointed out by Vilhu and Tuominen, because of rotational broadening, lines with equivalent widths less than 100 mÅ are likely to disappear in rapid rotators unless they are observed nearly pole-on!) *Fifth*, rapid rotation causes only small deviations from the usual relation between spectral type (measured by the ratio of equivalent widths of the lines He I 4121 and Si II 4130) and color index $(b - y)$ for B-type stars, particularly for aspect angles of $45°$ or less. In fact, the spread in colors (or effective temperatures) derived from spectral types and caused by a spread in turbulent velocities is greater than that due to uniform rotation alone. Thus, where spectral classification may be accomplished with reasonable accuracy, solid-body rotation by itself introduces very little change in the spectral-type-color relation. This result is due to Hardorp and Strittmatter.

Since neither theoretical considerations nor observations of the continuum can give a clear expectation for the surface angular velocity of a rapidly rotating star, we turn to the effect of differential rotation on the shape of spectral lines. As early as 1949, Slettebak concluded that such a second-order effect would be difficult to detect observationally. More recently, Stoeckley and his collaborators have presented a theoretical method for determining the aspect angle, as well as the presence or absence of differential rotation in the spectra of stars. According to these authors, enough good data exist to make possible a determination of this angle and the amount of differential rotation for fast rotators. *No results are available so far.* In this connection, let us also mention that Gray has measured line profiles in six A-type stars in order to detect differential rotation. His results seem to indicate that differential rotation does not exist or is small in early-type stars. As Gray pointed out, however, alternate conclusions are possible. Perhaps his sample of stars is too small and misrepresents A-type stars in general. Perhaps other effects are entering that cancel out or obliterate the spectroscopic characteristics of differential rotation. *More observations will be needed before the situation will be fully settled.*

To conclude, let us also mention the possibility of making interferometric observations of the surface distortion of rapidly rotating stars. According to Johnston and Wareing, however, until the present sensitivities increase considerably, it will be very difficult to observe the rotational distortion of any star, though with present sensitivities it should be possible to detect whether or not a star is in a state of extreme differential rotation. *No definite result is available so far.*

12.4. ROTATION OF THE Am- AND Ap-STARS

As was originally shown by Slettebak, the metallic-line (Am) and peculiar A-type (Ap) stars have small projected rotational velocities, $v_e \sin i$, relative to the means for normal stars of corresponding spectral types (see Figure 2.5). The first basic question is whether the Am- and Ap-stars are intrinsically slow rotators or normal stars observed pole-on. Since both groups of stars differ in many essential respects, we shall consider them in succession. The section concludes with some remarks on the relation between rotation and diffusion processes in the Am- and Ap-stars.

Abt's discovery that all, or most, Am stars are spectroscopic binaries leaves little doubt that the gross differences in mean rotational velocities for Am stars and normal A-type stars cannot be explained entirely by an aspect effect. Following Abt, this point can be argued in three different ways. *First*, the Am stars are usually or always members of spectroscopic binaries, while the normal A-type stars are only found occasionally in binaries and then only in systems with long periods ($P \gtrsim 300$ days). Since in closely spaced binaries the orbital and rotational axes will be tidally forced toward parallelism (cf. §§2.4 and 16.3), the spectroscopic-binary characteristics of these two groups of stars are just opposite to the expected ones if the aspect effect were valid: "pole-on" stars would tend to be in "pole-on" binaries, which would be more difficult to discover, whereas the Am stars are the ones frequently observed to be in close binaries. *Second*, the complete absence of binaries with periods less than 300 days among the normal A-type stars and the high frequency of short-period binaries among the Am stars would not occur if the difference between the two groups were merely an aspect effect. *Third*, if Am stars were pole-on aspects of normal A-type stars, there would be no short-period eclipsing Am binaries but many eclipsing A-type stars, whereas observations show that there are many eclipsing Am binaries and no known eclipsing binaries among normal A-type stars.

Given the foregoing arguments, we conclude with Abt that the Am stars are not necessarily observed pole-on, so that the sin i in the projected equatorial velocity does not dominate their distribution of v_e sin i values—although there always remains the possibility that aspect effects might contribute a small part to the differences between rotational velocities in Am stars and normal A-type stars. Assuming random distributions of the axes for both groups of stars, we can thus convert statistically from projected equatorial velocities (v_e sin i) to true equatorial velocities (v_e). Figure 12.16 depicts the frequency distributions of true equatorial velocities for representative samples of marked (Am) and marginal (Am:) metallic-line stars, and of normal A5-A9 (IV or V) stars. We observe that the distributions of metallic-line stars show little overlap with that of normal A-type stars. According to Abt, this overlap can be explained by differences in spectral classification. If so, then, the rotational velocity is a necessary *and* sufficient parameter to determine whether, within certain spectral- and age-ranges, a star will have a metallic-line or normal spectrum. (But it will not determine whether an abnormality in abundances will be marked or marginal.) Analysis of published data on binaries further indicates that the Am: stars have the same high frequency of short-period binaries as do the Am stars and that, in both cases, their low

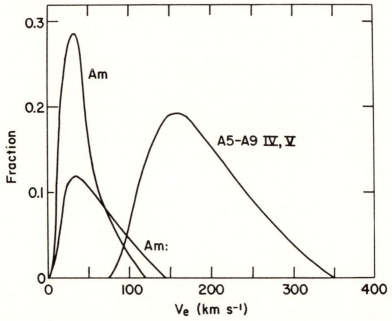

Figure 12.16. Frequency distributions of equatorial rotational velocities for Am, Am:, and normal A-type stars. The areas under the three curves are proportional to the numbers of stars in the samples, or 28 : 19 : 53 percent, respectively. Source: Abt, H. A., *Ap. J.* **195**, 405, 1975. (By permission of The University of Chicago Press. Copyright 1975 by the American Astronomical Society.)

rotational velocities are caused mostly by tendencies toward synchronous rotation in closely spaced binaries (see Figure 2.12). Putting it in another way, if we add the three curves in Figure 12.16, we obtain a bimodal distribution (with maxima at 35 and 150 km/sec, and a minimum at 80 km/sec), suggesting that we have two groups of late A-type stars whose present rotation rates have been produced by two different mechanisms: (i) members of binaries with small equatorial velocities due primarily to tidal interaction, and (ii) single stars (or members of widely spaced binaries) that have retained much of their initial angular momenta. The fact that slow rotation is the single most important factor in producing abnormal spectra in A-type stars is further confirmed by Strömgren's original observation that mild rotation reduces the apparent degree of metallicism in Am stars. Finally, as can be seen from Table 12.4, the mean values of the evolutionary luminosity increase, ΔM_V, above the zero-age main sequence also indicate that the Am and Am: stars are concentrated much closer to this sequence than the normal A-type stars. This result is in conformity

TABLE 12.4

Characteristics of Am, Am:, and normal A-type stars

Group	n	$\langle v_e \sin i \rangle$ (km/sec)	Maximum $v_e \sin i$ (km/sec)	$\langle v_e \rangle$ (km/sec)	$\langle \Delta M_V \rangle$ (mag)	Fraction with $M_V > 0.5$ mag
Am	66	33	95	42	0.38	36%
Am:	44	47	110	58	0.25	25%
A5–A9 IV, V	123	141	265	186	0.78	60%

SOURCE: Abt, H. A., *Ap. J.* **195**, 405, 1975. (By permission of The University of Chicago Press. Copyright 1975 by the American Astronomical Society.)

TABLE 12.5

Characteristics of Ap subgroups

Group	Number of stars	Mean Hydrogen Type & p.e. in the mean	Range in Type	$\langle v_e \sin i \rangle$ & p.e. in the mean (km/sec)	Range in $v_e \sin i$ (km/sec)
Hg, Mn	15	B8.0 ± 0.9	B5–A0	29 ± 5	5–100
Si	20	B7.8 ± 0.8	B5–B9	46 ± 4	16–105
Sr, Cr, Eu	24	A0.9 ± 0.4	B8–F0	30 ± 3	5–85

SOURCE: Abt, H. A., Chaffee, F. H., and Suffolk, G., *Ap. J.* **175**, 779, 1972. (By permission of The University of Chicago Press. Copyright 1972 by the American Astronomical Society.)

with Smith's result that the metallic-line phenomenon disappears as stars approach the giant region.

We now turn to the projected rotational velocities of the Ap stars. The first question is whether there are subgroups of Ap stars that are sufficiently different in rotational velocities to warrant their separate consideration. For the three subgroups summarized in Table 12.5, we note that the mean rotational velocities and ranges show marginal differences only, except that Guthrie's discovery of a deficiency of very narrow-lined Si stars ($v_e \sin i < 10$ km/sec) is confirmed and strengthened. Figure 12.17 illustrates the frequency distribution of rotational velocities for 49 Ap stars and 73 normal B7–A0 (IV or V) stars. The mean projected velocities are 42 and 178 km/sec for the Ap- and normal stars, respectively, indicating that the former have one-quarter the mean projected rotational velocity of the latter. However, in contrast to the Am stars, the distributions of $v_e \sin i$ values overlap—an indication that, if the rotational axes are oriented at random, there are statistically some normal stars with lower rotational velocities than for some of the Ap stars. This would imply that a low rotational velocity is not a sufficient condition for a late B-type star to exhibit abundance anomalies. The important question of whether the Ap stars could be normal stars observed pole-on (i.e., only the polar

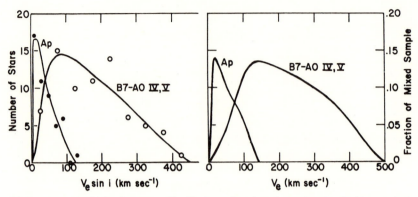

Figure 12.17. *Left*, the frequency distribution of projected rotational velocities, $v_e \sin i$, for 49 Ap stars with hydrogen types B7–A0 compared with the distribution for 73 normal A-type stars. The intervals used are 20 km/sec for the Ap stars (*dots*) and 50 km/sec for the normal stars (*circles*); the curves are free-hand averages. *Right*, the distributions in equatorial rotational velocities, v_e, normalized for a frequency of 20 percent Ap stars. Source: Abt, H. A., Chaffee, F. H., and Suffolk, G., *Ap. J.* **175**, 779, 1972. (By permission of The University of Chicago Press. Copyright 1972 by the American Astronomical Society.)

regions, where the rotational velocities are low, might have abundance anomalies) has been investigated (among others) by Preston and by Abt, Chaffee, and Suffolk. According to these authors, a comparison of projected rotational velocities with equatorial velocities computed from periods of variations in light, magnetic field, or spectrum leads strong support to the rigid-rotator model and to random orientations of the rotational axes (cf. §15.4).

On the evidence before us, it thus appears that the Ap stars are intrinsically slow rotators, but that rotation alone might not be a sufficient parameter to distinguish between Ap stars and normal stars—although there is the possibility that all the Ap stars have not been removed from the sample of normal stars, so that the two groups may not overlap in v_e. Moreover, the low rotational velocities of the Ap stars do not appear to be due to tidal interactions in binaries. Indeed, as was shown by Abt and Snowden, the frequency of visual binaries seems to be normal for all three subgroups of Ap stars. What is, then, the mechanism responsible for the abnormally low rotation rates of the Ap stars, when compared to the normal main-sequence stars of the same temperature? Of course, given the absence of extended subphotospheric convective zones and in view of their very short convective phases during the pre-main-sequence contraction, magnetic braking by thermally driven stellar winds cannot be invoked here (cf. §11.4). Havnes and Conti have proposed that Ap stars might accrete and spin up ionized matter of the surrounding interstellar medium at the boundary of their magnetospheres. From this assumption, a braking time of the order of 10^7 to 10^8 years is obtained, which is sufficiently short so that appreciable magnetic braking will occur to stars on the upper main sequence. Kulsrud has also investigated the possibility that magnetic stars might be rotationally decelerated solely by the radiation of hydromagnetic waves with no mass loss. Finally, Strittmatter and Norris have suggested that evolution toward the Ap state might critically depend on the initial ratio of magnetic energy to rotational kinetic energy. In their picture, if this ratio exceeds a critical value, thermally driven circulatory currents in the atmosphere will be suppressed by the magnetic field, so that the field lines always remain above the surface of the star. The field then continues to extend from the photosphere into the ambient medium to which it can transfer angular momentum through a centrifugal wind, as originally discussed by Mestel (cf. §11.4). Order-of-magnitude calculations show that angular momentum is lost on a time scale of the order of 10^7–10^9 years. Obviously, these various models for explaining the lower-than-average rotational velocities of the Ap stars are still very tentative, and all of them require further quantitative studies.

To the best of our knowledge, there is as yet no consensus about the actual mechanism that causes the abnormal abundances in the spectra of Am- and Ap-stars. As far as the Am stars are concerned, a well defined correlation exists between slow rotation and abnormal spectrum. A currently working hypothesis regarding these stars is that members of close binaries are, by tidal interaction, slowly rotating stars so that, given the lack of substantial mixing by circulatory currents in their atmospheres, they have abnormal spectra. Indeed, as was advocated by Michaud and his collaborators, *if a star has negligible macroscopic motions in its atmosphere* microscopic diffusion processes can separate the light elements from the heavy elements; that is to say, those elements absorbing more of the outward going radiative flux per atom move to the surface, while those absorbing less sink into the interior. The same mechanism has been propounded also to explain the abundance anomalies in the Ap stars: they are slow rotators, they have shallow surface convective zones, and some of them have strong magnetic fields. They are thus the stars most likely to possess stable envelopes in which gravitational sorting of the elements is possible.[4] Although no serious discrepancy between theory

---

[4] The magnetic-accretion process that purposely explains the slow rotation rates of the Ap stars has been also used by Havnes and Conti to predict abnormal surface concentrations in these stars. As they pointed out, however, it is not impossible that both an accretion mechanism and diffusion processes operate in magnetic Ap stars to enhance abundance peculiarities on their surfaces.

and observation appears definitively established at the present level of uncertainties, the above mechanism nevertheless presents a serious difficulty: for microscopic diffusion to be of importance, the various time scales for mixing (by circulation, convection or turbulence) or for replacement of material (by accretion or mass loss) should be much larger than the time scales for gravitational sorting to occur. The role of meridional currents has been especially considered by Schatzman, Baglin, and Vauclair; no firm conclusion can be drawn at this time, though, because self-consistent flow patterns near the surface of a rotating star are not yet available (cf. §§8.2–8.3). The importance of shear-flow instabilities in differentially rotating stars has been stressed also by Zahn, but no practical criterion for stability has emerged so far from this type of analysis (cf. §8.4). It would thus appear that, instead of looking for no-mixing conditions in slowly rotating stars, we should perhaps accept gravitational sorting as a most likely explanation and, thence, consider the diffusion velocities (<1 cm/sec) as a crude upper limit to circulation velocities (until reliable circulation flow patterns in radiative zones are eventually obtained).

12.5. STELLAR ROTATION IN OPEN CLUSTERS

The main observations pertaining to stellar rotation in open clusters and associations have been summarized in Section 2.3. The present section is confined entirely to a discussion of the following two questions: (i) what are the causes of the differences in projected rotational velocities between individual clusters and field stars? and (ii) how does rotation modify the age estimates of open clusters?

Figure 2.7 provides a comparison of $\langle v_e \sin i \rangle$ values in several open clusters and associations with the field stars. It seems apparent that, on the average, the projected rotational velocities in clusters do tend to be somewhat unique. The question immediately arises whether the $\langle v_e \sin i \rangle$ values of a given cluster are unusual because of high or low equatorial velocities, v_e, or because of preferential inclinations, i, of the rotational axes. Support for the view that the rotational axes in clusters are distributed at random has been set forth by Kraft. *First*, his investigation of the inclinations of orbital planes of visual binaries in the Hyades and Coma clusters shows no preferential orientation, and one might expect the rotational and orbital axes to be roughly aligned. *Second*, the Hyades and Coma clusters, which have many characteristics in common, also share an unusual dependence of $\langle v_e \sin i \rangle$ on spectral type (see Figure 2.7); since the directions of these two clusters from the Sun are very different ($b^{II} = -24°$ and $+84°$, respectively), it would be surprising if their v_e's and i's combined to give the same values of $v_e \sin i$. *On the contrary side*, arguments for a preferred orientation of the rotational axes in the cluster IC 4665 have been given recently by Ferrer and Jaschek, but the evidence is not strictly compelling. Admittedly, we do not know whether the axes of closely spaced binaries or rotational axes of stars in open clusters are oriented at random or whether there exists a preferential direction in some (if not all) clusters.

With regard to the causes of the differences in mean rotational velocities between clusters, three likely explanations have been examined at length by Abt; these are evolutionary expansion effects, the proportion of binaries, and the proportion of peculiar stars. We shall briefly summarize his discussion.

When a star leaves the zero-age main sequence and expands, its equatorial velocity decreases at a rate that is inversely proportional to the radius if the outer layers retain their initial angular momenta, or at a slower rate if angular momentum is transferred within the star so as to maintain approximate uniform rotation (cf. §12.6). In other words, the evolution of the brightest stars in a cluster causes them to have $v_e \sin i$'s that are lower than the comparable field values, because their radii are systematically larger than they would be for main-sequence stars of comparable absolute magnitudes. As can be observed in Figure 2.7, such an evolutionary effect would seem to explain the low rotational velocities of the brightest stars in, e.g., IC 4665. However, the fact that evolutionary expansion is *not* the main cause of abnormal rotational velocities in clusters is well illustrated by the α Persei cluster, where the evolved stars have larger, rather than smaller, mean rotational velocities than field stars! We conclude therefore that the post-main-sequence evolution and its consequent effect on the rotational velocities of stars should be considered, but this effect alone cannot explain the observed differences between individual clusters and field stars.

Abt suggested two other possibilities. As we recall from Sections 2.4 and 12.4, the initial rotational velocities of stars may be gradually modified by tidal interaction in closely spaced binaries (e.g., the Am stars) or by magnetic coupling in magnetic stars (e.g., the Ap stars). If some clusters differ in their number of spectroscopic binaries or Ap stars, we might expect that their mean rotational velocities will also depart significantly from the average rotational velocities of field stars. That this is indeed the case is illustrated in Figure 12.18, which shows the relation between (i) the frequency of binaries (with periods less than 100 and 10 days, respectively) plus Ap stars, and (ii) the mean rotational velocity relative to that of field stars of the same types. Taking into account the binaries with periods less than 100 days, we observe an approximate inverse correlation in the sense that clusters (α Persei, Pleiades) with rapidly rotating stars have

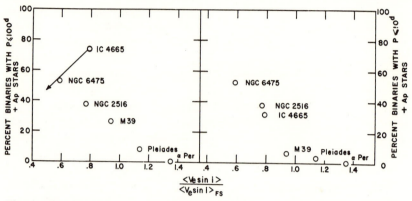

Figure 12.18. The percentage of spectroscopic binaries (*ordinates*) with periods less than 100 days (*left*) or less than 10 days (*right*) plus the Ap stars for open clusters with various mean rotational velocities (*abscissae*) relative to those of field stars of the same spectral types. Source: Abt, H. A., and Sanders, W. L., *Ap. J.* **186**, 177, 1973. (By permission of The University of Chicago Press. Copyright 1973 by the American Astronomical Society.)

far fewer binaries plus Ap stars than clusters (M 39) with normal rotational velocities, or, especially, than clusters (NGC 6475, NGC 2516, IC 4665) having low rotational velocities.[5] A much tighter correlation is obtained when only binaries with periods less than 10 days are used in the statistics. According to Abt and Sanders, this may be explained in two different ways: either duplicity with periods between 10 and 100 days is not very effective in reducing rotational velocities by tidal interaction, *or* there exists a preferred orientation of the rotational and orbital axes in some clusters. (The arrow in Figure 12.18, *left*, on the symbol for IC 4665 shows how that symbol is moved if the rotational and orbital inclinations are both changed so that the mean mass functions of IC 4665 and NGC 6475 are made to agree.)

In summary, it is apparent that tidal interaction and magnetic braking are very effective in reducing rotational velocities, so that a large part of the differences between clusters in their mean rotation rates can be assigned to these effects. However, as was pointed out by Abt, we have only succeeded in shifting the problem from trying to explain the various mean rotation rates in clusters to trying to explain the different binary- and Ap-star-frequencies. Furthermore, it also remains to see whether, after

---

[5] The very high binary frequency found by Abt and his collaborators in IC 4665 has been correctly questioned by Crampton, Hill, and Fisher. According to Abt, the evidence still shows an observed binary frequency of 50 percent or more for clusters with unusually narrow lines (M 39, IC 4665, NGC 2516, NGC 6475) and about half that for clusters with broad lines (α Persei, Pleiades).

allowance for these and other obvious effects, the mean rotational velocities in a given cluster specifically depend on the initial rotation rates of
the cloud from which the cluster has condensed, or whether all clusters
do have a common dependence of mean rotational velocity on spectral
type. A clear answer to this question would provide an important clue as
to how stars actually acquired their original angular momenta.

We now consider the modifications brought by rotation on the age
estimates of open clusters. As we know, the age of an open cluster is
usually obtained from its color-magnitude diagram by fitting the observed sequence in the turnoff region with isochronic lines derived from
nonrotating stellar models. The effects of rotation on age estimates are
essentially of two kinds: *aspect* effects on the color and magnitude of
each star belonging to the cluster, and *structural* effects on the models
that are used to draw the theoretical isochronic lines. Both aspects of the
problem have been considered by Maeder under the assumption of uniform rotation on and above the main sequence. The consistency of this
hypothesis is supported by various studies of observable properties, particularly those concerning the photometric effects of rotation on magnitudes and colors of A-type stars (cf. §12.3), as well as those concerning
the mean rotational velocities of A- and F-type stars on and above the
main sequence (cf. §12.6). As was convincingly shown by Maeder, the
structural effects of uniform rotation on age estimates are negligible in
comparison with the aspect effects. (The increase caused by uniform
rotation in the evolutionary time scale is a mere few percent only, whereas
the neglect of the aspect effects leads to an overestimate in age that may
reach up to 70 percent for clusters with the most rapidly rotating stars.)
Our main problem, therefore, is to obtain for each star in a cluster the
color and magnitude of the corresponding spherical star of the same total
luminosity. (These are not, in general, the color and magnitude the star
would have if it were nonrotating!) Fortunately, the effects of rotation on
absolute magnitude have little influence on age estimates, for the isochronic lines have an almost vertical tangent in the turnoff region (see
Figure 12.19). Moreover, by virtue of equation (3), the correction of the
color index essentially depends on the product $v_e \sin i$, and not on the
individual values of v_e and i. Individual corrections can thus be obtained
at once from the Maeder-Peytremann models discussed in Section 12.3.

Figure 12.19 illustrates the color-magnitude diagram for the α Persei
cluster with individual corrections in the turnoff region of the cluster.
Table 12.6 summarizes age estimates for this and six other clusters. These
ages were obtained (with and without rotational corrections) by using
isochronic lines derived by interpolation between the spherical models of
Kelsall and Strömgren ($X = 0.70$, $Y = 0.28$). We observe that the age

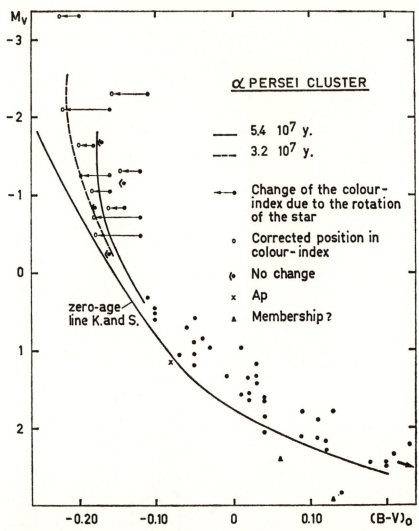

Figure 12.19. Color–magnitude diagram for the α Persei cluster with individual corrections for uniform rotation. (These are made only for stars in the turnoff region.) The isochronic lines correspond to the age with (*broken line*) and without (*continuous line*) rotational corrections. An arrow on the right-hand part of the main sequence indicates that it still extends to fainter stars. The spherical models are taken from: Kelsall, T., and Strömgren, B., *Vistas in Astronomy* (Beer, A., ed.) **8**, 159, 1966. Source: Maeder, A., *A. Ap.* **10**, 354, 1971.

TABLE 12.6

Age estimates of seven open clusters

Clusters	Without Rotation in 10^6y	With Rotation in 10^6y	Age Overestimate in %
α Persei	54	32	69
Pleiades	78	48	62
IC 4665	60	51	18
M39	230	200	15
Coma B.	560	500	12
Hyades	600	560	7
Praesepe	660	600	10

SOURCE: Maeder, A., *A. Ap.* **10**, 354, 1971.

overestimates caused by the neglect of rotation reach about 60–70 percent for the Pleiades- and α Persei-clusters; by contrast, the ages of the Hyades-, Coma-, and Praesepe-clusters (which have lower rotational velocities) undergo very little changes, approximately 10–20 percent. It is perhaps not inappropriate to recall at this place that these are orders of magnitude only. Indeed, in making use of the Kelsall-Strömgren models with $X = 0.70$ and $Y = 0.27$ (instead of $Y = 0.28$), Maeder obtains age estimates that are 22–28 percent smaller than those listed in Table 12.6; on the contrary, Iben's spherical models with $X = 0.71$ and $Y = 0.27$ give 12–35 percent higher values for ages between 10^8 and $7 \cdot 10^8$ years, and 0–12 percent higher values for ages less than 10^8 years! According to Maeder, due to the parallelism of the log (age) vs. log (T_{eff}) relation for the turnoff point of the various sets of existing models, his rotational corrections on the age estimates are identical for all of the various age scales. The third column in Table 12.6 should thus provide reliable results.

12.6. THE POST-MAIN-SEQUENCE PHASES

Among the many problems that beset the theory of rotating stars, the distribution of angular momentum in evolving stars is by far the least understood. Indeed, the post-main-sequence evolution of a star is accompanied by a strong contraction of its helium-rich core and by a corresponding expansion of the surrounding envelope. In the absence of any means of transport of angular momentum from the core to the envelope, it is obvious that the former has to spin up appreciably while at the same time the latter must spin down. To compute the gross changes caused by rotation on the post-main-sequence evolution of a star, we are therefore faced at once with two pressing questions: is the total angular momentum J of a star conserved or lost during its evolution? and is there an effective means to transfer angular momentum per unit mass from the core to the envelope? Given our complete ignorance in these matters, different limiting cases (mainly with constant J) have been considered in the literature. These are:

α) angular momentum per unit mass is conserved locally in radiative regions, and solid-body rotation prevails in convective regions;

β) angular momentum per unit mass is conserved locally in layers with stably stratified molecular weight only (cf. the "$\bar{\mu}$-barriers" of Section 8.3), and solid-body rotation prevails everywhere else;

γ) a star starting off the main sequence with solid-body rotation continues to rotate uniformly during its post-main-sequence evolution.

We shall now mainly report on the observational tests of the various assumptions about angular momentum transfer in evolving stars. Theoretical results and speculations about these matters are briefly discussed *in fine*.

OBSERVATIONAL TESTS

THE Be STARS. As we recall from Section 1.4, it has been originally conjectured by Struve that the rapidly rotating Be stars owe their emission properties to the presence of a disc of hot gas which is ejected when B-type stars find themselves rotating at their equatorial break-up velocities (cf. §2.3). Starting with the qualitative study of Crampin and Hoyle, many authors have computed evolutionary sequences of uniformly rotating stellar models including the effects of equatorial mass loss (i.e., case γ above). Detailed numerical results are now available for stars with initial masses in the range $5-10M_\odot$, mass loss occurring at the equator whenever the surface ratio (α_S, say) of the centrifugal force to gravitational attraction exceeds unity. The most important feature of these calculations lies in the fact that an early-type star evolving from the main sequence with moderate uniform rotation ($\alpha_S \gtrsim 0.2$) attains its equatorial break-up velocity during the hydrogen burning phase (from A to B in Figure 12.20), long before the second contraction phase following hydrogen exhaustion in the core (from B to C in Figure 12.20). Nobili and Secco's original calculations for a $7M_\odot$ star indicate that mass loss proceeds at a rate of about $10^{-9}M_\odot$/year during the hydrogen burning phase, and at about 10^{-7} to $10^{-6}M_\odot$/year during the hydrogen exhaustion phase. Similar rates of mass loss were subsequently obtained by Strittmatter, Robertson, and Faulkner for a $9M_\odot$ star; that is to say, $3\cdot10^{-9}M_\odot$/year during the first expansion phase, and $4\cdot10^{-7}M_\odot$/year during the second contraction

phase. (The total mass loss was found to be only 0.6 percent, the concomitant loss of angular momentum amounting to 23 percent of its initial
value.) More recently, Meyer-Hofmeister and Thomas have also computed the evolution of a $9M_\odot$ star by assuming uniform rotation at least
in the homogeneous radiative zone (i.e., case β above). In this case, two
distinct phases of mass loss were found. The first phase starts right on the
main sequence, because critical angular velocity was assumed at the
beginning, and it stops at the end of the first expansion phase. This phase
lasts for $2.6 \cdot 10^7$ years, with an average rate of mass loss of about $10^{-9} M_\odot/$
year. During the second contraction phase, however, no mass loss occurs
for about 10^6 years; then, it starts again and proceeds while the star is
expanding toward the red-giant region for $5 \cdot 10^4$ years, with a rate of
about $3 \cdot 10^{-7} M_\odot$/year. These mass-loss rates are thus very similar to those
obtained for models that maintain solid-body rotation all the way to the
red-giant region.

How do these theoretical mass-loss rates compare with the observed
mass losses from the Be stars? According to Strittmatter and his associates,
a theoretical mass-loss rate of about $3 \cdot 10^{-9} M_\odot$/year during the hydrogen
burning phases is in satisfactory agreement with various (indirect) measurements of mass loss from the Be stars, notably with those inferred
from the work of Burbidge and Burbidge. In their view, Struve's conjecture about these stars is strongly supported by the observational data.
As was pointed out by Gredley and Borra, however, other (indirect)
measurements of mass loss from the Be stars (notably those of Marlborough and Hutchings) lead to mass-loss rates that are one or two orders
of magnitude greater than those quoted by Strittmatter and his associates.
In particular, Hutchings finds that the mass-loss rates range from $10^{-7} M_\odot/$
year for early B-type stars to $2.5 \cdot 10^{-8} M_\odot$/year for late B-type stars. If
we accept these figures because they are derived from more detailed
numerical models that match the observed line profiles reasonably well,
we are thus led to choose between two alternatives: *either* the Be stars are
in the second contraction phase where the rates of mass loss can be much
greater than $10^{-9} M_\odot$/year *or* the assumption of rotational ejection of
matter is not entirely adequate to explain the emission properties of the
Be stars. A careful statistical analysis of these stars has been made by
Hardorp and Strittmatter. According to these authors, the conjecture
confining the Be stars to the second contraction phase cannot be maintained; hence, most of them must be placed into the first expansion phase,
their evolution probably proceeding with constant angular velocity at
least in the radiative zones. Given the serious discrepancies in the
"observed" rates of mass loss, it remains to be seen whether the Be stars
lose mass because they are rotating near (but probably not exactly at)

their break-up velocities, or because another ejection mechanism (such as a radiatively driven stellar wind, aided by centrifugal effects) is at play in these stars. In the latter case, rotation would then only decrease the effective gravity in the equatorial regions, thus favoring mass loss.

THE EVOLVED A- AND F- TYPE STARS. The problem of how angular momentum is redistributed in stellar interiors during evolution was originally discussed by Oke and Greenstein. In assuming that the total angular momentum of a star is conserved along its evolutionary path, they considered two limiting cases that might be thought to bracket the possibilities: (A) complete radial exchange of angular momentum (i.e., solid-body rotation), and (B) no radial exchange of angular momentum (i.e., each shell conserves its own angular momentum). In case B, we have $(v_e \sin i)/(v_e \sin i)_0 = R_0/R$, where R is the mean stellar radius and the subscript "0" refers to the initial state on the main sequence. In case A, however, the change in $v_e \sin i$ from its initial value must be further corrected for the change in the mass distribution within the evolving star; this correcting factor is of course related to the instantaneous moment of inertia of the star, and it must be computed from a theoretical evolutionary track. Because of increasing central concentration with stellar evolution, velocities for case A are always in excess of those for case B. The main goal of the analysis is now to determine whether angular momentum is indeed conserved as stars evolve and, if so, which of the two cases agrees more closely with the observational data.

The above method has been recently used by Danziger and Faber to discuss stellar rotation among evolved stars with spectral types between A5 and F9. The main points of their study may be summarized as follows. For stars in the main-sequence band, including stars of *bona fide* luminosity classes IV and V, the observations are much more compatible with the predictions of case A than with those of case B. It would thus appear that all stars that have expanded their radii by no more than a factor of 2 do rotate as solid bodies. (Note the slight discrepancy between case A and the Maeder-Peytremann results of Section 12.3, at least for stars later than spectral type A7!) Further complications arise with the analysis of giants and supergiants. As Table 12.7 shows, four different classes provide a reasonable grouping of these stars: Ia supergiants, Ib supergiants, class Ib II-II stars, and class III stars. The listed quantity K_{obs} is equal to the ratio $\langle v_e \sin i \rangle_{m.s.}/\langle v_e \sin i \rangle$, where $\langle v_e \sin i \rangle$ and $\langle v_e \sin i \rangle_{m.s.}$ designate, respectively, the mean value of $v_e \sin i$ for a group of stars and the mean value of $v_e \sin i$ this group had on the main sequence. The quantities K_A and K_B represent the theoretical values of this ratio computed for cases A and B, respectively; they are calculated for a mean model within each luminosity class by interpolating between Iben's theoretical tracks for

TABLE 12.7

Stars above the main-sequence band

Lum. Class	Spec. Type	No. of members	$\langle v_e \sin i \rangle$ (km/sec)	K_{obs}	K_A	K_B
Ia	A5–F2	3	26 ± 4	5.8 ± 0.8	3.6–5.7	21.0–32.9
Ib	A5–F2	7	17 ± 2	8.6 ± 0.8	3.1–5.1	6.8–11.3
Ib	F5–F8	5	12 ± 3	12.1 ± 2.3	4.0–6.7	8.9–14.9
Ib II–II	A5–F2	11	30 ± 8	5.7 ± 1.5	2.0–2.9	3.7–5.4
Ib II–II	F3–F8	10	30 ± 7	5.6 ± 1.3	2.3–3.4	4.3–6.3
III	A5–A9	6	49 ± 23	3.4 ± 1.6	1.2–1.4	2.0–2.3
III	F0–G0	17	87 ± 14	1.90 ± 0.3	1.4–1.6	2.2–2.5

SOURCE: Danziger, I. J., and Faber, S. M., *A. Ap.* **18**, 428, 1972.

spherical models. Errors in the computed values of K_A and K_B result from the uncertainties in placing stars in the $(M_{bol} - \log T_{eff})$-plane, which make the determination of the radius of expansion of the stars uncertain. For the less luminous stars, namely those belonging to class III, we note that the observational data do not conflict with the interpretation that such stars lie in a transition region between case A and case B. However, no firm conclusions about the actual mode of rotation of these giant stars can be drawn. Inspection of the results for class Ib II-II stars and class Ib stars reveals a situation clearly favoring case B. This may be a further indication that, as we go to situations where the radii are changing by large factors, the stars are no longer well coupled throughout their interiors so that solid-body rotation is no longer possible. Additional support for this conclusion is the fact that other sources of line-broadening (such as turbulence) would raise the values of K_{obs} further away from the values of K_A. (The data in Table 12.7 refer to the observed macroscopic broadening *uncorrected for turbulence*.) A similar comment might be made about the results for the class Ia supergiants, although in this case the shift would need to be much larger to change the observations from apparently good agreement with case A to agreement with case B. According to Danziger and Faber, a shift of this magnitude does not seem to be ruled out if we adopt the turbulent velocities derived by Rosendhal from his data on line-broadening in early-type supergiants. Furthermore, as was pointed out by Abt, it seems very likely that among the Ia supergiants the primary contributor to the line broadening is turbulence because, at each luminosity, the minimum broadening is large and the range in broadening is small, whereas for rotation and random orientation one would expect to observe some narrow-lined spectra and a large range in broadening. For all these reasons the striking agreement between solid-body rotation and observations for the Ia supergiants must probably be considered to be fortuitous.

THE G- AND K-TYPE GIANTS. A further tantalizing question is provided by the low rotational velocities observed among the G- and K-type giants in young clusters. Indeed, since the place of origin of these stars at the main sequence is fairly certain, it is possible to make a reliable comparison between theory and observation along the lines initiated by Oke and Greenstein. Their technique was applied by Kraft to the Hyades-, NGC 6633-, and NGC 2281-clusters, and by Strittmatter and Norris to the Hyades moving group. In all cases, the mean rotation rates of giant stars ($v_e \sin i \lesssim 10$ km/sec) are much lower than would be expected if they had evolved either (A) maintaining uniform rotation with total angular momentum conserved or (B) conserving the angular momentum of each element. Accordingly, unless these giant stars happened to have below-than-average rotational velocities at the main sequence, angular momentum must have been lost from at least their surface regions. To account for the low observed rotational velocities, Kraft has proposed that *either* the angular momentum somehow disappears from the surface layers into the deep interiors *or* it is lost from the surfaces by some appropriate form of mass loss in the Hertzsprung gap. The former solution gains support from Gough and Lynden-Bell's suggestion that angular momentum can be expelled from convective regions provided a suitable sink is available. However, their main piece of evidence is based on an Alka-Seltzer experiment only. An independent check by Strittmatter, Illingworth, and Freeman indicates that the experimental device used by Gough and Lynden-Bell should be significantly modified before any definitive conclusion can be drawn. For the present, the most likely explanation for the discrepancy seems to be the Schatzman mechanism, i.e., magnetic braking by flare emissions (or by stellar winds) from the outer convective layers of these giant stars (cf. 11.4). In this connection, let us note that Ca II emission is present in almost all stars to the right of late F-type stars in the Hertzsprung-Russell diagram. This result (due to Wilson and Bappu) leads strong support to the view that the anomalously slow rotation rates of the K-type giants is indeed a result of mass loss and operation of the Schatzman mechanism. However, because we do not know with any precision the rate of mass loss required to produce magnetic braking in these stars over periods of 10^5 to 10^6 years, a meaningful comparison with the observed mass-loss rates (i.e., 10^{-7} to 10^{-5} $M_\odot$/year) is not yet available. According to Strittmatter and Norris, differentiation between the two proposed mechanisms might be possible by using the Oke-Greenstein technique to late K- and early M-giants.

THE VARIABLE STARS. Since the regular intrinsic variables (such as the δ cepheids, the δ Scuti stars, and presumably the β cepheids) are encountered in post-main-sequence evolutionary stages, we shall briefly discuss their rotational velocities here. Preston has advanced the view that the δ cepheids have very sharp lines in their spectra because they are descended only from main-sequence progenitors with small equatorial velocities. (When interpreting their line widths as an effect of rotational broadening alone, Kraft finds $v_e \sin i$'s near 20 km/sec.) Another proposal that is more in line with the current knowledge of post-main-sequence evolution has been advanced by Kraft: (i) all δ cepheids are essentially post K-type supergiants, in accordance with Iben's evolutionary tracks, and (ii) their line profiles are a result of turbulence, not rotation. Putting this another way, the δ cepheids have formerly been in a part of the Hertzsprung-Russell diagram, where angular momentum may have been lost (presumably in the form of magnetically coupled winds) so that they descend from nonvariable stars of very slow rotation. The general idea that pulsation and rotation are physically incompatible phenomena has been proposed also for the β cepheids and for the δ Scuti stars. As far as the β cepheids are concerned, this early conjecture has not been confirmed by the recent observations of Hill. An examination of the rotational data for 41 of these stars shows that there is *no* preferred rotational velocity, the $v_e \sin i$'s falling in the range 0-300 km/sec (plus two β Cephei variables with $v_e \sin i > 300$ km/sec!) A similar statement can be made for the δ Scuti variables, for they also have a wide range of projected rotational velocities ($v_e \sin i \lesssim 200$ km/sec). However, as was shown by Breger, the relation between pulsation and rotation is rather complex for these stars. Following a more detailed discussion by Danziger and Faber, we can say that among the stars of lowest luminosity the δ Scuti variables have small amplitudes of pulsation and rotational velocities similar to those of nonvariable stars. On the contrary, the δ Scuti variables of higher luminosity have slower rotation rates than nonvariable stars, and for this group slow rotation is a prerequisite for large amplitudes. There are thus indications that the effect of rotation on the pulsation of δ Scuti stars may be dependent on their luminosities.

INTERACTION OF ROTATION AND EVOLUTION

To illustrate the influence of rotation on the post-main-sequence evolution of a star, we shall first consider the results obtained by Kippenhahn, Meyer-Hofmeister, and Thomas for a $9M_\odot$ star (with an initial chemical composition of $X = 0.739$ and $Y = 0.240$ on the main sequence). Since the actual redistribution of angular momentum within an evolving star (as well as its distribution of angular momentum on the main sequence!) is totally unknown, a prescription of how the angular velocity changes with time is necessary. For the initial main-sequence model, the maximum angular velocity possible for solid-body rotation was used; the evolutionary calculations were then made for the cases α and β described above (with schematic prescriptions for the rotational state of the convective and radiative regions, together with conservation of the total angular momentum). The computations were terminated in the phase where a carbon-oxygen core is surrounded by a helium-burning shell that produces most of the energy radiated from the surface of the model. For the sake of comparison, the evolution of a nonrotating, $9M_\odot$ star was also computed. Figure 12.20 depicts these evolutionary tracks in the Hertzsprung-Russell diagram. Table 12.8 provides the age, location, and equatorial

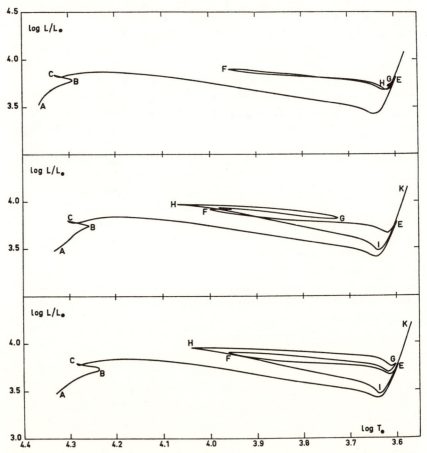

Figure 12.20. The Hertzsprung-Russell diagram with the evolutionary track for a model of $9M_\odot$. *Top*: nonrotating model, *middle*: rotating model under case α, *bottom*: rotating model under case β. Source: Kippenhahn, R., Meyer-Hofmeister, E., and Thomas, H. C., *A. Ap.* **5**, 155, 1970.

TABLE 12.8

Post-main-sequence models

age (10^{7}y)	$\log \dfrac{L}{L_\odot}$	$\log T_{\text{eff.}}$	v_e (km/sec)	α_c	
Nonrotating models					
0	3.5233	4.3655	0	0	A
2.5968	3.7836	4.2936	0	0	B
2.6733	3.8389	4.3315	0	0	C
2.7059	3.8220	3.5967	0	0	E
3.0512	3.8982	3.9563	0	0	F
3.1106	3.7774	3.6029	0	0	G
3.1187	3.7236	3.6144	0	0	H
Models with rotation, case (α)					
0	3.4906	4.3328	434	0.008	A
2.6593	3.7416	4.2586	306	0.008	B
2.7929	3.8047	4.3041	328	0.011	C
2.8307	3.7889	3.5986	4.5	0.044	E
3.2198	3.9194	3.9985	112	0.038	F
3.3124	3.8247	3.7226	37.6	0.081	G
3.3534	3.9735	4.0685	132	0.065	H
3.3693	3.4832	3.6339	14.8	0.134	I
3.3914	4.2029	3.5712	2.7	0.182	K
Models with rotation, case (β)					
0	3.4906	4.3328	434	0.008	A
2.6883	3.7552	4.2409	(381)	0.008	B
2.7526	3.8022	4.2888	(424)	0.011	C
2.7925	3.7609	3.6004	4.7	0.017	E
3.1743	3.9127	3.9613	132	0.017	F
3.2551	3.7877	3.6032	5.9	0.020	G
3.2970	3.9713	4.0432	183	0.015	H
3.3171	3.4725	3.6362	21.3	0.021	I
3.3365	4.2271	3.5696	2.5	0.021	K

SOURCE: Kippenhahn, R., Meyer-Hofmeister, E., and Thomas, H. C., *A. Ap.* **5**, 155, 1970.

velocity for certain characteristic stages of evolution. The fifth column lists the central ratio (α_c, say) of the centrifugal force to gravitational attraction for selected models. The letters in the sixth column give the correspondence to the models indicated in Figure 12.20. Two main results emerge from this analysis. *First*, since the lifting effect of the centrifugal force has a similar effect as reducing the stellar masses, the main-sequence lifetime is thus slightly increased by about 4 percent in cases α and β due to rotation. The same effect is more enhanced in the phase of central helium burning (13 percent in case α, and 11 percent in case β). This causes bigger loops in the Hertzsprung-Russell diagram. *Second*, regardless of whether case α or case β is used, a rapidly spinning core (with a period of about 60 seconds!) eventually develops between the exhaustion of helium and ignition of carbon in the core. This result is a direct consequence of the fact that transfer of angular momentum through regions of varying molecular weight was not allowed in the model calculations. (The subsequent evolution of such a core cannot be ascertained, though, for after helium burning the simplifications in the way of treating the centrifugal force break down in the deep interior.) Similar evolutionary tracks have been obtained by Meyer-Hofmeister for rotating models of $5M_\odot$ and $6M_\odot$ during the core helium-burning phase.

Under similar assumptions, Sackmann and Weidemann have further examined the effects of rotation in the $(\rho_c - T_c)$-plane during the pre-carbon-ignition phases of a $5M_\odot$ star. By starting with average uniform rotation on the main sequence ($v_e \approx 200$ km/sec) and then retaining the original angular momentum per unit mass during all later phases, they finally obtained a differentially rotating configuration that, to a first approximation, resembles the Ostriker-Bodenheimer white-dwarf models embedded in an extended envelope (cf. §13.2). Moreover, they found that, in the case of strict conservation of angular momentum, even an average amount of rotation on the main sequence suffices to prevent the carbon-oxygen core of a $5M_\odot$ star from igniting explosively. To be specific, carbon detonation is either prevented or, with smaller rotation rates on the main sequence, will occur at core masses much in excess of $1.4M_\odot$ (i.e., above the critical mass of detonating carbon-oxygen cores for nonrotating stellar models). According to Sackmann and Weidemann, the correspondingly increasing luminosity will then terminate this phase of evolution by mass loss. Their numerical results have been subsequently confirmed by analytical studies of Maeder. That is to say, (i) rotation allows the formation of much larger degenerate cores than in the absence of rotation, and (ii) at a given central density, the luminosity of a star is highly increased if the degree of differential rotation is sufficiently important in the core. More recently, further calculations of carbon detonations in rapidly rotating stellar cores have been made by Mahaffy and Hansen. By starting with differentially rotating, isothermal cores (with central densities ranging from $2 \cdot 10^9$ to 10^{10} g/cm^3 and an arbitrarily chosen angular momentum distribution), they found that the presence of rotation does not lower the central density required for leaving a bound remnant down to the required density range at which current evolutionary calculations predict the explosive ignition of carbon. On the basis of these preliminary calculations, it would appear that "reasonable" rotation rates cannot solve the immediate problem inherent in explaining Type II supernovae as detonations in the cores of stars within the mass range $4-8M_\odot$.

Most of these theoretical results have been obtained by assuming explicitly that *no* angular momentum can be transported from the core to the envelope of an evolving star. So we must now ask whether there are really no means of transferring angular momentum per unit mass through regions of varying molecular weight. In other words, does the real situation lie between case α and case β, or is it closer to case γ?

For the present, two rather speculative pieces of evidence support the occurrence of fast-spinning cores in evolving stars: (i) an excess of massive stars above the main sequence in the color-magnitude diagram of the open cluster NGC 6819, and (ii) the existence of pulsars with the observed periods. The *first* argument stems from the work of Maeder on the influence of rotation on the limiting isothermal core mass of upper main-sequence stars, i.e., the Schoenberg-Chandrasekhar mass. Indeed, whereas this limiting mass suffers a negligibly small decrease for a uniformly rotating model (3–4 percent at most), it is increased by about 10 percent if the core rotates 20 times as fast as the envelope. Hence, because an increase of the Schoenberg-Chandrasekhar mass leads to an increase in the lifetime of a star in the shell hydrogen-burning phase, extreme differential rotation would produce an excess of stars in this stage (compared to what one would expect from uniformly rotating or nonrotating models). According to Maeder, Burkhead's observations of NGC 6819 might thus be interpreted as favoring the existence of fast-spinning cores during the shell hydrogen-burning phase. The *second* argument is due to Fricke and Kippenhahn. In agreement with Gold's original idea, it assumes that the pulsars are rotating neutron stars that owe their high angular velocities to the angular momentum content of the cores of stars with initial masses in the range $4–10M_\odot$ (according to the Gunn-Ostriker analysis of the observational data). Thence, a simple calculation indicates that these cores would rotate *too slowly* to form pulsars with the observed periods, had they not decoupled rotationally from their surrounding envelopes *prior* to the core helium-burning phase. In Fricke and Kippenhahn's opinion, this strongly supports the hypothesis that rotational decoupling between the core and the envelope of an evolving star must have taken place earlier during evolution, leading to a core spinning perhaps 50 times as fast as the envelope during the core helium-burning phase.

On the contrary side, arguments favoring the view that the inner core of an evolving star exchanges a substantial amount of angular momentum with the envelope have also been presented. Particularly relevant are two hydrogen-line white dwarfs, 40 Eridani B and Wolf 1346, both of which show sharp absorption cores in the lines Hα and Hβ. As Greenstein and Peterson showed, the former appears to have a finite rotational velocity, $v_e \sin i = 40 \, (+20, \, -10)$ km/sec, and the latter has $v_e \sin i = 50 \, (+30,$

-20) km/sec. Unless these two stars are seen pole on, they are therefore slow rotators and have been able to dispose most of their angular momentum before reaching their final state. Following a qualitative discussion by Hardorp, this result indicates that rotational braking must somehow operate through the $\bar{\mu}$-barriers, unless one is willing to assume that the more massive white dwarfs now have fast-spinning cores underneath slowly rotating envelopes. In this connection, it is also interesting to note that a white dwarf with a strong magnetic field seems also to rotate extremely slowly. The star G 195-19 has been found by Angel and Landstreet to show cyclic changes of circular polarization with a period of 1.33 days. According to Greenstein and Peterson, if this arises from rotation, then it corresponds to $v_e \approx 0.4$ km/sec and suggests coupling of the core to the envelope through much later stages of evolution or a subsequent loss of angular momentum to the interstellar medium, or both. Finally, we note also that so far no DA white dwarf is known to have $v_e \sin i \gtrsim 150$ km/sec.

On the basis of the foregoing results, it appears that stellar cores are *not* perfectly isolated from their surrounding envelopes, i.e., as a core contracts, its spin-up is braked to some extent by the rest of the star. Semiquantitative studies of the removal of angular momentum from the interiors of rapidly rotating stellar cores by strong magnetic fieds are now available, but no definitive results have emerged so far from this type of analysis. Until such time as we *understand* the actual mechanisms that are responsible for angular momentum transfer within an evolving star (such as friction, magnetic fields, and meridional currents), we must thus question the purpose of calculating detailed post-main-sequence evolutionary tracks for rotating stellar models.

BIBLIOGRAPHICAL NOTES

A general account of the subject matter treated in this chapter will be found in:

1. Kraft, R. P., "Stellar Rotation" in *Stellar Astronomy* (Chiu, H.Y., Warasila, R. L., and Remo, J. L., eds.) **1**, pp. 317–367, New York: Gordon and Breach, 1969.

2. Kraft, R. P., "Stellar Rotation" in *Spectroscopic Astrophysics* (Herbig, G. H., ed.), pp. 385–422, Berkeley: Univ. of California Press, 1970.

See also references 1–3 of Chapter 7.

Section 12.2. Uniformly rotating models have been constructed in references 25–30, 32, 34, and 39 of Chapter 5; and in reference 26 of Chapter 16. See also:

3. Limber, D. N., and Roberts, P. H., *Astroph. J.* **141** (1965): 1439–1442.

4. Sackmann, I-J., and Anand, S. P. S., *ibid.* **155** (1969): 257–264; *ibid.* **162** (1970): 105–124.

5. Sackmann, I-J., *Astron. and Astroph.* **8** (1970): 76–84.

6. Sanderson, A. D., Smith, R. C., and Hazlehurst, J., *Astroph. J. Letters* **159** (1970): L69–L71.

Because some of the early models are either too crude or simply erroneous, the interested reader may well do starting his exploration of the literature with the following notes:

7. Jackson, S., Sanderson, A. D., Smith, R. C., and Hazlehurst, J., *Astroph. J.* **165** (1971): 223–224.

8. Whelan, J. A. J., Papaloizou, J. C. B., and Smith, R. C., *Monthly Notices Roy. Astron. Soc. London* **152** (1971): 9P–13P.

9. Aranda, J., and Thomas, H. C., *Astron. and Astroph.* **45** (1975): 441–442.

Differentially rotating models were originally discussed by Mark (reference 34 of Chapter 5). See also:

10. Bodenheimer, P., *Astroph. J.* **167** (1971): 153–163.

11. Monaghan, J. J., and Smart, N. C., *Monthly Notices Roy. Astron. Soc. London* **153** (1971): 195–204.

12. Tuominen, I. V., *Ann. Acad. Sci. Fennicae* (*A*), *VI. Phys.*, No 391 (1972): 1–49; *ibid.* No 392 (1972): 1–27; *ibid.* No 393 (1972): 1–20.

The presentation in the text closely follows references 5 and 10.

Section 12.3. The following general reference may be noted:

13. Collins, G. W., "The Effects of Rotation on the Atmospheres of Early-Type Main-Sequence Stars" in *Stellar Rotation* (Slettebak, A., ed.), pp. 85–109, New York: Gordon and Breach, 1970.

The early references on the effects of rotation on observable parameters are those of:

14. Collins, G. W., *Astroph. J.* **138** (1963): 1134–1146; *ibid.* **139** (1964): 1401; *ibid.* **142** (1965): 265–277; *ibid.* **146** (1966): 914–939.

15. Zhu, C.-S., *Acta Astron. Sinica* **11** (1963): 41–48 (English summary on p. 48).

See also the pioneering work of Sweet and Roy (reference 25 of Chapter 5). Further developments will be found in:

16. Roxburgh, I. W., and Strittmatter, P. A., *Zeit. f. Astroph.* **63** (1965): 15–21.

17. Ireland, J. G., *Publ. Roy. Observatory Edinburgh* **5** (1966): 63–88.

18. Rubin, R., *Monthly Notices Roy. Astron. Soc. London* **133** (1966): 339–344.
19. Ireland, J. G., *Zeit. f. Astroph.* **65** (1967): 123–132.
20. Hardorp, J., and Strittmatter, P. A., *Astroph. J.* **151** (1968): 1057–1073.
21. Jordahl, P. R., in *Stellar Rotation* (Slettebak, A., ed.), pp. 116–121, New York: Gordon and Breach, 1970.

See also reference 27 of Chapter 5. The most detailed studies are those of:

22. Maeder, A., and Peytremann, E., *Astron. and Astroph.* **7** (1970): 120–132; *ibid.* **21** (1972): 279–284.

These last papers provide results in the UBV-, uvby-, and Geneva-systems. Comparisons between theory and observation will be found in:

23. Kraft, R. P., and Wrubel, M. H., *Astroph. J.* **142** (1965): 703–711.
24. Roxburgh, I. W., Sargent, W. L. W., and Strittmatter, P. A., *The Observatory* **86** (1966): 118–120.
25. Strittmatter, P. A., *Astroph. J.* **144** (1966): 430–434.
26. Strittmatter, P. A., and Sargent, W. L. W., *ibid.* **145** (1966): 130–140.
27. Dickens, R. J., Kraft, R. P., and Krzeminski, W., *Astron. J.* **73** (1968): 6–13.
28. Golay, M., *Archives des Sciences (Genève)* **21** (1968): 105–124.
29. Maeder, A., *ibid.* **21** (1968): 125–132.
30. Smith, R. C., *Monthly Notices Roy. Astron. Soc. London* **151** (1971): 463–483.

See also references 21 and 22. Gravity darkening for various rotation laws has been considered by:

31. Smith, R. C., and Worley, R., *Monthly Notices Roy. Astron. Soc. London* **167** (1974): 199–213.

This last paper also discusses the changes in M_V for the Bodenheimer pseudo-barotropes (reference 10) and for the Roxburgh-Strittmatter baroclines (reference 31, pp. 345–357, of Chapter 7). Intrinsic polarization in the continuum of rapidly rotating, early-type stars has been discussed in:

32. Harrington, J. P., and Collins, G. W., *Astroph. J.* **151** (1968): 1051–1056.
33. Collins, G. W., *ibid.* **159** (1970): 583–591.
34. Cassinelli, J. P., and Haisch, B. M., *ibid.* **188** (1974): 101–104.
35. Peraiah, A., *Astron. and Astroph.* **46** (1976): 237–241.

The classical results on the rotational broadening of stellar lines have been summarized in references 24 and 25 of Chapter 2. Further discussions along similar lines will be found in:

36. Qu, Q.-Y., and Zhao, J.-B., *Acta Astron. Sinica* **12** (1964): 23–38 (English summary on p. 38).

37. Friedjung, M., *Astroph. J.* **151** (1968): 779–781.

38. Stoeckley, T. R., *Monthly Notices Roy. Astron. Soc. London* **140** (1968): 121–139, 141–147, 149–154.

39. Underhill, A. B., *Bull. Astron. Soc. Netherlands* **19** (1968): 526–536.

40. Wilson, A., *Monthly Notices Roy. Astron. Soc. London* **144** (1969): 325–332.

41. Collins, G. W., *Astroph. J.* **191** (1974): 157–164.

42. Delcroix, A., *Bull. Soc. Roy. Sci. Liège* **43** (1974): 223–248.

43. Stoeckley, T. R., and Morris, C. S., *Astroph. J.* **188** (1974): 579–594.

44. Buscombe, W., and Stoeckley, T. R., *Astroph. and Space Sci.* **37** (1975): 197–220.

See also reference 29 of Chapter 2. The effects of uniform rotation on line profiles and strengths have been originally considered by:

45. Collins, G. W., and Harrington, J. P., *Astroph. J.* **146** (1966): 152–176.

46. Collins, G. W., *ibid.* **151** (1968): 217–226; *ibid.* **152** (1968): 847–858.

See also reference 38 (pp. 149–154) and reference 41. An independent set of papers is due to:

47. Hardorp, J., and Strittmatter, P. A., *Astroph. J.* **153** (1968): 465–482.

48. Hardorp, J., and Scholz, M., *Astron. and Astroph.* **13** (1971): 353–358.

49. Hardorp, J., and Strittmatter, P. A., *ibid.* **17** (1972): 161–164.

50. Norris, J., and Scholz, M., *ibid.* **17** (1972): 182–189.

Other discussions are those of:

51. Tuominen, I. V., and Vilhu, O., in *Stellar Rotation* (Slettebak, A., ed.) pp. 110–114, New York: Gordon and Breach, 1970.

52. Tuominen, I. V., and Vilhu, O., *Astroph. Letters* **6** (1970): 143–145.

53. Vilhu, O., and Tuominen, I. V., *Astron. and Astroph.* **13** (1971): 136–146.

Most of the calculations on line profiles and strengths are based on the erroneous interiors of Roxburgh, Griffith, and Sweet (reference 26 of Chapter 5); it is suggested, therefore, to start the reading with references 48 and 53. The references to Guthrie and Balona are to their papers:

54. Guthrie, B. N. G., *Publ. Roy. Observatory Edinburgh* **3** (1963): 83–107.

55. Balona, L. A., *Monthly Notices Roy. Astron. Soc. London* **173** (1975): 449–464.

The effects of differential rotation have been especially examined in references 36, 38 (pp. 121–139), 43, and 44; see also the third reference 115. The following paper is also quoted in the text:

56. Johnston, I. D., and Wareing, N. C., *Monthly Notices Roy. Astron. Soc. London* **147** (1970): 47–58.

Section 12.4. The original measurements of $v_e \sin i$ values for Am- and Ap- stars were made by:

57. Slettebak, A., *Astroph. J.* **119** (1954): 146–167; *ibid.* **121** (1955): 653–669.

There is a wide literature on these stars; a recent survey of their properties will be found in:

58. Preston, G. W., "The Chemically Peculiar Stars of the Upper Main Sequence" in *Annual Review of Astronomy and Astrophysics* **12**, pp. 257–277, Palo Alto, California: Annual Reviews, Inc., 1974.

Among the many papers on the Am stars, let us mention:

59. Abt, H. A., and Moyd, K. I., *Astroph. J.* **182** (1973): 809–816.
60. Abt, H. A., and Levy, S. G., *ibid.* **188** (1974): 291–294.
61. Abt, H. A., *ibid.* **195** (1975): 405–409.

The presentation in the text largely follows reference 61; see also:

62. Strömgren, B., *Quart. J. Roy. Astron. Soc. London* **4** (1963): 8–36.
63. Smith, M. A., *Astroph. J. Supplements* **25** (1973): 277–314.

The rotational velocities of the Ap stars have been recently discussed by:

64. Preston, G. W., *Publ. Astron. Soc. Pacific* **83** (1971): 571–584.
65. Abt, H. A., Chaffee, F. H., and Suffolk, G., *Astroph. J.* **175** (1972): 779–785.
66. Abt, H. A., and Snowden, M. S., *Astroph. J. Supplements* **25** (1973): 137–162.

The following paper is also quoted in the text:

67. Guthrie, B. N. G., *Publ. Roy. Observatory Edinburgh* **5** (1965): 1–11.

Various braking mechanisms for the Ap stars will be found in:

68. Havnes, O., and Conti, P. S., *Astron. and Astroph.* **14** (1971): 1–11.
69. Kulsrud, R. M., *Astroph. J.* **163** (1971): 567–576.
70. Strittmatter, P. A., and Norris, J., *Astron. and Astroph.* **15** (1971): 239–250.

Recent contributions are due to:

71. Havnes, O., *ibid.* **32** (1974): 161–176.
72. Gedzelman, S. D., *Astroph. J.* **201** (1975): 509–520.
73. Wolff, S. C., *ibid.* **202** (1975): 121–126.

The present status of diffusion processes in the Am- and Ap-stars has been summarized by:

74. Michaud, G., "Abundance Anomalies, Diffusion and Constraints on F, A and B Stellar Envelope Models" in *Physics of Ap Stars* (Weiss, W. W., Jenkner, H., and Wood, H. J., eds.), pp. 81–107, Vienna, Austria: Universitätssternwarte Wien, 1975.

The magnetic-accretion mechanism has been propounded in reference 68. A critical survey of these and other theories will be found in reference 58. Attempts at a discussion of the importance of turbulence, circulation, and differential rotation in gravitational sorting are due to:

75. Schatzman, E., *Astron. and Astroph.* **3** (1969): 331–346.
76. Baglin, A., *ibid.* **19** (1972): 45–50.
77. Zahn, J. P., in *Journée d'étude sur les binaires serrées*, Nice, 1974.
78. Vauclair, G., *Astron. and Astroph.* **50** (1976): 435–444.

See also reference 70. Another viewpoint will be found in:

79. Mullan, D. J., *Astron. and Astroph.* **27** (1973): 379–381.

Section 12.5. The analysis in this section is taken from:

80. Abt, H. A., "Stellar Rotation in Open Clusters" in *Stellar Rotation* (Slettebak, A., ed.), pp. 193–203, New York: Gordon and Breach, 1970.
81. Abt, H. A., and Sanders, W. L., *Astroph. J.* **186** (1973): 177–183.

See, however:

82. Crampton, D., Hill, G., and Fisher, W. A., *ibid.* **204** (1976): 502–511.

See also references 1 and 2, as well as the following papers:

83. Kraft, R. P., *Astroph. J.* **142** (1965): 681–702.
84. Ferrer, O., and Jaschek, C., *Publ. Astron. Soc. Pacific* **85** (1973): 207–212.

The following paper may be noted also:

85. Guthrie, B. N. G., *The Observatory* **90** (1970): 233–236.

The influence of rotation on age estimates has been considered by:

86. Hazlehurst, J., and Thomas, H. C., *Monthly Notices Roy. Astron. Soc. London* **150** (1970): 311–323.

87. Maeder, A., *Astron. and Astroph.* **10** (1971): 354–361.

Other age estimates (without rotational corrections) may be traced to these papers.

Section 12.6. A general survey covering these and other matters will be found in reference 2 of Chapter 7. Recent papers lending qualitative support to Struve's conjecture are those of:

88. Limber, D. N., and Marlborough, J. M., *Astroph. J.* **152** (1968): 181–193.
89. Hardorp, J., and Strittmatter, P. A., in *Stellar Rotation* (Slettebak, A., ed.), pp. 48–59, New York: Gordon and Breach, 1970.

The following papers may also be noted:

90. Stothers, R., *Astroph. J.* **175** (1972): 431–452.
91. Maeder, A., *Astron. and Astroph.* **42** (1975): 471–473.

Theoretical studies of mass loss from Be stars were originally made by:

92. Crampin, J., and Hoyle, F., *Monthly Notices Roy. Astron. Soc. London* **120** (1960): 33–42.
93. Schmidt-Kaler, T., *Veröff. Astron. Inst. Univ. Bonn*, No 70 (1964): 1–43.

Detailed model calculations are due to:

94. Nobili, L., and Secco, L., in *Mass Loss from Stars* (Hack, M., ed.), pp. 135–138, New York: Springer-Verlag New York, Inc., 1969.
95. Kippenhahn, R., Meyer-Hofmeister, E., and Thomas, H. C., *Astron. and Astroph.* **5** (1970): 155–161.
96. Meyer-Hofmeister, E., and Thomas, H. C., *ibid.* **5** (1970): 490.
97. Strittmatter, P. A., Robertson, J. W., and Faulkner, D. J., *ibid.* **5** (1970): 426–430.
98. Gredley, P. R., and Borra, E. F., *Astroph. J.* **172** (1972): 609–614.

The most recent survey is due to:

99. Slettebak, A., "Be Stars as Rotating Stars: Observations" in *Be and Shell Stars* (Slettebak, A., ed.), pp. 123–136, Dordrecht: D. Reidel Publ. Co., 1976.

Other relevant papers will be found in this volume. The following studies are also quoted in the text:

100. Burbidge, E. M., and Burbidge, G. R., *Astroph. J.* **117** (1953): 407–430; *ibid.* **122** (1955): 89–94.
101. Marlborough, J. M., *ibid.* **156** (1969): 135–155.
102. Hutchings, J. B., *Monthly Notices Roy. Astron. Soc. London* **150** (1970): 55–66; *ibid.* **152** (1971): 109–119.

Detailed comparisons between theory and observation will be found in references 97–99. Rapidly rotating, early-type stars with radiatively driven stellar winds are discussed in:

103. Marlborough, J. M., and Zamir, M., *Astroph. J.* **195** (1975): 145–155.

Previous references on these winds may be traced to the foregoing work. See also references 55 and 99, as well as the following papers:

104. Massa, D., *Publ. Astron. Soc. Pacific* **87** (1975): 777–784.
105. Snow, T. P., and Marlborough, J. M., *Astroph. J. Letters* **203** (1976): L87–L90.

An entirely different approach to the Be-star problem is due to:

106. Kříž, S., and Harmanec, P., *Bull. Astron. Inst. Czechoslovakia* **26** (1975): 65–81.

Pioneering papers on the interaction between axial rotation and stellar evolution are those of:

107. Oke, J. B., and Greenstein, J. L., *Astroph. J.* **120** (1954): 384–390.
108. Sandage, A. R., *ibid.* **122** (1955): 263–270.
109. Abt, H. A., *ibid.* **126** (1957): 503–508; *ibid.* **127** (1958): 658–666.
110. Kraft, R. P., *ibid.* **144** (1966): 1008–1015.

See also references 1 and 2. The most recent contributions are due to:

111. Faber, S. M., and Danziger, I. J., in *Stellar Rotation* (Slettebak, A., ed.), pp. 39–47, New York: Gordon and Breach, 1970.
112. Strittmatter, P. A., and Norris, J., *Astron. and Astroph.* **11** (1971): 477–481.
113. Danziger, I. J., and Faber, S. M., *ibid.* **18** (1972): 428–443.

Early discussions of rotation and turbulence will be found in reference 25 of Chapter 2. The most recent contributions are due to:

114. Rosendhal, J. D., *Astroph. J.* **159** (1970): 107–118.
115. Gray, D. F., *ibid.* **184** (1973): 461–471; *ibid.* **202** (1975): 148–164; *ibid.* **211** (1977): 198–206.
116. Smith, M. A., and Parsons, S. B., *Astroph. J. Supplements* **29** (1975): 341–361.
117. Smith, M. A., *Astroph. J.* **208** (1976): 487–499.

See also Abt's comment in reference 111 (p. 47). The vorticity-expulsion mechanism is discussed at length in:

118. Gough, D. O., and Lynden-Bell, D., *J. Fluid Mechanics* **32** (1968): 437–447.
119. Strittmatter, P. A., Illingworth, G., and Freeman, K. C., *ibid.* **43** (1970): 539–544.

The following paper is also quoted in the text:

120. Wilson, O. C., and Bappu, M. K. Vainu, *Astroph. J.* **125** (1957): 661–683.

The view that stellar rotation might inhibit pulsation has been mainly advocated by:

121. Preston, G. W., *Kl. Veröff. Remeis-Sternwarte Bamberg* No 40 (1965): 155–163.

Rotation of the δ cepheids is discussed in reference 110. The most detailed observational studies of rotation among the β cepheids and δ Scuti variables are those of:

122. Hill, G., *Astroph. J. Supplements* **14** (1967): 263–300.
123. Breger, M., *ibid.* **19** (1969): 99–113.

See also references 113, 76, and 78. Previous studies may be traced to these papers. The present status of the post-main-sequence evolution of spherical models is discussed in:

124. Iben, I., Jr., "Post-Main-Sequence Evolution of Single Stars" in *Annual Review of Astronomy and Astrophysics* **12**, pp. 215–256, Palo Alto, California: Annual Reviews, Inc., 1974.

Detailed theoretical studies on the influence of rotation on post-main-sequence evolution are due to:

125. Kippenhahn, R., *Astron. and Astroph.* **8** (1970): 50–56.
126. LeBlanc, J. M., and Wilson, J. R., *Astroph. J.* **161** (1970): 541–551.
127. Maeder, A., *Astron. and Astroph.* **14** (1971): 351–358; *ibid.* **34** (1974): 409–414.
128. Meyer-Hofmeister, E., *ibid.* **16** (1972): 282–285.
129. Sackmann, I-J., and Weidemann, V., *Astroph. J.* **178** (1972): 427–432.
130. Mahaffy, J. H., and Hansen, C. J., *ibid.* **201** (1975): 695–704.
131. Zentsova, A. S., *Astron. Zh.* **52** (1975): 905–910.
132. Bisnovatyi-Kogan, G. S., Popov, Yu. P., and Samochin, A. A., *Astroph. and Space Sci.* **41** (1976): 287–320.
133. Meier, D. L., Epstein, R. I., Arnett, W. D., and Schramm, D. N., *Astroph. J.* **204** (1976): 869–878.
134. Mengel, J. G., and Gross, P. G., *Astroph. and Space Sci.* **41** (1976): 407–415.

See also reference 95. The following paper is further quoted in the text:

135. Burkhead, M. S., *Astron. J.* **76** (1971): 251–259.

Review papers on pulsars and neutron stars may be traced to reference 124. See, in particular:

136. Gold, T., *Nature* **218** (1968): 731–732.
137. Gunn, J. E., and Ostriker, J. P., *Astroph J.* **160** (1970): 979–1002.

Semiquantitative studies of braking during the post-main-sequence phases will be found in:

138. Morozov, V. N., *Astrofizika* **9** (1973): 567–579.
139. Hardorp, J., *Astron. and Astroph.* **32** (1974): 133–136.
140. Levy, E. H., and Rose, W. K., *Astroph. J.* **193** (1974): 419–427.

The importance of meridional currents in rotating degenerate objects has been also considered by Kippenhahn and Möllenhoff (reference 45 of Chapter 13). The slow rotation rates of some white dwarfs is discussed in:

141. Angel, J. R. P., and Landstreet, J. D., *Astroph. J. Letters* **165** (1971): L71–L75.
142. Greenstein, J. L., and Peterson, D. M., *Astron. and Astroph.* **25** (1973): 29–34.

13

Rotating White Dwarfs

13.1. INTRODUCTION

Despite our lack of knowledge about the exact manner in which the various possible endpoints of stellar evolution are attained, great progress has been made in our basic understanding of white dwarfs as cooling degenerate dwarfs. Since a number of comprehensive papers on the theory of spherical white-dwarf models now exist, a brief survey of their properties will suffice for our purpose. As we know, a white dwarf is largely supported against gravity by the pressure provided by the kinetic energy of the *degenerate* electrons; in contrast, its luminosity is almost entirely derived from the thermal energy of the *nondegenerate* ions, when nuclear processes no longer come into play and gravitational contraction has almost ceased. Because the ions have almost all of the thermal energy but provide almost none of the pressure, the mechanical and thermal properties of a white dwarf are, therefore, largely independent of each other.

Quite generally, the electronic pressure p_e of a white dwarf depends on the density ρ, temperature T, and chemical composition. However, because the electron gas in the main bulk of a white dwarf is so highly degenerate, it thus follows that in spite of possibly rather high internal temperatures ($T \approx 10^7 \, ^\circ K$, say) complete degeneracy ($T = 0 \, ^\circ K$) is nevertheless an excellent approximation in most cases, at least insofar as the *global* structure of the star is concerned. That is to say, to a first approximation, a white dwarf may be considered as a barotrope (cf. §4.2), and we can write

$$p_e = Af(x), \tag{1}$$

where

$$f(x) = x(2x^3 - 3)(x^2 + 1)^{1/2} + 3 \sinh^{-1} x, \tag{2}$$

and

$$x = (\rho/B)^{1/3}. \tag{3}$$

The quantity $x(= p_0/m_e c)$ is the maximum momentum in dimensionless form; letting $\bar{\mu}_e$ denote the mean molecular weight per electron, we also have $A = 6.002 \cdot 10^{22}$ dyne/cm^2 and $B/\bar{\mu}_e = 9.736 \cdot 10^5$ g/cm^3. In the limit of nonrelativistic electron velocities, equations (1)–(3) reduce to

$$p_e \to \frac{8}{5} \frac{A}{B^{5/3}} \rho^{5/3} \quad \text{(as } x \to 0\text{)},$$ (4)

whereas, in the limit of relativistic electron velocities, we similarly obtain

$$p_e \to 2 \frac{A}{B^{4/3}} \rho^{4/3} \quad \text{(as } x \to \infty\text{)}.$$ (5)

The existence of a maximum possible mass for spherical stars—the Chandrasekhar limit — can readily be seen from the following order-of-magnitude argument.[1] For sufficiently high masses, the electrons become relativistic; hence, by virtue of equation (5), we have $p_e \approx M^{4/3}/R^4$. Comparing next the pressure force (grad $p_e \approx M^{4/3}/R^5$) required to support the star against the gravitational force per unit volume ($\rho g \approx M^2/R^5$), we see that they exhibit the same dependence on R; hence, it is not at all certain that a small enough radius can be found to bring these two forces in balance. Detailed calculations indicate that the mass of a completely degenerate dwarf satisfying equations (1)–(3) must necessarily comply with the condition

$$M < \frac{5.75}{\bar{\mu}_e^2} M_\odot = M_3 \text{ (say)}.$$ (6)

Although various corrections to the equation of state, changes in $\bar{\mu}_e$ resulting from inverse β-decays, and modifications of the equation of hydrostatic equilibrium due to general-relativistic effects all slightly *reduce* the foregoing limit, Chandrasekhar's original calculations have largely withstood the passage of time.

To what extent are these results modified when allowance is made for rotation? Not surprisingly, observation provides no information as to what is the actual angular momentum distribution within a white dwarf. (As we recall from Section 12.6, a few white dwarfs are known to have very low $v_e \sin i$ values, but this does not preclude the existence of rapid rotators among these stars.) In the next section, we shall therefore consider the effects of both uniform and non-uniform rotation laws on the global structure, pulsation periods, and stability of white dwarfs. Evolutionary effects (such as viscous friction, crystallization, and meridional currents) will be discussed *in fine*.

---

[1] This result stems from the independent pioneering work of Chandrasekhar and Landau. See: Chandrasekhar, S., *M.N.* **91**, 456, 1931; *ibid.* **95**, 207, 1935; Landau, L. D., *Phys. Z. Sowjetunion* **1**, 285, 1932.

13.2.   MODELS IN A STATE OF PERMANENT ROTATION

It is apparent from equations (4) and (5) that zero-temperature, completely degenerate white dwarfs very much resemble polytropic configurations, ranging from the polytrope of index $n = 3/2$ (in the limit $M \to 0$) to the polytrope of index $n = 3$ (in the limit $M \to M_3$). By making use of the results described in Section 10.3, we may surmise that sequences of *uniformly rotating* white dwarfs (with fixed mass and increasing angular momentum) must necessarily terminate with models for which the effective gravity falls to zero at their equators, and that points of bifurcation (i.e., points at which non-axisymmetric figures of equilibrium branch off) cannot possibly occur along these sequences. In other words, because equations (1)–(3) are very similar to the polytropic equation $p = K\rho^{1 + 1/n}$ (with $3/2 \leqslant n \leqslant 3$) and because centrally condensed polytropes with $n \geqslant 0.808$ cannot store much angular momentum, the restriction of solid-body rotation should prevent the construction of white-dwarf models for which the ratio $\tau = K/|W|$ is greater than a mere few percent. (As usual, K and W designate, respectively, the rotational kinetic energy and the gravitational potential energy.) This, in turn, should imply that the Chandrasekhar limit is not much affected by the prescription of solid-body rotation, even for the most rapid rotators at their break-up velocities. That this is the case will become apparent from the discussion immediately below. Afterwards, we consider *differentially rotating* models with prescribed distributions of angular momentum for they exhibit quite different properties.

UNIFORMLY ROTATING MODELS

Many authors have calculated the effects of solid-body rotation on the global structure of white dwarfs obeying both the classical equation of state (1)—with $\bar{\mu}_e = 2$—and the Newtonian condition of mechanical equilibrium. Detailed numerical calculations have been presented by James, following a method similar to that used in the theory of uniformly rotating polytropes (cf. §10.3). White-dwarf structures have been determined also by means of various expansion techniques in the squared angular velocity Ω^2. The most detailed analytical study is due to Anand and Dubas, who considered second-order expansions that include all terms up to order of Ω^4. In all these studies, it is found that the global structure of white dwarfs is not significantly altered by uniform rotation because the total angular momenta consistent with equilibria having this special rotation law are severely limited by the restriction that centrifugal force be less than gravity at the equator. As a result, uniformly rotating white dwarfs do not greatly deviate from spherical symmetry, and the

Chandrasekhar limit is increased by only 3 percent. For example, by retaining all terms up to order of Ω^4, Anand found that inequality (6) must be replaced by

$$M < \frac{5.92}{\bar{\mu}_e^2} M_\odot, \tag{7}$$

in the extreme case of a white dwarf rotating at the maximum rate consistent with equilibrium and constant angular velocity. (To the first order in Ω^2, Roxburgh originally found the limit $5.96M_\odot/\bar{\mu}_e^2$.) The short-dashed curve in the lower left-hand portion of Figure 13.6 defines the small range of masses and angular momenta for which uniformly rotating white-dwarf models obeying equations (1)–(3) can be constructed.

Models of uniformly rotating white dwarfs pervaded by magnetic fields that vanish on their surfaces have been further constructed by Ostriker and Hartwick. By making use of the self-consistent-field method described in Section 5.5, they found that internal fields that have both poloidal and toroidal components can significantly increase the radii of white dwarfs while having their surface regions nearly spherical. To be specific, in the case of a $1.05M_\odot$ star, the fractional radius increase goes as $\exp[+3.5(M/|W|)]$, where M is the total magnetic energy of the configuration. (For the most extreme model considered, with $M/|W| \approx 0.1$, the central field approaches 10^{12} gauss but the poloidal components near the surface always remain several powers of 10 smaller than this value.) Following Mestel's original suggestion, strong internal magnetic fields might provide an explanation for the observation that the photometric radii of Sirius B, 40 Eridani B, and some other white dwarfs are apparently somewhat larger than the theoretical radii predicted by the classical zero-temperature mass-radius relation.[2] (As Ostriker and Hartwick pointed out, this discrepancy cannot be explained by rotation alone because, for the relatively low-mass stars considered here, the rotation velocities required to bring a significant change in the radii would produce spectral lines with broader cores than are observed.) For the present, the question of whether or not these white-dwarf models with large magnetic fields are stable is not settled. However, because the stability of a magnetized configuration usually depends on the ratio $M/|W|$ ($\leqslant 1$) rather than on the average field strength (cf. §15.2), the Ostriker-Hartwick models with $M/|W| \lesssim 0.1$ might well be stable after all. No firm conclusion can be drawn without a detailed study of their normal modes of oscillation.

---

[2] As was shown by Hubbard and Wagner, this discrepancy might also be explained by means of spherical models with *finite* internal temperatures, particularly for white dwarfs of low masses. This problem is outside the scope of the present work. See: Hubbard, W. B., and Wagner, R. L., *Ap. J.* **159**, 93, 1970 (and references quoted therein).

The above results specifically refer to models obeying the classical equation of state (1) and the Newtonian condition of mechanical equilibrium. At high densities, however, inverse β-decays and general relativistic corrections tend to reduce the maximum mass attainable by a spherical white-dwarf model. For example, in the case of Newtonian models satisfying the Hamada-Salpeter equation of state, the maximum mass ranges from $1.40 M_\odot$ for carbon to $1.11 M_\odot$ for iron. In contrast, in the case of general-relativistic models satisfying equations (1)–(3) with $\bar{\mu}_e = 2$, the maximum mass is found to be about $1.42 M_\odot$ (as opposed to the Newtonian limit $1.44 M_\odot$). The influence of uniform rotation on these limiting masses was originally discussed by Hartle and Thorne. Figure 13.1 illustrates the effects of solid-body rotation on the masses and mean radii of general-relativistic configurations obeying the Harrison-Wheeler equation of state. We observe that the maximum mass of white-dwarf

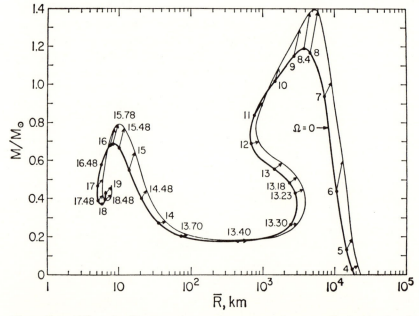

Figure 13.1. Effects of uniform rotation on the masses and mean radii of white dwarfs and neutron stars obeying the Harrison-Wheeler equation of state. The *thick* curve is a plot of mass vs. radius, parametrized by the logarithm of central density (in g/cm^3), for nonrotating models. The *thin* curve is an approximate plot of mass vs. radius, for configurations rotating with constant angular velocity, $\Omega = (GM/R^3)^{1/2}$. This angular velocity is approximately the amount needed to produce equatorial break-up. The small arrows indicate the displacement, with increasing angular velocity, of models with the same central densities. Source: Hartle, J. B., and Thorne, K. S., *Ap. J.* **153**, 807, 1968. (By permission of The University of Chicago Press. Copyright 1968 by the University of Chicago.)

models is increased by about 20 percent as a result of uniform rotation at the limit of equatorial break-up. (Although neutron stars are outside the scope of this book, notice that their global structures are little affected by uniform rotation!) Similar results have been obtained by Arutyunian, Papoian, Sedrakian, and Chubarian for general-relativistic white-dwarf models obeying the Saakian-Chubarian equation of state, which includes, in an approximate way, the effects of inverse β-decay. The nice feature of Figure 13.2 is that it allows at a glance an evaluation of the relative importance of the effects of general relativity, inverse β-decay, and uniform rotation. Evidently, solid-body rotation does not greatly modify the mass-radius relation, the sole difference being that a uniformly rotating con-figuration has a larger mass for the same central density (for $\rho_c \approx 10^9$ g/cm^3, the difference is of the order of 8 percent only). To summarize, as far as the global structure of white dwarfs is concerned, the *ad hoc* pre-scription of uniform rotation leads to only minor corrections, especially when these are compared to the changes brought on the models by inverse β-decay.

Let us now consider the stability of uniformly rotating white-dwarf models when they are subjected to infinitesimal disturbances. Because uniformly rotating white dwarfs do not greatly deviate from spherical symmetry, the fundamental pulsation mode of these stars can be obtained by suitably modifying the lowest radial mode of spherical stars (cf. §6.4: equation [41]). Indeed, as we shall see in Section 14.2, the lowest quasi-radial frequency of a white dwarf is given, to a good approximation, by the formula

$$\sigma_R{}^2 \propto \frac{|W|}{I}\left[(3\langle\Gamma_1\rangle - 4) + \alpha(5 - 3\langle\Gamma_1\rangle)\frac{K}{|W|} - \beta\frac{|W|}{Mc^2}\right], \qquad (8)$$

where α and β are *positive* constants of order unity.[3] As usual, I is the moment of inertia with respect to the center of mass, c is the speed of light, and $\langle\Gamma_1\rangle$ is the pressure-weighted average of the adiabatic exponent Γ_1. The first term in the bracket is for most nondegenerate stars the dominant term; the second term shows the stabilizing influence of rotation on the pulsations, and the third term, the destabilizing effect of general relativity. In the present instances, because $\langle\Gamma_1\rangle$ approaches the value 4/3 for high density, relativistic white dwarfs, the normally small "correction" terms might become important near the limiting mass.

---

[3] To be specific, rotation always couples the two lowest axisymmetric modes, i.e., those corresponding to the frequencies σ_R and σ_Z (cf. §6.4: equations [41] and [37]). However, although the frequency σ_Z may lead to periods as short as 0.25 seconds for uniformly rotating white dwarfs, this zonal mode never becomes unstable.

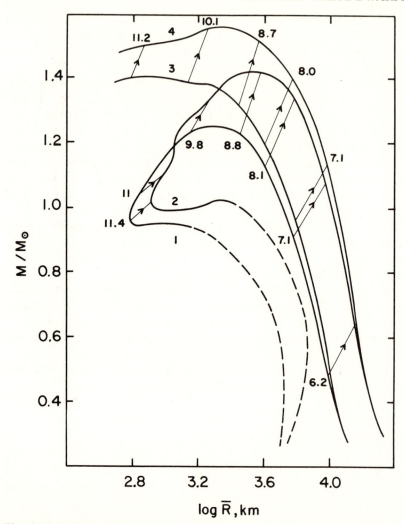

Figure 13.2. Effects of uniform rotation on the masses and mean radii of white-dwarf models obeying the Saakian-Chubarian equation of state. Curves 1 and 3 refer to spherical models with and without allowance for inverse β-decay, respectively; curves 2 and 4 refer to uniformly rotating models at their break-up velocities, with and without allowance for inverse β-decay, respectively. The figures along curves 1 and 4 designate the logarithm of central density (in g/cm^3). The arrows connect configurations having identical central densities. Source: Arutyunian, G. G., Sedrakian, D. M., and Chubarian, E. V., *Astrofizika* 7, 467, 1971.

TABLE 13.1

Pulsation periods of fully rotating white-dwarf
models obeying the Hamada-Salpeter equation of state[†]

$Log_{10}\rho_c$	R(10^7 cm)	M(10^{33} g)	$10^3\beta_{max}$	Ω^2(/sec^2)	Period (sec)
9.6	17.1	2.03	5.19	8.66	1.62[2.01]
9.8	15.2	2.00	4.57	12.1	1.46[1.90]
10.0	13.8	1.95	3.83	16.0	1.34[1.83]
10.2	12.3	1.88	3.26	21.6	1.23[1.79]
10.4	11.0	1.81	2.73	28.8	1.14[1.84]
10.6	9.69	1.72	2.42	40.4	1.07[2.22]
10.8	8.40	1.66	2.26	59.7	1.06[U]
11.0	7.46	1.60	1.96	82.0	1.07[U]
11.2	6.48	1.52	1.80	119	1.61[U]
11.4	5.93	1.47	1.43	151	U[U]

[†] The figures in brackets are the pulsation periods of the corresponding nonrotating models having the same central densities (Faulkner, J., and Gribbin, J. R., *Nature* **218**, 734, 1968).

SOURCE: Gribbin, J. R., *M. N.* **144**, 549, 1969.

Detailed numerical results have been presented by several authors. After Gribbin, Table 13.1 lists the approximate quasi-radial periods P_R ($= 2\pi/\sigma_R$) for relativistic models rotating at their break-up velocities and obeying the Hamada-Salpeter equation of state. The equilibrium models contain $^{56}Fe_{26}$ for $\log_{10}\rho < 7.15$ (in g/cm^3), and heavier, more neutron-rich nuclei for higher densities, with $^{120}Sr_{38}$ at the highest densities considered ($11.28 < \log_{10}\rho < 11.40$); they are further obtained from an equation that represents a combination of the relativistic structure equation for the corresponding nonrotating models, plus the Newtonian effects of solid-body rotation. (As was shown by Durney and Roxburgh, although this procedure is not strictly correct, the use of the Newtonian effects of rotation is nevertheless justifiable in the weak field, slow rotation approximation.) For the sake of completeness, Table 13.1 also lists the mean radius R, the mass M, the ratio $\beta_{max} = \Omega^2/2\pi G\rho_c$ (where ρ_c is the central density), and the squared angular velocity Ω^2. The figures given in brackets are the lowest pulsation periods of the corresponding spherical models having the same central densities. (The letter "U" designates an unstable model.) Two main results emerge from the linear calculations. *First*, for low-mass spherical models, the lowest radial periods decrease from about 20 seconds at 0.4M$_\odot$ to about 6 seconds at 1.0M$_\odot$ (not illustrated). The periods then reach a minimum (typically of the order of 2 seconds) and go to infinity for high-density models. Notice that dynamical instability sets in when the star's radius is still two orders of magnitude larger than its Schwarzschild radius R$_s$($= 2GM/c^2$). *Second*, because the second term in equation (8) is proportional to R^{-2} at high densities (as compared to R^{-1} for the third term), rotation is therefore more important than general relativitiy for uniformly rotating, high-density models. The stabilizing effect of solid-body rotation may readily be seen from Table 13.1.

DIFFERENTIALLY ROTATING MODELS

By virtue of the Poincaré-Wavre theorem, a barotropic configuration in a state of permanent rotation must necessarily comply with the condition $\Omega = \Omega(\varpi)$, where ϖ is the distance from the rotation axis (cf. §4.3). The particular case of $\Omega = constant$ has already been considered above, showing that solid-body rotation induces no substantial changes in the global structure of a degenerate dwarf. As we shall now see, a completely different picture emerges if we consider rotating white dwarfs for which we prescribe an angular momentum distribution $j(m_\varpi)$, where m_ϖ is the mass fraction interior to the cylinder of radius ϖ. This problem was considered first by Hoyle and by Roxburgh, who pointed out that disc-like equilibrium configurations could exist for masses arbitrarily greater than the Chandrasekhar limit. Indeed, because contraction without angular momentum loss forces the ratio $\tau = K/|W|$ to increase as R^{-1}, it should be possible to construct white-dwarf models of arbitrary mass for any (nonzero) angular momentum, the star being always able to find a radius small enough for the electronic pressure force and the centrifugal force to balance the gravitational attraction. Of course, the resulting configurations will necessarily be in a state of differential rotation, the angular velocity approaching Kepler's third law (i.e., $\Omega \propto \varpi^{-3/2}$) in the outer layers of rapidly rotating models. Detailed models of massive white dwarfs in fast non-uniform rotation have been constructed by Ostriker, Bodenheimer, and Lynden-Bell.

As we recall from Sections 4.2 and 4.3, when considering *idealized* configurations in a state of permanent rotation, we are free to *choose* any angular momentum distribution of the form $j = j(m_\varpi) = \Omega(\varpi)\varpi^2$, provided it satisfies the Høiland criterion (cf. §7.3). Following Ostriker and Bodenheimer, we shall *prescribe* an angular momentum distribution that is similar to that of a uniformly rotating polytrope of index $n = 3/2$ (cf. §10.4: equation [44]). Calculations that have been carried out with other forms for the function $j(m_\varpi)$ indicate that the main conclusions to be presented here are only slightly dependent on the actual distribution taken, so long as it remains within reasonable limits. The models were constructed by means of the self-consistent field method (cf. §5.5), all of them obeying equations (1)–(3) and the Newtonian condition of mechan-

TABLE 13.2

Properties of differentially rotating white dwarfs

	Model 1	Model 2	Model 3	Model 4	Model 5	Model 6
$M/M_\odot$	0.79	1.36	1.81	2.26	2.72	3.17
J	0.00	2.69(49)	2.31(50)	5.77(50)	8.85(50)	1.21(51)
R_e	7.27(8)	3.04(8)	6.17(8)	1.34(9)	1.67(9)	1.89(9)
K	0.00	1.16(49)	2.01(50)	2.86(50)	4.01(50)	5.37(50)
U_T	1.24(50)	1.38(51)	1.16(51)	7.42(50)	8.17(50)	9.64(50)
W	−2.11(50)	−1.81(51)	−2.03(51)	−1.77(51)	−2.18(51)	−2.73(51)
ρ_c	8.81(6)	4.84(8)	1.76(8)	4.07(7)	3.32(7)	3.25(7)
R_e/R_p	1.00	1.05	2.08	3.39	3.96	4.32
v_e	0.00	1.58(3)	5.01(3)	4.62(3)	4.75(3)	4.90(3)
Ω_e/Ω_c	0.00	0.476	0.350	0.274	0.254	0.241
$c\Delta\nu/\nu$	4.82(1)	2.06(2)	2.46(2)	2.12(2)	2.32(2)	2.58(2)
g_e	1.99(8)	1.88(9)	2.27(8)	3.71(6)	1.04(7)	1.28(7)
g_p	1.99(8)	2.12(9)	2.09(9)	1.26(9)	1.23(9)	1.28(9)
R_g/R_e	0.424	0.358	0.317	0.270	0.254	0.241

SOURCE: Ostriker, J. P., and Bodenheimer, P., *Ap. J.* **151**, 1089, 1968. (By permission of The University of Chicago Press. Copyright 1968 by the University of Chicago.)

ical equilibrium. Table 13.2 summarizes the main physical characteristics of six selected configurations. It includes a nonrotating model and five models of increasing mass, with J for each mass equal to the total angular momentum of a uniformly rotating star of the same mass on the main sequence. Figures 13.3 and 13.4 provide the detailed structures of Models 3 and 6, respectively. The fourteen lines in Table 13.2 give, respectively, the mass M, the total angular momentum J, the equatorial radius R_e, the rotational kinetic energy K, the internal energy U_T, the gravitational potential energy W, the central density ρ_c, the ratio of equatorial radius to polar radius R_e/R_p, the equatorial velocity v_e (in km/sec), the ratio Ω_e/Ω_c of angular velocity at the equator and at the center, the gravitational redshift $c\,\Delta\nu/\nu$ at the poles (in km/sec), the surface gravity at the equator g_e, the surface gravity at the poles g_p, and the ratio R_g/R_e of the radius of gyration to the equatorial radius. (All units are c.g.s. unless otherwise specified; an integer in parentheses gives the power of ten by which the corresponding entry should be multiplied.) Because $\log_{10} \rho_c \lesssim 9$, these zero-temperature, completely degenerate configurations are thus stable to inverse β-decay. For the very same reason, the assumed equation of state should provide a good approximation also (if the chemical composition is mainly formed out C, O, Mg, and Si), and the (neglected) corrections caused by general relativity should not substantially affect the models. The limitations brought by viscous friction and meridional circulation on massive white-dwarf models in a state of permanent rotation will be considered in Section 13.3.

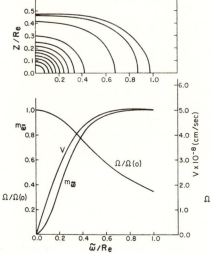

Figure 13.3.   Detailed structure of Model 3. The upper portion shows the isopycnic surfaces of densities 0.9, 0.8, 0.7, 0.6, 0.5, 0.4, 0.3, 0.2, 0.1, 0.01, 0.001, and 0.0001 times the central density. The radius in the equatorial plane is indicated by ϖ, that along the rotation axis by z, while R_e is the total equatorial radius. The lower portion shows the ratio of the angular velocity Ω to its central value $\Omega(0)$, the fraction m_ϖ of the total mass interior to the cylinder of radius ϖ, and the circular velocity v_e. Source: Ostriker, J. P., and Bodenheimer, P., *Ap. J.* **151**, 1089, 1968. (By permission of The University of Chicago Press. Copyright 1968 by the University of Chicago.)

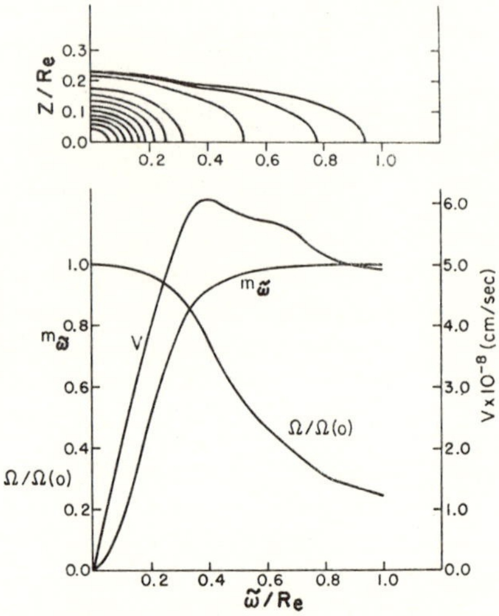

Figure 13.4.   Detailed structure of Model 6. All curves have the same meaning as the corresponding ones in Figure 13.3. Source: Ostriker, J. P., and Bodenheimer, P., *Ap. J.* **151**, 1089, 1968. (By permission of The University of Chicago Press. Copyright 1968 by the University of Chicago.)

The main features of these idealized configurations have been summarized by Ostriker and Bodenheimer. *First*, in all cases, we have $K < U_T$, indicating that the models—considered as a whole—are supported primarily by pressure, not centrifugal force. *Second*, the binding energy always remains large and negative; in all models, we have $E = K + W + U_T < -K$. *Third*, in no case is the differential rotation very extreme, as indicated by the fact that $\Omega_e/\Omega_c > 0.2$ for all configurations presented. *Fourth*, the value of the gravitational redshift becomes rather insensitive to the mass for the models with $M \geqslant 1.3 M_\odot$. *Fifth*, the equatorial velocities of the rotating models with $M > M_3$ lie in the range 3–$7 \cdot 10^3$ km/sec, and the radii fall between $5 \cdot 10^8$ and $3 \cdot 10^9$ cm. The equatorial radii of rapidly rotating models thus lie in the range of the radii of nonrotating white dwarfs with $M < M_3$. *Sixth*, comparison of Figures 13.3 and 13.4 shows that both configurations have fairly spherical, fairly uniformly rotating, high-density central regions; however, the outer regions of Model 6 (that with the higher J) are considerably more extended and flattened than those of Model 3. *Seventh*, the decrease in velocity as a function of radius in the outer regions of Model 6 indicates that pressure effects are so small there that the material is essentially in Keplerian orbits, with centrifugal force rather than the pressure gradient providing the main part of the balance against gravity in the direction perpendicular to the rotation axis.

By comparing Figures 13.3–4 to Figure 10.15, we readily notice the striking resemblance between differentially rotating polytropes and massive white-dwarf models having about the same values of the ratio $\tau = K/|W|$. Accordingly, since polytropes that greatly depart from spherical symmetry may become unstable with respect to non-axisymmetric disturbances, a detailed stability analysis of the Ostriker-Bodenheimer models is required. The general strategy to obtain approximate values for the lowest frequencies of oscillation has been described at length in Section 6.7, and it has been used to discuss the stability of differentially rotating polytropes in Section 10.4. As we recall, the second-order virial equations provide the most likely modes to determine the stability of configurations in non-uniform rotation. These are: the lowest *zonal* modes (σ_R and σ_Z), the lowest *tesseral* modes (σ_{+1}, σ_{-1}, and σ_0), and the lowest *sectorial* modes (σ_{+2} and σ_{-2}). In the limit of zero rotation, these modes reduce to the five f-modes belonging to the spherical harmonics $l = 2$, the lowest p-mode, and one zero-mode. Figure 13.5 illustrates the seven corresponding frequencies along a sequence of differentially rotating white-dwarf models of $1.13M_\odot$, the angular momentum distribution being that of a uniformly rotating, homogeneous spheroid (cf. §10.4: equation [44]). The oscillation periods of $2.26M_\odot$ models are given in Table 13.3. Figure 13.6 summarizes computations made for various sequences.

As we expected, no instability occurs with respect to the axisymmetric modes. The reason is that, for all our models, we have $\langle \Gamma_1 \rangle > 4/3$. Had we included inverse β-decay and general relativity, we would have found that such an instability may indeed occur, at least for small values of the ratio $\tau = K/|W|$ (see Table 13.1). Particular attention must be paid to the *sectorial* modes, however, for they transform an axisymmetric configuration into another one which has a genuine triaxial symmetry. (At this stage, it is worth comparing Figure 13.5 to Figures 10.3, 10.16, and 10.17!) As in the case of Maclaurin spheroids and differentially rotating polytropes, a neutral mode (i.e., $\sigma_{-2} = 0$) occurs when $\tau \approx 0.14$, thus implying the existence of massive triaxial white dwarfs beyond this limit. To be specific,

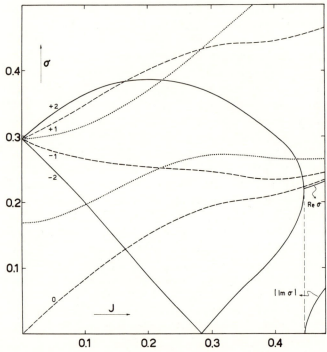

Figure 13.5. Frequencies of oscillation along a sequence of differentially rotating white dwarfs ($M = 1.13M_\odot$). The angular momentum distribution is that of a uniformly rotating, homogeneous spheroid. J and σ are given, respectively, in the units of $M^{5/3}G^{1/2}\rho_c^{-1/6}$ and $(4\pi G\rho_c)^{1/2}$. Sectorial, tesseral, and zonal modes are represented, respectively, by full-line-, dashed-, and dotted-curves. The upper dotted curve represents a zonal mode. See Table 13.3 for the labeling of the remaining curves. Note that the frequency of the "radial" mode (*lower dotted curve*) increases with increasing J, i.e., rotation decreases the pulsation period for given density. Note also the neutral mode at $J \approx 0.27$ and the onset of dynamical instability at $J \approx 0.46$. Source: Ostriker, J. P., and Tassoul, J. L., *Ap. J.* **155**, 987, 1969. (By permission of The University of Chicago Press. Copyright 1969 by the University of Chicago.)

the value $\tau \approx 0.14$ defines an *approximate* upper limit for secular stability to gravitational radiation-reaction (cf. §10.4). Further along a sequence of axisymmetric configurations, we also note that the models become dynamically unstable past the limit $\tau \approx 0.26$, the instability growing on a time scale of the order of a few seconds (see Figure 13.5). Calculations made for other sequences indicate that, within the accuracy of the numerical work, the neutral mode and the onset of dynamical instability always occur for $\tau \approx 0.14$ and $\tau \approx 0.26$, respectively, *almost independently of the total mass and angular momentum distribution*. Figure 13.6 illustrates these limits in the (M–J)-plane. On the reasonable assumption that the limit $\sigma_{-2} = 0$

TABLE 13.3

Properties and periods of differentially rotating white dwarfs[†]

$(M = 2.26 M_\odot)$

J	2.31(50)	3.40(50)	3.85(50)	5.39(50)	6.93(50)	8.47(50)	1.00(51)	1.12(51)	1.15(51)
R_e	2.51(8)	4.33(8)	5.09(8)	7.70(8)	1.04(9)	1.33(9)	1.64(9)	1.90(9)	1.97(9)
R_e/R_p	2.02	2.12	2.20	2.46	2.77	3.10	3.45	3.71	3.78
v_e	5.10(8)	4.35(8)	4.19(8)	3.87(8)	3.68(8)	3.52(8)	3.37(8)	3.25(8)	3.21(8)
Ω_c/Ω_e	6.23	5.21	4.91	4.21	3.81	3.58	3.42	3.32	3.29
ρ_c	2.71(9)	4.26(8)	2.45(8)	6.17(7)	2.39(7)	1.14(7)	6.12(6)	4.08(6)	3.68(6)
U_T	4.61(51)	2.17(51)	1.71(51)	9.13(50)	5.74(50)	3.92(50)	2.81(50)	2.25(50)	2.13(50)
K	8.83(50)	5.45(50)	4.80(50)	3.62(50)	3.05(50)	2.66(50)	2.37(50)	2.18(50)	2.14(50)
$|W|$	7.27(51)	3.95(51)	3.30(51)	2.10(51)	1.53(51)	1.20(51)	9.71(50)	8.45(50)	8.15(50)
P_{-2}	6.56	∞	4.69(1)	2.05(1)	1.87(1)	1.89(1)	1.96(1)	1.93(1)	*1.77(1)*
P_2	4.29(−1)	*1.08*	1.42	2.87	4.79	7.35	1.09(1)	1.50(1)	*1.77(1)*
P_{-1}	6.37(−1)	*1.59*	2.09	4.09	6.51	9.38	1.27(1)	1.55(1)	*1.63(1)*
P_1	3.81(−1)	9.25(−1)	1.20	2.27	3.53	5.00	6.71	8.17	*8.60*
P_0	8.89(−1)	2.11	2.70	4.96	7.34	1.05(1)	1.38(1)	1.65(1)	*1.73(1)*
P_z	3.97(−1)	9.46(−1)	1.21	2.20	3.23	4.34	5.52	6.45	*6.72*
P_R	1.18	2.35	2.86	4.69	6.78	9.29	1.23(1)	1.49(1)	*1.57(1)*

[†] The calculations are for $\bar{\mu}_e = 2$. For the other values of $\bar{\mu}_e$, P should be multiplied by $(2/\bar{\mu}_e)^{1/2}$. Units are in the c.g.s. system. The first and second columns in italics pertain to the models at the point where $\sigma_{-2} = 0$ and at the point of dynamical over-stability.

SOURCE: Ostriker, J. P., and Tassoul, J. L., *Ap. J.* **155**, 987, 1969. (By permission of The University of Chicago Press. Copyright 1969 by the University of Chicago.)

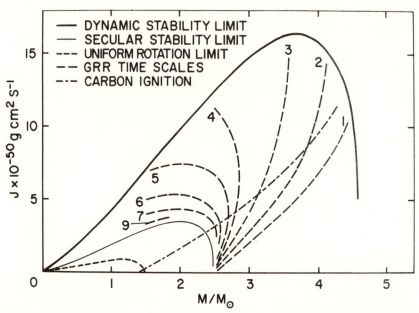

Figure 13.6. Stability limits in the (M–J)-plane. The angular momentum distribution is that of a uniformly rotating, homogeneous spheroid. For $M \lesssim 2M_\odot$, the curves are taken from the work of Ostriker and Tassoul (1969). Curves (*long dash*) of constant e-folding time τ_s of the secular instability due to gravitational radiation-reaction are labeled in $\log_{10} \tau_s$ (years). The short-dashed curve bounds the region to which uniformly rotating white-dwarf models are confined according to James (1964). Carbon ignites in the centers of carbon-oxygen models below the dot-dash curve. Source: Durisen, R. H., *Ap. J.* **199**, 179, 1975. (By permission of The University of Chicago Press. Copyright 1975 by the American Astronomical Society.)

also defines the onset of secular instability, we notice that upper mass limits beyond which the models become secularly and dynamically unstable are found at $2.5M_\odot$ and $4.6M_\odot$, respectively. As was shown by Durisen, for $3.5M_\odot \lesssim M \lesssim 4.6M_\odot$, the secular instability is mainly due to gravitational radiation-reaction (cf. §10.2), the instability growing on a time scale of the order of 10^3 to 10^1 years in this mass range.

In summary, differentially rotating white-dwarf models that are secularly and dynamically stable against non-axisymmetric perturbations can be constructed over a certain range of mass and total angular momentum. Note that Figure 13.6 also depicts the line below which models with central densities exceeding $3 \cdot 10^9$ g/cm^3 are likely to become unstable to detonation or implosion. In spite of these severe limitations in the (M–J)-plane, the rapidly rotating, massive white dwarf can be considered a possible final state of stellar evolution for a star of mass greater than M_3. In particular, it is no longer necessary to invoke mass loss from a massive star in order to bring it to the white-dwarf stage. However, because the equatorial velocities would be expected to be in the range 10^3–10^4 km/sec and because of the relatively narrow cores of lines in many observed white dwarfs, it is probable that the rapidly rotating, massive white dwarf is an exceptional case in nature. The white dwarfs of class DC—those which have no observable lines—are possible candidates. To test this suggestion, Milton has calculated the emergent spectra of hydrogen-rich, differentially rotating white-dwarf models. One significant conclusion of this work is that it is possible to generate white-dwarf models that have no detectable lines. However, since many of the models with "lineless" spectra have (U–V) color indices significantly more negative than any known DC white dwarf, it seems unlikely that these models can explain the class of DC white dwarfs as a whole.

13.3. EVOLUTIONARY EFFECTS

The major conclusion obtained in the previous section is that fully degenerate white dwarfs in rapid differential rotation can have masses considerably greater than the Chandrasekhar limiting mass. However, because this result was demonstrated by means of idealized configurations with arbitrarily prescribed distributions of angular momentum, we must now consider the difficult problem of the evolution of these models during the phase of gradual cooling. In other words, do rapidly rotating, massive white-dwarf models represent as truly a "final" stage of stellar evolution as do nonrotating white dwarfs, or do substantial changes in their mechanical and thermal properties occur over the cooling period? Since so many factors may affect the velocity field of an evolving white dwarf, it is not surprising that different pictures of the evolution of differentially rotating white dwarfs have been sketched in the literature. Only Durisen has attempted a *quantitative* discussion. His work refers to the detailed changes brought by viscous friction on the Ostriker-Bodenheimer models. For the sake of clarity, we shall consider the results pertaining to viscous friction first, proceeding next to some brief comments on the possible effects of meridional currents, crystallization, and magnetic fields on the final evolution of massive white dwarfs in non-uniform rotation.

As can readily be seen from Table 13.2, the amount of kinetic energy stored in differential motions is comparable to the total gravitational energy of a massive white dwarf. Accordingly, because viscous friction continuously generates thermal energy from non-uniform motions and because the viscous time scale of these objects is estimated to be 10^9 years or larger, both the residual thermal energy of the nondegenerate ions and the rate of viscous dissipation of kinetic energy will contribute to the total luminosity of the Ostriker-Bodenheimer models throughout the cooling phase. In other words, if rapidly rotating white dwarfs exist in nature, they will be hot and bright due to viscous dissipation. Durisen has demonstrated that this is the case. For this purpose, evolutionary sequences of differentially rotating, zero-temperature white-dwarf models were constructed, taking detailed account of angular momentum transport by a nonconstant isotropic viscosity. The principal physical assumption involved is the constancy of the angular momentum per unit mass on cylinders about the rotation axis, i.e., $j = j(\varpi,t)$ with some prescribed distribution for $j = j(\varpi,t_0) = j_0(\varpi)$ at some initial instant $t = t_0$. The relativistically corrected coefficient of viscosity used in the calculations is that of the degenerate electrons for a carbon-oxygen mixture. An evolutionary sequence, with fixed values of M and J, is thus generated by calculating the change with time of the angular momentum per unit mass due to viscous friction alone. The "initial" models at $t = t_0$ are those constructed by Ostriker and Bodenheimer, the function $j_0(\varpi)$ being that of uniformly rotating polytropes of index $n = 0$ or $n = 3/2$ (cf. §10.4: equation [44]).

Apart from details, the important physical conclusion of Durisen's calculations is that, contrary to common opinion, *viscous friction in rapidly rotating white-dwarf models does not lead to overall uniform rotation in many interesting cases*. Indeed, since uniformly rotating models exist only for $J \leqslant J_u(M)$ and $M < M_u \approx 1.5 M_\odot$, qualitatively different behaviors are obtained, depending on how the M and J values along a given sequence compare with the M and J values for which uniformly rotating models do exist (see Figure 13.6). Following Durisen, we must distinguish between three regions in the (M–J)-plane:

(i) $M < M_u$ and $J < J_u(M)$. A differentially rotating model lying just inside this small region evolves toward the strictly uniformly rotating model having the same values for M and J.

(ii) $M < M_u$ and $J > J_u(M)$. A contracting core grows in mass and approaches solid-body rotation by transferring angular momentum to the outermost mass elements. The final model thus consists of a slowly contracting, uniformly rotating core ($\Omega \propto constant$) containing most of the mass and growing in mass fraction, an expanding flattening Keplerian outer envelope ($\Omega \propto \varpi^{-3/2}$) comprising the outermost few percent of the mass, and a transitional envelope region ($\Omega \propto \varpi^{-s}, 0 < s < 3/2$) between them.

(iii) $M > M_u$. These evolutions show a smooth change in behavior with increasing M. In all cases, a contracting core grows in mass by transfer of angular momentum to an expanding envelope, which becomes Keplerian. For $M_u < M < 1.9 M_\odot$, the evolutionary sequences are similar over times considered to those for models lying in region (ii). For $1.9 M_\odot < M < 3 M_\odot$, the envelope becomes thicker with time. Finally, for $M > 3M_\odot$, the core does *not* appear to approach solid-body rotation over any sizable portion of the central regions. (As we recall from Section 13.2, however, all models with $M \gtrsim 2.5M_\odot$ are secularly unstable to non-axisymmetric disturbances!)

In summary, for atomic number $Z \leqslant 10$ and $M \gtrsim 1.5M_\odot$, the viscous evolutionary time scales of the Ostriker-Bodenheimer models are of the order of 10^9 years or larger. In addition, viscous dissipation alone is sufficient to maintain the luminosity of these massive models at more than $10^{-1}L_\odot$ and to keep their internal temperatures high enough so as to prevent crystallization over these times (unless the configurations are made of a very high-Z material, such as iron). However, because the central density is an increasing function of time in all evolutionary sequences considered, the long time scales for some values of M and J strongly suggest that these configurations could eventually reach carbon-ignition densities in 10^{10} years. According to Durisen, such models may thus represent Type I supernova precursors.

Let us now turn to the possible influence of meridional currents on the evolution of massive white dwarfs. As we recall from Section 8.1, meridional circulation occurs in the radiative regions of *nondegenerate* stars because the condition of mechanical equilibrium is not compatible with the condition of thermal radiative equilibrium. That is to say, whereas the rate of energy released—$\rho\epsilon_{Nuc}$—is constant on the level surfaces of a chemically homogeneous star, the energy transported by radiation or conduction—div $\mathscr{F}$—is not constant on such surfaces. Meridional motions are therefore necessary to insure conservation of energy. As was pointed out by Kippenhahn and Möllenhoff, circulatory currents must also prevail in *degenerate* stars, because thermal conduction alone cannot transport all the energy released by cooling and by friction, i.e., both sources of thermal energy must necessarily generate meridional currents in differentially rotating, massive white dwarfs. As a matter of fact, circulation caused by cooling is very similar to the circulation in the radiative zones of nondegenerate stars, since the thermal energy released by cooling is constant on the level surfaces. In the case of thermal energy released by friction—$\rho\epsilon$ (say)—the situation is quite different, for neither $\rho\epsilon$ nor div $\mathscr{F}$ are then constant on the level surfaces. According to Kippenhahn and Möllenhoff, if white dwarfs produce their luminosity by cooling of the nearly isothermal degenerate interior, the time scale for this kind of circulation might be shorter than the time scale for cooling. In using a similar order-of-magnitude argument, they have shown also that the circulation necessary to carry away the heat released by viscous dissipation might be more important for the evolution of the angular momentum than the transport of momentum by friction. However, because all these estimates are based on formulae derived for slowly rotating, slightly distorted configurations, further work is needed to evaluate the actual role of circulatory currents in massive white dwarfs during the phase of gradual cooling.

To conclude, let us now consider the possible influence of crystallization on the structure of the Ostriker-Bodenheimer models. As we know, because of thermal energy losses, the nearly isothermal interior of a non-rotating white-dwarf model must cool and finally, proceeding outward from the center, form a crystalline lattice. According to Schwartz and Africk, although the viscous dissipation of rotational kinetic energy may in some cases lengthen the cooling time scale of differentially rotating white dwarfs, it cannot prevent this final state. In brief, their qualitative picture of the endpoints of the evolution of the Ostriker-Bodenheimer models is the following. Any part of a degenerate star that forms a lattice must be in rigid rotation for, if it were not, the energy dissipated by differential motions would immediately melt the lattice. It thus follows that the star cools until the center is at its melting temperature $T_m (\propto \rho^{1/3} Z^{5/3})$, the angular momentum being redistributed throughout the interior in order to enable the core to rotate rigidly. Then, the solidification of the star proceeds outward from the center, with the internal temperature of the nearly isothermal star given by the local melting temperature at the density boundary of the solid core. Eventually, this solid core grows to a large enough mass so that it can no longer support itself in rigid rotation; and a catastrophic collapse must ensue, perhaps leading to a supernova explosion. In Koester's opinion, however, crystallization may lead *either* to the critical central density for supernova explosions *or* to mass loss with subsequent formation of stable, uniformly rotating white dwarfs. In this connection, a further interesting possibility was recently proposed by Ostriker. If these differentially rotating white dwarfs exist, and if some have significant magnetic fields, then it may be argued that they can lose rotational energy via magnetic dipole radiation in the same way that rotating neutron stars are thought to lose energy. This process would also lead to collapse, then resulting in the production of energy emitted as cosmic rays rather than as thermal luminosity.

BIBLIOGRAPHICAL NOTES

Section 13.1. Detailed derivation and discussion of equations (1)–(6) will be found, e.g., in references 7 and 8 of Chapter 3. The following reviews are particularly worth noting:

   1. Mestel, L., "The Theory of White Dwarfs" in *Stellar Structure* (Aller, L. H., and McLaughlin, D. B., eds.), pp. 297–325, Chicago: The Univ. of Chicago Press, 1965.

   2. Ostriker, J. P., "Recent Developments in the Theory of Degenerate Dwarfs" in *Annual Review of Astronomy and Astrophysics* **9**, pp. 353–366, Palo Alto, California: Annual Reviews, Inc., 1971.
   3. Saakian, G. S., *Equilibrium Configurations of Degenerate Gaseous Masses*, New York: Wiley and Sons, 1974.

Reference 2 contains a clear and concise presentation of the role of rotation in white dwarfs. The rotational velocities of some white dwarfs have been discussed in references 141 and 142 of Chapter 12; see also references 40 and 41 below.

Section 13.2. The influence of uniform rotation on the global structure of white-dwarf models has been originally considered by:

   4. Chandrasekhar, S., in *Novae and White Dwarfs* (Shaler, A. J., ed.), *Colloque International d'Astrophysique* **3**, pp. 239–248, Paris: Hermann et Cie, 1941.
   5. Bandyopadhyay, G., *Bull. Calcutta Math. Soc.* **44** (1952): 89–91.
   6. Suda, K., *Sci. Reports Tôhoku Univ.* (I) **37** (1953): 307–319.
   7. Krishan, S., and Kushwaha, R. S., *Publ. Astron. Soc. Japan* **15** (1963): 253–255.

The most detailed models are due to James (reference 13 of Chapter 10). Further contributions are due to:

   8. Anand, S. P. S., *Proc. Nat. Acad. Sci. U.S.A.* **54** (1965): 23–26.
   9. Roxburgh, I. W., *Zeit. f. Astroph.* **62** (1965): 134–142.
   10. Monaghan, J. J., *Monthly Notices Roy. Astron. Soc. London* **132** (1966): 305–316.
   11. Anand, S. P. S., and Dubas, O. V., *Astroph. and Space Sci.* **2** (1968): 520–528.

Inequality (7) was derived by Anand in reference 21 of Chapter 10. The effects of inverse β-decay are discussed in:

   12. Papoian, V. V., Sedrakian, D. M., and Chubarian, E. V., *Comm. Burakan Observatory* **39** (1968): 101–112; *ibid.* **40** (1969): 86–97.
   13. Papoian, V. V., Sedrakian, D. M., and Chubarian, E. V., *Astrofizika* **7** (1971): 95–105.

See also reference 19 of Chapter 5. The role of magnetic fields has been considered by:

   14. Ostriker, J. P., and Hartwick, F.D.A., *Astroph. J.* **153** (1968): 797–806.
   15. Raadu, M. A., *Astroph. and Space Sci.* **14** (1971): 464–472.

See, however:

   16. O'Connell, R. F., *Astroph. J.* **195** (1975): 751–752.

Uniformly rotating models which include the effects of general relativity and inverse β-decay are due to:

17. Hartle, J. B., and Thorne, K. S., *ibid.* **153** (1968): 807–834.
18. Papoian, V. V., Sedrakian, D. M., and Chubarian, E. V., *Astrofizika* **5** (1969): 415–424.
19. Arutyunian, G. G., Sedrakian, D. M., and Chubarian, E. V., *ibid.* **7** (1971): 467–479.

The following stability analysis of uniformly rotating white dwarfs may be noted:

20. Roxburgh, I. W., and Durney, B. R., *Zeit. f. Astroph.* **64** (1966): 504–511.
21. Durney, B. R., Faulkner, J., Gribbin, J. R., and Roxburgh, I. W., *Nature* **219** (1968): 20–21.
22. Ostriker, J. P., and Tassoul, J. L., *ibid.* **219** (1968): 577–579, 1091.
23. Gribbin, J. R., *Monthly Notices Roy. Astron. Soc. London* **144** (1969): 549–552.
24. Imshennik, V. S., and Seidov, Z. F., *Astrofizika* **6** (1970): 301–307.
25. Vartanian, Yu. L., and Ovsepian, A. V., *ibid.* **6** (1970): 601–613.
26. Papoian, V. V., Sedrakian, D. M., and Chubarian, E. V., *ibid.*, **7** (1971): 643–649.
27. Arutyunian, G. G., Sedrakian, D. M., and Chubarian, E. V., *Astron. Zh.* **49** (1972): 1216–1220.
28. Vartanian, Yu. L., *Astrofizika* **8** (1972): 413–418.
29. Vartanian, Yu. L., Ovsepian, A. V., and Adzhian, G. S., *Astron. Zh.* **50** (1973): 989–995.
30. Shapiro, S. L., and Teukolsky, S. A., *Astroph. J.* **203** (1976): 697–700.

See also reference 35. Application of the static method was also made by Bisnovatyi-Kogan and Blinnikov (reference 58 of Chapter 6). The absence of critical mass for rapidly rotating white dwarfs has been originally pointed out by:

31. Hoyle, F., *Monthly Notices Roy. Astron. Soc. London* **107** (1947): 231–236.

See also reference 9. Detailed models of zero-temperature white dwarfs in non-uniform rotation will be found in:

32. Ostriker, J. P., Bodenheimer, P., and Lynden-Bell, D., *Phys. Rev. Letters* **17** (1966): 816–818.
33. Ostriker, J. P., and Bodenheimer, P., *Astroph. J.* **151** (1968): 1089–1098.

34. Smart, N. C., and Monaghan, J. J., *Monthly Notices Roy. Astron. Soc. London* **151** (1971): 427–435.

See also reference 36 of Chapter 5. The oscillations and stability of these models have been originally considered by:

35. Ostriker, J. P., and Tassoul, J. L., *Astroph. J.* **155** (1969): 987–997.

The approximate stability limits $\tau \approx 0.14$ and $\tau \approx 0.26$ were first derived in this paper; more refined calculations are due to:

36. Durisen, R. H., *ibid.* **199** (1975): 179–183.

See, however, reference 36 of Chapter 10. Papers of related interest are those of:

37. Gribbin, J. R., *Astroph. Letters* **4** (1969): 77–79.
38. Basko, M. M., and Imshennik, V. S., *Astrofizika* **8** (1972): 387–391.
39. Wiita, P. J., and Press, W. H., *Astroph. J.* **208** (1976): 525–533.

The effects of rapid rotation on the spectra of white dwarfs are considered in:

40. Wickramasinghe, D. T., and Strittmatter, P. A., *Monthly Notices Roy. Astron. Soc. London* **147** (1970): 123–131.
41. Milton, R. L., *Astroph. J.* **189** (1974): 543–554.

Analytical approximations to some of the mass-radius relations discussed in this section will be found in:

42. Nauenberg, M., *ibid.* **175** (1972): 417–430.

Section 13.3. Various factors influencing the evolution of the Ostriker-Bodenheimer models have been considered by:

43. Schwartz, R., and Africk, S., *Astroph. Letters* **5** (1970): 141–144.
44. Durisen, R. H., *Astroph. J.* **183** (1973): 205–214, 215–231; *ibid.* **195** (1975): 483–492.
45. Kippenhahn, R., and Möllenhoff, C., *Astroph. and Space Sci.* **31** (1974): 117–141.
46. Koester, D., *Astron. and Astroph.* **35** (1974): 77–80.
47. Van Horn, H. M., in *Physics of Dense Matter* (Hansen, C. J., ed.), pp. 251–264, Dordrecht: D. Reidel Publ. Co. 1974.

See also reference 2. Other contributions are due to:

48. Vartanian, Yu. L., and Ovsepian, A. V., *Astrofizika* **7** (1971): 107–119.
49. Ovsepian, A. V., *ibid.* **10** (1974): 99–107.

Papers of related interest are those of:

50. Finzi, A., and Wolf, R. A., *Astroph. J.* **150** (1967): 115–129.

51. Hansen, C. J., and Wheeler, J. C., *Astroph. and Space Sci.* **3** (1969): 464–474.

52. Brecher, K., and Morrison, P., *Astroph. J. Letters* **180** (1973): L107–L112.

53. Lamb, D. Q., and Van Horn, H. M., *Astroph. J.* **183** (1973): 959–966.

54. Durisen, R. H., *Astron. and Astroph.* **44** (1975): 473–476.

The effects of rotation on gravitational sorting in white dwarfs are briefly discussed in:

55. Strittmatter, P. A., and Wickramasinghe, D. T., *Monthly Notices Roy. Astron. Soc. London* **152** (1971): 47–73.

56. Baglin, A., in *European Workshop on White Dwarfs*, Kiel, G. F. R., October 1974.

14

Oscillations and Stability

14.1. INTRODUCTION

As we recall from Section 6.4, the natural oscillation frequencies of a spherical star in hydrostatic equilibrium can be classified according to several types: the (radial and nonradial) p-modes, the (nonradial) f-modes, and the (nonradial) g-modes. In the absence of any dissipative mechanism, the possible modes of instability of a spherical star are these: (i) the lowest radial mode that becomes dynamically unstable when the adiabatic exponent is less than $4/3$ in a large portion of the star, and (ii) the g-oscillations that become convectively unstable when the Schwarzschild criterion for stability is violated in some part of the star (i.e., when the local temperature gradient is steeper than the corresponding adiabatic lapse rate).

The situation is quite different in the case of a rotating star for, then, the pattern of instabilities may be altered in two ways. *First*, rotation modifies the criterion for the onset of dynamical instability with respect to the lowest "radial" mode, and it alters the growth rates of convectively unstable motions. *Second*, further instabilities may be introduced by the presence of rotation through the f-oscillations which, as we know, are always stable in the limit of zero rotation (cf. §6.4). The latter instabilities have been discussed already in the context of rapidly rotating, severely distorted polytropes and white-dwarf models (cf. §§10.5 and 13.2); they correspond to non-axisymmetric disturbances and occur only in configurations that greatly depart from spherical symmetry. The main purpose of this chapter is to find the effects of rotation on the oscillation frequencies belonging to the lowest p- and f-modes, as well as those belonging to the g-modes.

In Section 14.2, we discuss the dynamical stability of slowly rotating, slightly distorted stars when subject to pseudo-radial motions; nonrelativistic stars and relativistic stars are considered in succession. In Section 14.3, we examine the resonances of selected modes in a rotating star as plausible explanations of the beat phenomenon occurring in some of the β Cephei variables. The next section is devoted to the effects of rotation on the vibrational stability of upper main-sequence stars against axisymmetric disturbances. We conclude the chapter with a brief discussion of convectively unstable motions in rotating bodies. The modifications brought by magnetic fields and tidal disturbances on the natural frequencies of a rotating star are considered in the two subsequent chapters.

14.2. PSEUDO-RADIAL MOTIONS

The pseudo-radial isentropic pulsations of a uniformly rotating, slightly distorted star were first considered, in the linear approximation, by Ledoux in 1945. For a slowly rotating, nonrelativistic configuration, Ledoux's formula for the square of the characteristic frequency of oscillation is

$$\sigma_R^2 = (3\gamma - 4)\frac{|W|}{I} + \tfrac{2}{3}(5 - 3\gamma)\Omega^2, \tag{1}$$

where γ is the (constant) ratio of specific heats, and can be modified to include the effects of radiation pressure, partial ionization, and degeneracy (cf. §3.4). As usual, Ω is the (constant) angular velocity, W is the gravitational potential energy, and I is the moment of inertia about the center of mass. (Note that in equation [1] both W and I refer to the *rotating* configuration and therefore include, implicitly, terms of order Ω^2.) By virtue of equation (1), the critical value γ_c below which dynamical instability sets in is thus reduced from 4/3—the critical value for a spherical star—to

$$\gamma_c(\text{Rot.}) = \frac{4}{3} - \frac{2}{9}\frac{\Omega^2 I}{|W|}. \tag{2}$$

As was recently shown by Lebovitz, the stabilizing influence of a slow rotational motion holds not only for solid-body rotation but for *any* rotation law whatever, certainly if γ is constant and probably in many cases where γ is variable.[1]

At this stage, the question naturally arises whether the stabilizing influence of rotation on the lowest p-mode remains true in the case of rapidly rotating, severely distorted configurations. This problem has received comparatively little attention so far, because rapid rotation couples the two lowest axisymmetric modes which, in the limit of hydrostatic equilibrium, are independent; that is, the fundamental radial mode—σ_R—and the lowest axisymmetric f-mode—σ_Z (cf. §6.7: equations [120]–[123]). On the evidence before us, there are strong reasons to believe that the universally stabilizing influence of rotation (in the context of a constant value of γ) will persist in the case of large departures from sphericity. Figure 14.1 illustrates the critical adiabatic exponent γ_c along various sequences of differentially rotating polytropes, as a function of the ratio $\tau = K/|W|$ of the rotational kinetic energy to the gravitational potential energy. (The remaining symbols are defined in Section 10.4.) The sequence

---

[1] In case γ is not constant, it is generally believed that equations (1) and (2) still hold true, provided γ and γ_c are replaced by their pressure-weighted averages. Because this conclusion has never been properly demonstrated for an arbitrary γ, we shall restrict our subsequent discussion to the case $\gamma = constant$.

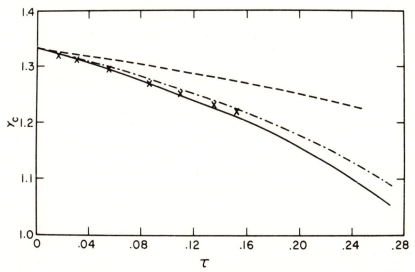

Figure 14.1. The critical value of γ for dynamical instability as a function of $\tau = K/|W|$ for four sequences: *solid line*, $n = 0$, $n' = 0$; *crosses*, $n = 1.5$, $n' = 1.5$; *dash-dot line*, $n = 1.5$, $n' = 0$; *dashed line*, $n = 3$, $n' = 0$. Source: Ostriker, J. P., and Bodenheimer, P., *Ap. J.* **180**, 171, 1973. (By permission of The University of Chicago Press. Copyright 1973 by the American Astronomical Society.)

$(n,n') = (0,0)$ was originally discussed by Chandrasekhar and Lebovitz; it refers to the compressible Maclaurin spheroids and provides an *exact* limit of dynamical stability against axisymmetric pulsations. These configurations are, of course, atypical since their mass distribution is highly unrealistic. The remaining sequences were obtained by Ostriker and Bodenheimer on the basis of the second-order virial equations and an *approximate* expression for the Lagrangian displacement (cf. §6.7). We note that rapid rotation has some effect on the stabilization of the lowest *p*-mode, although the degree of stabilization depends on n as well as τ. For configurations near the point of bifurcation ($\tau \approx 0.14$), the critical value γ_c is reduced from $4/3$ to 1.23 for the case $n = 1.5$, and to 1.28 for the case $n = 3$.

The above result is of particular importance when effects besides rotation alter the critical adiabatic exponent γ_c from its "classical value" $4/3$. Indeed, it is well known that the effects of general relativity have a destabilizing influence on the radial pulsations of a spherical body and that, to first approximation,

$$\gamma_c(\text{G.R.}) = \frac{4}{3} + k\,\frac{R_s}{R}, \tag{3}$$

where k is a certain (*positive*) constant that depends on the model, $R_s = 2GM/c^2$ is the Schwarzschild radius, and R is the radius of the spherical body. This result—which derives from the independent work of Kaplan, Chandrasekhar, and Fowler—is a direct consequence of the existence of the critical length R_s. Now, if the effects of solid-body rotation and of general relativity are both present, and are both considered as first-order effects, then the critical value γ_c below which dynamical instability sets in becomes, to first approximation,

$$\gamma_c(\text{Rot.} + \text{G.R.}) = \frac{4}{3} - \frac{2}{9}\frac{\Omega^2 I}{|W|} + k\,\frac{R_s}{R}. \tag{4}$$

This last result was originally obtained by Fowler, and by Durney and Roxburgh, in the post-Newtonian approximation of general relativity. Subsequent refined treatments of the problem have largely confirmed the conflicting influences of rotation and general relativity, at least insofar as infinitesimal disturbances are concerned.

For the small-amplitude motions considered above, the origin of the stabilizing influence of rotation on the lowest pseudo-radial mode can be traced to the conservation of angular momentum of *each* fluid element expelled from its equilibrium position. However, it is not at all obvious that this physical argument will remain valid when a slowly rotating configuration is subject to finite-amplitude, axisymmetric disturbances. Indeed, as we shall now see, stability toward finite-amplitude motions crucially depends on the adiabatic exponent γ, on the rotation parameter τ, on the ratio R_s/R, *and on the amount of energy available to sustain the oscillations.* Moreover, the importance of the latter factor becomes paramount near the cutoff value γ_c predicted by a linear stability analysis (cf. equation [4]). For the sake of simplicity, we shall mainly discuss the case of nonrelativistic stars, passing next to the results for the case of relativistic stars in the post-Newtonian approximation of general relativity.

THE NEWTONIAN CASE

In the linear approximation, a compressible Maclaurin spheroid admits of axisymmetric motions for which the Lagrangian displacement $\xi(x,t)$ is a linear function of the Eulerian variables **x** (cf. §6.7: equations [109], [120], and [121]). It is a simple matter to verify that such an idealized configuration admits of a wider class of finite-amplitude motions that, in the limit of infinitesimal disturbances, reduce to the two (nontrivial) *zonal* modes discussed in Section 6.7. For this purpose, let us view the spheroid in a nonrotating frame of reference and introduce the Lagrangian variables **X** (cf. §3.2). At some initial instant $t = 0$ (say), the Maclaurin spheroid rotates with the constant angular velocity Ω_0 about the X_3-axis; in our Lagrangian representation, such a motion has the form

$$x_1 = X_1 \cos \Omega_0 t - X_2 \sin \Omega_0 t, \tag{5}$$

$$x_2 = X_2 \cos \Omega_0 t + X_1 \sin \Omega_0 t, \tag{6}$$

$$x_3 = X_3. \tag{7}$$

Then, if all dissipative effects and electromagnetic forces are neglected altogether, it can be shown that the simplest finite-amplitude motions are

$$x_1 = v(t)(X_1 \cos \phi - X_2 \sin \phi), \tag{8}$$

$$x_2 = v(t)(X_2 \cos \phi + X_1 \sin \phi), \tag{9}$$

$$x_3 = w(t)X_3, \tag{10}$$

where

$$\phi(t) = \int_0^t \Omega(t) \, dt, \tag{11}$$

and $\Omega(t)$ is the instantaneous angular velocity of the homogeneous spheroid. The functions $v(t)$ and $w(t)$ satisfy a system of ordinary differential equations that may be obtained by substituting equations (8)–(11) into the basic equations of hydrodynamics. (Let us recall that $\mathbf{x} = \mathbf{X}$ and $\Omega = \Omega_0$ at $t = 0$, so that $v \equiv w \equiv 1$ at this instant.) Equations (8)–(11) provide an *exact* homological solution, which represents, in an inertial frame, expanding and contracting motions, together with a concomitant variation in the rotation rate so as to preserve the total angular momentum of the configuration. In particular, if we restrict ourselves to a slowly rotating spheroid, we can further simplify this solution by letting

$$v(t) \equiv w(t). \tag{12}$$

Equations (8)–(12) accurately describe the simplest pseudo-radial pulsation of a homogeneous spheroid, having the Ledoux frequency (1) in the limit of infinitesimal disturbances.[2]

Proceeding along the same lines as those followed in Section 6.7, we shall now use the virial theorem and consider equations (8)–(12) as a "trial solution" for the simplest axisymmetric motions of a slowly rotating, *centrally condensed* star. Hence, in both the linear and nonlinear regimes, the larger the departure from uniformity in angular velocity and mass

---

[2] The presence of two time-dependent functions—$v(t)$ and $w(t)$—in equations (8)–(10) implies a *coupling* between the two zonal modes which, in the limit of hydrostatic equilibrium, reduce to the lowest radial mode (σ_R) and the lowest axisymmetric f-mode (σ_Z). This coupling is already apparent in the linear approximation (cf. §6.7: equation [121]), and it also occurs in the nonlinear regime. For the sake of simplicity, we shall restrict our discussion to the case of slowly rotating stars for, then, it is a simple matter to isolate the pseudo-radial motions.

distribution, the larger the error in the description of the motion will be. Quite generally, the scalar virial theorem has the form

$$\frac{1}{2}\frac{d^2}{dt^2}\int_{\mathscr{V}} x_k x_k \, dm = \int_{\mathscr{V}} \frac{\partial x_k}{\partial t}\frac{\partial x_k}{\partial t}\, dm + W + 3\int_{\mathscr{V}} \frac{p}{\rho}\, dm, \qquad (13)$$

where, as usual, a summation over repeated indices must always be understood; we also have

$$W = -\tfrac{1}{2}G\int_{\mathscr{V}}\int_{\mathscr{V}'} \frac{dm\, dm'}{|\mathbf{x} - \mathbf{x}'|} \qquad (14)$$

(cf. §3.7). At time $t = 0$, the body is assumed to be in a state of permanent rotation so that equations (5)–(7) and (13) then imply the condition

$$\int_{\mathscr{V}} \Omega_0{}^2(X_1{}^2 + X_2{}^2)\, dm + W_0 + 3\int_{\mathscr{V}} \frac{p_0}{\rho_0}\, dm = 0, \qquad (15)$$

where a subscript "0" denotes an initial value at $t = 0$. Because of mass conservation, let us note that the mass element dm can be written equally as $dm = \rho_0\, d\mathbf{X} = \rho\, d\mathbf{x}$; hence, we must have

$$\rho J = \rho_0, \qquad (16)$$

where J is the Jacobian of the transformation that relates the dependent variables $\mathbf{x}$ to the independent variables $\mathbf{X}$ (cf. §3.3: equation [22]). Because we restrict our analysis to isentropic motions only, it also follows that

$$p = p_0\left(\frac{\rho}{\rho_0}\right)^{\gamma}, \qquad (17)$$

where the adiabatic exponent γ is taken as a constant throughout the whole mass. Finally, we must impose the condition

$$\int_{\mathscr{V}} \Omega(x_1{}^2 + x_2{}^2)\, dm = \int_{\mathscr{V}} \Omega_0(X_1{}^2 + X_2{}^2)\, dm, \qquad (18)$$

expressing conservation of angular momentum.

Let us now evaluate the various terms occurring in equation (13). In making use of equations (8)–(12), we can first write

$$\frac{1}{2}\frac{d^2}{dt^2}\int_{\mathscr{V}} x_k x_k\, dm = \left[\left(\frac{dw}{dt}\right)^2 + w\frac{d^2w}{dt^2}\right]I_0, \qquad (19)$$

where I_0 is the moment of inertia with respect to the center of mass of the equilibrium configuration at $t = 0$. Similarly, because conservation of angular momentum implies the condition $\Omega = \Omega_0/w^2$, we further obtain

$$\int_{\mathscr{V}} \frac{\partial x_k}{\partial t}\frac{\partial x_k}{\partial t}\, dm = \left(\frac{dw}{dt}\right)^2 I_0 + \frac{2K_0}{w^2}, \qquad (20)$$

where K_0 is the rotational kinetic energy at time $t = 0$. In making use next of equation (14) and applying equations (8)–(12) to both unprimed and primed coordinates, we obtain $|\mathbf{x} - \mathbf{x}'| = w|\mathbf{X} - \mathbf{X}'|$, in which $\mathbf{x} = \mathbf{X}$ and $\mathbf{x}' = \mathbf{X}'$ at $t = 0$; it thus follows that

$$W = \frac{W_0}{w}. \tag{21}$$

Finally, if we insert equations (8)–(12) into equations (16) and (17), we can write

$$3 \int_{\mathscr{V}} \frac{p}{\rho}\, dm = \frac{3}{w^{3\gamma - 3}} \int_{\mathscr{V}} \frac{p_0}{\rho_0}\, dm = \frac{1}{w^{3\gamma - 3}} (|W_0| - 2K_0). \tag{22}$$

In establishing this last equality, we made use of the equilibrium condition (15) to eliminate the integral over the pressure.

Inserting now equations (19)–(22) into the scalar virial theorem (13), we obtain at once the required condition on the function $w(t)$; that is

$$\frac{d^2 w}{dt^2} = \frac{1}{w^2}\left(\frac{1}{w^{3\gamma - 4}} - 1\right)\frac{|W_0|}{I_0} + \frac{1}{w^3}\left(1 - \frac{1}{w^{3\gamma - 5}}\right)\frac{2K_0}{I_0}, \tag{23}$$

where I_0, K_0, and W_0 refer to the *rotating* configuration at time $t = 0$. Hence, both coefficients on the r.h.s. of equation (23) include terms that are proportional to Ω_0^2 (for a uniformly rotating body) or to some *mean* squared angular velocity $\langle \Omega_0^2 \rangle$ (for a differentially rotating body). In the latter case, we expect greater errors in our description of the motion, but the *qualitative* features of the results should be trustworthy. Equation (23) must now be solved with appropriate initial conditions, e.g., $w = 1$ and $dw/dt = \dot{w}_0$(say) at $t = 0$. No confusion should arise if we henceforth suppress the subscript "0" attached to the quantities I_0, K_0, W_0, and Ω_0.

Before discussing equation (23), let us first consider the case of a uniformly rotating star when subject to infinitesimal disturbances. Obviously, we can then replace equation (23) by its Taylor expansion near the value $w = 1$ which defines the state of equilibrium. Letting

$$w(t) = 1 + \epsilon(t), \tag{24}$$

and neglecting the second and higher powers of ϵ, we thus obtain

$$\frac{d^2 \epsilon}{dt^2} + \sigma_R^2 \epsilon = 0, \tag{25}$$

where σ_R^2 is given by equation (1) since $2K/I$ is very closely equal to $2\Omega^2/3$. Equation (25) yields at once the solution

$$\epsilon(t) = \epsilon_0 \sin \sigma_R t, \tag{26}$$

where $\epsilon_0 = \dot{w}_0/|\sigma_R|$ (when $\sigma_R^2 \neq 0$). The above equation readily demonstrates that the critical value γ_c given by equation (2) indeed separates stable ($\sigma_R^2 > 0$) from unstable ($\sigma_R^2 < 0$) motions in the linear approximation. This is the main content of Ledoux's classical result.

Let us now turn to the finite-amplitude, pseudo-radial motions described by equations (8)–(12) and (23). It is convenient to rewrite this last equation in the dimensionless form

$$\frac{d^2w}{dt^2} = F(w; \gamma, \tau), \tag{27}$$

where

$$F(w; \gamma, \tau) = \frac{1}{w^2}\left(\frac{1}{w^{3\gamma-4}} - 1\right) + \frac{2}{w^3}\left(1 - \frac{1}{w^{3\gamma-5}}\right)\tau, \tag{28}$$

the variable t being here given in the units of $(|W|/I)^{-1/2}$. These pseudo-radial motions are thus characterized by a time scale of the order of $(G\bar{\rho})^{-1/2}$, where $\bar{\rho}$ designates the mean density of the configuration at time $t = 0$. A first integral of equation (27) can be obtained at once; we have

$$\frac{1}{2}\left(\frac{dw}{dt}\right)^2 + \Phi(w; \gamma, \tau) = E_0, \tag{29}$$

where E_0 is a constant. The potential Φ associated with the motion has the form

$$\Phi(w; \gamma, \tau) = \frac{\tau - w}{w^2} + \frac{1 - 2\tau}{3(\gamma - 1)}\frac{1}{w^{3\gamma-3}}, \tag{30}$$

when $\gamma \neq 1$, and

$$\Phi(w; 1, \tau) = \frac{\tau - w}{w^2} - (1 - 2\tau)\log w. \tag{31}$$

In accordance with equation (29), we note that the constants $\dot{w}_0$ and E_0 are *not* independent quantities, but are related by the equation

$$(\dot{w}_0)^2 = 2[E_0 - \Phi(1; \gamma, \tau)]. \tag{32}$$

A second quadrature eventually leads to the implicit solution of equation (27); that is

$$t = \int_w^1 \frac{dw}{[2(E_0 - \Phi)]^{1/2}}, \tag{33}$$

which, for periodic motions, is a multivalued function.

Equation (29) merely expresses the fact that conservation of energy does obtain in our model of axisymmetric pulsations. Accordingly, the

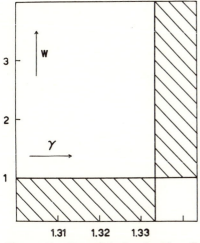

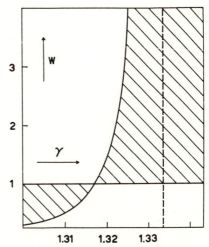

Figure 14.2. Zeros of the function $F(w; \gamma, 0)$ in the (w,γ)-plane for a nonrotating gaseous sphere. The acceleration is negative in shaded areas, and positive in unshaded areas. After Ledoux (1952). Source: Tassoul, J. L., *Ap. J.* **160**, 1031, 1970. (By permission of The University of Chicago Press. Copyright 1970 by the University of Chicago.)

Figure 14.3. Zeros of the function $F(w; \gamma, \tau)$ in the (w,γ)-plane for a uniformly rotating spheroid with $\tau = 0.022874$. The (*dashed*) line $\gamma = 4/3$ is an asymptote for the vertical branch of the curve $F(w; \gamma, \tau) = 0$. Shaded and unshaded areas have the same meanings as in Figure 14.2 ($\gamma_c = 1.3174$). Source: Tassoul, J. L., *Ap. J.* **160**, 1031, 1970. (By permission of The University of Chicago Press. Copyright 1970 by the University of Chicago.)

pseudo-radial motions described by equations (8)–(12) are either oscillatory or monotonically increasing with time. (Damped or overstable motions would have required a term which is proportional to dw/dt in equation [27]!) Hence, for every stable (i.e., periodic) solution, the acceleration d^2w/dt^2 must be negative during expansion ($w > 1$) and positive during contraction ($w < 1$). Conditions under which this stability criterion is satisfied can readily be deduced by plotting, for a given rotation rate τ, the zeros of the function $F(w; \gamma, \tau)$ in the (w,γ)-plane.[3] This approach (originally due to Poincaré) is illustrated in Figures 14.2 and 14.3. Further physical insight is provided, in the usual fashion, by a plot of the potential $\Phi(w; \gamma, \tau)$ as a function of w, for given values of γ and τ (see Figures 14.4 and 14.5).

It is immediately apparent from Figure 14.2 that, in the absence of rotation, stable motions exist only when $\gamma > 4/3$ and that unstable motions

[3] See, e.g., Andronov, A. A., Vitt, A. A., and Khaikin, S. E., *Theory of Oscillators*, New York: Pergamon Press, 1966.

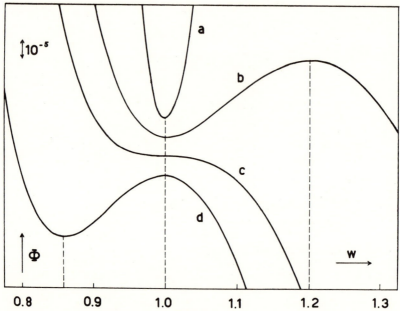

Figure 14.4. Potential $\Phi(w; \gamma, \tau)$ of a uniformly rotating spheroid with $\tau = 0.022874$. Curves labeled a, b, c, and d correspond, respectively, to $\gamma = 4/3$, $\gamma = 1.3188$, $\gamma = \gamma_c$, and $\gamma = 1.3160$. Origin of vertical scale is arbitrary. Source: Tassoul, J. L., *Ap. J.* **160**, 1031, 1970. (By permission of The University of Chicago Press. Copyright 1970 by the University of Chicago.)

exist only when $\gamma < 4/3$. In that case, the equilibrium position $w = 1$ corresponds to an absolute minimum (for stable motions) or maximum (for unstable motions) of the potential Φ. When $\gamma = 4/3$, the displacement $w(t)$ is a linear function of time, for the acceleration then identically vanishes; hence, unless the initial velocity $\dot{w}_0$ is rigorously zero, the marginal state corresponding to $\gamma = 4/3$ is always unstable. This result is originally due to Ledoux.

These conclusions are substantially modified when a slow rotational motion is taken into account. Indeed, Figure 14.3 clearly shows that the zeros of the function F are now given by the line $w = 1$ *and a curve that admits the line $\gamma = 4/3$ as a vertical asymptote*. Their common intersection defines, of course, the critical adiabatic exponent γ_c provided by a linear stability analysis (cf. equation [2]). For a general discussion, we must thus distinguish between three possible situations:

(i) $\gamma \geqslant 4/3$. All motions are *stable*, for the equilibrium state $w = 1$ then corresponds to an absolute minimum of the potential (see Figure 14.4:

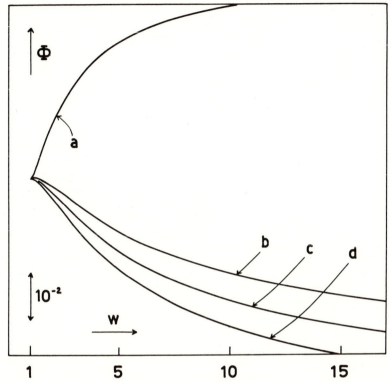

Figure 14.5. Continuation of the four curves illustrated in Figure 14.4, for very large values of w. Origin of vertical scale is arbitrary. Values for $w < 1$ have not been reproduced (see Figure 14.4). Note that the potential becomes independent of the displacement when $w \gtrsim 15$. Source: Tassoul, J. L., *Ap. J.* **160**, 1031, 1970. (By permission of The University of Chicago Press. Copyright 1970 by the University of Chicago.)

curve labeled *a*). Accordingly, irrespective of the sign and magnitude of $\dot{w}_0$, the slowly rotating body will always oscillate about the mean position $w = 1$;

(ii) $\gamma_c < \gamma < 4/3$. The system is *metastable*, i.e., growing motions always occur whenever the energy available to sustain the pulsations becomes too large. In other words, for oscillatory motions to prevail, the largest value of w during a cycle, w_M(say), must always remain smaller than a critical amplitude, w_{max}(say), which depends on the values of γ and τ (see Figure 14.3). This case is further illustrated in Figure 14.4 (curve labeled *b*), which shows that the potential Φ now has *two* extrema, i.e., $w = 1$ and $w = w_{max}$. Thus, whenever $\dot{w}_0$, γ, and τ are such that w_M becomes exactly equal to w_{max}, the rotating system tends toward a new state of equilibrium. This (unstable) marginal state is depicted in Figures 14.6 and 14.7 (dashed

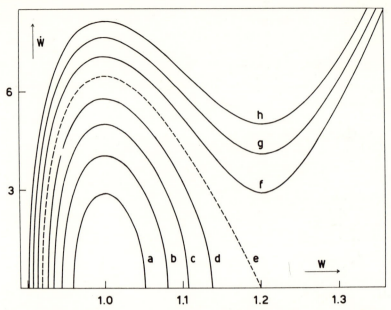

Figure 14.6. Pseudo-radial motion of a uniformly rotating spheroid with $\tau = 0.022874$, in the $(w, dw/dt)$-plane. The various curves correspond to $\gamma = 1.3188$ and different initial velocities (case [ii] discussed in the text). The curves are labeled to correspond to those illustrated in Figure 14.7. The vertical scale should be multiplied by 0.001. The dashed curve separates stable from unstable motions. For stable pulsations, the graph is symmetric with respect to the line $dw/dt = 0$; for unstable motions, this symmetry is physically relevant only for $w \leqslant 1$. Pseudo-radial expansion becomes uniform in time when $w \gg 10$ (see Figure 14.5). Source: Tassoul, J. L., *Ap. J.* **160**, 1031, 1970. (By permission of The University of Chicago Press. Copyright 1970 by the University of Chicago.)

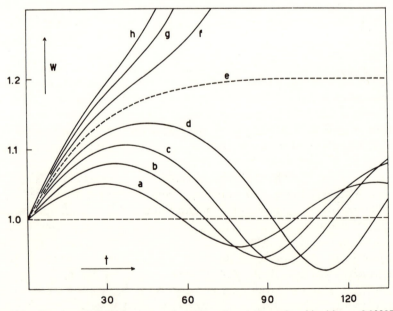

Figure 14.7. Pseudo-radial displacement of a uniformly rotating spheroid with $\tau = 0.022874$, as a function of time (in units of $[|W|/I]^{-1/2}$). The various curves correspond to $\gamma = 1.3188$ and different initial velocities (case [ii] discussed in the text); see Figure 14.6 for labeling of the curves. For very large values of t (*not illustrated*) unstable displacements become a linear function of time (see Figure 14.5). Note that the marginal state (*dashed curve*) rapidly becomes independent of time. Source: Tassoul, J. L., *Ap. J.* **160**, 1031, 1970. (By permission of The University of Chicago Press. Copyright 1970 by the University of Chicago.)

curves), and clearly separates oscillatory motions from expanding motions;

(iii) $\gamma \leqslant \gamma_c$. The configuration is thoroughly *unstable* (see Figure 14.4: curves labeled c and d). When $\gamma = \gamma_c$, the equilibrium state $w = 1$ corresponds to an inflection point of the potential Φ. Hence, any initial disturbance will necessarily grow in time. When $\gamma < \gamma_c$, the potential Φ has *two* extrema, i.e., $w = w_{min}$(say) and $w = 1$. At first glance, this situation appears to be very similar to that of case (ii); however, given any initial velocity, the slowly rotating body will now become unstable, for the position $w = 1$ is a relative maximum of the potential Φ. (In case [ii], $w = 1$ is a relative minimum!) Moreover, because of energy conservation, it also follows that these instabilities are always associated in the long run with expanding motions.

In summary, although a linear stability analysis predicts a critical value γ_c that unequivocally separates stable from unstable configurations with respect to pseudo-radial disturbances, *no such clear-cut distinction exists when allowance is made for finite-amplitude oscillations.* That is to say, in the range $\gamma_c \leqslant \gamma \leqslant 4/3$, stability depends on the values of γ and τ, as well as on the amount of energy available to sustain the pseudo-radial motions. This is, in essence, the result summarized in Figures 14.3–14.7. It should be borne in mind, however, that the illustrative examples depicted in Figures 14.4–14.7 correspond to an adiabatic exponent γ and a rotation rate τ such that the critical amplitude of the oscillations is equal to 20 percent of the mean radius of the slowly rotating configuration (i.e., $w_{max} = 1.2$). Had we taken γ slightly larger than γ_c, for a given τ, w_{max} would then have been slightly larger than unity only (see Figure 14.3), thereby allowing for the growth of initial disturbances with physically small amplitudes (and this in spite of the fact that a linear analysis predicts stable pseudo-radial motions in the slowly rotating body!). In other words, although equation (2) provides a stability limit with respect to infinitesimal disturbances, this limit cannot be used in the nonlinear regime, *even allowing for initial motions with small- but finite-amplitudes only.*

THE POST-NEWTONIAN CASE

The foregoing discussion clearly exhibits the distinction between stability with respect to infinitesimal disturbances (i.e., motions that are small in the *mathematical* sense) and stability with respect to small- but finite-amplitude disturbances (i.e., motions that are small in the *physical* sense). Now, as can readily be seen from equation (4), the conflicting effects of rotation and general relativity become of paramount importance in configurations for which the adiabatic exponent γ is very closely equal to

4/3. Accordingly, because allowance for finite-amplitude motions reduces the stabilizing effects of rotation, we must evaluate the changes brought about by these motions on the stability limit (4). Again, we shall use the scalar virial theorem and equations (8)–(12) to obtain an approximate description of the pseudo-radial oscillations of a slowly rotating star within the framework of general relativity. For the sake of simplicity, however, we shall restrict our discussion to the case of slowly rotating, homogeneous bodies for which the Schwarzschild radius R_s is a small fraction of the mean radius R only; that is to say, we shall consider

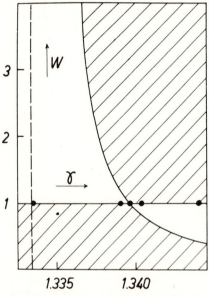

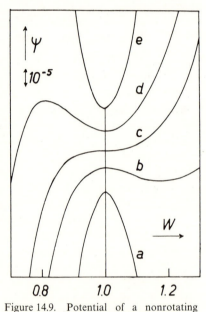

Figure 14.8. Zeros of the acceleration in the (w,γ)-plane for a nonrotating sphere in the post-Newtonian approximation ($R_s/R = 0.007$, $\gamma_c = 1.33967$). Acceleration is negative in shaded areas, and positive in unshaded areas. The values of γ used in Figure 14.9 are depicted by small circles on the line $w = 1$. The (dashed) line $\gamma = 4/3$ is an asymptote for the vertical branch of the curve. Source: Dedic, H., and Tassoul, J. L., Ap. J. **188**, 173, 1974. (By permission of The University of Chicago Press. Copyright 1974 by the American Astronomical Society.)

Figure 14.9. Potential of a nonrotating sphere in the post-Newtonian approximation ($R_s/R = 0.007$, $\gamma_c = 1.33967$). Origin of vertical scale is arbitrary. Curves correspond to the five particular values of γ located on Figure 14.8. [(a), $\gamma = 1.33350$; (b), $\gamma = 1.33905$; (c), $\gamma = \gamma_c$; (d), $\gamma = 1.34035$; (e), $\gamma = 1.34400$]. Curves a and e are similar to b and d; their secondary extremum occurs far outside the frame of the figure. However, in view of the post-Newtonian approximation, the description becomes invalid whenever w is substantially smaller than 1. Source: Dedic, H., and Tassoul, J. L., Ap. J. **188**, 173, 1974. (By permission of The University of Chicago Press. Copyright 1974 by the American Astronomical Society.)

structures for which we may neglect all terms of order $\Omega^2 R^2/c^2$, while retaining quantities of order R_s/R only.

Figures 14.8 and 14.9 illustrate the behavior of a nonrotating configuration. In contrast to a spherical body in the Newtonian limit, we observe that the zeros of the acceleration d^2w/dt^2 are now given by the line $w \approx 1$, and a curve that admits the vertical line $\gamma = 4/3$ as an asymptote (compare with Figure 14.2); their intersection corresponds to the critical value γ_c defined in equation (3). It is immediately apparent that all motions are definitively unstable when $\gamma \leqslant \gamma_c$; however, in the range $\gamma > \gamma_c$, instability may also occur when the amplitudes of the initial disturbances become too large. This is again a case of metastability.

The conflicting roles of rotation and general relativity are depicted in Figures 14.10–14.15. When the latter dominate, it is apparent that $\gamma_c > 4/3$; and, as in the nonrotating case, γ_c then separates unstable from metastable

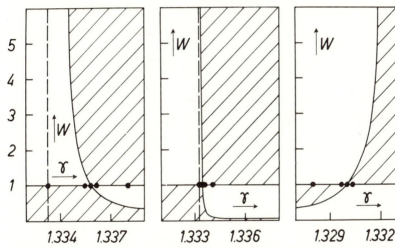

1.334   1.337   1.333   1.336   1.329   1.332

Figure 14.10. Zeros of the acceleration in the (w,γ)-plane for a slowly rotating spheroid ($\tau = 0.0057185$, $R_s/R = 0.007$, $\gamma_c = 1.33584$). Conventions and comments as in Figure 14.8; see also Figure 14.13. Source: Dedic, H., and Tassoul, J. L., *Ap. J.* **188**, 173, 1974. (By permission of The University of Chicago Press. Copyright 1974 by the American Astronomical Society.)

Figure 14.11. Zeros of the acceleration in the (w,γ)-plane for a slowly rotating spheroid ($\tau = 0.0008$, $R_s/R = 0.0007$, $\gamma_c = 1.33343$). Conventions and comments as in Figure 14.8; see also Figure 14.14. Source: Dedic, H., and Tassoul, J. L., *Ap. J.* **188**, 173, 1974. (By permission of The University of Chicago Press. Copyright 1974 by the American Astronomical Society.)

Figure 14.12. Zeros of the acceleration in the (w,γ)-plane for a slowly rotating spheroid ($\tau = 0.0057185$, $R_s/R = 0.0007$, $\gamma_c = 1.33007$). Conventions and comments as in Figure 14.8; see also Figure 14.15. Source: Dedic, H., and Tassoul, J. L., *Ap. J.* **188**, 173, 1974. (By permission of The University of Chicago Press. Copyright 1974 by the American Astronomical Society.)

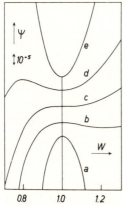

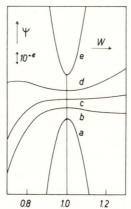

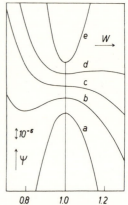

08   1.0   1.2   08   1.0   1.2   08   1.0   1.2

Figure 14.13. Potential of a slowly rotating spheroid (τ = 0.0057185, R_s/R = 0.007, γ_c = 1.33584). Conventions as in Figure 14.9; see Figure 14.10 [(a), γ = 4/3; (b), γ = 1.33559; (c), γ = γ_c; (d), γ = 1.33620; (e), γ = 1.33800]. Curve a has an absolute maximum at w = 1. Curve e is similar to d; its secondary extremum occurs far outside the frame of the figure. Source: Dedic, H., and Tassoul, J. L., *Ap. J.* **188**, 173, 1974. (By permission of The University of Chicago Press. Copyright 1974 by the American Astronomical Society.)

Figure 14.14. Potential of a slowly rotating spheroid (τ = 0.0008, R_s/R = 0.0007, γ_c = 1.33343). Conventions as in Figure 14.9; see Figure 14.11 [(a), γ = 4/3; (b), γ = 1.33342; (c), γ = γ_c; (d), γ = 1.33345; (e), γ = 1.33400]. Comments as in Figure 14.13. Source: Dedic, H., and Tassoul, J. L., *Ap. J.* **188**, 173, 1974. (By permission of The University of Chicago Press. Copyright 1974 by the American Astronomical Society.)

Figure 14.15. Potential of a slowly rotating spheroid (τ = 0.0057185, R_s/R = 0.0007, γ_c = 1.33007). Conventions as in Figure 14.9; see Figure 14.12 [(a), γ = 1.32800; (b), γ = 1.32970; (c), γ = γ_c; (d), γ = 1.33038; (e), γ = 4/3]. Curve a is similar to curve b; its secondary extremum occurs far outside the frame of the figure. Curve e has an absolute minimum at w = 1. Source: Dedic, H., and Tassoul, J. L., *Ap. J.* **188**, 173, 1974. (By permission of The University of Chicago Press. Copyright 1974 by the American Astronomical Society.)

motions. However, if we allow for a sufficiently large increase in the rotation rate, γ_c then becomes smaller than 4/3 (see Figures 14.12 and 14.15); and it is this change that substantially modifies the whole picture provided by a linear stability analysis! As in the Newtonian case, we must now distinguish between three cases:

(i) $\gamma \leqslant \gamma_c$. The configuration is always *unstable* with respect to pseudo-radial motions;

(ii) $\gamma_c < \gamma < 4/3$. The slowly rotating body becomes *metastable*, in spite of the fact that a linear stability analysis predicts stable pseudo-radial motions only;

(iii) $\gamma \geqslant 4/3$. The configuration is completely *stable*.

Thus, although the conflicting effects of rotation and general relativity are already apparent in a linear stability analysis, allowance for finite-amplitude motions modifies the entire situation. In particular, it considerably reduces the stabilizing role of a slow rotational motion. Unconditional stability with respect to finite-amplitude, pseudo-radial oscillations now requires *two* stringent conditions: $\gamma \geqslant 4/3$ and $\gamma_c < 4/3$. For polytropic models, the latter condition may be rewritten in the form

$$\frac{K}{|W|} > 10^{-6} A_n \frac{M/M_\odot}{R/R_\odot}, \tag{34}$$

where A_n depends on the polytropic index ($A_0 = 2.88$, $A_3 = 7.15$). Accordingly, even if $\gamma \geqslant 4/3$, great care should be exercised if condition (34) is not met, since the actual behavior of a metastable star primarily depends on the amount of energy available to sustain the pseudo-radial disturbances. A detailed study of this problem in the exact framework of general relativity would be most interesting.

14.3. THE BEAT PHENOMENON IN THE β CEPHEIDS

The β Cephei (β Canis Majoris) variables are hot stars of mean spectral types between B0.5 and B2 that are confined to a narrow strip lying somewhat above and to the right of the upper main sequence. (Note, however, that they share this region with nonvariable B-type stars!) The β cepheids are the hottest variable stars known, with effective temperatures T_{eff} in the range 20,000–25,000°K, and periods of only a few hours. The theoretical interpretation of these stars is still in a very unsatisfactory state, particularly with respect to the mechanism responsible for their variability. As a matter of fact, even the precise nature of the motions involved is not yet known with certainty. (The radial-velocity curves of some β Cephei variables are quite complex, and do not always admit of a simple interpretation in terms of radial pulsations.) Moreover, excitation of pulsations in these stars by the usual envelope ionization mechanisms, which apparently account for the instability of the cooler variables, can be ruled out by their high surface temperatures. (In this respect, let us also mention the phase relation between the light- and radial-velocity curves which, according to the simplest interpretation, implies that the phase lag of maximum luminosity behind minimum radius is approximately *zero* for the β cepheids.) The outstanding properties of several of the β Cephei variables are the beat phenomenon (somewhat as in the case of the RR Lyrae and δ Scuti variables) and the periodic line widening and/or line splitting. These observational facts have prompted many

authors to suggest that nonradial pulsations, possibly coupled with rotation and/or tidal interaction with a companion, must be involved in the
β cepheids. We shall now briefly report on two of these mechanisms.

THE LEDOUX-OSAKI APPROACH.   As we recall from Section 6.4, the free
oscillations of a nonrotating star correspond to the surface spherical
harmonics $Y_l^m(\theta,\varphi)$, with $|m| \leq l$, the characteristic frequencies σ_l being
then $(2l + 1)$-fold degenerate. In the presence of a slow rotational motion,
however, the degeneracy of the f-, p-, and g-modes is lifted and, to first
order in the angular velocity Ω, the frequencies belonging to given values
of l and m become

$$\sigma_{l,m} = \sigma_l \pm m\beta_l\Omega \quad (m = 0, 1, 2, \ldots, l) \tag{35}$$

where σ_l is the frequency of the corresponding star in hydrostatic equilibrium, and β_l are certain calculable constants that depend on the model.
In particular, for the f-modes belonging to the second-order harmonics
$l = 2$, we are left with only five frequencies. The zonal mode $(m = 0)$
corresponds to standing oscillations that are symmetrical with respect
to the axis of rotation; modes with positive m describe waves traveling
in the opposite direction as the rotation, and those with negative m in
the same direction as the rotation. The sectorial modes $(|m| = 2)$ are
symmetrical with respect to the equatorial plane, whereas the tesseral
modes $(|m| = 1)$ are antisymmetrical with respect to this plane.

Ledoux's original theory postulates that the beat phenomenon of the
β cepheids is the result of a coupling in the presence of rotation of two
different nonradial oscillations, both having the form of second-order
spherical harmonics. Since the frequencies of two such modes would be
equal in the absence of rotation, Ledoux thus assumes that the pulsating
star has a small angular velocity, which may be determined by the observed
beat frequency. On the basis of this conjecture, Osaki subsequently computed the variable line profiles and radial-velocity curves of various
models, assuming the presence of a wave traveling in the same direction
as the rotation of the models. His calculations confirm Christy's finding
that such a wave may indeed explain the variable line widths and discontinuous radial-velocity curves in some of the β cepheids. By identifying
the two oscillations with nearly equal periods as a wave traveling in the
same direction as the rotation and a standing oscillation, respectively,
Osaki has further estimated the equatorial rotational velocities from the
beat periods and compared these with the rotational velocities obtained
from the observed line widths. Since the agreement between theory and
observation is fairly good, Osaki thus concludes that nonradial oscillations
might well explain most of the properties of the β Cephei variables.

Now, what excitation mechanism is responsible for preferentially se-
lecting from all the modes a wave traveling around the equator in the
same direction as the rotation? According to Osaki, such a mechanism
could perhaps involve overstable g^--modes in the (assumed) rapidly
rotating cores of massive, slightly evolved, core hydrogen-burning stars
(cf. §14.5). Indeed, by suitably adjusting the rotation rate of the convective
core of a 12$M_\odot$ star, Osaki finds that it is possible to achieve resonance
of large-scale convective motions with some of the lowest g^+-modes in
the radiative envelope of the star. The most likely mode to be excited by
this mechanism seems to be the sectorial mode, which is a wave traveling
around the equator in the same direction as the rotation; and, as we have
seen, this seems to be in reasonable agreement with the observations.
However, given the many uncertainties entailed in this approach, further
studies are needed to decide whether this is indeed the relevant excitation
mechanism for the β Cephei variables, and whether the Ledoux-Osaki
mechanism is indeed the correct explanation of these stars.

THE CHANDRASEKHAR-LEBOVITZ APPROACH.   As we have shown in Section 6.7, rotation couples two modes of oscillation that are, in the absence
of rotation, purely radial and purely nonradial. In the limit of hydrostatic
equilibrium, the corresponding characteristic frequencies are, to first
approximation,

$$\sigma_R{}^2 = (3\gamma - 4)\frac{|W|}{I}, \tag{36}$$

and

$$\sigma_Z{}^2 = \frac{4}{5}\frac{|W|}{I} \tag{37}$$

(cf. §6.4: equations [41] and [37]). Therefore, when $\gamma = 1.6$, these two
frequencies coincide and we have a case of degeneracy. Accordingly, one
might expect to observe beats if (i) both kind of modes are excited and
(ii) γ is near 1.6. (More refined calculations by Hurley, Roberts, and
Wright indicate that this accidental degeneracy always occurs at about
$\gamma = 1.6$ in centrally condensed bodies, so long as $\rho_c/\bar{\rho} \lesssim 50$, where ρ_c is
the central density and $\bar{\rho}$ the mean density.) Given these results, we can
now picture the Chandrasekhar-Lebovitz mechanism as follows. It is
first supposed that the physical conditions prevalent in the β cepheids
(in particular, the ratio of specific heats) are such that a degeneracy occurs
between the frequencies σ_R and σ_Z. According to the theory, rotation
then removes this degeneracy and gives rise to two nonradial modes
characterized by slightly different frequencies, the difference being proportional to the squared angular velocity Ω^2.

This hypothesis as applied to the β Cephei variables has the advantage that the lowest axisymmetric f-mode is *naturally* coupled by rotation to the lowest radial mode to produce two distinct normal modes, both of which are nonradial. Further support for the Chandrasekhar-Lebovitz mechanism derives from the extended numerical computations of Harper and Rose, as well as from those of Deupree. In brief, the Harper-Rose calculations indicate that when spherical models of $10M_\odot$ have evolved to about 1 magnitude above the main sequence (i) their pulsation frequencies σ_R closely agree with those of the β cepheids and (ii) the theoretical frequencies σ_R and σ_Z are then nearly in resonance. A similar agreement was found by Deupree between his realistic calculations and a number of the β cepheids showing the beat phenomenon. However, as we know from the work of Davey, no mechanism has yet been found that excites these two modes in a spherical model. In addition to this important difficulty, the Chandrasekhar-Lebovitz approach has also raised the following objection: on the assumption of *uniform* rotation, their explanation of the beat phenomenon in the β cepheids leads to angular velocities that are 3 or 4 times larger than those observed at the surfaces of these stars. According to Clement, however, the discrepancy that exists between the observations and the calculations based on the assumption of solid-body rotation may be removed by suitably choosing models in *non-uniform* rotation. Finally, as was shown by Osaki, the use of two axisymmetric modes also meets with difficulties in explaining the observed variability in line profiles. Until such time as we understand the actual mechanism that maintains the multiple periodicity in several of the β cepheids, the validity of this model will thus remain an open question.

14.4.   VIBRATIONAL STABILITY

Questions pertaining to the *dynamical* stability of rotating stars have already been discussed (cf. §6.1). As we know, for all such discussions it is sufficient to consider small isentropic disturbances, i.e., motions for which the entropy of each mass element is preserved along its path. Quite generally, the Lagrangian variations in pressure and density are then related by the condition

$$\frac{\Delta p}{p} = \Gamma_1 \frac{\Delta \rho}{\rho} = -\Gamma_1 \operatorname{div} \xi, \tag{38}$$

where Γ_1 is the first generalized adiabatic exponent (cf. §3.4). Although equation (38) does represent a reasonable assumption in most cases, there are situations for which departure from isentropy may play an essential role (e.g., upper main-sequence stars in which the radiation pressure dominates over the gaseous pressure). To the best of our knowledge, the question of the *vibrational* stability of rotating stars has not yet been considered, at least insofar as the *global* structure of these stars is concerned. (Departure from isentropy was considered in Section 7.4 for the problem of thermal imbalance in baroclines; however, as we recall, only *local* disturbances were taken into account.) Following Aizenman and Cox, we first present a formal solution of the problem of vibrational stability for stars in a state of permanent rotation (cf. §4.2). We conclude this section with some brief comments on the maximum mass above which main-sequence stars might become vibrationally unstable with respect to axisymmetric pulsations.

In Section 6.6, we presented a variational principle governing the dynamical stability of a barocline when subject to axisymmetric motions satisfying equation (38). When departure from isentropy is taken into account, this equation no longer applies; and we must then deal with the full energy equation (cf. §3.4: equation [63]). Neglecting viscous dissipation and heat transfer, we shall first rewrite this conservation principle in the convenient form

$$\frac{Dp}{Dt} - \Gamma_1 \frac{p}{\rho} \frac{D\rho}{Dt} = (\Gamma_3 - 1)\rho\left(\epsilon_{Nuc} - \frac{1}{\rho} \operatorname{div} \mathscr{F}\right), \tag{39}$$

in which we write $\mathscr{F}$ for $\mathscr{F}_r$ without confusion. Hence, because the operations of Δ and total derivation do commute in the linear approximation, it follows that

$$\frac{D}{Dt}\Delta p - \frac{\Gamma_1 p}{\rho}\frac{D}{Dt}\Delta\rho = (\Gamma_3 - 1)\rho\,\Delta\left(\epsilon_{Nuc} - \frac{1}{\rho}\operatorname{div}\mathscr{F}\right) \tag{40}$$

(cf. §6.2: equation [12]). This equation is generally valid if we assume that the unperturbed state is everywhere in thermal equilibrium; we thus ignore altogether the possible influence of meridional currents (cf. §8.1). Now, by using cylindrical coordinates (ϖ,φ,z), we have

$$\frac{D}{Dt} = \frac{\partial}{\partial t} + v.\operatorname{grad} = \frac{\partial}{\partial t} + \Omega\,\frac{\partial}{\partial\varphi}, \tag{41}$$

whenever the operator D/Dt is applied to a *scalar* quantity. For axisymmetric disturbances, equation (40) therefore reduces to

$$\frac{\partial}{\partial t}\Delta p - \Gamma_1\frac{p}{\rho}\frac{\partial}{\partial t}\Delta\rho = (\Gamma_3 - 1)\rho\,\Delta\left(\epsilon_{Nuc} - \frac{1}{\rho}\operatorname{div}\mathscr{F}\right), \tag{42}$$

which now replaces equation (38).

Following the method used in Section 6.6, we now redefine the Lagrangian displacement $\xi(\varpi,z,t)$ to be a two-dimensional vector with components

ξ_{ϖ} and ξ_z; next combining equations (74)–(77) of Section 6.6 with equation (42) above, we eventually obtain

$$\rho \frac{\partial^3 \xi}{\partial t^3} = \mathbf{L} \frac{\partial \xi}{\partial t} - (\Gamma_3 - 1) \, \mathrm{grad} \left[\rho \, \Delta \left(\epsilon_{Nuc} - \frac{1}{\rho} \, \mathrm{div} \, \mathscr{F} \right) \right], \qquad (43)$$

where the linear operator $\mathbf{L}$ has the *same* meaning as in Section 6.6. As usual, we further assume a dependence of the two-dimensional vector ξ on time of the form

$$\xi(\varpi, z, t) = \xi(\varpi, z) e^{i\sigma t}. \qquad (44)$$

By multiplying equation (43) by ξ and integrating over the total volume $\mathscr{V}$ of the rotating star, we can thus write

$$\sigma^3 \int_{\mathscr{V}} \rho \xi \cdot \xi \, \mathbf{dx} + \sigma \int_{\mathscr{V}} \xi \cdot \mathbf{L} \xi \, \mathbf{dx}$$

$$+ i \int_{\mathscr{V}} (\Gamma_3 - 1) \xi \cdot \mathrm{grad} \left[\rho \, \Delta \left(\epsilon_{Nuc} - \frac{1}{\rho} \, \mathrm{div} \, \mathscr{F} \right) \right] \mathbf{dx} = 0. \quad (45)$$

If we perform an integration by parts of the third integral and make use next of the conditions $\Delta p = \Delta \rho = 0$ on the outer surface $\mathscr{S}$, equations (38) and (45) may be combined to give the cubic equation

$$\sigma^3 - \sigma_a^2 \sigma - 2i\sigma_a^2 \kappa = 0, \qquad (46)$$

where

$$\sigma_a^2 = -\int_{\mathscr{V}} \xi \cdot \mathbf{L} \xi \, \mathbf{dx} \bigg/ \int_{\mathscr{V}} \rho \xi \cdot \xi \, \mathbf{dx}, \qquad (47)$$

and

$$\kappa = -\int_{\mathscr{V}} (\Gamma_3 - 1) \, \Delta \rho \, \Delta \left(\epsilon_{Nuc} - \frac{1}{\rho} \, \mathrm{div} \, \mathscr{F} \right) \mathbf{dx} \bigg/ 2\sigma_a^2 \int_{\mathscr{V}} \rho \xi \cdot \xi \, \mathbf{dx}. \quad (48)$$

As we expected, if we ignore possible departures from isentropy (i.e., $\kappa = 0$), equation (46) readily leads to the Fjørtoft-Lebovitz necessary condition for dynamical stability, i.e., $\sigma_a^2 > 0$ with the present notation (cf. §§6.6 and 6.8).

In principle, for equation (46) to give the *exact* eigenvalues σ of the problem, the coefficients σ_a and κ must be evaluated by inserting the *exact* eigenfunctions $\xi(\varpi, z)$. Since these (*complex*) quantities are simply not known, we shall thus restrict our consideration to systems that only slightly depart from isentropy, i.e., rotating stars for which the condition $|\kappa/\sigma| \ll 1$ holds true. Under this assumption, the independent term in equation (46) represents a small contribution; that is to say, we may now obtain the frequencies σ by merely inserting the (*real*) eigenfunctions $\xi_a(\varpi, z)$ of the corresponding isentropic problem into the coefficient κ, the coefficient $\sigma_a{}^2$ being then equal to the (*real*) squared frequency associated with the function $\xi_a(\varpi,z)$. On the basis of the foregoing approximation, two solutions of equation (46) can be written in the form

$$\sigma = \pm\sigma_a + i\kappa, \tag{49}$$

the third (real) root presenting no interest for the pulsation problem. Under the same assumption, the two-dimensional vector $\xi(\varpi,z,t)$ becomes

$$\xi(\varpi,z,t) = \xi_a(\varpi,z)e^{\pm i\sigma_a t}e^{-\kappa t}. \tag{50}$$

Given these *approximate* results, it is apparent that small-amplitude, quasi-isentropic pulsations are normally oscillatory, with periods equal to $2\pi/\sigma_a$, but their amplitudes either grow or slowly decay in time. To be specific, when $\kappa \geqslant 0$, the oscillations are stable; they are overstable when $\kappa < 0$. For this reason, the quantity κ is usually called the coefficient of vibrational stability.

The solution given by equations (49) and (50) is surprisingly simple, and it bears much resemblance to the formal solution derived by Eddington in 1919 for the case of spherical stars. In the present instances, however, equation (48) must be evaluated by integrating over the total volume of the rotating star and by using the two-dimensional displacement $\xi_a(\varpi,z)$ or, if not available, approximate trial functions in place of this function (cf. §6.6). The physical meaning of the coefficient κ can now be seen as follows. For the usual thermonuclear reactions, we have $\epsilon_{Nuc} = \epsilon_0\rho T^\nu$; neglecting any possible phase delays due to abundance variations, we can thus relate the Lagrangian variation $\Delta\epsilon_{Nuc}$ to the vector ξ_a by the relation

$$\frac{\Delta\epsilon_{Nuc}}{\epsilon_{Nuc}} = \frac{\Delta\rho}{\rho} + \nu\frac{\Delta T}{T} = [1 + \nu(\Gamma_3 - 1)]\frac{\Delta\rho}{\rho} \tag{51}$$

$$= -[1 + \nu(\Gamma_3 - 1)] \operatorname{div} \xi_a.$$

Since $\Delta\epsilon_{Nuc}$ is always of the same sign as $\Delta\rho$, it tends to render κ negative, thus contributing to the vibrational instability of the star. Obviously, this *nuclear driving* mechanism becomes particularly effective whenever the central pulsation amplitude is relatively large. (This may be caused by $\langle\Gamma_1\rangle$ being lowered to a value near 4/3 by the action of radiation pressure.) In contrast, the part of the stability integral (48) that arises from energy transfer may either stabilize or destabilize the pulsations of a rotating star. In the latter case, the destabilizing effect results from a modulation of the flux via "valve-type" mechanisms associated with ionization zones of abundant elements (such as hydrogen and helium) in the outer layers of the star. These *envelope ionization* mechanisms have been mainly invoked to explain the variability of the common types of pulsating stars. Because these stars are thought to be unusually slow rotators, the "valve-type" mechanisms are thus outside the scope of this book.

Detailed calculations of the vibrational stability of nonrotating main-sequence stars with normal Population I compositions were first carried out by Ledoux in 1941. As we know, the dominance of radiation pressure over gaseous pressure in upper main-sequence stars results in an upper mass limit to pulsationally stable, chemically homogeneous stars. This theoretical mass limit for spherical main-sequence stars derives from the fact that, as $\langle \Gamma_1 \rangle$ approaches 4/3, the lowest radial mode becomes nearly homologous (i.e., $\xi_r \propto r$ in spherical coordinates); hence, the pulsation amplitude of this mode has a finite value, thus allowing the nuclear-driven mechanism in the core to overcome the damping effects of energy transfer in the envelope. Other linear stability analysis based on more realistic models lead to an upper mass limit in the range $60–120 M_\odot$ for likely Population I compositions. Until recently, the presumption has been that vibrational instability would result in a build-up of the pulsation amplitudes to such large values that mass loss would necessarily ensue, and that the mass of a pulsationally unstable main-sequence star would eventually be reduced to a value below the upper mass limit. Unfortunately, detailed nonlinear calculations have *not* confirmed this early conjecture. That is to say, in considering finite-amplitude motions, many authors have found that the radial pulsations of spherical stars with $M \simeq 100–300 M_\odot$ reach a limiting amplitude and have a much less disruptive effect on the stars than the linear stability analysis would predict. In other words, spherical models in the mass range $100–300 M_\odot$ may well be able to survive the nuclear-driven pulsational instability without any appreciable mass loss. Hence, the Ledoux mechanism may therefore not be effective in limiting the masses of upper main-sequence stars. Moreover, because these stars might store a large amount of angular momentum below their surfaces, it also remains to investigate whether rotation may raise the upper mass limit well above the limit derived for nonrotating stellar models.

In connection with these problems, let us mention that Larson and Starrfield have presented arguments, based on theories of star formation with negligible angular momentum, to the effect that stars having masses in excess of $60–120 M_\odot$ may not be able to condense out of the present interstellar medium. Several factors may operate to set the above limit, whose precise value depends on the initial conditions. *First*, the time required for a star to form by accretion at the center of a collapsing protostellar cloud becomes larger than the main-sequence evolutionary lifetime for $M \gtrsim 60–120 M_\odot$, depending on the conditions that prevail in the cloud. *Second*, radiation pressure acting on the outer parts of a proto-stellar cloud may become sufficient to halt the collapse at a mass of the order of $50-100M_\odot$. *Third*, the strongest limit on the mass of a main-sequence star is provided by the formation of an HII region in the collapsing protostellar envelope surrounding a forming star; this occurs at a mass of the order of $25-60M_\odot$, depending on the initial conditions. For stars containing little or no heavy elements, however, much larger masses may be possible. According to Conti and Burnichon, given the many uncertainties in the parameters involved, these arguments do not conflict with their recent finding of stars that are in the vicinity of the zero-age main sequence and that have masses in the range $60-120M_\odot$.

As far as the effects of rotation on the Ledoux critical mass are concerned, very little is known with any certainty. Preliminary calculations by Stothers seem to indicate that the theoretical limiting mass for rapidly rotating, main-sequence stars might be much larger than in the case of spherical stars. (Note, however, that this result does not stem from an explicit evaluation of the stability coefficient κ!) This conclusion, if correct, lends support to the Larson-Starrfield arguments. As was pointed out by Stothers, their suggested mechanisms might be aided, perhaps crucially, by fast rotation. In particular, approximate conservation of angular momentum during the collapse of the original gas cloud could further prevent the formation of very massive individual stars if the fission mechanism is indeed operative (cf. §11.3). Not enough is known, however, about these early stages of stellar evolution to predict a meaningful upper mass limit for pre-main-sequence stars.

14.5. CONVECTIVE MOTIONS

In spite of many independent contributions, the interaction between rotation and convection continues to be one of the most frustrating problems of the theory of rotating stars. Actually, two distinct questions need to be considered. One is whether the condition for the onset of convection is altered by the presence of rotation; the second is whether convection, once established in a rotating star, is seriously different in its general properties from that in a nonrotating star. We shall mainly comment on the first question, stability with respect to small isentropic displacements being here studied.

As we recall from Section 6.4, convection occurs in a spherically symmetric star whenever the square of the local Brunt-Väisälä frequency

$$N^2 = \frac{1}{\rho} \operatorname{grad} p \cdot \left(\frac{1}{\rho} \operatorname{grad} \rho - \frac{1}{\gamma p} \operatorname{grad} p \right) \qquad (52)$$

becomes negative. On the basis of simple considerations, it is tempting to believe that rotation would generally tend to inhibit convection. To illustrate this view, let us consider a uniformly rotating star. Then, the angular momentum per unit mass $\Omega\varpi^2$ regularly increases outward; and, for axisymmetric motions only, $\Omega\varpi^2$ is constant for any fluid parcel during its displacement. An element of mass carried outward from the axis of rotation by these motions will thus lack angular momentum relative to its surroundings and will tend to slow down; similarly, matter carried inward will have an excess of angular momentum and will likewise tend to slow down (cf. §7.3). For example, Randers obtained the following *local* stability criterion for cells extending radially from the center of the star:

$$N^2 + 4\Omega^2 \sin^2 \theta > 0, \tag{53}$$

where θ is the colatitude.[4] As was pointed out by Cowling, however, if non-axisymmetric motions are considered, azimuthal pressure gradients insure that $\Omega\varpi^2$ does not remain constant; accordingly, the stabilizing effects of rotation on convection might be lost for these displacements. That this is indeed the case was convincingly shown by Cowling by means of a linear stability analysis, with local disturbances of the form

$$\exp[i(k_\varpi\varpi + k_z z) + im\varphi + i\sigma t]. \tag{54}$$

Under these assumptions, the condition for convective stability in a uniformly rotating star becomes

$$N^2\left[\left(\frac{m}{\varpi}\right)^2 + (k_\varpi \cos\alpha - k_z \sin\alpha)^2\right] + 4\Omega^2 k_z^2 > 0, \tag{55}$$

where α is the angle (nearly equal to our earlier θ) that the effective gravity **g** makes with the polar axis. This *local* stability condition embodies the main effects of rotation on the onset of convection when both axisymmetric and non-axisymmetric displacements are considered.

In the absence of rotation, inequality (55) reduces to $N^2 > 0$, and gives the usual condition that the actual temperature gradient does not exceed the adiabatic lapse rate in a star in radiative equilibrium. Also, if displacements symmetric about the axis of rotation alone are considered, we must set $m = 0$ in equation (55). In particular, if $m = 0$ and $k_\varpi \sin\theta + k_z \cos\theta = 0$ (corresponding to radial displacements), the Cowling criterion reduces to

$$N^2 \cos^2(\alpha - \theta) + 4\Omega^2 \sin^2 \theta > 0, \tag{56}$$

which is essentially equivalent to Randers's criterion (53), since $\alpha - \theta$ is the small angle between **g** and the radial direction. Similarly, if we let $m = 0$

[4] In the equatorial plane (i.e., $\theta = \pi/2$), the stabilizing effect of rotation on axisymmetric convective motions can readily be seen from equations (41) and (42) of Section 7.3.

and $k_\varpi = 0$, inequality (55) reduces to Walén's stability criterion

$$N^2 \sin^2 \alpha + 4\Omega^2 > 0, \tag{57}$$

for displacements perpendicular to the axis of rotation. In general, equation (55) indicates that N^2 must be comparable with $4\Omega^2$, so that the onset of convection is put off until the temperature gradient exceeds the adiabatic lapse rate by an amount that, though small, is not negligible; this shows the stabilizing influence of uniform rotation for a large class of displacements. For certain disturbances, however, the Cowling criterion (55) indicates that the adiabatic lapse rate needs be exceeded by only a very small amount to insure the existence of convective motions; these are the disturbances in which k_z is small compared with one or both of k_ϖ and m/ϖ. Accordingly, *convective motions must occur in a uniformly rotating star whenever the actual temperature gradient exceeds the adiabatic lapse rate*, but this is limited to certain types of displacements unless the adiabatic lapse rate is markedly exceeded. In other words, the Schwarzschild criterion remains unaltered by the presence of uniform rotation, the effects of rotation being simply to stabilize *some* of the disturbances that would be unstable in the absence of rotation.

The above conclusions were originally obtained by Cowling on the basis of a *local* stability analysis of a uniformly rotating star; thus the convected elements were assumed to have small dimensions compared with the scales of variation of the macroscopic variables (such as pressure and density), and any fluctuations in gravity due to the disturbances were ignored altogether. Further attempts to study the influence of rotation on the onset of convection in a star have focused on simple configurations that allow a *global* stability analysis of their p-, f-, and g-modes. Three such systems will now be considered; all of them confirm the rough but convincing analysis of Cowling.

The simplest of these models is the semi-infinite, isothermal atmosphere that, in the absence of rotation, becomes convectively unstable when $N^2 < 0$, i.e., $\gamma < 1$. (The physical contradiction implied by this inequality does not impair the qualitative conclusions drawn from using it.) When rotation is present, there are two sets of modes for the isothermal atmosphere: the (high-frequency) p-modes and the (low-frequency) g-modes. As was shown by Lebovitz, the characteristic frequencies associated to these isentropic motions are:

$$\sigma^2 = \tfrac{1}{2}(4\Omega^2 + k^2 + k_z^2 + \tfrac{1}{4}) \pm \tfrac{1}{2}\left\{(4\Omega^2 + k^2 + k_z^2 + \tfrac{1}{4})^2 \right.$$

$$\left. - 4\left[4\Omega^2(k_z^2 + \tfrac{1}{4}) + \frac{\gamma - 1}{\gamma^2}k^2\right]\right\}^{1/2}, \tag{58}$$

where k_z and k are, respectively, the "vertical" and "horizontal" wavenumbers. (All variables are made dimensionless by choosing the scale height H as the unit of length and $[H/\gamma g]^{1/2}$ as the unit of time.) Stable modes are those for which $\sigma^2 \geqslant 0$. Hence, for the convective modes, we clearly find instability if

$$4\Omega^2(k_z^2 + \tfrac{1}{4}) + \frac{\gamma - 1}{\gamma^2} k^2 < 0. \tag{59}$$

A necessary condition for convective instability is, then, that $\gamma < 1$. Further, this condition is also sufficient, as one sees by taking k "large enough" and k_z "small enough." This agrees with Cowling's conclusions; that is to say, whenever $N^2 < 0$, there exist displacements, of sufficiently large horizontal wavenumber, that are of an unstable type.

A further confirmation of Cowling's analysis is provided by a global study of the nonradial oscillations of uniformly rotating cylinders, of finite radius and infinite length, with pressure and density related by the polytropic equation $p = K\rho^{1 + 1/n}$. (In the absence of rotation, these configurations become convectively unstable when $\gamma < 1 + 1/n$.) For these purely nonradial motions, the Lagrangian displacement becomes, therefore,

$$\xi(\varpi,\varphi,t) = \xi(\varpi)e^{i(m\varphi + \sigma t)}. \tag{60}$$

The limiting case $n = 0$ (i.e., the homogeneous cylinder) is particularly instructive, since it admits of *exact* solutions that are expressible in terms of Jacobi polynomials; furthermore, if we choose $(2\pi G\rho)^{-1/2}$ as the unit of time, it can be shown that the p- and g-modes are solutions of the quartic equation

$$\sigma^4 - 2\Delta_v\sigma^2 - 4m\Omega(1 - \Omega^2)\sigma - m^2(1 - \Omega^2)^2 = 0, \tag{61}$$

where

$$\Delta_v = \gamma(1 - \Omega^2)(v + 1)(v + |m| + 1) + 2\Omega^2 - 1 \tag{62}$$

$(v = 0, 1, 2, \dots)$. As far as we know, this is the only consistent configuration for which the complete sets of acoustic and convective modes have been obtained.[5] Figure 14.16 illustrates the lowest g-modes as functions of the rotation rate Ω^2. (In order to comply with the condition of mechanical equilibrium, here we must have $0 \leqslant \Omega^2 \leqslant 1$.) We observe, *first*, that

---

[5] The uniformly rotating, homogeneous cylinder also admits of a set of (stable) f-modes; these are $\sigma = \pm\Omega \pm (|m| - 1)^{1/2}(1 - \Omega^2)^{1/2}$, where $|m| \geqslant 2$ and $\Omega^2 \leqslant 1$. In the absence of rotation, these f-modes and the p- and g-modes defined in equation (61) are very similar to those obtained by Pekeris for the homogeneous sphere (cf. §6.4: equations [35], [38], and [42]). See: Tassoul, J. L., *Ann. Ap.* **26**, 444, 1963; Ostriker, J. P., *Ap. J.* **140**, 1529, 1964.

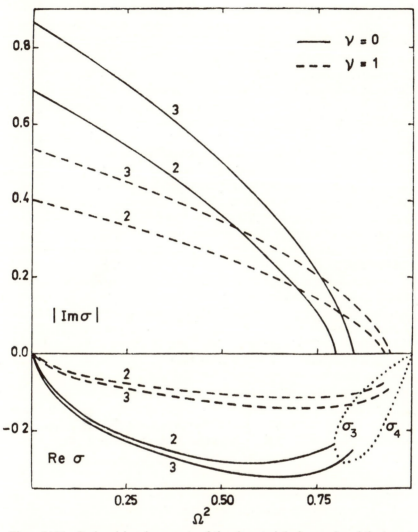

Figure 14.16. Real and imaginary parts of the characteristic frequencies of the lowest *g*-modes, as functions of Ω^2, for homogeneous columns in uniform rotation. The curves are labeled by the value of *m* to which they refer ($\gamma = 5/3$). σ and Ω are given in units of $(2\pi G\rho)^{1/2}$. The real frequencies (σ_3 and σ_4) are represented for $m = 2$ and $v = 0$ only (*dotted curves*). Source: Cretin, M., et Tassoul, J. L., *Ann. Ap.* **28**, 982, 1965.

rotation lifts the twofold degeneracy that prevails in the limit of hydro-static equilibrium. *Second*, whereas all *g*-modes are dynamically unstable in this limit, convective motions set in as oscillations of increasing ampli-tude when rotation is present. *Third*, rotation reduces the growth rates of these overstable oscillations. *Fourth*, although rotation is able to stabi-lize in sequence some of the *g*-modes, vibrational instability is nevertheless possible for *g*-modes having sufficiently large values of $|m|$ and v. Similar conclusions were drawn by Robe for axially condensed cylinders of indices $n = 1$ and $n = 3$, and $\gamma = 5/3$. As was shown by this author, however, the eigenvalues of all *g*-modes always remain complex, whatever the rotation rate of the axially condensed cylinder may be. The complete stabilization of some of the *g*-modes (as depicted in Figure 14.16) might thus be an artifact of polytropic models with index equal (or nearly equal) to zero. In any event, both works clearly show that (overstable) convective motions always occur in uniformly rotating polytropic cylinders when $\gamma < 1 + 1/n$. This conclusion lends further support to Cowling's local stability analysis.

The complexity of our problem is perhaps best illustrated by the case of compressible Maclaurin spheroids. (Although these rapidly rotating systems are highly unrealistic, they nevertheless provide the only examples of *bounded* systems that were thoroughly examined from the viewpoint of convective instability.) Figures 14.17 and 14.18 depict the three con-vective modes that, in the limit of hydrostatic equilibrium, degenerate into the (triple) *g*-mode belonging to $v = 0$ and $l = 1$ (cf. §6.4: equation [42]). For the sake of clarity, the frequencies are plotted as functions of the flattening $f[= (1 - a_3/a_1)]$, where a_3/a_1 is the ratio of the semi-axes $(0 \leqslant f \leqslant 1)$. In the present case, the components of the Lagrangian dis-placement associated with these motions can be given in *exact* form; we have

$$\xi_i(\mathbf{x},t) = (L_{i,jk}x_jx_k + L_i)e^{i\sigma t}, \tag{63}$$

where a summation over repeated indices must be understood. The $L_{i,jk}$'s and L_i's are calculable constants; they depend on the mode, since rotation lifts the threefold degeneracy that one observes in the limit $f = 0$. As can be seen in Figure 14.17, the characteristic frequency belonging to the *axisymmetric g*-mode remains at first imaginary; further along the Mac-laurin sequence, however, it becomes real. (Note that the precise location of the exchange of stability strongly depends on the value of γ.) Hence, overstability never occurs for this mode; and it may even become com-pletely stable when departure from spherical symmetry is quite large. In contrast, the two *non-axisymmetric g*-modes (labeled g_1 and g_2 in Figure 14.18) become at first overstable, their growth rates being somewhat reduced by rotation. For slow rotation rates $(f \ll 1)$, these results are thus

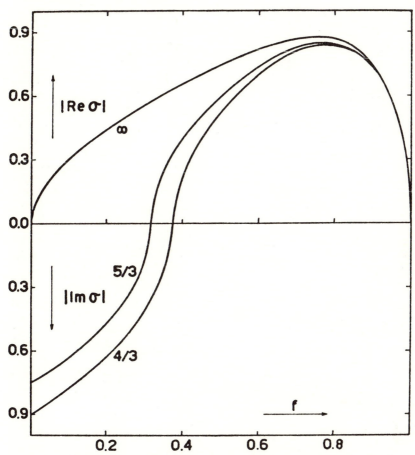

Figure 14.17.   Characteristic frequency belonging to the lowest axisymmetric g-mode, as a function of the flattening f, for the homogeneous but compressible Maclaurin spheroids. Curves are labeled by the value of γ to which they refer. σ is given in units of $(\pi G\rho)^{1/2}$. Source: Tassoul, M., and Tassoul, J. L., *Ap. J.* **150**, 1031, 1967. (By permission of The University of Chicago Press. Copyright 1967 by the University of Chicago.)

very similar to those obtained by Smeyers and Denis for higher modes. (A similar behavior was also observed by Clement, and by Durney and Skumanich for the lowest g-modes of slowly rotating, slightly distorted polytropes.) For larger rotation rates, however, the behavior of the modes g_1 and g_2 becomes rather complex, as indicated on Figure 14.18, which shows the existence of a coupling between these two modes and another mode which, in the limit of hydrostatic equilibrium, has the frequency $\sigma = 0$. (When $\gamma \lesssim 1.5$, a fourth mode comes into play; that is, an f-mode belonging to the spherical harmonics $l = 3$ in the limit $f = 0$!) Hence,

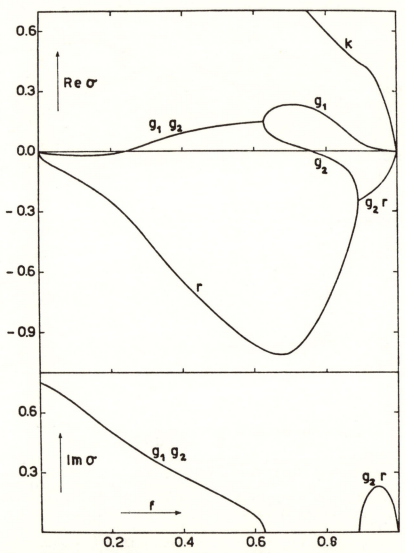

Figure 14.18. Characteristic frequencies belonging to the lowest non-axisymmetric g-modes, as functions of the flattening f, for the homogeneous but compressible Maclaurin spheroids ($\gamma = 5/3$). The imaginary part $Im\ \sigma$ is represented in absolute value. σ is given in units of $(\pi G \rho)^{1/2}$. Source: Tassoul, M., and Tassoul, J. L., *Ap. J.* **150**, 1031, 1967. (By permission of The University of Chicago Press. Copyright 1967 by the University of Chicago.)

although the complete stabilization of the modes g_1 and g_2 at intermediate rotation rates might be once more an artifact of the models, our exact solutions strongly suggest that the derivation of higher g-modes in rapidly rotating, severely distorted bodies might eventually prove an impossible task. Further progress along these lines would require the study of differentially rotating, centrally condensed configurations, perhaps using equation (63) and higher-order polynomials as trial functions for the Lagrangian displacement. No results are available so far.

BIBLIOGRAPHICAL NOTES

Section 14.2. The pseudo-radial oscillations of a slowly rotating star were originally discussed by Ledoux (reference 45 of Chapter 6). The following papers may also be noted:

1. Bhatnagar, P. L., *Bull. Calcutta Math. Soc.* **38** (1946): 93–95.
2. Cowling, T. G., and Newing, R. A., *Astroph. J.* **109** (1949): 149–158.
3. Huang, S. S., *Ann. Astroph.* **16** (1953): 315–320.
4. Chandrasekhar, S., and Lebovitz, N. R., *Astroph. J.* **135** (1962): 248–260; *ibid.* **136** (1962): 1069–1081, 1082–1104; *ibid.* **152** (1968): 267–291.
5. Clement, M. J., *ibid.* **141** (1965): 210–231.
6. Zahn, J. P., *C. R. Acad. Sci. Paris* **263B** (1966): 1077–1079.
7. Occhionero, F., *Ann. Astroph.* **30** (1967): 761–771; *ibid.* **31** (1968): 1–4, 345–347.
8. Dyachenko, V. F., Zeldovich, Ya. B., Imshennik, V. S., and Paleichik, V. V., *Astrofizika* **4** (1968): 159–180.
9. Simon, R., *Astron. and Astroph.* **2** (1969): 390–397.
10. Simon, R., in *Stellar Rotation* (Slettebak, A., ed.), pp. 37–38, New York: Gordon and Breach, 1970.

The most detailed discussion of this problem is by Lebovitz (reference 12 of Chapter 5); this paper also contains pertinent comments on the literature. A further contribution will be found in:

11. Lebovitz, N. R., and Russell, G. W., *Astroph. J.* **171** (1972): 103–105.

See also reference 44 of Chapter 10. The destabilizing effects of general relativity on the radial pulsations of a spherical star have been considered, independently, by:

12. Kaplan, S. A., *Uchenye Zapiski Lvovskovo Univ.* **15** (1949): 109–115.
13. Chandrasekhar, S., *Astroph. J.* **140** (1964): 417–433, 1342; *ibid.* **142** (1965): 1519–1540.
14. Fowler, W. A., *Rev. Modern Phys.* **36** (1964): 545–555, 1104.

The combined effects of rotation and general relativity were first discussed by:

15. Fowler, W. A., *Astroph. J.* **144** (1966): 180–200.
16. Durney, B. R., and Roxburgh, I. W., *Proc. Roy. Soc. London, A,* **296** (1967): 189–200.

See also:

17. Chandrasekhar, S., and Friedman, J. L., *Astroph. J.* **175** (1972): 379–405; *ibid.* **176** (1972): 745–768; *ibid.* **181** (1973): 481–495.
18. Hartle, J. B., Thorne, K. S., and Chitre, S. M., *ibid.* **176** (1972): 177–194.
19. Will, C. M., *ibid.* **190** (1974): 403–410.
20. Hartle, J. B., and Friedman, J. L., *ibid.* **196** (1975): 653–660.
21. Hartle, J. B., and Munn, M. W., *ibid.* **198** (1975): 467–476.

Independent contributions are due to:

22. Vartanian, Yu. L., Ovsepian, A. V., and Adzhian, G. S., *Astrofizika* **7** (1971): 624–641.
23. Arutyunian, G. G., and Sedrakian, D. M., *ibid.* **8** (1972): 419–423.
24. Papoian, V. V., Sedrakian, D. M., and Chubarian, E. V., *ibid.* **8** (1972): 405–412.
25. Papoian, V. V., Sedrakian, D. M., and Chubarian, E. V., *Astron. Zh.* **49** (1972): 750–755.
26. Vartanian, Yu. L., Ovsepian, A. V., and Adzhian, G. S., *ibid.* **50** (1973): 48–59.

See also references 20–30 of Chapter 13. Application of the scalar virial theorem to the finite-amplitude, radial oscillations of a spherical star was first made by:

27. Ledoux, P., *Bull. Acad. Roy. Belgique: Classe des Sciences* (5) **38** (1952): 352–365.

Finite-amplitude motions of rotating bodies are considered in:

28. Tassoul, J. L., *Astroph. J.* **160** (1970): 1031–1042.
29. Tassoul, J. L., and Tassoul, M., *ibid.* **166** (1971): 621–625.
30. Dedic, H., and Tassoul, J. L., *ibid.* **188** (1974): 173–179.

The presentation of the problem in the text largely follows references 28 and 30. Reference 29 contains a discussion of rapidly rotating, severely distorted bodies.

Section 14.3. Most of the relevant papers on the β cepheids may be traced to:

31. Cox, J. P., "Pulsating Stars" in *Rep. Progress Phys.* **37** (1974): 563–698.

32. Cox, J. P., "Stellar Oscillations, Stellar Stability and Applications to Variable Stars" in *Mém. Soc. Roy. Sci. Liège* (6) **8** (1975): 129–159.

The Ledoux-Osaki approach is mainly discussed in:

33. Ledoux, P., *Astroph. J.* **114** (1951): 373–384.
34. Christy, R. F., *Astron. J.* **72** (1967): 293–294.
35. Osaki, Y., *Publ. Astron. Soc. Japan* **23** (1971): 485–502; *ibid.* **27** (1975): 237–258.
36. Osaki, Y., *Astroph. J.* **189** (1974): 469–477.

Conflicting reports on the Chandrasekhar-Lebovitz approach will be found in:

37. Chandrasekhar, S., and Lebovitz, N. R., *ibid.* **135** (1962): 305–306; *ibid.* **136** (1962): 1105–1107.
38. Böhm-Vitense, E., *Publ. Astron. Soc. Pacific* **75** (1963): 154–161.
39. van Hoof, A., *Zeit. f. Astroph.* **60** (1964): 184–189.
40. Clement, M. J., *Astroph. J.* **141** (1965): 1443–1454; *ibid.* **144** (1966): 841–842; *ibid.* **150** (1967): 589–605.
41. Hurley, M., Roberts, P. H., and Wright, K., *ibid.* **143** (1966): 535–551.

See also:

42. Vandakurov, Yu. V., *Astron. Zh.* **43** (1966): 1009–1017; *ibid.* **44** (1967): 786–797.
43. Vandakurov, Yu. V., *Astroph. J.* **149** (1967): 435–438.

The following papers are also quoted in the text:

44. Harper, R. Van R., and Rose, W. K., *ibid.* **162** (1970): 963–969.
45. Davey, W. R., *ibid.* **179** (1973): 235–240.
46. Deupree, R. G., *ibid.* **190** (1974): 631–636; *ibid.* **194** (1974): 393–401.

The effects of tidal interaction with a companion have been considered by:

47. Fitch, W. S., *ibid.* **148** (1967): 481–496; *ibid.* **158** (1969): 269–280.
48. Kato, S., *Publ. Astron. Soc. Japan* **26** (1974): 341–353.
49. Denis, J., and Smeyers, P., *Astron. and Astroph.* **40** (1975): 411–414.

Section 14.4. The formal solution derived in the text is largely due to:

50. Aizenman, M. L., and Cox, J. P., *Astroph. J.* **202** (1975): 137–147.

Most of the papers devoted to the Ledoux critical mass may be traced to reference 31; further contributions are due to:

51. Aizenman, M. L., Hansen, C. J., and Ross, R. R., *Astroph. J.* **201** (1975): 387–391.
52. Conti, P. S., and Burnichon, M.-L., *Astron. and Astroph.* **38** (1975): 467–470.

The following papers are also quoted in the text:

53. Larson, R. B., and Starrfield, S., *ibid.* **13** (1971): 190–197.
54. Stothers, R., *Astroph. J.* **192** (1974): 145–148.

Section 14.5. A general account of these and related matters will be found in:

55. Spiegel, E. A., "Convection in Stars. II. Special Effects" in *Annual Review of Astronomy and Astrophysics* **10**, pp. 261–304, Palo Alto, California: Annual Reviews, Inc., 1972.

The earliest contributions are due to Randers (reference 14 of Chapter 7), Walén (reference 26 of Chapter 3), and Cowling:

56. Cowling, T. G., *Astroph. J.* **114** (1951): 272–286.

The effects of a slow rotation on the onset of convection in gaseous masses have been considered by:

57. Clement, M. J., *ibid.* **142** (1965): 243–252.
58. Durney, B. R., and Skumanich, A., *ibid.* **152** (1968): 255–266.
59. Smeyers, P., *Astron. and Astroph.* **7** (1970): 204–209.
60. Smeyers, P., and Denis, J., *ibid.* **14** (1971): 311–318.

Technical matters pertaining to this and other problems are discussed in reference 9, as well as in the following paper:

61. Denis, J., and Smeyers, P., *Med. Kon. Acad. België: Klasse der Wetenschappen* **37**, No 1 (1975): 1–18.

A work of related interest is:

62. Wolff, C. L., *Astroph. J.* **193** (1974): 721–727.

Explicit evaluations of the *g*-modes in rapidly rotating bodies will be found in:

63. Cretin, M., et Tassoul, J. L., *Ann. Astroph.* **28** (1965): 982–991.
64. Lebovitz, N. R., *Astroph. J.* **150** (1967): 203–212.
65. Tassoul, M., and Tassoul, J. L., *ibid.* **150** (1967): 1031–1040.
66. Robe, H., *Ann. Astroph.* **31** (1968): 549–558.

See also:

67. Kochhar, R. K., and Trehan, S. K., *Astroph. and Space Sci.* **26** (1974): 271–287.

The semi-infinite isothermal atmosphere is considered in reference 64. References 63 and 66 treat the corresponding problem in cylindrical geometry. In reference 65, the lowest convective modes of a compressible Maclaurin spheroid are obtained in a rotating frame of reference; the same problem was subsequently treated by Kochhar and Trehan in a nonrotating frame (reference 67). When corrected for algebraic errors, the Kochhar-Trehan equations lead to the results derived in reference 65; that is to say, their (*corrected*) frequencies may be obtained at once from those derived in reference 65 by merely substituting $\sigma \pm m\Omega$ in place of σ (with $m = 0, 1, 2,$ or 3). In this connection, see:

68. Tassoul, M., *Astroph. and Space Sci.* **48** (1977): 89–102.

The complicated behavior of the non-axisymmetric modes depicted in Figure 14.18 is further discussed in this paper.

15

Stellar Magnetism and Rotation

15.1. INTRODUCTION

The role of magnetic fields in rotating stars has been considered at various places in this book. For example, in Section 11.4 we showed that magnetically controlled winds provide an efficient means for extracting angular momentum from rotating stars with convective envelopes. It is the purpose of this chapter to outline the major theoretical problems of interest in the study of rotating stars containing large-scale magnetic fields. It will in no way be an exhaustive study; however, references will be given for further pursuit. In Section 15.2, we discuss the gross changes brought by large-scale magnetic fields on the structure of nonrotating stellar models. The current ideas on the origin of stellar magnetic fields are summarized in Section 15.3. The next section explores the outstanding features of the strongly magnetic main-sequence stars. We conclude the chapter with brief comments on dynamo action and differential rotation in the solar convection zone.

15.2. SOME GENERAL PROPERTIES

The basic equations for magnetic stars have been discussed in Section 3.6; these are, of course, the Maxwell equations (with the displacement currents ignored) and the equations of hydrodynamics suitably modified to include the Lorentz force and Ohmic dissipation. Here we shall briefly consider some general properties of nonrotating stellar models containing large-scale magnetic fields.

Quite generally, the changes in the magnetic field $\mathbf{H}$ are given by

$$\frac{\partial \mathbf{H}}{\partial t} = \operatorname{curl}(\mathbf{v} \times \mathbf{H}) - \operatorname{curl}(v_m \operatorname{curl} \mathbf{H}), \tag{1}$$

where $\mathbf{v}$ is the macroscopic velocity and $v_m(=1/4\pi\sigma_e)$ is the coefficient of magnetic diffusivity (cf. §3.6: equations [125] and [126]). In particular, treating the coefficient of electrical conductivity σ_e as constant and remembering that div $\mathbf{H} = 0$, we obtain

$$\frac{\partial \mathbf{H}}{\partial t} = \operatorname{curl}(\mathbf{v} \times \mathbf{H}) + v_m \nabla^2 \mathbf{H}. \tag{2}$$

Now, if L is a measure of the linear dimension of the mass, the two terms on the r.h.s. are comparable to vH/L and $v_m H/L^2$, respectively. (Here, v and H are quantities of the order of the velocity and of the magnetic field, respectively.) Accordingly, if $v_m \ll vL$, the second term in the right of equation (2) becomes small compared to the first, so that equation (2) approximates to

$$\frac{\partial \mathbf{H}}{\partial t} = \text{curl}(\mathbf{v} \times \mathbf{H}), \tag{3}$$

the exact form it would take if σ_e were infinite. Given the discussion in Section 3.6, this means that the magnetic lines of force are then dragged along with the velocity $\mathbf{v}$ of the material. That is, the material can stream freely along the lines of force; but, when it moves perpendicular to the lines of force, the material must carry these with it or be carried along by them. As we shall see in Section 15.3, in electrically conducting masses of the size of a star, the lines of force leak so slowly through the material that, for most purposes, they can be regarded as frozen into it.

A stellar magnetic field can be regarded as a sort of "strain," possessing a potential energy M given by

$$M = \frac{1}{8\pi} \int |\mathbf{H}|^2 \, \mathbf{dx}, \tag{4}$$

where the volume integral extends over all parts of space occupied by the star *and* field. As we recall from Section 3.6, the field produces mechanical effects equivalent to the action of a magnetic pressure $|\mathbf{H}|^2/8\pi$, which is uniform in all directions, and a tension $|\mathbf{H}|^2/4\pi$ along the lines of force. As was shown by Chandrasekhar and Fermi in 1953, the scalar virial theorem can be used to fix an upper limit for the magnetic energy of a star. In the present instance, it is a simple matter to prove that

$$\frac{1}{2}\frac{d^2 I}{dt^2} = 2K + M - |W| + 3(\gamma - 1)U_T, \tag{5}$$

where all symbols have their standard meanings (cf. §3.7: equations [148]–[150]). In the case of a nonrotating stellar model in equilibrium, we thus have

$$M - |W| + 3(\gamma - 1)U_T = 0. \tag{6}$$

This relation expresses, in integral form, the fact that a star is supported against gravity partly by the magnetic force, and partly by the pressure. Equation (6) readily provides an upper limit to the magnetic energy consistent with equilibrium, since it implies that

$$M < |W|. \tag{7}$$

Thus, for the existence of an equilibrium it is necessary that the total magnetic energy of a nonrotating star does not exceed the absolute value of the gravitational potential energy.

The upper limit given by equation (7) is, however, unlikely even to be approached for actual stars. Indeed, by writing

$$M = \tfrac{1}{6}R^3 H_0{}^2 \quad \text{and} \quad |W| = k\,\frac{GM^2}{R}, \tag{8}$$

where H_0 is an average value of the field and k is a numerical constant of order unity, we can rewrite inequality (7) in the form

$$H_0 < (6kG)^{1/2}\,\frac{M}{R^2}. \tag{9}$$

For a uniform spherical star of mass $M \simeq 2M_\odot$ and radius $R \simeq 3R_\odot$, with a uniform field H_0 in the interior and a dipole field outside, condition (9) gives

$$H_0 < 4 \cdot 10^7 \; gauss. \tag{10}$$

Comparing this upper limit with the observed surface fields of typical magnetic stars, a few thousand gauss in magnitude, we see that the limit $M = |W|$ could be approached only if the inner fields are inordinately greater than those observed on their surfaces (cf. §15.4). Thus, provided a star is not to contain very strong subsurface fields, the magnetic energy is a negligible fraction of the gravitational potential energy. The above result thus leads to the suggestion that magnetic fields are not likely, in general, to produce serious modifications in the internal structure of a star.

An extensive literature deals with the effects of various axisymmetric fields on the shape of nonrotating self-gravitating bodies (with $\sigma_e = \infty$). In addition to the distention caused by the magnetic pressure, a stellar model is, in general, distorted from a spherical form. That is to say, a field in meridian planes through a diameter (i.e., a poloidal field) corresponds, in general, to an *oblate* form of equilibrium, whereas one whose lines of force are circles with a diameter as axis (i.e., a toroidal field) corresponds, in general, to a *prolate* form of equilibrium. With appropriate combinations of poloidal and toroidal fields, oblate, prolate, or spherical forms of the free surface are possible. Various perturbation techniques have been used to construct sequences of nonrotating, centrally condensed bodies pervaded by poloidal and/or toroidal magnetic fields. In all these studies it is found that the surface distortion produced by a magnetic field is small if $M/|W|$ is small, the ellipticity of the outer surface being then of the order of $M/|W|$. The structure of nonrotating polytropes with very strong poloidal fields of the dipolar type has been also determined by Monaghan (using a perturbation technique) and by Miketinac (integrating the exact equations governing internal structure). For such large fields, with $M/|W|$ ranging up to 0.25, their calculations show that all centrally condensed, axially symmetric polytropes are grossly distorted when the magnetic energy is several percent of the gravitational potential energy. It is also found that the magnetic field is a fairly insensitive function of the polytropic index. By virtue of equation (10), all of these results provide no reason to modify the view that magnetic fields are unable to distort a normal star appreciably or to exert any decisive influence on its global properties.

In the special case when only magnetic fields are present, the scalar virial theorem can also be used to discuss stability. Indeed, the total energy E of a nonrotating star in equilibrium is

$$E = U_T + M - |W|, \tag{11}$$

or, using equation (6),

$$E = -(3\gamma - 4)U_T = -\frac{3\gamma - 4}{3(\gamma - 1)}(|W| - M). \tag{12}$$

By virtue of equation (12), because the total thermal energy U_T is always positive and because stable configurations must have a total energy which is negative, it is thus necessary for stability that the ratio of specific heats γ must be greater than 4/3. This is the condition valid in the absence of magnetic field (cf. §6.4). However, in the stable case $\gamma > 4/3$, E increases through negative values toward zero as M increases toward its upper limit $|W|$. Hence, in a certain sense, one may say that the stability of a nonrotating star decreases as its magnetic energy increases. As a matter of fact, if such a star pulsates more or less radially, the fundamental frequency of oscillation is reduced by the presence of a magnetic field. As was shown by Chandrasekhar and Limber, this frequency becomes, to first approximation.

$$\sigma^2 = (3\gamma - 4)\frac{|W|}{I}\left(1 - \frac{M}{|W|}\right). \tag{13}$$

This illustrates the decrease in dynamical stability in a star as its magnetic field increases. Similar calculations further indicate that all natural frequencies of oscillation of nonrotating polytropes containing weak magnetic fields are only slightly altered by these fields. Roughly speaking, when $M/|W| \ll 1$, the various squared frequencies discussed in Section 6.4 are then modified by factors of the form $(1 + \alpha M/|W|)$, where α is a (positive or negative) constant of order unity which depends on the mode. (Also obtained are modes with squared frequencies $\sigma^2 = \beta M/|W|$, where the β's are positive constants.) Hence, unless the magnetic energy is a substantial fraction of the gravitational potential energy, the presence of a magnetic field is not likely to endanger the global stability of a star.

15.3. ORIGIN OF STELLAR MAGNETIC FIELDS

No completely satisfactory theory of stellar magnetic fields has been given so far. Only the theories of fossil magnetism and dynamo maintenance have undergone recent development. According to the "fossil" theory, the magnetic field existing in a star is the slowly decaying relic of the field present in the gas from which the star formed. On the other hand, the dynamo theory supposes that the motion of material across the magnetic lines of force of an existing field generates electric currents that maintain this "seed" field. As we shall see in Section 15.5, the discovery that the general solar field reverses with the solar cycle lends support to the idea of a contemporary dynamo process rather than a "fossil" explanation of the surface fields of the Sun and solar-type stars. By contrast, the long decay-time of the fields of the strongly magnetic stars does suggest that these fields may be slowly decaying relics of a distant past—*either* remnants of the interstellar field present in the primeval gas cloud from which these stars formed *or* survival of dynamo-generated fields from an earlier epoch, e.g., the convective phase of pre-main-sequence contraction. Here we shall comment mainly on the "fossil" theory of stellar magnetism.

Consider a primeval field in a star without mass motions, and so without any energy source to replace the energy dissipated by the large but finite electrical conductivity of the body. Under these conditions, equation (2) reduces to the diffusion equation

$$\frac{\partial \mathbf{H}}{\partial t} = \frac{1}{4\pi\sigma_e} \nabla^2 \mathbf{H}, \tag{14}$$

thus indicating that a magnetic field decays by leaking toward regions where the field is in the opposite direction and being neutralized there. So far as orders of magnitude are concerned, this equation indicates that the characteristic decay time t_d of a local field varying appreciably over the length L is

$$t_d = 4\pi\sigma_e L^2. \tag{15}$$

As was originally shown by Cowling in 1945, this time of free decay of a stellar magnetic field is comparable with 10^{10} years, being thus of the same order as, or larger than, the lifetime of a star. In making explicit use of equation (14), Wrubel subsequently reduced to $4 \cdot 10^9$ years the e-folding time for the slowest decaying dipole component of a quasi-steady solar field. Hence, *provided there are no mass motions*, free decay is unlikely to reduce a star's magnetic field appreciably if it is born with such a field. This result is the basis of the "fossil" theory of stellar magnetism.

Given the many lines of evidence converging on a background galactic magnetic field in the range $3-5 \cdot 10^{-6}$ gauss, the "fossil" theory is thus intimately related to the problem of stellar formation in the presence of rotation and magnetic fields. In particular, as we recall from Section 11.1, both effects prevent the gravitational collapse of protostellar clouds with masses less than $10^3-10^4 M_\odot$. Moreover, if the lines of force are *strictly* frozen into the material, the total magnetic flux through the mass remains constant during the contraction (cf. §3.6: equation [131]); because of mass conservation, it follows that the magnetic field should increase as $\rho^{2/3}$. (Assuming $H \approx 10^{-6}$ gauss and $\rho \approx 10^{-24}$ g/cm^3 in the collapsing cloud, this would imply a final field of 10^{10} gauss!) The first difficulty might be partly overcome by supposing that more of the contraction takes place along the lines of force than perpendicular to them. According to Mestel and his associates, because such an anisotropy in the gravitational collapse allows the density to increase with very small changes in the magnetic field, this could well lead to a process of successive fragmentations, thus yielding protostellar fragments with masses of the right order of magnitude. However, since these protostars would then have magnetic energies comparable to their gravitational energies, the "fossil" theory still presents many tantalizing problems: (i) Why are the strong surface fields observed only in an apparent minority of main-sequence stars, which appear to be *almost* co-extensive with the Ap stars? (ii) Why are the strongest observed stellar fields not even stronger? and (iii) Why is the weak general solar field—due probably to a contemporary dynamo process—not masked by a strong primeval field emanating from the bulk of the Sun?

The "fossil" theory owes its appeal to the fact that the Ohmic decay time of a magnetic field in a star with no mass motions is long compared with the time scale of evolution of the system. However, if most stars are not to contain very strong magnetic fields, an efficient process of flux destruction must be found. Various means for disposing of unwanted flux have been proposed. For example, Mestel and Spitzer have suggested that the magnetic field and the low density ionized component of a collapsing cloud would drift together out of the gas early in its contraction phase. More recent work strongly suggests, however, that approximate flux-freezing holds in dusty HI clouds until much higher densities. Hence, although it is plausible that some protostars lose most of their primeval magnetic flux by ambipolar diffusion, we may still ask what happens to protostars that do manage to reach their pre-main-sequence while retaining most of their primeval flux.

It is well known that a nonmagnetic star is completely (or almost completely) convective at some stage of its quasi-static contraction toward the main sequence. Unfortunately, the interaction between turbulent convection and prescribed magnetic fields is still poorly understood. According to Gough, Moss, and Tayler, it seems probable that in a star whose magnetic energy exceeds the turbulent energy of the eddying motions (and certainly in any star with comparable magnetic and gravitational energies) the flux can survive the slow contraction phase—unless of course other mechanisms (such as hydromagnetic instabilities) lead to rapid flux-losses. In the case of a weak primeval field with energy much less than the turbulent energy, however, Sweet and Elsasser have pointed out that convective motions might then cause a sufficient reduction in the length scale of the magnetic field so that much more rapid Ohmic decay would occur (just as molecular diffusion can be supplemented by a much larger mixing effect due to turbulence). On the contrary side, Spitzer has argued that, though turbulence may well lead to a reduction of a weak magnetic field in a turbulent region, it leads to no appreciable change in the total magnetic moment of a star. That is to say, the flux will be concentrated in the surface regions of the star, and the decay rate of the dipolar component of the field will remain essentially unchanged; once the convective phase is over, the field will then leak back through the star into a more normal dipolar-type structure. If this argument is correct, weakly magnetic stars would thus reach the main sequence with their primeval flux intact. Given the many uncertainties in the parameters involved, no firm conclusion can be drawn at this time.

Another possible mechanism for disposing of unwanted magnetic flux in a star is that many magnetic field configurations *might* be unstable on a short time scale and that these local instabilities *might* lead to enhanced field decay. This approach has been mainly discussed by Markey and Tayler, and by Wright. They have shown that these instabilities are essentially the same as those which occur in laboratory experiments involving cylindrical and toroidal pinched gas discharges, and which have proved so troublesome in attempts to produce controlled thermonuclear reactions. In particular, it is found that, in a star containing purely poloidal fields (with closed field lines inside the star), instability is likely to occur in the neighborhood of the neutral line of the field, with a growth rate very much shorter than the thermal time scale of the star. The motions driven by these local instabilities are almost entirely parallel to the level surfaces. Tayler has also obtained criteria for the stability of purely toroidal fields and has shown that a wide class of these is unstable; instability then occurs close to the axis of symmetry of the star. According to Wright, a mixed poloidal-toroidal field may be stable if the toroidal component is sufficiently large, but it is not yet possible to state whether a wide range of such configurations is completely stable to short time scale disturbances. The ultimate result of these local instabilities is still unclear. At present, only the occurrence of instability in nonrotating stars has been demonstrated, and no discussion has taken place concerning the nonlinear growth of the perturbations. It has been conjectured that this will lead to flux destruction, but has not been properly demonstrated. In addition, even though some magnetic field configurations may be stable to isentropic perturbations, finite thermal conductivity and finite electrical conductivity may promote additional instabilities. The study of these non-isentropic processes is still in its infancy.

In summary, although it is not yet clear whether or not a large-scale primeval field threading a protostellar cloud could survive the pre-main-sequence phases, observation reveals that a main-sequence star must have been able to reduce its magnetic flux to well below the maximum limit (7) that the scalar virial theorem allows. This may occur during the star formation process—by ambipolar diffusion—as well as after the star is formed, either by enhanced Ohmic decay or by hydromagnetic instabilities. However, even if we accept the "fossil" theory for the strongly magnetic stars, we must still explain why the high-intensity surface fields are found only in a small portion of the main sequence. It has often been argued that any primeval magnetic flux retained by a late-type main-sequence star is concentrated deep down by its surface convection zone, and so is unobservable. (If this is correct, the surface fields of the Sun and solar-type stars would then arise via a contemporary dynamo process, having no connection with the primeval field deep down in the radiative core.) Similarly, it has long been surmised that the meridional currents that necessarily flow in the radiative zone of a rapidly rotating, upper main-sequence star will distort a given primeval field by dragging its lines of force beneath the surface, thus leaving the rapid rotator apparently as a nonmagnetic body. (Such a dragging mechanism would be less efficient for the slowly rotating, upper main-sequence stars, e.g., the Ap stars; hence, only sharp-lined stars would exhibit high-intensity surface fields.) Circulation patterns that can indeed trap magnetic lines of force within a star have been constructed by Mestel and Moss (cf. §8.3).

Since there is as yet no consensus about the origin of magnetic fields in main-sequence stars, we conclude by noting that recent work (notably by Braginski and by Steenbeck, Krause, and Rädler) strongly suggests that turbulence with a sufficient degree of anisotropy can act as a dynamo (cf. §15.5). Accordingly, even if a star approaches the main sequence with little memory of its primeval field, it will almost certainly acquire a dynamo-built field during the convective portion of its pre-main-sequence contraction. Detailed numerical calculations by Schüssler show that in a slowly contracting, uniformly rotating star (with a weak primeval field and $M \simeq 2M_\odot$) this dynamo process indeed can build up a new magnetic field that can survive the transition from the fully convective phase to the main-sequence stage. On the basis of this result, Schüssler further speculated that no dynamo-built field can occur during the contraction of the O- and B-type stars; that is to say, because stars earlier than spectral type A do not pass through a fully convective phase, none of them can exhibit the remnant of a dynamo-built field when it arrives on the main sequence. In this connection, Wolff and Wolff have recently shown that the sharp-lined B-type star HR 7129—a helium spectrum variable—has a longitudinal magnetic field that varies between the limits $+7,000$ gauss and $-5,000$ gauss in a period of 3.670 days (cf. §15.4). Since the surface temperature of this star is approximately $20,000°K$, it represents an extension to much higher temperatures of the phenomena that are observed in the magnetic Ap stars. In other words, unless we are willing to accept the (quite implausible) idea that the surface magnetic field of this early B-type star has not the same origin as those observed in the Ap stars, we are thus forced back to the "fossil" theory of stellar magnetism for the upper main-sequence stars.

15.4. MAGNETIC STARS

Stellar magnetic fields are currently detected with methods derived from the photographic technique developed by Hale in 1908 for the study of sunspot fields, and first applied by Babcock in 1947 to the study of stellar magnetism. These fields are measured spectroscopically by the Zeeman effect, which gives the mean field directed toward the observer, averaged over the whole disc; hence, the measurements do not show the distribution of the field, but indicate only that it extends over a large portion of the stellar disc. As we pointed out previously, *almost* all high-intensity magnetic fields known in nondegenerate stars are found in the peculiar A-type (Ap) stars, i.e., stars lying on or near the main sequence, with effective temperatures approximately in the range $8,000-15,000°K$ and characterized by abnormal abundances in their spectra (cf. §12.4).[1] Except for the manganese stars, all Ap stars with spectral lines not too greatly broadened by rotation possess surface-averaged fields that range from

---

[1] Following the recent discovery by Wolff and Wolff of a large magnetic field in the star HR 7129, with $T_e \approx 20,000°K$, it now appears that the so-called "Ap phenomenon" must be taken in an extended sense so as to cover also the early B-type, helium variables. Other known magnetic stars with effective temperatures in this region are: HD 215441 (Babcock's star) with $T_e \approx 16,000-18,000°K$, a Cen with $T_e \approx 19,400°K$ (Wolff and Morrison), and 3 Sco with $T_e \approx 16,800°K$ (Landstreet, unpublished).

the limit of detectability to 34,000 gauss.[2] A search for strong magnetic fields ($\gtrsim 1,000$ gauss) in "rapidly" rotating Ap stars (i.e., stars with projected equatorial velocities $v_e \sin i \gtrsim 30$ km/sec) has been made by Landstreet, Borra, Angel, and Illing. Observations of 16 stars reveal that no definite fields can be found, which indicates that longitudinal fields in excess of 1,000 gauss are significantly less common among "rapidly" rotating Ap stars than among slowly rotating ones. To summarize the outstanding properties of the magnetic Ap stars very briefly: (i) magnetic fields are variable and, in many stars, proceed as far as actual reversals of polarity; (ii) the actual moment of reversal is often accompanied by the cross-over effect discovered by Babcock; that is, a systematic difference in width between the right-hand and left-hand circularly polarized components into which a line is split by an analyzer; (iii) in all Ap stars with well-determined periods, the variabilities of magnetic field, spectrum, and luminosity exhibit the same periodic behavior; typical periods are of the order of 5–9 days, but periods considerably shorter and longer are observed. (Note that many of the magnetic stars that were previously classified as irregular variables are now being reclassified as regular variables!); and (iv) the line widths of periodic Ap stars tend to vary inversely as their periods—an effect originally noticed by Deutsch.

Three theories have been proposed to explain the magnetic Ap stars: (i) the magnetic oscillator, (ii) the "stellar-cycle" model, and (iii) the oblique rotator. Very few people still believe in the magnetic-oscillation theory since it seems incapable of yielding the observed periods and fields that reverse (cf. §15.2). The second theory, which attempts to explain magnetic variations by analogy with the solar cycle, also presents difficulties and is much less definitely formulated than its contenders. Here we shall mainly consider the oblique rotator, for it is generally agreed that the magnetic Ap stars find a natural explanation in terms of this model. Following the pioneering work of Babcock, Stibbs, and Deutsch, we shall assume that a magnetic Ap star rotates uniformly about an axis inclined at some angle to the line of sight, its magnetic axis being also inclined at some angle to the rotation axis. (Note that this model is independent of the fossil vs. dynamo issue: all that we require is a fair degree of permanence in the surface magnetic field, so that what we observe is a field that varies as a consequence of the uniform rotation of the star.)

---

[2] Let us note that much greater magnetic fields have now been found in some white dwarfs; these fields lie in the range $1-50 \cdot 10^6$ gauss. (These observed values lie well below the upper limit of about 10^{12} gauss which is imposed by condition (9), with $M \approx M_\odot$ and $R \approx 10^{-2} R_\odot$!) See: Landstreet, J. D., in *Planets, Stars and Nebulae Studied with Photopolarimetry* (Gehrels, T., ed.), p. 981, Tucson: Univ. of Arizona Press, 1974; Ingham, W. H., Brecher, K., and Wasserman, I., *Ap. J.* **207**, 518, 1976 (and references quoted therein).

The strongest support for the oblique rotator comes from Deutsch's finding that the line widths of periodic Ap stars tend to vary inversely as their periods. Indeed, assuming that the line widths are due mainly to the star's rotation and that the radius does not vary much among the Ap stars, we thus have $v_e \propto 1/P$; to be specific, we can write

$$v_e = \frac{50.6\bar{R}}{P}, \tag{16}$$

if the mean radius $\bar{R}$, equatorial velocity v_e, and rotational period P are expressed in solar radii, km/sec, and days, respectively. Because the stars are viewed at unknown inclinations i, equation (16) should form the upper envelope of the $v_e \sin i$ values *if the oblique rotator is the correct explanation*. According to Preston, we have $\bar{R} \simeq 3.2 R_\odot$ for all Ap stars, with no evidence that $\bar{R}$ may increase with T_{eff} among these stars. The solid line in Figure 15.1 is based on this mean value. Clearly, this is consistent with the magnetic period being equal to the rotation period.

The oblique-rotator model also accounts satisfactorily for the reversals of magnetic polarity that are commonly observed, and it explains the order of magnitude, sign convention, and phases of the cross-over effect. That is, when the polarity is changed from (say) positive to negative, the widths of the Zeeman components of a line are consistent with the simultaneous presence, on the visible disc, of spectroscopic and/or magnetic patches, of which those of positive polarity are receding while those of negative polarity are approaching. Obviously, this effect is consistent with the oblique-rotator model, for it ascribes changes of polarity to the motion of patches of different polarity from the visible disc to the invisible disc, and conversely. Finally, let us mention that another independent confirmation of the oblique-rotator model has recently been obtained by Borra and Vaughan, who observed the transverse Zeeman effect in the line Fe II 4520.2 in the spectrum of β Coronae Borealis and found that the linear polarization rotates through $360°$ with the star's magnetic period.

All detailed studies of the oblique-rotator model of magnetic Ap stars are necessarily *phenomenological*: they ask which distribution of magnetic field and which angle of inclination between the magnetic and rotation axes will reproduce the observed field variations in a given star. In its simplest form, first discussed by Stibbs, the model consists of a magnetic dipole placed at the center of the star and inclined at some angle with respect to the rotation axis. Following the recent work of Landstreet, it is now believed that a *decentered* poloidal field may be more adequate for almost all magnetic variables. In accordance with Preston's statistical analysis, it is also believed that the rotation and magnetic axes are approximately

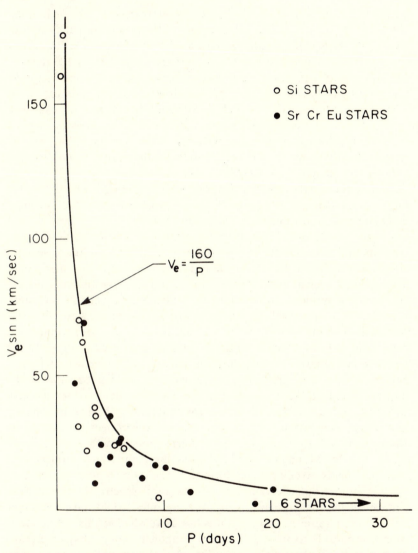

Figure 15.1. A plot of projected equatorial velocity vs. magnetic period for periodic Ap stars. The curve represents the relation between equatorial velocity and rotational period for stars with radius $R = 3.2R_\odot$, as discussed in the text. Note that the curve is a reasonable upper envelope for the plotted points. Source: Preston, G. W., *P.A.S.P.* **83**, 571, 1971.

orthogonal in most (but not all) Ap stars. Nevertheless, as Preston pointed out, there remains one disquieting feature of the observational data from the point of view of the oblique-rotator model: there are now a few known or suspected magnetic Ap stars with periods ranging from 100 days to several years. Wolff's more recent observations, however, strongly support the hypothesis that the periods of variation of the Ap stars are to be identified in all cases with the periods of rotation, and that the existence of Ap stars with extremely long periods is due to the loss of angular momentum through some form of magnetic braking (cf. §12.4). As was emphasized by Cowling, "the validity of the oblique-rotator model must in large part be left to the observers to settle."

Despite the obvious requirement of *no* axial symmetry with respect to the rotation axis, almost all attempts to construct models of strongly magnetic main-sequence stars have concentrated on the case of axial symmetry—both for the case of strict radiative equilibrium with no large-scale meridional currents in the radiative zone and for the case with circulation included. Since magnetic-star models with slow meridional currents have been already presented in Section 8.3, here we shall consider axisymmetric models containing magnetic fields that generate a circulation that *exactly* cancels the rotationally driven currents, resulting in uniformly rotating configurations that are in strict radiative equilibrium near their surfaces.[3] Stellar models of this type with weak dipole fields and slow uniform rotation were first considered, independently, by Davies and by Wright. Both authors have shown that, for a fixed value of the magnetic field at the center of the star, rapidly rotating models have weaker surface fields than the slowly rotating ones with the same opacity and energy generation. A similar result was subsequently obtained by Monaghan and Robson for axisymmetric models with strong dipole fields, and by Monaghan and by Moss for comparable models with multipole fields. Arguing solely from this result, one would predict that, of those upper main-sequence stars of a given spectral type that have surface magnetic fields, the slowly rotating ones would have the larger fields at their surfaces. Although it is not clear whether this prediction is in fact applicable to strongly magnetic stars with no axial symmetry (as the observational material suggests in most—but not all—cases), such a result would offer an alternative explanation to the observed properties of the Ap stars and normal A-type stars. (As Section 15.3 noted, it is quite plausible that strong magnetic fields are not observed in rapidly rotating, upper main-sequence stars because large-scale meridional cur-

---

[3] These models are thus somewhat similar (at least in their principle) to the nonmagnetic, circulation-free models in which the angular velocity $\Omega(\varpi, z)$ is chosen so as to prevent rotationally driven meridional currents (cf. §7.5).

rents have dragged the field lines beneath the surface of these stars.)
In any event, the Davies-Wright result is only suggestive, as the fields
studied are all purely poloidal and so are not dynamically stable (cf. §15.3).
Therefore, Moss has also constructed models appropriate to uniformly
rotating, upper main-sequence stars that contain mixed poloidal-toroidal
fields. The equilibrium models are found to have comparable poloidal
and toroidal fluxes, so that it is plausible that some of them are stabilized
against the local instabilities experienced by purely poloidal fields. Since
the fields are assumed to be symmetric about the rotation axis, detailed
comparison with the observational data does not seem possible either.
A preliminary attempt to correct this deficiency has been made by
Monaghan, assuming again uniform rotation and a magnetic field that
is not symmetric with respect to the rotation axis. Interpreting the field
in terms of an inclined dipole, Monaghan found that the ratio of the
central field to the surface field decreases as the angle between the rotation
and magnetic axes increases. Magnetic stars with a large angle between
the rotation and magnetic axes would thus tend to have the larger surface
fields. However, because the computations fail to converge when the
ratio of rotational energy to magnetic energy increases, these results must
be extrapolated with caution until more extensive calculations are made.

To conclude, let us mention two alternative mechanisms that may
spontaneously yield a large angle χ between the magnetic and rotation
axes, as required by the oblique-rotator model and the reversals of polarity
observed in most of the magnetic Ap stars. The *first* mechanism, an extension of the magnetic braking process discussed in Section 11.4, has
been worked out by Mestel and Selley. They postulate an oblique rotator
and a stellar wind, so that the torques resulting from the loss of angular
momentum cause the instantaneous axis of rotation to precess through
the star and the magnetic axis to rotate in space. Approximate calculations
indicate that the change in the obliquity χ is always small, if the braking
by a stellar wind is not to be excessive. The *second* explanation, which
ignores coupling with the external world altogether, was subsequently
proposed by Mestel and Takhar. For the sake of simplicity, the frozen-in
magnetic field is taken to be symmetric about an axis, which is inclined
at an angle χ to the angular-momentum vector. Because of the effect of
both centrifugal and magnetic forces, the body will thus have three unequal axes of inertia; and if it were rigid, its motion could be described
approximately as a combination of rotation, Eulerian nutation and a
periodic rocking in space of the magnetic axis. But a gaseous star cannot
support any nonhydrostatic stresses, and so it will react to the nutation
by a complex system of internal motions. As was originally noted by
Spitzer, these motions must be subject to dissipative processes that drain

energy from the rotation field, and so steadily change the obliquity χ until the star is finally rotating about a principal axis. According to the calculations by Mestel and Takhar, if a star is prolate with respect to the magnetic axis (possibly because of a mixed poloidal-toroidal field), the combined effect of nutation and dissipation within the star leads to the angular-momentum vector becoming parallel to the maximum principal axis. In this case, the rotation axis is thus ultimately perpendicular to the magnetic axis. The estimated dissipation rates suggest that the large obliquity required for most of the magnetic Ap stars can be achieved well within a stellar lifetime.

15.5. THE SOLAR DYNAMO AND DIFFERENTIAL ROTATION

The Sun is observed to have a mottled magnetic face, with general fields of only a few gauss over most of the surface, and strong localized fields (up to 4,000 gauss) in sunspots and in active regions. A weak field of about 1 gauss dominates the northern polar region above heliocentric latitude $\phi = 55°$, and a roughly equivalent field of opposite polarity dominates the corresponding region of the southern hemisphere. A reversal of the polarities of these fields has been observed near the peak of the sunspot activity of the last two cycles. However, it is not yet certain whether these fields always maintain dominant polarities according to a neat schedule synchronized to the phase of the 22-year solar cycle. Similarly, although the detailed variation of the magnetic fields in the broad equatorial region is immensely complicated, the overall pattern of these fields nevertheless follows the solar cycle, but this regularity is only statistical. The reversal of sunspot magnetic polarities at the beginning of a new 11-year half of the solar cycle and the longitude pattern in the photospheric magnetic fields are two examples of this statistical regularity. The weaker fields of the Sun are extended by the solar wind, filling the entire solar system to distances of 30 AU or more.

Although we can only infer from the surface fields what fields the Sun's interior may contain, it is generally believed that the observed fields extend out from the hydrogen convection zone and that the sunspots and bipolar magnetic regions are evidence of a general toroidal field of several hundred gauss at some depth beneath the photosphere. Furthermore, since these toroidal bands are observed to migrate from latitudes of about 45° to the equator in each 11-year half of the solar cycle, it is now commonly assumed that the maintenance of the solar fields is an *active* process from which all traces of the primeval field have long since vanished. Further pieces of evidence that support a migratory dynamo action in the solar convective zone are these: (i) turbulent convection in this surface layer would cause decay of a primeval field in a decade or so, and (ii) the observed reversals of the polar fields during the last two sunspot cycles are in contradiction to the existence of a primeval field. In what follows, we shall thus assume that the migratory fields of the Sun are maintained against Ohmic decay by the combined effect of differential rotation and turbulent motions in the convection zone. Because the literature contains many comprehensive discussions of the solar-dynamo problem, a brief survey will suffice for our purpose, which is to derive information about the (*unobservable*) state of differential rotation within the subphotospheric convective shell from the (*observed*) properties of the Sun's activity cycle.

The principle of dynamo action was first suggested by Larmor in 1919; it requires that the conducting fluid moves in such a way as to induce electromotive forces capable of maintaining a "seed" field against Ohmic decay. This is by no means a perpetual-motion engine, however, because energy must be supplied by the forces driving the fluid flow and it must be converted into magnetic energy. Most studies of the Sun have been devoted to the so-called *kinematic dynamo*, in which a fluid flow is specified in the convection zone, and the back reaction of the magnetic field, which opposes the motion, is ignored. Such an approximation is quite acceptable for the solar problem: (i) dynamo excitation must occur for arbitrary small seed fields, and (ii) the average magnetic energy density is small compared with the kinetic energy density at the photosphere. It is naturally convenient, therefore, to consider smoothed fields and to treat the solar cycle in a statistical manner. In contrast, both large-scale motions (such as differential rotation and meridional currents) and irregular small-scale motions (such as turbulence) are needed to produce dynamo action and migratory magnetic fields in the solar convection zone and in the photosphere.

All early efforts to build a solar dynamo have been hindered by Cowling's discovery, in 1934, that *a steady magnetic field symmetric about an axis cannot be maintained by a dynamo effect*. Indeed, the poloidal part of such a field must vanish on a circle around the axis, and so dynamo action can generate no electromotive force and no currents around this neutral line. Hence, the lines of force near this circle must contract toward the circle, and a steady field cannot be maintained in its vicinity. A more general approach to Cowling's theorem has been adopted by Weiss. Consider first axisymmetric fields and velocities within a differentially rotating body. They can be separated into poloidal and toroidal parts:

$$\mathbf{H} = \mathbf{H}_p + H_\varphi \mathbf{1}_\varphi \quad \text{and} \quad \mathbf{v} = \mathbf{v}_p + v_\varphi \mathbf{1}_\varphi, \tag{17}$$

where 1_φ is the unit vector in the azimuthal direction. (Note that $\mathbf{v}_p$ describes meridional currents whereas $v_\varphi = \Omega\varpi$ is the rotational velocity.) Then, the φ-component of equation (2) becomes

$$\frac{\partial}{\partial t}\left(\frac{H_\varphi}{\varpi}\right) + \operatorname{div}\left(\frac{H_\varphi}{\varpi}\mathbf{v}_p\right) = \mathbf{H}_p.\operatorname{grad}\left(\frac{v_\varphi}{\varpi}\right) + \frac{v_m}{\varpi}\left(\nabla^2 - \frac{1}{\varpi^2}\right)H_\varphi. \quad (18)$$

In this equation, the second term on the l.h.s. represents the advection of the toroidal field by the fluid flow; on the r.h.s., however, the creation of toroidal flux from the poloidal field as a result of differential rotation can compensate for the Ohmic losses. Now, the poloidal field can be expressed in terms of a toroidal vector potential; we thus have

$$\mathbf{H}_p = \operatorname{curl}(A 1_\varphi), \quad (19)$$

so that equation (2) can be integrated to give

$$\frac{\partial A}{\partial t} + \frac{1}{\varpi}\mathbf{v}_p.\operatorname{grad}(\varpi A) = v_m\left(\nabla^2 - \frac{1}{\varpi^2}\right)A. \quad (20)$$

In contrast with equation (18), the r.h.s. of equation (20) is purely dissipative. Hence, this equation does not allow the regeneration of poloidal flux from the toroidal field: $\mathbf{H}_p$ must ultimately decay, followed by H_φ. In other words, rotationally symmetric velocity fields—such as differential rotation—will not suffice to maintain rotationally symmetric magnetic fields against Ohmic decay.

A realistic dynamo model must therefore be non-axisymmetric, with an azimuthally averaged field **H**, whose poloidal part is maintained by a toroidal electromotive force

$$\overline{(\mathbf{v} \times \mathbf{H})_\varphi}. \quad (21)$$

As we shall see below, the currently accepted models incorporate *two* scales of motion. Averaging over the azimuthal coordinate φ then produces a simplified expression of this mean electromotive force which can be used to solve the induction equation for the mean field **H**. The major problem is, therefore, to find the appropriate motions that can maintain such a field against Ohmic decay.

It is apparent from equation (18) that it is easy to produce a toroidal field from a poloidal field by differential rotation; indeed, such a large-scale motion necessarily shears any poloidal field and draws out the lines of force into an azimuthal field. However, in order to regenerate the initially poloidal field from the azimuthal component, it is necessary that the motions should not be axially symmetric. *This is certainly true of the turbulent convection below the solar surface.* As Parker originally pointed out in 1955, the regeneration mechanism may be understood as follows.

The Coriolis force acting on convection causes it to be cyclonic, with a rising cell of fluid rotating and carrying the lines of force of the toroidal field into loops with nonvanishing projection on the meridian plane; a large number of such loops then coalesce to regenerate the poloidal field. According to Parker, the Coriolis force thus shapes the convection, producing the non-uniform rotation (which generates the toroidal field from the poloidal field) and producing the cyclonic rotation of the individual convective cells (which generate the poloidal field from the toroidal field). On the basis of this *topological* discussion, Parker argued heuristically that the rate of regeneration of the poloidal field is proportional to the toroidal field strength, i.e., cyclonic convection produces the mean electromotive force

$$e_\varphi = \Gamma H_\varphi, \tag{22}$$

where the (*variable*) coefficient Γ is a measure of the mean rate and strength of the local cyclonic convection. (For the Sun, owing to the Coriolis force, these motions are clockwise in the northern hemisphere, and counterclockwise in the south.) Accordingly, equation (20) must now be replaced by

$$\frac{\partial A}{\partial t} + \frac{1}{\varpi} \mathbf{v}_p.\mathrm{grad}(\varpi A) = \Gamma H_\varphi + v_m \left(\nabla^2 - \frac{1}{\varpi^2} \right) A. \tag{23}$$

Because of the extra term in equation (23), dynamo action is thus possible.

More recently, Steenbeck, Krause, and Rädler have shown through *formal calculation* that cyclonic turbulence (actually any turbulence that lacks reflection symmetry) can indeed regenerate magnetic fields; that is to say, the net effect of this small-scale, anisotropic turbulence on the *mean* field $\mathbf{H}$ is given by the mean electromotive force

$$\mathbf{e} = \alpha \mathbf{H} - \beta \, \mathrm{curl} \, \mathbf{H}, \tag{24}$$

with

$$\alpha = -\tfrac{1}{3}\tau \overline{\mathbf{v}'.\mathrm{curl} \, \mathbf{v}'} \quad \text{and} \quad \beta = \tfrac{1}{3}\tau \overline{\mathbf{v}'.\mathbf{v}'}, \tag{25}$$

where τ is the correlation time for a turbulent eddy, and $\mathbf{v}'$ is the turbulent velocity. Since the (*variable*) coefficient α vanishes for homogeneous, isotropic turbulence, the existence of this electromotive force (the α-effect) thus depends on having nonzero helicity, corresponding to twisted motions with some preferred sense of rotation. According to Krause, the helicity of the turbulence in the solar convection zone is furnished by the Coriolis force, so that the function α takes the form

$$\alpha = \frac{\tau^2 v_t^2 \Omega}{H_\rho} \cos \theta, \tag{26}$$

where v_t^2 is the r.m.s. value of the turbulent velocity, Ω the angular velocity, H_ρ the density scale height, and θ the colatitude. Given these results, equation (1) can now be used to relate the *mean* fields v and H; we obtain

$$\frac{\partial \mathbf{H}}{\partial t} = \text{curl}(\mathbf{v} \times \mathbf{H} + \alpha\mathbf{H}) - \text{curl}(\beta \text{ curl } \mathbf{H}), \qquad (27)$$

since the diffusivity v_m is largely dominated by the *eddy* diffusivity β (cf. §8.4). Equation (27)—with prescribed values for α, β, and v—embodies the kinematic $\alpha\omega$-dynamo, in which the shear of the non-uniform rotation (*the ω-effect*) winds up an initially poloidal field into a toroidal field, and a reversed poloidal field is generated from the toroidal field by cyclonic turbulence (*the α-effect*) in the solar convection zone.[4] When appropriate boundary conditions are imposed, the solar-dynamo problem becomes an eigenvalue problem: it involves finding eigenvalues λ (say) for which the corresponding solutions $\mathbf{H}(\mathbf{x},t) = \mathbf{H}_0(\mathbf{x}) \exp(\lambda t)$ grow with time, or at least do not decay.

Parker was the first to suggest that the mean magnetic field governing the 22-year solar cycle is an oscillatory solution of an $\alpha\omega$-dynamo. There are now many dynamo models that reasonably explain the maintenance and the variation of the solar fields, beginning with the semi-empirical model of Leighton and the more theoretical model of Steenbeck and Krause. Broadly speaking, all of these models involve the production of toroidal flux by non-uniform rotation in the solar convection zone and a global effect of non-axisymmetric motions that regenerate the poloidal field. The major uncertainties of these kinematic models are these: (i) how does the large-scale velocity field v vary with depth and colatitude in the solar convection zone, and (ii) which of the observed or inferred convective motions is the seat of dynamo regeneration of the poloidal field? Since our knowledge in these matters is very meagre indeed, the functions $\alpha(r,\theta)$, $\beta(r,\theta)$, $\Omega(r,\theta)$, and $\mathbf{v}_p(r,\theta)$ are usually chosen so as to obtain solutions that simulate the periodic variations associated with the solar cycle. It is found, in particular, that the solar-dynamo models are very sensitive to the values of $\partial\Omega/\partial\theta$ and $\partial\Omega/\partial r$. This is quite fortunate since the dependence of Ω on depth and colatitude must now satisfy *two* independent constraints: (i) the values of $\Omega(r,\theta)$ must agree with the observed angular velocity $\Omega(R_\odot,\theta)$ at the solar surface (cf. §2.2), and (ii) this dependence must be such that the dynamo theories reproduce the observed

---

[4] Other mean-field dynamos have been proposed: (i) the α^2-dynamo in which the α-effect alone provides self-maintained solutions of the induction equation for the mean magnetic field (Steenbeck and Krause), and (ii) the $\Omega \times \mathbf{j}$-dynamo, which does not depend on the helicity of turbulence; instead, it makes use of a mean electromotive force perpendicular to both the rotation vector, Ω, and the mean electric current density, $\mathbf{j} = \text{curl } \mathbf{H}$ (Rädler).

properties of the solar activity cycle. The theories of differential rotation (as discussed in Chapter 9) and the present dynamo theories are thus closely linked, each putting severe constraints on the other.

One particular result of the $\alpha\omega$-dynamo is that the mean toroidal field propagates from high latitudes toward the solar equator when the product $\alpha . \partial\Omega/\partial r$ is negative in the northern hemisphere and positive in the southern hemisphere. (The direction of propagation is reversed when the sign of $\alpha . \partial\Omega/\partial r$ is reversed.) Hence, most of the dynamo models of the solar cycle require that the angular velocity *increases* with depth ($\partial\Omega/\partial r < 0$), while most of the theories of differential rotation predict that the angular velocity should *decrease* with depth ($\partial\Omega/\partial r > 0$) because rotational constraints tend to make the angular velocity constant on cylinders centered on the rotation axis. According to Yoshimura, however, the α-term (due to the global convection) changes sign in the lower part of the solar convection zone and is therefore negative. Hence, the constancy of Ω on cylinders could be compatible with Yoshimura's model of the solar cycle, *if the observed toroidal fields are generated in the lower part of the convection zone*, as has recently been argued by Parker.[5] As Stix pointed out, however, observation reveals a phase relation between the toroidal and poloidal components of the mean field that could not be satisfied if Ω decreases with depth. In this context, let us also mention Köhler's finding that the angular velocity must increase with depth for circulatory flows which rise at the poles and sink at the equator, i.e., for meridional currents that give rise to differential rotation with equatorial acceleration. For the moment, it is not yet clear whether real contradictions exist between all of these results, since the approximations used in the (independent) theories of differential rotation and of dynamo action are not always justified for the Sun.

BIBLIOGRAPHICAL NOTES

No attempt at a complete bibliography has been made in this chapter. The following general references may be noted:

1. Cowling, T. G., "Magnetic Stars" in *Stellar Structure* (Aller, L. H., and McLaughlin, D. B., eds.), pp. 425–463, Chicago: The Univ. of Chicago Press, 1965.

[5] According to Parker, magnetic buoyancy would bring a toroidal field of 100 gauss through much of the solar convection zone in a time rather less than 10 years, thus implying that the dynamo operates principally in the very lowest levels of the convective shell. As was subsequently shown by Unno and Ribes, however, eddy viscosity may balance magnetic buoyancy, so that the toroidal field of 100 gauss can be retained long enough for the solar dynamo to operate *in* the solar convection zone.

2. Ledoux, P., and Renson, P., "Magnetic Stars" in *Annual Review of Astronomy and Astrophysics* 4, pp. 293–352, Palo Alto, California: Annual Reviews, Inc., 1966.

3. Mestel, L., "The Theory of Magnetic Stars" in *Quart. J. Roy. Astron. Soc. London* 12 (1971): 402–416.

4. Mestel, L., "Stellar Magnetism and Rotation" in *Stellar Evolution* (Chiu, H. Y., and Muriel, A., eds.), pp. 643–734, Cambridge, Massachusetts: The M.I.T. Press, 1972.

5. Mestel, L., "Effects of Magnetic Fields" in *Mém. Soc. Roy. Sci. Liège* (6) 8 (1975): 79–91.

Section 15.2. The classical paper on the subject matter treated in this section is due to:

6. Chandrasekhar, S., and Fermi, E., *Astroph. J.* 118 (1953): 116–141; *ibid.* 122 (1955): 208.

The pioneering work of Ferraro, Prendergast, Wentzel, and Woltjer may be traced to references 1 and 2. Among the many contributions on the effects of a magnetic field on the structure of self-gravitating systems, we shall only mention those which refer to nonrotating, *centrally condensed* bodies; these are:

7. Monaghan, J. J., *Monthly Notices Roy. Astron. Soc. London* 131 (1965): 105–119; *ibid.* 132 (1966): 1–14; *ibid.* 134 (1966): 275–285.

8. Roxburgh, I. W., *ibid.* 132 (1966): 347–358.

9. Van der Borght, R., *Australian J. Phys.* 20 (1967): 643–650.

10. Monaghan, J. J., *Zeit. f. Astroph.* 69 (1968): 146–153.

11. Sinha, N. K., *Australian J. Phys.* 21 (1968): 283–291.

12. Chiam, T. C., and Monaghan, J. J., *Monthly Notices Roy. Astron. Soc. London* 155 (1971): 153–167.

13. Miketinac, M. J., *Astroph. and Space Sci.* 22 (1973): 413–419; *ibid.* 35 (1975): 349–362.

The problem of the global oscillations and stability of a magnetized body was originally clarified by:

14. Chandrasekhar, S., and Limber, D. N., *Astroph. J.* 119 (1954): 10–13.

15. Ledoux, P., et Simon, R., *Ann. Astroph.* 20 (1957): 185–195.

16. Wentzel, D. G., *Astroph. J.* 135 (1962): 593–598.

17. Woltjer, L., *ibid.* 135 (1962): 235–237.

Detailed evaluations of the effects of various magnetic fields on the natural frequencies of nonrotating, *compressible* configurations will be found in:

18. Roxburgh, I. W., and Durney, B. R., *Monthly Notices Roy. Astron. Soc. London* **135** (1967): 329–337.
19. Tassoul, J. L., *ibid.* **138** (1968): 123–136.
20. Anand, S. P. S., *Astroph. and Space Sci.* **4** (1969): 255–274.
21. Singh, S., and Tandon, J. N., *Astron. and Astroph.* **2** (1969): 266–273.
22. Fahlman, G. G., *Astroph. and Space Sci.* **12** (1971): 424–455.
23. Sood, N. K., and Trehan, S. K., *ibid.* **10** (1971): 393–401; *ibid.* **16** (1972): 451–464; *ibid.* **19** (1972): 441–467; *ibid.* **33** (1975): 165–172.
24. Trehan, S. K., and Billings, D. F., *Astroph. J.* **169** (1971): 567–584.
25. Goossens, M., *Astroph. and Space Sci.* **16** (1972): 386–404; *ibid.* **43** (1976): 9–18.
26. Trehan, S. K., and Uberoi, M. S., *Astroph. J.* **175** (1972): 161–169.
27. Billings, D. F., Singh, M., and Trehan, S. K., *Astroph. and Space Sci.* **25** (1973): 457–469.
28. Grover, S. D., Singh, S., and Tandon, J. N., *Astron. and Astroph.* **22** (1973): 133–137.
29. Miketinac, M. J., *Astroph. and Space Sci.* **28** (1974): 193–203.
30. Goossens, M., Smeyers, P., and Denis, J., *ibid.* **39** (1976): 257–272.
31. Sobouti, Y., *Astron. and Astroph.* **55** (1977): 339–346.

Section 15.3. The presentation in the text largely follows references 3–5. Most of the relevant papers may be traced to references 1–6. The "fossil" theory was originally discussed by:

32. Cowling, T. G., *Monthly Notices Roy. Astron. Soc. London* **105** (1945): 166–174.

See also:

33. Wrubel, M. H., *Astroph. J.* **116** (1952): 291–298.
34. Smith, T. S., *ibid.* **139** (1964): 767–772.

The theory of star formation in magnetic dusty clouds is due to:

35. Mestel, L., and Spitzer, L., Jr., *Monthly Notices Roy. Astron. Soc. London* **116** (1956): 503–514.

See also references 1 (pp. 238–242), 9, and 10 of Chapter 11. Among the pioneering papers on the effects of turbulence on a magnetic field, reference may be made to:

36. Sweet, P. A., *Monthly Notices Roy. Astron. Soc. London* **110** (1950): 69–83.
37. Elsasser, W. M., *Rev. Mod. Phys.* **28** (1956): 135–163.

See also:

38. Spitzer, L., Jr., *Astroph. J.* **125** (1957): 525–534.

39. Parker, E. N., *ibid.* **138** (1963): 552–575.
40. Weiss, N. O., *Proc. Roy. Soc. London, A*, **293** (1966): 310–328.

The influence of a magnetic field on the Schwarzschild criterion has been discussed by:

41. Gough, D. O., and Tayler, R. J., *Monthly Notices Roy. Astron. Soc. London* **133** (1966): 85–98.
42. Kovetz, A., *ibid.* **137** (1967): 169–174.
43. Moss, D. L., and Tayler, R. J., *ibid.* **145** (1969): 217–240; *ibid.* **147** (1970): 133–138.

Various local instabilities are considered in:

44. Schubert, G., *Astroph. J.* **151** (1968): 1099–1110.
45. Vandakurov, Yu. V., *Astron. Zh.* **49** (1972): 324–333.
46. Markey, P., and Tayler, R. J., *Monthly Notices Roy. Astron. Soc. London* **163** (1973): 77–91; *ibid.* **168** (1974): 505–514.
47. Tayler, R. J., *ibid.* **161** (1973): 365–380.
48. Wright, G. A. E., *ibid.* **162** (1973): 339–358.

The following papers are further quoted in the text:

49. Braginski, S. I., *Zh. Exper. i Theoret. Fiz. (U.S.S.R.)* **47** (1964): 1084–1098, 2178–2193.
50. Steenbeck, M., Krause, F., und Rädler, K.-H., *Zeit. f. Naturforschg.* **21a** (1966): 369–376.
51. Schüssler, M., *Astron. and Astroph.* **38** (1975): 263–270.

See also reference 55. A paper of related interest is:

52. Drobyshevski, E. M., *Astrofizika* **9** (1973): 119–138.

Section 15.4. Recent measurements of strong magnetic fields have been reported by:

53. Wolff, S. C., and Morrison, N. D., *Publ. Astron. Soc. Pacific* **86** (1974): 935–939.
54. Landstreet, J. D., Borra, E. F., Angel, J. R. P., and Illing, R. M. E., *Astroph. J.* **201** (1975): 624–629.
55. Wolff, R. J., and Wolff, S. C., *ibid.* **203** (1976): 171–176.
56. Wolff, S. C., and Hagen, W., *Publ. Astron. Soc. Pacific* **88** (1976): 119–121.

The oblique-rotator model is discussed at length in:

57. Preston, G. W., "The Rotation of the Ap Stars from the Point of View of the Rigid Rotator Model" in *Stellar Rotation* (Slettebak, A., ed.), pp. 254–263, New York: Gordon and Breach, 1970.

The pioneering work of Babcock, Stibbs, and Deutsch may be traced to this paper; see also reference 64 of Chapter 12. Recent studies of the oblique-rotator model will be found in:

58. Landstreet, J. D., *Astroph. J.* **159** (1970): 1001–1007.

59. Hockey, M. S., *Monthly Notices Roy. Astron. Soc. London* **152** (1971): 97–107.

60. Huchra, J., *Astroph. J.* **174** (1972): 435–438.

61. Borra, E. F., *ibid.* **187** (1974): 271–274; *ibid.* **188** (1974): 287–290; *ibid.* **193** (1974): 699–703.

62. Monaghan, J. J., *Monthly Notices Roy. Astron. Soc. London* **167** (1974): 163–168.

63. Stift, M. J., *ibid.* **169** (1974): 471–476; *ibid.* **172** (1975): 133–139.

64. Wolff, S. C., *Astroph. J.* **202** (1975): 127–136.

65. Borra, E. F., and Vaughan, A. H., *Astroph. J. Letters* **210** (1976): L145–L147.

66. Borra, E. F., and Landstreet, J. D., *Astroph. J.* **212** (1977): 141–148.

An interesting survey of the theoretical work on equilibrium magnetic fields in rotating stars will be found in reference 5. See also:

67. Davies, G. F., *Australian J. Phys.* **21** (1968): 293–305.

68. Vandakurov, Yu. V., *Astron. Zh.* **45** (1968): 103–114, 813–822.

69. Wright, G. A. E., *Monthly Notices Roy. Astron. Soc. London* **146** (1969): 197–212.

70. Trasco, J. D., *Astroph. J.* **161** (1970): 633–642.

71. Monaghan, J. J., and Robson, K. W., *Monthly Notices Roy. Astron. Soc. London* **155** (1971): 231–247.

72. Maheswaran, M., and de Silva, H. A. B. M., *ibid.* **162** (1973): 289–293.

73. Monaghan, J. J., *Astroph. J.* **186** (1973): 631–642.

74. Monaghan, J. J., *Monthly Notices Roy. Astron. Soc. London* **163** (1973): 423–435.

75. Moss, D. L., *ibid.* **164** (1973): 33–51; *ibid.* **171** (1975): 303–309; *ibid.* **173** (1975): 141–160; *ibid.* **178** (1977): 61–70.

76. Smith, R. C., *ibid.* **173** (1975): 97–102.

77. Das, M. K., and Tandon, J. N., *Astroph. J.* **209** (1976): 233–242.

78. Monaghan, J. J., *Astroph. and Space Sci.* **40** (1976): 385–391.

Theoretical work on the oblique-rotator model is due to:

79. Spitzer, L., Jr., in *Electromagnetic Phenomena in Cosmical Physics* (Lehnert, B., ed.), pp. 169–181, Cambridge: At the Univ. Press, 1958.

80. Mestel, L., and Takhar, H. S., *Monthly Notices Roy. Astron. Soc. London* **156** (1972): 419–436.

See also reference 78 of Chapter 11.

Section 15.5. There is a wide literature on dynamo action in celestial bodies. Most of the early references may be traced to:

81. Parker, E. N., "The Origin of Solar Magnetic Fields" in *Annual Review of Astronomy and Astrophysics* **8**, pp. 1–30, Palo Alto, California: Annual Reviews, Inc., 1970.

82. Weiss, N. O., "Theories of Large Scale Fields and the Magnetic Activity Cycle" in *Solar Magnetic Fields* (Howard, R., ed.), pp. 757–769, Dordrecht: D. Reidel Publ. Co., 1971.

83. Gubbins, D., "Theories of the Geomagnetic and Solar Dynamos" in *Reviews of Geophysics and Space Physics* **12** (1974): 137–154.

Other relevant papers will be found in:

84. Bumba, V., and Kleczek, J., eds., *Basic Mechanisms of Solar Activity*, Dordrecht: D. Reidel Publ. Co., 1976.

See especially the final discussion (pp. 479–481) by Durney, Gilman, and Stix. Observations of the solar magnetic fields are discussed in:

85. Stenflo, J. O., *Solar Phys.* **23** (1972): 307–339.

86. Gillespie, B., Harvey, J., Livingston, W., and Harvey, K., *Astroph. J. Letters* **186** (1973): L85–L86.

Among the many theoretical papers on dynamo action and the solar cycle, reference may be made to:

87. Parker, E. N., *Astroph. J.* **122** (1955): 293–314; *ibid.* **164** (1971): 491–509; *ibid.* **198** (1975): 205–209.

88. Leighton, R. B., *ibid.* **156** (1969): 1–26.

89. Steenbeck, M., und Krause, F., *Astron. Nachr.* **291** (1969): 49–84.

90. Iroshnikov, R. S., *Astron. Zh.* **47** (1970): 1253–1267.

91. Deinzer, W., and Stix, M., *Astron. and Astroph.* **12** (1971): 111–119.

92. Stix, M., *ibid.* **13** (1971): 203–208; *ibid.* **20** (1972): 9–12; *ibid.* **24** (1973): 275–281; *ibid.* **37** (1974): 121–133; *ibid.* **47** (1976): 243–254.

93. Lerche, I., and Parker, E. N., *Astroph. J.* **176** (1972): 213–223.

94. Roberts, P. H., *Phil. Trans. Roy. Soc. London, A,* **272** (1972): 663–698.

95. Roberts, P. H., and Stix, M., *Astron. and Astroph.* **18** (1972): 453–466.

96. Yoshimura, H., *Astroph. J.* **178** (1972): 863–886; *ibid.* **201** (1975): 740–748.

97. Köhler, H., *Astron. and Astroph.* **25** (1973): 467–476.

98. Deinzer, W., v. Kusserow, H.-U., and Stix, M., *ibid.* **36** (1974): 69–78.

99. Jepps, S. A., *J. Fluid Mechanics* **67** (1975): 625–646.

100. Yoshimura, H., *Astroph. J. Supplements* **29** (1975): 467–494.

101. Unno, W., and Ribes, E., *Astroph. J.* **208** (1976): 222–223.
102. Yoshimura, H., *Solar Phys.* **50** (1976): 3–23.

Differential rotation, meridional currents, and dynamo action are considered in:

103. Tuominen, I. V., and Tuominen, J., *Astroph. Letters* **1** (1968): 95–97.
104. Köhler, H., *Solar Phys.* **34** (1974): 11–14.
105. Durney, B. R., *Astroph. J.* **199** (1975): 761–764.

Papers of related interest are:

106. Piddington, J. H., *Astroph. and Space Sci.* **38** (1975): 157–166.
107. Vandakurov, Yu. V., *Astron. Zh.* **52** (1975): 351–358.
108. Wolff, C. L., *Astroph. J.* **205** (1976): 612–621.

16

Rotation in Close Binaries

16.1. INTRODUCTION

The effect of axial rotation upon the structure, evolution, and stability of a single star has been the main subject matter of this book. In the preceding chapter, we showed that the presence of magnetic fields actuates extra degrees of freedom that introduce a number of new elements into the problem of axial rotation. Further challenging problems arise from the study of double stars whose components are close enough to induce tidal distortion on each other. (The main observational data pertaining to these matters was presented in Section 2.4.) Since many comprehensive surveys of the double-star problem are now available in the literature, our concern with close binaries will be only to the extent that they bear on the general theory of axial rotation and stellar hydrodynamics. The major ideas on the formation of double and multiple systems have already been discussed in Chapter 11; here we shall consider mainly the combined effect of axial rotation and tidal action on the gross properties of components of close binaries. In Section 16.2, the problem of determining the internal structure of a synchronously rotating binary component is discussed on the basis of idealized models. Various physical mechanisms that may cause synchronization between the periods of axial rotation and orbital revolution in close binaries are considered further in Section 16.3. The chapter concludes with a brief discussion of the gas outflow from a mass-losing component of a close binary.

16.2. SYNCHRONOUSLY ROTATING MODELS

If we ignore possible disturbances caused by the presence of magnetic fields, the components of a close binary are subject to two types of distortion: a *polar flattening* due to rotation of the components about axes passing through their centers of mass, and *tides* raised by one component on the other. Moreover, since both these effects cause departure from spherical symmetry, large-scale *meridional currents* must also be taken into account to comply with the simultaneous conditions of mechanical and thermal equilibrium in radiative zones. So far, almost all studies of the internal structure of rotating stars deformed by the gravitational field of a close companion have implicitly assumed that these stars remain, at all times, in mechanical equilibrium; and meridional circulation has never been properly taken into account. The approximation of mechanical equilibrium is certainly justified when the angular velocity of rotation of the binary component is equal to the orbital angular velocity, both in magnitude and in direction (i.e., when the uniform rotation is perfectly synchronized with the circular orbital motion so that the tidal bulges are steady in time). Postponing to Section 16.3 the study of some time-dependent problems, we shall first consider the highly idealized Roche ellipsoids, for they illustrate quite interesting features of the double-star problem; thence, we shall proceed with centrally condensed, tidally distorted models, concluding the section with a brief discussion of the so-called Roche model, which, for most purposes, provides a simple and useful representation of actual close binaries.

As we have seen in Section 10.2, the Maclaurin spheroids adequately illustrate the general properties of axially rotating bodies. Hence, we shall first consider the equilibrium and stability of a class of ellipsoidal figures that arise when, in addition to axial rotation, the disturbances caused by a neighboring mass are included in a first approximation. The simplest problem, which was originally dealt with by Roche in 1850, is concerned with a homogeneous ellipsoid ("the primary") rotating about a rigid sphere ("the secondary") in such a way that their relative dispositions remain the same, with no internal motions in the primary.

Let the masses of the primary and the secondary be M and M′, respectively; let the distance between their centers of mass be d; and let the angular velocity of rotation about their common center of mass be Ω. Next choose a system of reference in which the origin is at the center of mass of the primary; for convenience, the x_1-axis points toward the center of mass of the secondary, and the x_3-axis is parallel to the direction of Ω. Then, the equation of the rotation axis, which of course passes through the center of mass of the two bodies, is

$$x_1 = \frac{M'}{M + M'} d \quad \text{and} \quad x_2 = 0. \tag{1}$$

Accordingly, the centrifugal force acting on the mass M may be derived from the potential

$$-\tfrac{1}{2}\Omega^2 \left[\left(x_1 - \frac{M'}{M + M'} d \right)^2 + x_2^2 \right]. \tag{2}$$

In Roche's particular problem, the secondary is treated as a rigid sphere; hence, over the primary, the tide-generating potential can be expanded in the form

$$-\frac{GM'}{d} \left(1 + \frac{x_1}{d} + \frac{x_1^2 - \tfrac{1}{2}x_2^2 - \tfrac{1}{2}x_3^2}{d^2} + \cdots \right). \tag{3}$$

The approximation that underlies the theory is to omit all terms beyond those written down. On this assumption, we find that, apart from its own gravitation, the primary may be supposed to be acted upon by a total field of force derived from the potential

$$-\tfrac{1}{2}\Omega^2(x_1{}^2 + x_2{}^2) - \mu(x_1{}^2 - \tfrac{1}{2}x_2{}^2 - \tfrac{1}{2}x_3{}^2) - \left(\mu - \frac{M'}{M + M'}\,\Omega^2\right)dx_1, \quad (4)$$

where

$$\mu = \frac{GM'}{d^3}. \tag{5}$$

Further letting Ω^2 have its "Keplerian value"

$$\Omega^2 = \frac{G(M + M')}{d^3}, \tag{6}$$

we can thus write the conditions of relative equilibrium for the primary in the form

$$\frac{1}{\rho}\,\mathrm{grad}\,p = -\mathrm{grad}[V - \tfrac{1}{2}\Omega^2(x_1{}^2 + x_2{}^2) - \mu(x_1{}^2 - \tfrac{1}{2}x_2{}^2 - \tfrac{1}{2}x_3{}^2)], \quad (7)$$

where V is the self-gravitating potential of the primary.

Now, the function V is quadratic in the coordinates for a homogeneous ellipsoid (cf. §10.2: equations [4]–[6]); hence, it is clear at once that such a configuration is consistent with equation (7) as well as the condition that requires the pressure to be constant over the bounding surface. Detailed numerical solutions of this problem have been presented by Chandrasekhar for various values of the ratio

$$q = \frac{M}{M'}. \tag{8}$$

As was expected, when $q \to \infty$, the term in μ in equation (7) becomes negligible, so we then recover the pure rotationally distorted configurations of Maclaurin and Jacobi (cf. §10.2). In the range $0 \leqslant q < \infty$, the figures of equilibrium are ellipsoids with semi-axes such that $a_1 > a_2 > a_3$; that is, for a given value of q, each series begins with a sphere and ends with an infinitely long prolate spheroid. Figure 16.1 depicts the variations of Ω^2, in units of $\pi G\rho$, along various Roche sequences. Solutions of equation (7) for the illustrative case $q = 1$ are given in Table 16.1; in this table, in addition to the principal constants (a_2/a_1, a_3/a_1, $\Omega^2/\pi G\rho$, and $\mu/\pi G\rho$), the semi-axes of the ellipsoids, in units of $(a_1 a_2 a_3)^{1/3}$, are also listed. It is apparent that along each series the normalized quantities Ω^2 and μ at first increase then, after passing through maximum values,

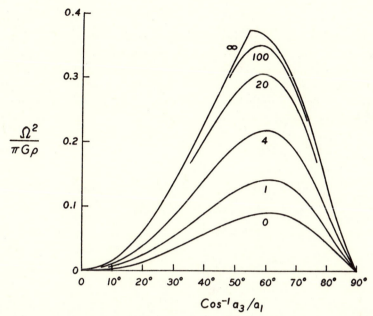

Figure 16.1. The variations of Ω^2 along the Roche sequences. The curves are labeled by the values of q to which they belong; and the curve labeled "∞" belongs to the combined Maclaurin-Jacobi sequence. The maxima of the curves define the Roche limit. Source: Chandrasekhar, S., *Ap. J.* **138**, 1182, 1963. (By permission of The University of Chicago Press. Copyright 1963 by the University of Chicago.)

TABLE 16.1

The properties of the Roche ellipsoids for $q = 1$[†]

$\cos^{-1}a_3/a_1$	a_2/a_1	a_3/a_1	Ω^2	μ	a_1	a_2	a_3
12°	0.98660	0.97815	0.009293	0.004647	1.0119	0.9984	0.9898
24	.94376	.91355	.036152	.018076	1.0507	.9916	.9598
36	.86345	.80902	.076342	.038171	1.1270	.9731	.9118
48	.73454	.66913	.118726	.059363	1.2671	.9308	.8479
54	.64956	.58779	.134284	.067142	1.3784	.8954	.8102
59	.56892	.51504	.140854	.070427	1.5056	.8566	.7754
60	.55186	.50000	.141250	.070625	1.5360	.8477	.7680
61	.53451	.48481	.141298	.070649	1.5685	.8384	.7604
66	.44429	.40674	.135785	.067892	1.7688	.7859	.7194
69	.38813	.35837	.127424	.063712	1.9300	.7491	.6917
71	.35022	.32557	.119625	.059812	2.0622	.7222	.6714
72	.33119	.30902	.115054	.057527	2.1379	.7080	.6606
73	.31213	.29237	.110044	.055022	2.2211	.6933	.6494
75	.27405	.25882	.098753	.049376	2.4158	.6620	.6252
78	.21726	.20791	.078934	.039467	2.8079	.6100	.5838
81	0.16126	0.15643	0.056499	0.028249	3.4097	0.5498	0.5334

[†] Ω^2 and μ are listed in units of $\pi G\rho$, the a_i's in units of $(a_1 a_2 a_3)^{1/3}$.

SOURCE: Chandrasekhar, S., *Ap. J.* **138**, 1182, 1963. (By permission of The University of Chicago Press. Copyright 1963 by the University of Chicago.)

decrease. Actually, both Ω^2 and μ attain maxima, simultaneously, at some determinate point along a Roche sequence. *Accordingly, for distances d smaller than what corresponds to the maximum value of μ attained, no ellipsoidal figures of equilibrium are possible.* This limit, which defines the distance of closest approach for equilibrium to be possible, is commonly spoken of as Roche's limit. The points along various Roche sequences where Ω^2 and μ attain their maxima are listed in Table 16.2.

TABLE 16.2

The Roche limit and the constants of the critical ellipsoid[†]

q	Ω^2_{max}	μ_{max}	a_1	a_2	a_3
0	0.090068	0.090068	1.5947	0.8151	0.7693
1	.141322	.070661	1.5565	0.8418	.7632
4	.216861	.043372	1.4944	0.8913	.7508
20	.306396	.014590	1.3989	0.9821	.7279
100	.350562	0.003471	1.3224	1.0642	.7106
∞	0.374230	0	1.1972	1.1972	0.6977

[†] Ω^2_{max} and μ_{max} are listed in units of $\pi G \rho$, the a_i's in units of $(a_1 a_2 a_3)^{1/3}$.

SOURCE: Chandrasekhar, S., *Ap. J.* **138**, 1182, 1963. (By permission of The University of Chicago Press. Copyright 1963 by the University of Chicago.)

It was generally believed that the non-existence of figures of equilibrium below the Roche limit implies the onset of some sort of instability at the point where μ attains its maximum value. This question has been recently revived by Chandrasekhar. According to this author, a Roche ellipsoid does not become dynamically unstable with respect to second-order disturbances at the point where Ω^2 and μ attain their maxima; rather, *dynamical* instability sets in only at a subsequent point along the sequence where the distance d is somewhat greater than the distance of closest approach (see Table 16.3). As Robe subsequently demonstrated, however, a Roche ellipsoid does become *secularly* unstable, by a purely viscous mode belonging to second-order harmonics, between the Roche limit and the point of onset of dynamical instability. Such a situation is somewhat reminiscent of the case of the Maclaurin spheroids, which become secularly unstable first, dynamical instability occurring further along the sequence (cf. §10.2).[1] In the present instance, however, the onset of secular

---

[1] Still, one may rightfully ask whether the existence of two distinct stability limits for the Roche ellipsoids is physically significant, or whether this is an unwanted consequence of the fact that, because of the truncated expansion (3), the *total* angular momentum is not exactly preserved during the oscillations.

TABLE 16.3

The point at which dynamical instability
sets in along the Roche sequences[†]

q	Ω^2	μ	a_1	a_2	a_3
0	0.090034	0.090034	1.611	0.8108	0.7655
1	.141229	.070614	1.581	0.8350	.7577
4	.216581	.043316	1.526	0.8808	.7439
20	.305937	.014568	1.430	0.9683	.7222
100	.350303	0.003468	1.343	1.0523	.7078
∞	0.374230	0	1.197	1.1972	0.6977

[†] Ω^2 and μ are listed in units of $\pi G\rho$, the a_i's in units of $(a_1 a_2 a_3)^{1/3}$.

SOURCE: Chandrasekhar, S., *Ap. J.* **138**, 1182, 1963. (By permission of The University of Chicago Press. Copyright 1963 by the University of Chicago.)

instability at the Roche limit is not associated with the presence of a point of bifurcation, since along a Roche sequence there is no point of bifurcation where another sequence of ellipsoidal figures branches off.

From the absence of equilibrium figures inside the Roche limit (established, as we have stated, only when the orbits are circular) it has been also inferred that if the primary describing, for example, an eccentric orbit should come inside the Roche limit, it would break up into pieces. As was shown by Nduka, a consistent formulation of the motion of the primary when it transgresses the Roche limit requires that we allow internal motions of uniform vorticity as well as the possibility of varying orientation of its ellipsoidal figure. According to Nduka, so long as the orbit remains outside the Roche limiting circle, the figure and the orientation of the primary vary only by moderate amounts; in contrast, if the orbit intersects this circle, then the ratio a_3/a_1 decreases and the primary begins to change into a disc-like figure. No results are so far available for the case where third- and higher-order harmonics distort the primary when it transgresses the Roche limit.

Let us now discuss the effect of compressibility on the natural frequencies of a rotationally and tidally distorted body. As we recall from Section 14.2, rotation alone couples the two lowest zonal modes which, in the limit of hydrostatic equilibrium, reduce to the lowest radial mode (σ_R) and the lowest axisymmetric f-mode (σ_Z); moreover, in the absence of tidal interaction, rotation reduces the critical adiabatic exponent γ_c from its "classical value" 4/3 (see Figure 14.1). The very same stability analysis has been extended by Chandrasekhar to determine the influence of compressibility on the stability of the Roche ellipsoids. Specifically, the problem to be considered is that of the small isentropic oscillations belonging to second-order harmonics. The assumption of *homogeneity* insures that, in the equilibrium state, the configurations will be indistinguishable from the incompressible Roche ellipsoids. But the assumption that they are *gaseous* has the consequence that the Lagrangian displacement describing a deformation can no longer be volume preserving; instead, we must apply the laws appropriate to a gas that is subject to isentropic changes. The main results of Chandrasekhar's calculations are depicted in Figure 16.2. (As usual, γ_c denotes the critical adiabatic exponent at which the

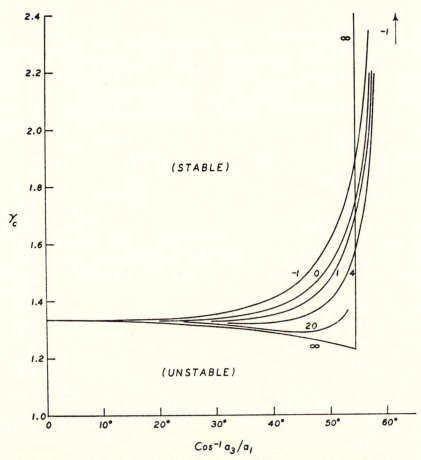

Figure 16.2.   The critical values γ_c at which marginal stability occurs for the homogeneous but compressible Roche ellipsoids. Except for the curve labeled "−1," which refers to the Jeans spheroids, the curves are labeled by the values of q to which they belong. Source: Chandrasekhar, S., *Ap. J.* **138**, 1182, 1963. (By permission of The University of Chicago Press. Copyright 1963 by the University of Chicago.)

squared frequency σ^2 vanishes; the curve labeled "-1" describes the so-called Jeans sequence, i.e., nonrotating prolate spheroids that depart from spherical symmetry by the sole tidal action of a secondary.) *We observe that gaseous configurations that are normally stable when $\gamma > 4/3$ may become dynamically unstable under the combined influence of uniform rotation and tidal action.* Since rotation alone has a stabilizing influence on the lowest zonal modes, it follows that tidal action destabilizes these isentropic motions. So far, however, the accentuation of the instability of a gaseous configuration by a close companion has been established only for homogeneous ellipsoids. Although it would not seem that the tidally induced instability of the lowest zonal modes is peculiar to them, further calculations are needed to demonstrate that centrally condensed, tidally distorted bodies do imitate the Roche ellipsoids.

We now turn to the question of the internal structure of a centrally condensed, synchronously rotating member of a close binary. For the sake of clarity, we shall first consider the so-called Roche model, in which practically all the mass of each component is concentrated in a central point surrounded by a tenuous envelope of vanishing density. This double-star model (originally proposed by Roche in 1873) stands at the opposite of the homogeneous Roche ellipsoids. Its importance stems from the fact that it not only provides an *exact* expression for the level surfaces of a binary whose components can be regarded as mass points, but also a *good approximation* for components whose degree of central condensation— though finite—happens to be high.

As before, let M and M′ denote the masses of the two components, and let d be their mutual separation. Suppose further that the positions of the two masses are referred to a Cartesian frame of reference—with the origin at the center of mass of the primary—the x_1-axis of which coincides with the line joining the centers of the two masses, while the x_3-axis is perpendicular to the plane of the orbit. If so, then, the rotation axis is given by equation (1), and the total potential Ψ of the combined forces acting at an arbitrary point $P(\mathbf{x})$ becomes expressible as

$$\Psi = -G\frac{M}{r} - G\frac{M'}{r'} - \tfrac{1}{2}\Omega^2\left[\left(x_1 - \frac{M'}{M+M'}d\right)^2 + x_2{}^2\right], \qquad (9)$$

where r and r' represent, respectively, the distances of the point P from the centers of mass of the two components. The first term on the r.h.s. of equation (9) represents the potential arising from the mass M, the second, the disturbing potential of its companion of mass M′, and the third, the potential arising from the centrifugal force. If we assume further that the angular velocity occurring in equation (9) is identical with the Keplerian angular velocity (6), the level surfaces may be expressed as

$$\Phi = \frac{1}{1+M'}\frac{1}{r} + \frac{M'}{1+M'}\left(\frac{1}{r'} - x_1\right) + \tfrac{1}{2}(x_1{}^2 + x_2{}^2) = constant, \quad (10)$$

with

$$\Phi = -\frac{1}{1+M'}\Psi - \frac{1}{2}\frac{M'^2}{(1+M')^2}, \quad (11)$$

where we have adopted a system of units such that $G = M = d = 1$. Now, as can readily be seen from the analysis of Section 4.3, the level surfaces $\Phi = constant$ coincide with the surfaces of constant density, as well as with those of constant pressure. Equation (10) provides, therefore, a complete description of a close-binary system whose synchronously rotating components may be regarded as mass points.

Figure 16.3 represents a section of the level surfaces cut by the orbital plane for a mass ratio $M'/M = 0.215$. Quite generally, level surfaces corresponding to high values of the parameter Φ form separate lobes enclosing each one of the two centers of mass and differ little from spheres. With diminishing values of Φ, the lobes defined by equation (10) become increasingly elongated in the direction of the center of mass of the system until, for a certain critical value $\Phi = \Phi_c$ characteristic of each mass ratio M'/M, both ovals will come into contact at a single point on the axis joining the centers of mass of the two components to form a dumb-bell-like configuration. It will henceforth be called the Roche critical surface; and its two lobes will be called the Roche critical lobes. When $\Phi < \Phi_c$, the connecting part of the dumb-bell would open up so that single level surfaces enclose both bodies, thus depriving us of the possibility of regarding equation (10) as the representation of a detached double star. For each $\Phi > \Phi_c$, equation (10) will, however, define two detached sets of level surfaces that approximate the forms of centrally condensed components of close binaries. As far as the stability of the masses M and M' is concerned, each will remain stable as long as its volume remains smaller than that of its Roche critical lobe. As we shall see in Section 16.4, however, if the volume of the primary (say) increases because of evolutionary expansion, then instability will arise as soon as the primary fills its Roche critical lobe, and matter will be lost through the inner Lagrangian point L_1 (see Figure 16.3).

Because actual stars have fairly large central condensations, the Roche model is generally believed to be useful for the discussion of the evolution of close binaries. In order to check the validity of this assumption, various authors have considered the combined effect of rotation and tidal action on the internal structure of polytropes and realistic stellar models. The

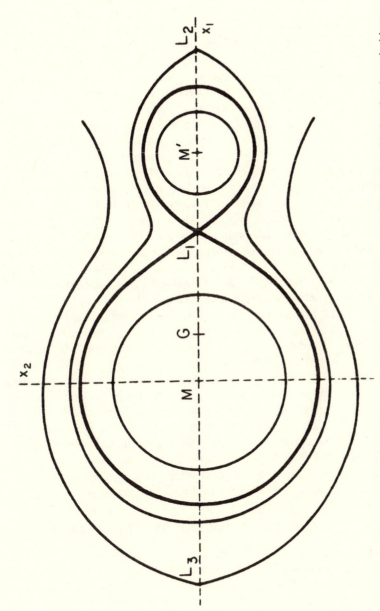

Figure 16.3. Level surfaces of the Roche model in the orbital plane ($M'/M = 0.215$). The Roche critical surface is marked by a heavy line. Source: Plavec, M., *Adv. A. Ap.* **6**, 201, 1968.

principal techniques employed in dealing with synchronously rotating components of close binaries merely generalize the methods of constructing rotationally distorted stars (see Chapter 5). In all these studies (notably those by Jackson, by Kippenhahn and Thomas, and by Naylor and Anand) tidal forces have been found to have almost no effect on the binary component's luminosity and are several times less effective than uniform rotation in increasing the mean radius. Moreover, because the rotation is assumed to be synchronous, the binary component cannot rotate as fast as a single star. Consequently, the reduction of luminosity, which is due almost entirely to rotation, is correspondingly less than for a single star. As was originally surmised by Dziembowski, it thus appears that a synchronously rotating component of a close binary may be treated, indeed, as a spherically symmetric star, *even if it fills up its Roche critical lobe in the course of its evolution.* Because the small departures from spherical symmetry caused by axial rotation and tidal action do not play a decisive role in binary star evolution (at least insofar as synchronously rotating models are concerned), these matters are thus outside the scope of this book.

Several authors (notably Plavec, Kruszewski, and Limber) have also examined the problem of *nonsynchronous* close binaries under the explicit assumption that the gravitational potential of each star could be represented by that of a mass point, i.e., the Roche model. Recently, Naylor has calculated models of uniformly rotating, nonsynchronous binary polytropes and compared these with the results Limber obtained from the Roche model. He found that the relative dimensions of nonsynchronous systems are adequately portrayed by the Roche model; however, the degree of central condensation affects the minimum separation by an amount that is probably not insignificant. In other words, whenever axial rotation is not synchronous with the orbital revolution, the Roche model is applicable providing only the relative dimensions are of interest; but if the radii and separation are required separately, then the effects of the finite central condensation of real stars should be considered.

16.3.   TIDAL INTERACTION AND SYNCHRONIZATION

In Section 2.4, we showed that there is substantial evidence of a tendency toward synchronization between the axial and orbital motions among a large number of close binaries. Observation reveals, *first*, that the components of close binaries have small projected rotational velocities, $v_e \sin i$, relative to the means for single stars of corresponding spectral types (see Figure 2.11). *Second*, when comparing the equatorial velocities v_e with the computed equatorial velocities $v(\text{syn})(= 2\pi R/P_0)$ that would be synchronous with the orbital period P_0, we observe that synchronization obtains in most close binaries of short periods, either perfectly or approximately (see Figure 2.12); however, roughly at least 10 percent of close binary systems rotate too rapidly for synchronization to occur, and this in spite of their short periods. *Third*, as Levato recently demonstrated, the periods below which the axial and orbital motions are synchronized are systematically larger for evolved main-sequence stars than for the ones near the zero-age main sequence. In the light of these observational facts, it is believed that tidal interaction between the components of a close binary system is responsible for the required rotational braking, *synchronization (for a given period) depending on how long the tidal forces have been acting on the components.* For the very same reason, it is now commonly accepted that the slow rotation rates of the marked (Am) and marginal (Am:) metallic-line stars are caused mostly by tidal interaction in closely spaced systems (cf. §12.4). Here we shall briefly comment on some physical mechanisms that may cause synchronization in close binaries.

As far back as 1879, Darwin outlined the evolutionary processes that may produce secular changes in the axial rotation of a member of a close binary, and established that—for systems exhibiting no mass loss or exchange—the principal causes of such an evolution are tides lagging in phase behind the external field of force on account of the viscosity of stellar matter. As we said previously, if the relative orbit of the component of a double star is circular, and the uniform rotation synchronized with revolution, the tidal bulges raised by one component upon another will remain steady in time; hence, they will give rise to no motion relative to a frame of reference rotating with each star. On the contrary, if none of these circumstances prevails, an observer "stationed" on one of the components will see the position of the companion changing in time. As long as the material may be regarded as inviscid, the shape of each component will almost instantaneously adjust itself to the external field of force (i.e., on a dynamical time scale); *but this does not imply that the rotation rate will adjust itself as well.* In the presence of viscosity, however, the response of the individual tidal bulges will no longer be instantaneous—but will advance, or lag behind the disturbing body, depending on whether the angular velocity of rotation of the component is greater or smaller than the instantaneous angular velocity of its companion. Accordingly, the asymmetrical position of the tidal bulges with respect to the line joining the two centers of mass will introduce a net *torque* between the components of a close binary. This tidal torque will cause, in turn, a secular decrease in spin angular momentum of the individual components—the effects of which will be reflected in secular changes in the orbital elements of the binary. In addition, the relative motion of material of finite viscosity within each component will gradually degrade the kinetic energy of internal motion into heat through the process of *viscous friction*. In formulating these ideas, however, Darwin treated the case of a homogeneous deformable body in which tides are produced by a companion that may be regarded as possessing a negligible amount of spin angular momentum. But it was not until 1966 that Zahn revived the tidal-lag and viscous-friction phenomena in a context that was more appropriate to the double-star problem. Subsequently, Kopal and his associates developed an elegant formalism for treating the evolution of a close binary due to the viscous tidal interaction between its components.

Whenever a star senses a variable external gravitational field, it becomes liable to oscillatory motions that may be described as both an "equilibrium tide" and a "dynamical tide." The former is just the instantaneous shape obtained by assuming that mechanical equilibrium does obtain, even though the forcing potential depends on time; that is to say, it is assumed that the natural frequencies of oscillation of the star are rapidly damped out and do not affect the "equilibrium distortion." On the contrary, when the period of the tidal disturbance is comparable to some of the natural periods of oscillation of the star, one cannot rule out *a priori* the possibility of a resonance or at least of an enhancement of the tides due to the outer gravitational potential; this is the dynamical tide.

Let us first consider the equilibrium tide. As was originally shown by Zahn, its effects are relatively large if the tidally distorted star has an extended outer convective zone, since *eddy* viscosity is fairly large in such a region. According to Kopal, however, the tidal torque predicted for stars lacking a convective envelope is much too weak to account for the observed synchronism in close binaries whose components have spectral types F or earlier.[2] In other words, so long as *plasma* viscosity and *radiative* viscosity constitute the sole sources of dissipation, tidal friction is an extremely inefficient process. (To be specific, the demand that tidal friction be effective in the main-sequence lifetime of binaries with spectral types B0 -A5 requires a coefficient of dynamical viscosity in the range 10^{11}–10^{15} g/cm sec, whereas typical coefficients of plasma or radiative viscosity in these stars are only of the order 10^2–10^4 g/cm sec.) Sutantyo has also

---

[2] Of course, as was suggested by Plavec, it may well be that synchronization between axial and orbital motions actually occurs during the pre-main-sequence contraction, at least in those stars that pass through a turbulent convective phase prior to their arrival on the main sequence. This suggestion seems to run counter to Levato's finding that synchronization is a function of age, however, for this strongly suggests that synchronism occurs gradually during the pre-main-sequence contraction, *as well as during the main-sequence lifetime of a binary.*

invoked the possibility of tidal energy dissipation in stars having magnetic fields of the order of 1 gauss. But (as was first suggested, independently, by Horedt and by Press, Wiita, and Smarr) another natural mechanism seems to deserve further attention: *the viscosity of fully developed turbulence driven by tidal shear*. According to these authors, the Reynolds number (Re) associated with this shear is many orders of magnitude larger than the critical value, Re_c, for the onset of fully developed turbulence, and the time scale for the establishment of tidally induced turbulence is much shorter than the Kelvin-Helmholtz time. However, as they correctly pointed out, the inequality $Re \gg Re_c$ is a necessary (but not sufficient) condition for the onset of turbulence in a stratified medium (cf. §8.4). Further studies based on the Richardson criterion are thus needed to ascertain whether tidal shear does lead to turbulence on a scale, and with the eddy viscosity, required to cause synchronism.

The problem of the dynamical tide in a close binary has received, comparatively, much less attention. Cowling was the first to describe the forced oscillations of a star, but he restricted attention to the possible resonances of the lowest g^+-modes only (cf. §6.4). According to this author, the tidal distortion in a star may be expected not to differ seriously from the equilibrium distortion, save, perhaps, in rare cases. The reasons Cowling gave for this were (i) the apparently small contribution of a resonant oscillation to the equilibrium distortion of a star, needing a very close resonance to enhance the total tide, and (ii) the fact that such a close resonance would be destroyed rapidly by second-order effects due to the large horizontal displacements associated with the high-order g^+-modes. Zahn extended Cowling's work to these overtones and found that it is not necessary to come very close to a resonance with a high-order g^+-mode to achieve tides with large surface amplitudes. But, as Zahn pointed out, both results are questionable since they were obtained in the isentropic approximation, which becomes very poor near the surface of a star. In a subsequent work, Zahn further included the effects of *radiative damping* in the description of the forced oscillations of a nonrotating star possessing a convective core and a radiative envelope. It was concluded that the resonances of the g^+-modes are damped out by radiative dissipation, which operates in a relatively thin region below the surface of the star. But the most interesting results bear on the excited oscillations of a star that is a member of a close binary. Zahn's conclusions may be summarized as follows. *First*, due to radiative dissipation near the surface of the star, the dynamical tide does not have the same symmetry properties as the forcing potential. Accordingly, a net torque is applied to the star, which tends to synchronize its axial rotation with the orbital motion. Moreover, this torque is strong enough in relatively close binaries to achieve synchronism in a time that is short compared to the nuclear lifetime of the star. *Second*, whenever the axial rotation of a component of a binary departs significantly from synchronism, the brightness distribution over the star's surface is generally shifted with respect to the companion, thus modifying both the luminosity and the radial velocity that would be observed. On the basis of these preliminary results, it appears that the dynamical tide with radiative dissipation is a most efficient process for synchronizing close binaries, when the components have no outer convective zones. However, Zahn's results were derived for *nonrotating* models only. It would be highly desirable, therefore, to extend this non-isentropic analysis to rotating models as well, but to do this is not a trivial matter.

16.4. GAS DYNAMICS OF CLOSE BINARIES

According to the currently accepted theory of stellar evolution, the post-main-sequence evolution of a star is accompanied by a strong contraction of its helium-rich core and by a corresponding expansion of the surrounding envelope. If a double star is wide enough, each component cannot significantly influence its companion, so that the two stars should evolve like single stars. The situation is quite different when a secularly expanding star has a closely orbiting companion, for the growth of such a star must then come to a halt when it fills up its Roche critical lobe (cf. §16.2). Once this maximum distension permissible on dynamical grounds has been attained, a continuing tendency to expand should induce a secular loss of mass from the star, either to its companion or from the system altogether. That this is indeed the case has been confirmed by numerous observations, which reveal features interpretable as arising in gas streams in many types of evolved close binaries. Various theoretical studies have been devoted to the evolution of a mass-losing component of a close binary, though none of them has ever considered the details of the actual mass transfer that must take place when the expanding body fills up its Roche critical lobe. Because so many *ad hoc* assumptions have been made concerning the transfer rates of mass and angular momentum in a close binary, the reader is referred to the literature for a detailed description.

As we know, if we regard the synchronously rotating members of a close binary as two mass points, the motion of a gas particle whose gravitational attraction may be neglected will be governed by the equations of the restricted three-body problem. If we further restrict attention to motions in the orbital plane of the system, we can thus write

$$\tfrac{1}{2}|\mathbf{v}|^2 + \Phi = constant, \tag{12}$$

where $\mathbf{v}$ is the velocity of the gas particle in the rotating frame of reference.

(The potential Φ is defined in equation [10].) This first integral gives the speed $|\mathbf{v}|$ at any permissible point, once the constant occurring in equation (12) has been determined by initial conditions. In particular, if we allow $|\mathbf{v}|$ to vanish identically, the resulting equation defines the zero-velocity surfaces as $\Phi = constant$, which are in fact identical with the level surfaces of the Roche model (see equation [10]). Of particular importance are the Lagrangian points (designated in Figure 16.3 as L_1, L_2, and L_3) for they correspond to locations where both velocities and accelerations vanish simultaneously; two other Lagrangian points, L_4 and L_5, occur at the vertices of the equilateral triangles built on the mutual separation of the two mass points. On the basis of these classical results, considerable effort has been devoted to the elaboration of mass loss following Kuiper's original suggestion that the inner Lagrangian point L_1 must play an essential role in this problem. Indeed, when a star fills up its Roche critical lobe, matter lying at the point L_1 is not subject to any acceleration, so that any small outward initial velocity will suffice to cause mass outflow. Particle-trajectory calculations by Kopal and others have shown that, with suitable initial conditions, families of orbits could be found that run from the point L_1 to the mass-accreting companion and form a ring around it. However, as was first pointed out by Prendergast, the correct approach to the problem of gas motions in a close binary must be a *hydrodynamical* one, since the mean free path of a gas particle is generally smaller than the linear dimension of the system. Detailed hydrodynamical flows that may be relevant to our problem are now available. It is to some of these solutions that we now turn.

Detailed numerical calculations that simulate the gas flow in a close binary, with as few restrictive assumptions as possible, have been presented by Prendergast and Taam and by Flannery. Both works allow, therefore, for the presence of shock waves, pressure and gravitational effects in the context of the Roche model, and radiative cooling. However, since the forces perpendicular to the orbital plane tend to confine the gas flow in this plane, hydrostatic equilibrium and isothermality in the x_3-direction are thus assumed (see, however, the closing remark of this section). It is also assumed that the flow thickness out of the orbital plane is much smaller than the mutual separation of the two mass points. The numerical technique used by Prendergast and Taam starts from the observation that the equations of hydrodynamics are moment equations of the Boltzmann equation. Hence, their method does not involve the solution of a set of difference equations (as in Flannery's work), but instead simulates a solution of the Boltzmann equation; that is, the gas is treated on a microscopic level as comprising a large number of individual particles, and they make use of the strong tendency of collisions to bring about a Maxwellian velocity distribution. This approach occupies, therefore, a middle ground between Kuiper's particle approach and a pure hydrodynamical approach.

The Prendergast-Taam method was used to study gas motions beyond the inner Lagrangian point in a binary-star system modeled to resemble U Cephei. The steady-state solutions for three separate cases have been considered, viz., synchronous rotation of both components, nonsynchronous rotation of the less massive component, and nonsynchronous rotation of the more massive component. In all cases, a shock wave forms on the surface of the more massive component; in addition, a turbulent region probably forms from a shock resulting from the collision of the returning stream (i.e., the stream that completes a full orbit around the more massive component) with the main stream between the components. It is also found that nonsynchronous rotation of the less massive component leads to the formation of a ring around its companion, with the gas flowing along nearly circular streamlines; in this case, a density protuberance is also seen in the vicinity of the advancing hemisphere of the more massive star. In contrast, a detachment of the stream from the surface of this star (in the same vicinity of its advancing hemisphere) results in the other two cases, the streamlines in this region being then more elliptical in shape.

Flannery's numerical scheme was applied to examine mass transfer in a system resembling the short-period, cataclysmic variable stars. With the inclusion of hydrodynamical effects, it is found that rapid energy dissipation, via radiative cooling in a shock front formed between the stream from the point L_1 and the returning gas that has circuited the star, quickly transforms the flow into a stream-ring configuration. Moreover, an optically thick, confined shock forms at the intersection of the stream and ring. According to Flannery, the shock's location, temperature, and energy release are in substantial agreement with the observed properties of cataclysmic variables stars, and confirm the major elements of the "hot spot" model proposed to account for the observations of these stars.

The main disadvantage of these numerical techniques for simulating mass exchange in close binaries is the enormous computing time they require on large machines. Moreover, as Lubow and Shu pointed out, it is quite possible that the grid sizes used in such calculations are still too coarse to give reliable details because the pressure effects may be important over lengths that are comparable to the mesh spacing. Therefore, these authors have presented a semianalytical analysis of the gas flow in a semidetached system within the context of the Roche model. For the sake of simplicity, it is assumed that the components rotate synchronously and that the flow occurs isothermally, with the sound speed being a small fraction ϵ of the relative orbital speed, Ωd, of the two stars. On the basis of these hypotheses, the steady flow can then be formulated in terms of a problem with multiple length scales. The main results obtained by Lubow and Shu may be summarized as follows. *First*, the escape of material from the surface of the contact component is accomplished by a highly non-isotropic flow that reaches sonic velocities in a neighborhood of the point L_1, of size ϵ in comparison with the distance d. *Second*, this flow throttles into a narrow stream of material, which makes a prescribed angle with respect to the line joining the two mass points, ranging from $19°\!.5$ to $28°\!.4$ for the full range of possible mass ratios. *Third*, the stream width remains nearly constant over the part of the stream that is nearly straight, and narrows somewhat as the stream curves toward the detached companion. *Fourth*, if the detached companion is smaller than a certain specified size, the stream results in the formation of a disc of material of prescribed size orbiting the detached component in a direct sense; this is in apparent agreement with the observations.

To conclude, let us mention that Lubow and Shu have also considered the three-dimensional dynamics of the stream in semidetached binaries undergoing Roche lobe overflow, thus generalizing the above results to include the dynamical effects in the direction perpendicular to the orbital plane. They found that the scale height of the stream characteristically exceeds its corresponding hydrostatic value by a significant factor, because the inertia of the gas prevents it from responding instantaneously to the changing gravitational field. According to Lubow and Shu, this effect is important for the interpretation of the observations relating to stream-disc impacts in cataclysmic variables and in binary X-ray sources of low total mass.

BIBLIOGRAPHICAL NOTES

No attempt at a complete bibliography has been made in this chapter. The classical references in the subject matter treated in this chapter are those of:

1. Kopal, Z., *Close Binary Systems*, London: Chapman and Hall, Ltd, 1959.
2. Batten, A. H., *Binary and Multiple Systems of Stars*, Oxford: Pergamon Press, 1973.

The pioneering work of Roche (1850–1873) and Darwin (1879–1906) may be traced to reference 1. The following survey is particularly noteworthy in the present context:

3. Plavec, M., "Rotation in Close Binaries" in *Stellar Rotation* (Slettebak, A., ed.), pp. 133–146, New York: Gordon and Breach, 1970.

Related matters not treated in the text are contained in:

4. Eggleton, P., Mitton, S., and Whelan, J., eds., *Structure and Evolution of Close Binary Systems*, Dordrecht: D. Reidel Publ. Co., 1976.

Section 16.2. A modern account of the equilibrium and stability of the Roche ellipsoids will be found in reference 28 (pp. 189–240) of Chapter 3. Chandrasekhar's stability analysis of the congruent Darwin ellipsoids is erroneous; the proper treatment of this problem within the framework of ellipsoidal figures of equilibrium has been given by Monique Tassoul:

5. Tassoul, M., *Astroph. J.* **202** (1975): 803–808.

The following extension of her results to nonsynchronous oscillations may also be noted:

6. Chandrasekhar, S., *ibid.* **202** (1975): 809–814.

The nature of the instability of the Roche ellipsoids beyond the Roche limit has been clarified by Robe (reference 5 of Chapter 10). See also:

7. Nduka, A., *Astroph. J.* **170** (1971): 131–142.

A paper of related interest is:

8. Kopal, Z., *Astroph. and Space Sci.* **27** (1974): 389–418.

The influence of tidal action due to a companion on the natural frequencies of compressible bodies has been determined by:

9. Chandrasekhar, S., and Lebovitz, N. R., *Astroph. J.* **137** (1963): 1172–1184.

10. Chandrasekhar, S., *ibid.* **138** (1963): 1182–1213.

11. Tassoul, M., and Tassoul, J. L., *ibid.* **150** (1967): 213–221.

12. Denis, J., *Astron. and Astroph.* **20** (1972): 151–155.

13. Denis, J., and Smeyers, P., *Med. Kon. Acad. België: Klasse der Wetenschappen* **38**, No 4 (1976): 1–19.

See also reference 5. The following paper may also be noted:

14. Papaloizou, J. C. B., and Bath, G. T., *Monthly Notices Roy. Astron. Soc. London* **172** (1975): 339–357.

The Roche model has been discussed at length by Kopal:

15. Kopal, Z., "The Roche Model and Close Binary Systems" in *Advances in Astronomy and Astrophysics* **9** (1972): 1–65.

Various independent techniques to construct inhomogeneous, tidally distorted models will be found in:

16. Chandrasekhar, S., *Monthly Notices Roy. Astron. Soc. London* **93** (1933): 462–471.

17. Agostinelli, C., *Rendiconti Seminario Mat., Univ. e Politecnico di Torino* **9** (1949): 179–224.

18. Evrard, L., *Ann. Astroph.* **14** (1951): 17–39.

These papers generalize, respectively, the Chandrasekhar-Milne method (§5.3), the Clairaut-Legendre method (§5.2, and reference 1), and the Wavre method (reference 2 of Chapter 4). Polytropic models of close binaries have been constructed by:

19. Orlov, A. A., *Astron. Zh.* **37** (1960): 902–907.
20. Durney, B. R., and Roxburgh, I. W., *Monthly Notices Roy. Astron. Soc. London* **148** (1970): 239–247.
21. Martin, P. G., *Astroph. and Space Sci.* **7** (1970): 119–138.
22. Naylor, M. D. T., and Anand, S. P. S., in *Stellar Rotation* (Slettebak, A., ed.), pp. 157–164, New York: Gordon and Breach, 1970.
23. Green, L. C., and Kolchin, E. K., *Astroph. and Space Sci.* **21** (1973): 285–288.
24. Green, L. C., and Kolchin, E. K., *Astroph. J. Supplements* **28** (1974): 449–463.

The validity of the Roche model has been originally discussed by:

25. Dziembowski, W., *Acta Astron.* **13** (1963): 157–164.

Detailed interior models of uniformly rotating binaries have been first constructed, independently, by Kippenhahn and Thomas (reference 28 of Chapter 5) and by Jackson:

26. Jackson, S., *Astroph. J.* **160** (1970): 685–699.

Further models are due to:

27. Naylor, M. D. T., and Anand, S. P. S., *Astroph. and Space Sci.* **16** (1972): 137–150; *ibid.* **18** (1972): 59–84.
28. Whelan, J. A. J., *Monthly Notices Roy. Astron. Soc. London* **160** (1972): 63–77.

Departure from synchronism is discussed in:

29. Plavec, M., *Mém. Soc. Roy. Sci. Liège* (4) **20** (1958): 411–420.
30. Kruszewski, A., *Acta Astron.* **13** (1963): 106–117.
31. Limber, D. N., *Astroph. J.* **138** (1963): 1112–1133.
32. Naylor, M. D. T., *Astroph. and Space Sci.* **18** (1972): 85–88.

Section 16.3. A penetrating study of tidal interaction in close binaries is due to:

33. Zahn, J.P., *Ann. Astroph.* **29** (1966): 313–330, 489–506, 565–591.

Refined mathematical treatments of these and related matters will be found in:

34. Kopal, Z., *Astroph. and Space Sci.* **1** (1968): 179–215, 284–300, 411–423; *ibid.* **16** (1972): 3–51; *ibid.* **17** (1972): 161–185; *ibid.* **18** (1972): 287–305.

35. Roach, G. F., *ibid.* **1** (1968): 32–67.

36. Alexander, M. E., *ibid.* **23** (1973): 459–510.

37. Haynes, W. E., *ibid.* **22** (1973): 165–192.

38. Tokis, J. N., *ibid.* **26** (1974): 447–476, 477–495; *ibid.* **36** (1975): 427–438.

39. Dolginov, A. Z., and Yakovlev, D. G. *ibid.* **36** (1975): 31–78.

Other discussions are those of:

40. Dziembowski, W., *Comm. Obs. Roy. Belgique, B,* **17** (1967): 105–108.

41. James, R., *ibid.* **17** (1967): 99–105.

42. Monaghan, J. J., *Astron. and Astroph.* **6** (1970): 464–467.

43. Stothers, R., *Publ. Astron. Soc. Pacific* **85** (1973): 363–378.

Further relevant papers may be traced to reference 43. Magnetic fields have been invoked by:

44. Sutantyo, W., *Astron. and Astroph.* **35** (1974): 251–257.

Synchronization by tidally induced turbulence is discussed in:

45. Horedt, G., *ibid.* **44** (1975): 461–463.

46. Press, W. H., Wiita, P. J., and Smarr, L. L., *Astroph. J. Letters* **202** (1975): L135–L137.

See also the third reference 49, as well as the following papers:

47. Lecar, M., Wheeler, J. C., and McKee, C. F., *Astroph. J.* **205** (1976): 556–562.

48. Seguin, F. H., *ibid.* **207** (1976): 848–859.

The dynamical tide has been considered by Cowling (reference 12 of Chapter 6) and by Zahn:

49. Zahn, J. P., *Astron. and Astroph.* **4** (1970): 452–461; *ibid.* **41** (1975): 329–344; *ibid.* **57** (1977): 383–394.

Section 16.4. A comprehensive survey of the particle approach will be found in:

50. Kruszewski, A., "Exchange of Matter and Period Changes in Close Binary Systems" in *Advances in Astronomy and Astrophysics* **4** (1966): 233–299.

Other relevant papers may be traced to reference 4. The hydrodynamical approach was originally discussed by:

51. Prendergast, K. H., *Astroph. J.* **132** (1960): 162–174; *ibid.* **133** (1961): 732.

See also:

52. Huang, S. S., *ibid.* **141** (1965): 201–209.

Simple models of mass exchange in close binaries will be found in:

53. Sobouti, Y., *Astron. and Astroph.* **5** (1970): 149–154.
54. Biermann, P., *ibid.* **10** (1971): 205–212.

The most detailed discussions are those of:

55. Prendergast, K. H., and Taam, R. E., *Astroph. J.* **189** (1974): 125–136.
56. Flannery, B. P., *ibid.* **201** (1975): 661–694.
57. Lubow, S. H., and Shu, F. H., *ibid.* **198** (1975): 383–405.
58. Lubow, S. H., and Shu, F. H., *Astroph. J. Letters* **207** (1976): L53–L55.

See also:

59. Sørensen, S. A., Matsuda, T., and Sakurai, T., *Astroph. and Space Sci.* **33** (1975): 465–480.

The presentation in the text largely follows references 55–58.

Epilogue

The theory of rotating stars had its beginnings in the Scientific Revolution, which took place in Europe during the seventeenth century, and it has since aroused the interest of many distinguished scientists. Yet, most (if not all) problems of stellar rotation belong to what Kuhn has described as "normal science," that is to say, a highly convergent or consensus-bound activity that is firmly based upon one or more major scientific achievements of the past and that results in a cumulative progression of new ideas developing from antecedent ideas in a logical sequence. Indeed, as can be inferred from the study by Crane, the recent growth of research on rotating stars reflects a kind of social diffusion process in which early adepts influence later adepts, the growth of knowledge in the field representing the accretion of many small innovations. It is not surprising, therefore, that expanding enrollments in universities during the 1960s together with enhancement of interest in astronomy following the space race have produced, *for a time*, a period of exponential growth in the bulk of literature on stellar rotation.[1] And yet, despite the fact that some important results have emerged from this large and varied harvest of papers, research on rotating stars lives with inadequacies, incompleteness, and a sense that it has often fallen short of its goal. In what follows, I will attempt to delineate the rapidly moving frontier of knowledge in the field as of late 1977, commenting briefly on some promising avenues of research for the coming decade or so.

Since much of the research on rotating stars is an attempt to adjust theory and observation in order to bring the two in closer agreement, I shall comment first on the existing measurements of rotation rates for single, nonmagnetic stars. Obviously, the main shortcoming of this kind of material lies in the fact that a study of line broadening interpreted as due to axial rotation provides only the *projected* equatorial velocity, $v_e \sin i$, whereas it is the true equatorial velocity, v_e, and the aspect angle, i, in which we are interested. Moreover, because of differences in the methods used to derive the $v_e \sin i$ values, the observational material that can now

---

[1] It is not my purpose to go into the many historical and sociological studies in the growth of scientific knowledge. But the interested reader is referred to Kuhn, T. S., *The Structure of Scientific Revolutions*, 2nd edn., Chicago: The University of Chicago Press, 1970. Other viewpoints—notably the frankly Darwinian models of Holton and Toulmin—and further pertinent results will be found in Crane, D., *Invisible Colleges*, Chicago: The University of Chicago Press, 1972.

be found in catalogues is far from being homogeneous, and the possibility of appreciable systematic errors for some groups of stars should not be excluded. New measurements based on a standard calibration are thus needed to improve the existing statistical studies of axial rotation as functions of spectral type and luminosity class. The recently published atlas of rotationally broadened spectra based on model-atmosphere calculations by Slettebak and his associates may provide the right instrument for this purpose (cf. §2.3). As far as the determination of v_e and i for single, nonmagnetic stars is concerned, however, very little has been achieved beyond the trial stage. Yet, Heap and Hutchings have recently found that in hot, rapidly rotating stars the rotational broadening of spectral lines is a function of wavelength.[2] According to these authors, measurements and detailed calculations of specific line profiles in the visual and far ultraviolet regions would offer a method of determining v_e and i for a large number of stars. Particularly interesting are the various Fourier-transform techniques that have already proved to be quite useful for obtaining detailed information from high-resolution spectra. Indeed, as was recently shown by Gray, a Fourier dissection of broadened line profiles allows in principle the separation of rotation from macroturbulence, and possibly the detection of small departures from solid-body rotation (cf. §12.3).

On the theoretical side, it is apparent from the discussions in the preceding chapters that we are still very far from having an adequate picture of the structure and evolution of rotating stars. This is especially true for the pre-main-sequence contraction phases, which are still a matter of considerable debate. For example, we briefly commented in Section 11.2 on the so-called angular momentum problem, but we have not discussed this important question in any detail because it has so far eluded a definitive answer. It is my belief that more observations of interstellar clouds are needed before we will be able to produce a reliable model that explains how the angular momentum of a protostellar cloud is dissipated so as to form the observed main-sequence stars. Recent discussions of the observational data on this and related problems have been made by Burki and Maeder, and by Mouschovias.[3] Perhaps more promising for the near future is the problem of double-star formation—a question that has seen advances on both the observational and theoretical fronts. As we said in Section 11.2, however, it is not yet entirely clear from the observational data whether all binaries have a common origin or whether we must look

---

[2] Heap, S. R., in *Be and Shell Stars* (Slettebak, A., ed.), p. 165, Dordrecht: D. Reidel and Co., 1976; Hutchings, J. B., *P.A.S.P.* **88**, 5, 1976; Hutchings, J. B., and Stoeckley, T. R., *ibid.* **89**, 19, 1977; Sonneborn, G. H., and Collins, G. W., *Ap. J.* **213**, 787, 1977.

[3] Burki, G., and Maeder, A., *A. Ap.* **57**, 401, 1977; Mouschovias, T. Ch., *Ap. J.* **211**, 147, 1977 (and observational work quoted therein).

for a separate mode of formation for the wide and close binaries (e.g., capture and fission). Further observations of double stars having separations in the region where the incidence of selection is the largest are thus badly needed.

Although the detailed processes of double-star formation remain largely unknown, substantial theoretical progress has been made concerning the so-called fission problem (cf. §11.3). Indeed, given some simplifying assumptions, it is now generally agreed that the free-fall, axisymmetric collapse of a rotating cloud leads to the formation of a ring-shaped structure, the ensuing instabilities in this rotating ring probably resulting in fragmentation and the formation of two or more pieces in orbit around each other (cf. §11.2). The fission hypothesis that has been advocated so often to explain binary formation thus looks very promising again, though the proper treatment of the real problem will obviously require a full three-dimensional discussion of the collapse of a rotating cloud. Moreover, because of the many uncertainties that necessarily entail long-run calculations on a large computer, independent runs on different machines will be needed to confirm the results. Hence, given the small number of theoreticians who can afford to engage in long-range research projects, it is obvious that quite some time will elapse before the validity and limitations of the fission hypothesis will have been thoroughly ascertained for the case of both wide and close binaries.

Let us consider next rotation among main-sequence stars. There is now ample evidence that the remarkable decline in rotational velocities in the middle F's along the main sequence is caused by a stellar wind and/or flare activity on stars that retain an outer convection zone all the way from the Hayashi phase to the main sequence, with concomitant corpuscular emission and magnetically enhanced loss of angular momentum (cf. §11.4). Because the surface rotation rates of main-sequence stars crucially depend on the presence or absence of an outer convective shell, the study of rotation among high-mass and low-mass stars leads to entirely distinct problems.

Concerning the Sun, which is probably a typical lower-main-sequence star, it is now commonly accepted that its *surface* rotation rate is somehow caused by the interaction of rotation with turbulent convection in its external layers (cf. §9.4). But here, as we know, there are further complications because of the solar cycle and variations of the mean surface rotation rate over the past centuries.[4] It is generally assumed that the migratory magnetic fields of the Sun are maintained against Ohmic decay by the

---

[4] Eddy, J. A., *Science* **192**, 1189, 1976; Eddy, J. A., Gilman, P. A., and Trotter, D. E., *Solar Phys.* **46**, 3, 1976; Link, F., *A. Ap.* **54**, 857, 1977.

combined effect of differential rotation and turbulent convection below the surface (cf. §15.5). Although detailed kinematic-dynamo models have been proposed, it is becoming apparent that further substantial progress will require a dynamical-dynamo model, taking fully into account the coupling between the velocity field and the magnetic field as well as the highly compressible nature of the solar convection zone. Until we understand turbulent convection *without* rotation and magnetic field, I thus seriously question the purpose of making further theoretical models of the solar cycle and the solar surface rotation rate.

As far as the *inner* rotation rates of the Sun and low-mass main-sequence stars are concerned, the situation is also quite frustrating for a theoretician. Of course, it is possible that these stars approach the main sequence with a core that rotates much more rapidly than their outer layers. For the moment, however, nothing is known with any degree of certainty about rapidly rotating cores and transport of angular momentum deep inside lower main-sequence stars.

Not surprisingly, the question of the angular momentum distribution within upper-main-sequence stars also remains largely unresolved. As a matter of fact, because no such star is located at a sufficiently close distance from the Sun, we do not even know what surface rotation law these stars actually follow (cf. §12.3). Hence, in all models proposed to date, the rotation rate is always specified in an *ad hoc* manner. Although such a procedure is certainly acceptable to evaluate the *gross* changes caused by rotation on stellar models, it is well known that pseudo-barotropic models in a state of permanent rotation cannot be used to describe rotating stars in *strict* radiative equilibrium (cf. §7.2). For this reason, it is now becoming necessary to construct models with large-scale meridional currents superimposed on a pure rotational motion.[5] Randers's finding that self-consistent flow patterns *cannot* be found in steady, inviscid, and nonmagnetic models should no longer be overlooked, however, as it was in the past (cf. §8.2). Recently, Mestel and Moss have presented models of uniformly rotating stars with steady-state circulation and large-scale magnetic fields (cf. §8.3). Further work along these lines is certainly necessary to understand the structure of magnetic stars. Since it is not proved that all upper-main-sequence stars are pervaded by a sizable magnetic field, independent work should also be done to obtain unsteady solutions as well as those that take fully into account the role of viscosity in the free boundary layer of a rotating star in radiative equilibrium.

---

[5] Steady-state baroclinic models for which the rotation rate adjusts itself so as to prevent meridional currents can be found in the literature. In my opinion, very little can be gained from such models, mainly because they are thermally unstable (cf. §§7.5 and 8.2). See also: Sharp, C. M., Smith, R. C., and Moss, D. L., *M. N.* **179**, 699, 1977.

To conclude, let us recall that the influence of meridional currents on the evolution of rotating stars is generally ignored in stellar evolution calculations. Rather, it is often assumed that nuclear reactions produce gradients of mean molecular weight, which, under certain circumstances, will tend to inhibit the circulation and protect the evolving core from rotational mixing. The recent semiquantitative analysis by Huppert and Spiegel seems to indicate, however, that in spite of a $\bar{\mu}$-barrier in the core, mixing may well occur at a nonnegligible rate.[6] Also poorly understood is the final fate of a rotating stellar model, given as usual its total mass, total angular momentum, and a prescribed angular momentum distribution. The main reason for this is that we ignore to what extent angular momentum is outwardly transferred and then lost during the post-main-sequence phases of stellar evolution (cf. §12.6). On the basis of the most recent observational data on rotating white dwarfs and pulsars, it now appears that a very effective transport of angular momentum has to be postulated.[7] This again lends strong support to the idea that $\bar{\mu}$-barriers are ineffective in inhibiting angular momentum transfer from the core to the envelope of an evolved star. Clearly, more theoretical discussions of meridional currents, angular momentum transfer, and mass loss in realistic stellar models will be needed before we can present a coherent picture of the very late phases of stellar evolution. Since most of these effects will require a treatment within the framework of general relativity, I shall leave this task to a better qualified individual.

[6] Huppert, H. E., and Spiegel, E. A., *Ap. J.* **213**, 157, 1977.

[7] Greenstein, J. L., Boksenberg, A., Carswell, R., and Shortridge, K., *Ap. J.* **212**, 186, 1977; Liebert, J., Angel, J. R. P., Stockman, H. S., Spinrad, H., and Beaver, E. A., *ibid.* **214**, 457, 1977; Weidemann, V., *A. Ap.* **59**, 411, 1977.

Appendix A

A TABLE OF PHYSICAL AND ASTRONOMICAL CONSTANTS

Digits in parentheses correspond to one standard deviation uncertainty in the final digits of the quoted value. Sources: Allen, C. W., *Astrophysical Quantities*, 3rd edition, London: The Athlone Press, 1973; Cohen, E. R., and Taylor, B. N., *Journal of Physical and Chemical Reference Data* **2**, 663, 1973.

Speed of light in vacuo:
$$c = 2.99792458(1.2) \cdot 10^{10} \text{ cm/sec}$$

Constant of gravitation:
$$G = 6.6720(41) \cdot 10^{-8} \text{ dyn cm}^2/\text{g}^2$$

Gas constant:
$$\mathscr{R} = 8.31441(26) \cdot 10^7 \text{ erg/deg mole}$$

Electron charge:
$$e = 4.8032431(46) \cdot 10^{-10} \text{ e.s.u.}$$

Electron rest mass:
$$m_e = 9.109534(47) \cdot 10^{-28} \text{ g}$$

Ratio, proton mass to electron mass:
$$m_p/m_e = 1836.15152(70)$$

Planck constant:
$$h = 2\pi\hbar = 6.626176(36) \cdot 10^{-27} \text{ erg sec}$$

Boltzmann constant:
$$k = 1.380662(44) \cdot 10^{-16} \text{ erg/deg}$$

Stefan-Boltzmann constant:
$$\sigma = 5.67032(71) \cdot 10^{-5} \text{ erg/cm}^2 \text{ deg}^4 \text{ sec}$$

Radiation pressure constant:
$$a = 4\sigma/c = 7.56566(71) \cdot 10^{-15} \text{ erg/cm}^3 \text{ deg}^4$$

Solar luminosity:
$$L_\odot = 3.826(8) \cdot 10^{33} \text{ erg/sec}$$

Solar mass:
$$M_\odot = 1.989(2) \cdot 10^{33} \text{ g}$$

Solar radius:
$$R_\odot = 6.9599(7) \cdot 10^{10} \text{ cm}$$

Solar surface gravity:

$$g_\odot = 2.7398(4) \cdot 10^4 \text{ cm/sec}^2$$

Solar effective temperature:

$$T_{\text{eff},\odot} = 5,770°\text{K}$$

Solar absolute bolometric magnitude:

$$M_{b,\odot} = +4.75$$

Astronomical unit:

$$\text{AU} = 1.495979(1) \cdot 10^{13} \text{ cm}$$

Parsec:

$$pc = 206,264.806 \text{ AU} = 3.085678 \cdot 10^{18} \text{ cm}$$

Appendix B

THE HYDRODYNAMICAL EQUATIONS IN CYLINDRICAL AND SPHERICAL COORDINATES

The operations D/Dt and ∇^2 listed in equations (B.1), (B.5), (B.18), and (B.22) are for operations on a scalar; these are not the same as operation on a vector (except in Cartesian coordinates). In this appendix, equations (B.1) and (B.18) thus de*fine* the symbol D/Dt occurring in the components of the Navier-Stokes equations.

CYLINDRICAL COORDINATES (ϖ, φ, z)

Operators:

$$\frac{D}{Dt} = \frac{\partial}{\partial t} + v_\varpi \frac{\partial}{\partial \varpi} + \frac{v_\varphi}{\varpi} \frac{\partial}{\partial \varphi} + v_z \frac{\partial}{\partial z}, \tag{B.1}$$

$$\operatorname{div} \mathbf{\Phi} = \frac{1}{\varpi} \frac{\partial}{\partial \varpi} (\varpi \Phi_\varpi) + \frac{1}{\varpi} \frac{\partial \Phi_\varphi}{\partial \varphi} + \frac{\partial \Phi_z}{\partial z}, \tag{B.2}$$

$$\operatorname{grad} \Phi = \frac{\partial \Phi}{\partial \varpi} \mathbf{1}_\varpi + \frac{1}{\varpi} \frac{\partial \Phi}{\partial \varphi} \mathbf{1}_\varphi + \frac{\partial \Phi}{\partial z} \mathbf{1}_z, \tag{B.3}$$

$$\operatorname{curl} \mathbf{\Phi} = \left(\frac{1}{\varpi} \frac{\partial \Phi_z}{\partial \varphi} - \frac{\partial \Phi_\varphi}{\partial z} \right) \mathbf{1}_\varpi + \left(\frac{\partial \Phi_\varpi}{\partial z} - \frac{\partial \Phi_z}{\partial \varpi} \right) \mathbf{1}_\varphi$$
$$+ \frac{1}{\varpi} \left[\frac{\partial}{\partial \varpi} (\varpi \Phi_\varphi) - \frac{\partial \Phi_\varpi}{\partial \varphi} \right] \mathbf{1}_z, \tag{B.4}$$

$$\nabla^2 = \frac{1}{\varpi} \frac{\partial}{\partial \varpi} \left(\varpi \frac{\partial}{\partial \varpi} \right) + \frac{1}{\varpi^2} \frac{\partial^2}{\partial \varphi^2} + \frac{\partial^2}{\partial z^2}. \tag{B.5}$$

Conservation of mass:

$$\frac{\partial \rho}{\partial t} + \frac{1}{\varpi} \frac{\partial}{\partial \varpi} (\varpi \rho v_\varpi) + \frac{1}{\varpi} \frac{\partial}{\partial \varphi} (\rho v_\varphi) + \frac{\partial}{\partial z} (\rho v_z) = 0. \tag{B.6}$$

Components of the viscous stress tensor:

$$\tau_{\varpi\varpi} = 2(\mu + \mu_r) \left[\frac{\partial v_\varpi}{\partial \varpi} \right] + (\mu_\vartheta + \mu_r - \tfrac{2}{3}\mu) \operatorname{div} \mathbf{v}, \tag{B.7}$$

$$\tau_{\varphi\varphi} = 2(\mu + \mu_r) \left[\frac{1}{\varpi} \frac{\partial v_\varphi}{\partial \varphi} + \frac{v_\varpi}{\varpi} \right] + (\mu_\vartheta + \mu_r - \tfrac{2}{3}\mu) \operatorname{div} \mathbf{v}, \tag{B.8}$$

$$\tau_{zz} = 2(\mu + \mu_r)\left[\frac{\partial v_z}{\partial z}\right] + (\mu_9 + \mu_r - \tfrac{2}{3}\mu) \text{ div } \mathbf{v}, \tag{B.9}$$

$$\tau_{\varpi\varphi} = \tau_{\varphi\varpi} = 2(\mu + \mu_r)\left[\frac{1}{2}\left(\frac{1}{\varpi}\frac{\partial v_\varpi}{\partial \varphi} + \varpi\frac{\partial}{\partial \varpi}\frac{v_\varphi}{\varpi}\right)\right], \tag{B.10}$$

$$\tau_{\varphi z} = \tau_{z\varphi} = 2(\mu + \mu_r)\left[\frac{1}{2}\left(\frac{\partial v_\varphi}{\partial z} + \frac{1}{\varpi}\frac{\partial v_z}{\partial \varphi}\right)\right], \tag{B.11}$$

$$\tau_{z\varpi} = \tau_{\varpi z} = 2(\mu + \mu_r)\left[\frac{1}{2}\left(\frac{\partial v_z}{\partial \varpi} + \frac{\partial v_\varpi}{\partial z}\right)\right]. \tag{B.12}$$

Note that the components of the deformation tensor **D** are the quantities in square brackets, e.g., $D_{\varpi\varpi} = \partial v_\varpi/\partial \varpi$, etc. (cf. §3.3: equations [43] and [47]).

Equations of motion:

$$\rho\left(\frac{Dv_\varpi}{Dt} - \frac{v_\varphi{}^2}{\varpi}\right) = -\rho\frac{\partial V}{\partial \varpi} - \frac{\partial p}{\partial \varpi} + \frac{1}{\varpi}\frac{\partial}{\partial \varpi}(\varpi\tau_{\varpi\varpi})$$

$$+ \frac{1}{\varpi}\frac{\partial \tau_{\varpi\varphi}}{\partial \varphi} + \frac{\partial \tau_{\varpi z}}{\partial z} - \frac{\tau_{\varphi\varphi}}{\varpi}, \tag{B.13}$$

$$\rho\left(\frac{Dv_\varphi}{Dt} + \frac{v_\varpi v_\varphi}{\varpi}\right) = -\frac{\rho}{\varpi}\frac{\partial V}{\partial \varphi} - \frac{1}{\varpi}\frac{\partial p}{\partial \varphi} + \frac{1}{\varpi}\frac{\partial}{\partial \varpi}(\varpi\tau_{\varphi\varpi})$$

$$+ \frac{1}{\varpi}\frac{\partial \tau_{\varphi\varphi}}{\partial \varphi} + \frac{\partial \tau_{\varphi z}}{\partial z} + \frac{\tau_{\varpi\varphi}}{\varpi}, \tag{B.14}$$

$$\rho\frac{Dv_z}{Dt} = -\rho\frac{\partial V}{\partial z} - \frac{\partial p}{\partial z} + \frac{1}{\varpi}\frac{\partial}{\partial \varpi}(\varpi\tau_{z\varpi})$$

$$+ \frac{1}{\varpi}\frac{\partial \tau_{z\varphi}}{\partial \varphi} + \frac{\partial \tau_{zz}}{\partial z}. \tag{B.15}$$

Dissipation function:

$$\Phi_v = 2(\mu + \mu_r)(D_{\varpi\varpi}^2 + D_{\varphi\varphi}^2 + D_{zz}^2 + 2D_{\varpi\varphi}^2 + 2D_{\varphi z}^2 + 2D_{z\varpi}^2)$$

$$+ (\mu_9 + \mu_r - \tfrac{2}{3}\mu)(\text{div } \mathbf{v})^2. \tag{B.16}$$

Conservation of thermal energy:

$$\rho T\frac{DS}{Dt} = \Phi_v + \rho\epsilon_{Nuc} + \frac{1}{\varpi}\frac{\partial}{\partial \varpi}\left[(\chi + \chi_r)\varpi\frac{\partial T}{\partial \varpi}\right]$$

$$+ \frac{1}{\varpi^2}\frac{\partial}{\partial \varphi}\left[(\chi + \chi_r)\frac{\partial T}{\partial \varphi}\right] + \frac{\partial}{\partial z}\left[(\chi + \chi_r)\frac{\partial T}{\partial z}\right]. \tag{B.17}$$

SPHERICAL COORDINATES (r, θ, φ)

Operators:

$$\frac{D}{Dt} = \frac{\partial}{\partial t} + v_r \frac{\partial}{\partial r} + \frac{v_\theta}{r} \frac{\partial}{\partial \theta} + \frac{v_\varphi}{r \sin \theta} \frac{\partial}{\partial \varphi}, \tag{B.18}$$

$$\text{div } \mathbf{\Phi} = \frac{1}{r^2} \frac{\partial}{\partial r} (r^2 \Phi_r) + \frac{1}{r \sin \theta} \frac{\partial}{\partial \theta} (\Phi_\theta \sin \theta) + \frac{1}{r \sin \theta} \frac{\partial \Phi_\varphi}{\partial \varphi}, \tag{B.19}$$

$$\text{grad } \Phi = \frac{\partial \Phi}{\partial r} \mathbf{1}_r + \frac{1}{r} \frac{\partial \Phi}{\partial \theta} \mathbf{1}_\theta + \frac{1}{r \sin \theta} \frac{\partial \Phi}{\partial \varphi} \mathbf{1}_\varphi, \tag{B.20}$$

$$\text{curl } \mathbf{\Phi} = \frac{1}{r \sin \theta} \left[\frac{\partial}{\partial \theta} (\Phi_\varphi \sin \theta) - \frac{\partial \Phi_\theta}{\partial \varphi} \right] \mathbf{1}_r$$

$$+ \frac{1}{r \sin \theta} \left[\frac{\partial \Phi_r}{\partial \varphi} - \frac{\partial}{\partial r} (r \Phi_\varphi \sin \theta) \right] \mathbf{1}_\theta$$

$$+ \frac{1}{r} \left[\frac{\partial}{\partial r} (r \Phi_\theta) - \frac{\partial \Phi_r}{\partial \theta} \right] \mathbf{1}_\varphi, \tag{B.21}$$

$$\nabla^2 = \frac{1}{r^2} \frac{\partial}{\partial r} \left(r^2 \frac{\partial}{\partial r} \right) + \frac{1}{r^2 \sin \theta} \frac{\partial}{\partial \theta} \left(\sin \theta \frac{\partial}{\partial \theta} \right) + \frac{1}{r^2 \sin^2 \theta} \frac{\partial^2}{\partial \varphi^2}. \tag{B.22}$$

Conservation of mass:

$$\frac{\partial \rho}{\partial t} + \frac{1}{r^2} \frac{\partial}{\partial r} (r^2 \rho v_r) + \frac{1}{r \sin \theta} \frac{\partial}{\partial \theta} (\rho v_\theta \sin \theta) + \frac{1}{r \sin \theta} \frac{\partial}{\partial \varphi} (\rho v_\varphi) = 0. \tag{B.23}$$

Components of the viscous stress tensor:

$$\tau_{rr} = 2(\mu + \mu_r) \left[\frac{\partial v_r}{\partial r} \right] + (\mu_\vartheta + \mu_r - \tfrac{2}{3}\mu) \text{ div } \mathbf{v}, \tag{B.24}$$

$$\tau_{\theta\theta} = 2(\mu + \mu_r) \left[\frac{1}{r} \frac{\partial v_\theta}{\partial \theta} + \frac{v_r}{r} \right] + (\mu_\vartheta + \mu_r - \tfrac{2}{3}\mu) \text{ div } \mathbf{v}, \tag{B.25}$$

$$\tau_{\varphi\varphi} = 2(\mu + \mu_r) \left[\frac{1}{r \sin \theta} \frac{\partial v_\varphi}{\partial \varphi} + \frac{v_r}{r} + \frac{v_\theta \cot \theta}{r} \right]$$

$$+ (\mu_\vartheta + \mu_r - \tfrac{2}{3}\mu) \text{ div } \mathbf{v}, \tag{B.26}$$

$$\tau_{r\theta} = \tau_{\theta r} = 2(\mu + \mu_r) \left[\frac{1}{2} \left(\frac{1}{r} \frac{\partial v_r}{\partial \theta} + r \frac{\partial}{\partial r} \frac{v_\theta}{r} \right) \right], \tag{B.27}$$

$$\tau_{\theta\varphi} = \tau_{\varphi\theta} = 2(\mu + \mu_r) \left[\frac{1}{2} \left(\frac{1}{r \sin \theta} \frac{\partial v_\theta}{\partial \varphi} + \frac{\sin \theta}{r} \frac{\partial}{\partial \theta} \frac{v_\varphi}{\sin \theta} \right) \right], \tag{B.28}$$

$$\tau_{\varphi r} = \tau_{r\varphi} = 2(\mu + \mu_r)\left[\frac{1}{2}\left(r\frac{\partial}{\partial r}\frac{v_\varphi}{r} + \frac{1}{r\sin\theta}\frac{\partial v_r}{\partial\varphi}\right)\right]. \tag{B.29}$$

Note that the components of the deformation tensor **D** are the quantities in square brackets, e.g. $D_{rr} = \partial v_r/\partial r$, etc. (cf. §3.3: equations [43] and [47]).

Equations of motion:

$$\rho\left(\frac{Dv_r}{Dt} - \frac{v_\theta{}^2 + v_\varphi{}^2}{r}\right)$$

$$= -\rho\frac{\partial V}{\partial r} - \frac{\partial p}{\partial r} + \frac{1}{r\sin\theta}\left[\frac{\sin\theta}{r}\frac{\partial}{\partial r}(r^2\tau_{rr}) + \frac{\partial}{\partial\theta}(\tau_{r\theta}\sin\theta) + \frac{\partial\tau_{r\varphi}}{\partial\varphi}\right]$$

$$- \frac{\tau_{\theta\theta} + \tau_{\varphi\varphi}}{r}, \tag{B.30}$$

$$\rho\left(\frac{Dv_\theta}{Dt} + \frac{v_r v_\theta}{r} - \frac{v_\varphi{}^2\cot\theta}{r}\right)$$

$$= -\frac{\rho}{r}\frac{\partial V}{\partial\theta} - \frac{1}{r}\frac{\partial p}{\partial\theta} + \frac{1}{r\sin\theta}\left[\frac{\sin\theta}{r}\frac{\partial}{\partial r}(r^2\tau_{\theta r}) + \frac{\partial}{\partial\theta}(\tau_{\theta\theta}\sin\theta) + \frac{\partial\tau_{\theta\varphi}}{\partial\varphi}\right]$$

$$+ \frac{\tau_{r\theta}}{r} - \frac{\tau_{\varphi\varphi}\cot\theta}{r}, \tag{B.31}$$

$$\rho\left(\frac{Dv_\varphi}{Dt} + \frac{v_r v_\varphi}{r} + \frac{v_\theta v_\varphi\cot\theta}{r}\right)$$

$$= -\frac{\rho}{r\sin\theta}\frac{\partial V}{\partial\varphi} - \frac{1}{r\sin\theta}\frac{\partial p}{\partial\varphi} + \frac{1}{r\sin\theta}\left[\frac{\sin\theta}{r}\frac{\partial}{\partial r}(r^2\tau_{\varphi r}) + \frac{\partial}{\partial\theta}(\tau_{\varphi\theta}\sin\theta)\right.$$

$$\left. + \frac{\partial\tau_{\varphi\varphi}}{\partial\varphi}\right] + \frac{\tau_{r\varphi}}{r} + \frac{\tau_{\theta\varphi}\cot\theta}{r}. \tag{B.32}$$

Dissipation function:

$$\Phi_v = 2(\mu + \mu_r)(D_{rr}^2 + D_{\theta\theta}^2 + D_{\varphi\varphi}^2 + 2D_{r\theta}^2 + 2D_{\theta\varphi}^2 + 2D_{\varphi r}^2)$$

$$+ (\mu_{\mathfrak{s}} + \mu_r - \tfrac{2}{3}\mu)(\text{div } \mathbf{v})^2. \tag{B.33}$$

Conservation of thermal energy:

$$\rho T\frac{DS}{Dt} = \Phi_v + \rho\epsilon_{Nuc} + \frac{1}{r^2}\frac{\partial}{\partial r}\left[(\chi + \chi_r)r^2\frac{\partial T}{\partial r}\right]$$

$$+ \frac{1}{r^2\sin\theta}\frac{\partial}{\partial\theta}\left[(\chi + \chi_r)\frac{\partial T}{\partial\theta}\sin\theta\right] + \frac{1}{r^2\sin^2\theta}\frac{\partial}{\partial\varphi}\left[(\chi + \chi_r)\frac{\partial T}{\partial\varphi}\right]. \tag{B.34}$$

Appendix C

THE SPHERICAL HARMONICS

The fundamental equation of potential theory, $\nabla^2\Phi = 0$, can be written in the form

$$\frac{1}{r^2}\frac{\partial}{\partial r}\left(r^2\frac{\partial\Phi}{\partial r}\right) + \frac{1}{r^2\sin\theta}\frac{\partial}{\partial\theta}\left(\frac{\partial\Phi}{\partial\theta}\sin\theta\right) + \frac{1}{r^2\sin^2\theta}\frac{\partial^2\Phi}{\partial\varphi^2} = 0, \quad (C.1)$$

where r, θ, and φ are the usual spherical coordinates. It is a simple matter to verify that this equation is separable and admits of the *particular* solutions

$$\Phi_{nm}(r,\theta,\varphi) = \begin{Bmatrix} r^n \\ r^{-n-1} \end{Bmatrix} P_n^m(\cos\theta) \begin{Bmatrix} \sin m\varphi \\ \cos m\varphi \end{Bmatrix}, \quad (C.2)$$

where n is a positive integer or zero, and m is a positive or negative integer or zero. The Legendre associated functions $P_n^m(\cos\theta)$ are defined by the differential equation

$$\frac{1}{\sin\theta}\frac{d}{d\theta}\left(\sin\theta\frac{dP_n^m}{d\theta}\right) + \left[n(n+1) - \frac{m^2}{\sin^2\theta}\right]P_n^m = 0, \quad (C.3)$$

or

$$(1-\mu^2)\frac{d^2P_n^m}{d\mu^2} - 2\mu\frac{dP_n^m}{d\mu} + \left[n(n+1) - \frac{m^2}{1-\mu^2}\right]P_n^m = 0, \quad (C.4)$$

where $\mu = \cos\theta$. Quite generally, it can be shown that

$$P_n^m(\mu) \propto \frac{(1-\mu^2)^{|m|/2}}{2^n n!}\frac{d^{n+|m|}(\mu^2-1)^n}{d\mu^{n+|m|}}. \quad (C.5)$$

Hence, we have $P_n^m(\mu) \equiv 0$ for $|m| > n$ since then the order of differentiation in equation (C.5) is greater than the degree of the differentiated polynomial. Moreover, because equation (C.4) depends only on m^2, it follows that the function P_n^m for negative values of m can differ from the function $P_n^{|m|}$ only by a constant factor. Finally, we perceive at once that for even values of m the function $P_n^m(\mu)$ is a polynomial of degree n, whereas for odd values of m the function $P_n^m(\mu)$ is $(1-\mu^2)^{1/2}$ times a polynomial of degree $(n-1)$. In particular, we see from equations (C.4) and (C.5) that

$$P_n^0(\mu) \equiv P_n(\mu), \quad (C.6)$$

where the $P_n(\mu)$'s are the Legendre polynomials.

The Legendre associated functions $P_n{}^m(\mu)$ belonging to the lowest values of n and m are:

$$P_0(\mu) = 1, \tag{C.7}$$

$$P_1(\mu) = \mu = \cos\theta, \tag{C.8}$$

$$P_1{}^1(\mu) = (1 - \mu^2)^{1/2} = \sin\theta, \tag{C.9}$$

$$P_2(\mu) = \tfrac{1}{2}(3\mu^2 - 1) = \tfrac{1}{4}(1 + 3\cos 2\theta), \tag{C.10}$$

$$P_2{}^1(\mu) = 3(1 - \mu^2)^{1/2}\mu = \tfrac{3}{2}\sin 2\theta, \tag{C.11}$$

$$P_2{}^2(\mu) = 3(1 - \mu^2) = \tfrac{3}{2}(1 - \cos 2\theta), \tag{C.12}$$

$$P_3(\mu) = \tfrac{1}{2}(5\mu^3 - 3\mu) = \tfrac{1}{8}(3\cos\theta + 5\cos 3\theta), \tag{C.13}$$

$$P_3{}^1(\mu) = \tfrac{3}{2}(1 - \mu^2)^{1/2}(5\mu^2 - 1) = \tfrac{3}{8}(\sin\theta + 5\sin 3\theta), \tag{C.14}$$

$$P_3{}^2(\mu) = 15(1 - \mu^2)\mu = \tfrac{15}{4}(\cos\theta - \cos 3\,\theta), \tag{C.15}$$

$$P_3{}^3(\mu) = 15(1 - \mu^2)^{3/2} = \tfrac{15}{4}(3\sin\theta - \sin 3\theta). \tag{C.16}$$

The surface spherical harmonics $P_n{}^m(\cos\theta)\sin m\varphi$ and $P_n{}^m(\cos\theta)\cos m\varphi$ are periodic on the surface of a unit sphere, and the indices n and m determine the number of nodal lines. Indeed, when $m = 0$, the lines of zeros divide the surface of the unit sphere into latitudinal regions of different signs. In particular, when $n = 1$ and $m = 0$, there is a single nodal line at the equator $\theta = 90°$, along which the function is zero. When $n = 2$ and $m = 0$, there are two nodal lines following the parallels of latitude at $\theta = 55°$ and $\theta = 125°$, approximately. In general, there are n nodal lines and $(n + 1)$ zones within which the function $P_n{}^0(\cos\theta)$ is alternately positive and negative. Accordingly, the Legendre polynomials are often called *zonal* harmonics. In contrast, when $m \neq 0$, the Legendre associated function $P_n{}^m(\cos\theta)$ is zero at the poles due to the factor $(1 - \mu^2)^{1/2}$, and the number of nodal lines parallel to the equator is $n - m$; moreover, this function also vanishes along meridians determined by the roots of $\sin m\varphi$ and $\cos m\varphi$. Hence, the lines of zeros then divide the surface of the unit sphere into quadrangles (tesserae) of different signs that are bounded by circles of constant latitude and longitude. In particular, the functions $P_n{}^n(\cos\theta)\sin n\varphi$ and $P_n{}^n(\cos\theta)\cos n\varphi$ divide the spherical surface into sectorial domains within which these functions are alternately positive and negative. Therefore, they are often called *tesseral* and *sectorial* harmonics, respectively. Obviously, there are $(2n - 2)$ tesseral harmonics of degree n and two sectorial harmonics of the same degree.

Appendix D

THE MACLAURIN AND JACOBI ELLIPSOIDS

In Tables D.3 and D.4, the number in parentheses following an entry is the power of ten by which that entry must be multiplied. The a_i's are listed in units of $\bar{a} = (a_1 a_2 a_3)^{1/3}$, J in units of $(GM^3\bar{a})^{1/2}$, Ω^2 in units of $2\pi G\rho$, K and W in units of $(4/3)\pi G\rho M\bar{a}^2$. *By comparing the sixth column of Tables D.2 and D.4, we observe that for equal values of the total angular momentum J the total mechanical energy $K + W$ is smaller in the Jacobi ellipsoid than in the Maclaurin spheroid having the same values of M and $\bar{a}$.* Courtesy Mr. Y. Charland (for the Jacobi ellipsoids) and Dr. (Mrs.) M. Tassoul (for the Maclaurin spheroids).

TABLE D.1

Properties of the Maclaurin Spheroids

τ	e	a_1	a_3	A_1	A_3
0.00	0.	1.000000	1.000000	0.666667	0.666667
0.02	0.373909	1.025418	0.951039	0.646382	0.707236
0.04	0.511028	1.051737	0.904035	0.625530	0.748939
0.06	0.605447	1.079070	0.858817	0.604117	0.791766
0.08	0.676908	1.107544	0.815227	0.582147	0.835706
0.10	0.733413	1.137304	0.773121	0.559625	0.880750
0.12	0.779224	1.168520	0.732365	0.536555	0.926889
0.14	0.816962	1.201394	0.692834	0.512943	0.974114
0.16	0.848380	1.236162	0.654409	0.488792	1.022415
0.18	0.874723	1.273107	0.616978	0.464108	1.071784
0.20	0.896912	1.312575	0.580433	0.438894	1.122211
0.22	0.915652	1.354987	0.544666	0.413156	1.173688
0.24	0.931495	1.400870	0.509571	0.386897	1.226206
0.26	0.944881	1.450890	0.475041	0.360122	1.279756
0.28	0.956167	1.505909	0.440963	0.332835	1.334331
0.30	0.965646	1.567062	0.407218	0.305039	1.389921
0.32	0.973562	1.635890	0.373673	0.276740	1.446520
0.34	0.980120	1.714542	0.340176	0.247941	1.504119
0.36	0.985492	1.806147	0.306545	0.218645	1.562710
0.38	0.989825	1.915476	0.272550	0.188857	1.622285
0.40	0.993246	2.050287	0.237887	0.158581	1.682837
0.42	0.995863	2.224338	0.202115	0.127820	1.744359
0.44	0.997770	2.465304	0.164535	0.096578	1.806844
0.46	0.999049	2.841411	0.123860	0.064858	1.870283
0.48	0.999772	3.604033	0.076988	0.032664	1.934671
0.50	1.000000	∞	0.	0.	2.000000

TABLE D.2

Properties of the Maclaurin Spheroids

τ	J	Ω^2	K	$|W|$	$K + W$
0.00	0.	0.	0.	0.600000	-0.600000
0.02	0.100445	0.038022	0.011994	0.599698	-0.587704
0.04	0.145586	0.072177	0.023951	0.598787	-0.574835
0.06	0.182704	0.102586	0.035835	0.597252	-0.561417
0.08	0.216141	0.129366	0.047606	0.595077	-0.547471
0.10	0.247555	0.152625	0.059224	0.592243	-0.533018
0.12	0.277797	0.172464	0.070647	0.588724	-0.518077
0.14	0.307386	0.188979	0.081829	0.584491	-0.502662
0.16	0.336674	0.202259	0.092721	0.579508	-0.486787
0.18	0.365933	0.212388	0.103272	0.573733	-0.470461
0.20	0.395384	0.219447	0.113423	0.567113	-0.453690
0.22	0.425231	0.223511	0.123109	0.559586	-0.436477
0.24	0.455674	0.224650	0.132258	0.551076	-0.418818
0.26	0.486924	0.222933	0.140787	0.541490	-0.400703
0.28	0.519221	0.218423	0.148599	0.530711	-0.382112
0.30	0.552848	0.211181	0.155578	0.518594	-0.363016
0.32	0.588164	0.201265	0.161584	0.504950	-0.343366
0.34	0.625639	0.188731	0.166441	0.489533	-0.323092
0.36	0.665924	0.173630	0.169923	0.472009	-0.302086
0.38	0.709969	0.156013	0.171726	0.451909	-0.280184
0.40	0.759255	0.135927	0.171418	0.428544	-0.257127
0.42	0.816297	0.113418	0.168347	0.400826	-0.232479
0.44	0.885914	0.088530	0.161418	0.366860	-0.205441
0.46	0.979309	0.061304	0.148485	0.322793	-0.174308
0.48	1.134414	0.031782	0.123844	0.258009	-0.134165
0.50	∞	0.	0.	0.	$-0.$

TABLE D.3
Properties of the Jacobi Ellipsoids

τ	a_1	a_2	a_3	A_1	A_2	A_3
0.1375	1.197234	1.197234	6.976571(−1)	5.158905(−1)	5.158905(−1)	9.682190(−1)
0.14	1.362619	1.058823	6.931102(−1)	4.331591(−1)	5.976933(−1)	9.691476(−1)
0.15	1.623843	9.128330(−1)	6.746283(−1)	3.319254(−1)	6.953345(−1)	9.727401(−1)
0.16	1.829508	8.332346(−1)	6.559916(−1)	2.722174(−1)	7.517110(−1)	9.760716(−1)
0.17	2.026971	7.742668(−1)	6.371794(−1)	2.272467(−1)	7.936073(−1)	9.791460(−1)
0.18	2.228015	7.260630(−1)	6.181695(−1)	1.909075(−1)	8.271249(−1)	9.819676(−1)
0.19	2.439037	6.845402(−1)	5.989390(−1)	1.605615(−1)	8.548969(−1)	9.845416(−1)
0.20	2.665006	6.475539(−1)	5.794633(−1)	1.347689(−1)	8.783570(−1)	9.868740(−1)
0.21	2.910679	6.138146(−1)	5.597169(−1)	1.126344(−1)	8.983940(−1)	9.889716(−1)
0.22	3.181212	5.824744(−1)	5.396727(−1)	9.355011(−2)	9.156077(−1)	9.908422(−1)
0.23	3.482603	5.529374(−1)	5.193021(−1)	7.707683(−2)	9.304287(−1)	9.924944(−1)
0.24	3.822142	5.247622(−1)	4.985752(−1)	6.288051(−2)	9.431813(−1)	9.939382(−1)
0.25	4.208964	4.976078(−1)	4.774606(−1)	5.069614(−2)	9.541193(−1)	9.951845(−1)
0.26	4.654786	4.712008(−1)	4.559259(−1)	4.030532(−2)	9.634493(−1)	9.962454(−1)
0.27	5.174943	4.453148(−1)	4.339376(−1)	3.152166(−2)	9.713443(−1)	9.971340(−1)
0.28	5.789902	4.197578(−1)	4.114622(−1)	2.418071(−2)	9.779546(−1)	9.978647(−1)
0.29	6.527538	3.943635(−1)	3.884668(−1)	1.813288(−2)	9.834143(−1)	9.984528(−1)
0.30	7.426615	3.689867(−1)	3.649206(−1)	1.323824(−2)	9.878473(−1)	9.989145(−1)
0.31	8.542318	3.435010(−1)	3.407974(−1)	9.362768(−3)	9.913709(−1)	9.992663(−1)
0.32	9.955242	3.177994(−1)	3.160786(−1)	6.375882(−3)	9.940987(−1)	9.995254(−1)
0.33	1.178655(+1)	2.917972(−1)	2.907582(−1)	4.149021(−3)	9.961425(−1)	9.997085(−1)
0.34	1.422451(+1)	2.654381(−1)	2.648497(−1)	2.555540(−3)	9.976129(−1)	9.998316(−1)
0.36	2.234398(+1)	2.116265(−1)	2.114801(−1)	7.805414(−4)	9.992638(−1)	9.999557(−1)
0.38	4.060607(+1)	1.569406(−1)	1.569183(−1)	1.567998(−4)	9.998506(−1)	9.999926(−1)
0.40	9.348701(+1)	1.034255(−1)	1.034240(−1)	1.591045(−5)	9.999847(−1)	9.999994(−1)
0.45	1.387580(+4)	8.489281(−3)	8.489281(−3)	1.048054(−11)	1.000000	1.000000
0.50	∞	0.	0.	0.	1.000000	1.000000

TABLE D.4

Properties of the Jacobi Ellipsoids

τ	J	Ω^2	K	$|W|$	$K + W$
0.1375	3.037510(−1)	1.871148(−1)	8.046137(−2)	5.850539(−1)	−5.045925(−1)
0.14	3.115277(−1)	1.824065(−1)	8.147653(−2)	5.819752(−1)	−5.004987(−1)
0.15	3.442578(−1)	1.640304(−1)	8.538107(−2)	5.692071(−1)	−4.838261(−1)
0.16	3.791934(−1)	1.467273(−1)	8.894712(−2)	5.559195(−1)	−4.669724(−1)
0.17	4.165932(−1)	1.304912(−1)	9.215489(−2)	5.420876(−1)	−4.499327(−1)
0.18	4.567600(−1)	1.153154(−1)	9.498333(−2)	5.276852(−1)	−4.327018(−1)
0.19	5.000515(−1)	1.011921(−1)	9.741002(−2)	5.126843(−1)	−4.152743(−1)
0.20	5.468926(−1)	8.811188(−2)	9.941110(−2)	4.970555(−1)	−3.976444(−1)
0.21	5.977918(−1)	7.606377(−2)	1.009612(−1)	4.807677(−1)	−3.798064(−1)
0.22	6.533628(−1)	6.503467(−2)	1.020334(−1)	4.637884(−1)	−3.617549(−1)
0.23	7.143523(−1)	5.500897(−2)	1.025994(−1)	4.460843(−1)	−3.434849(−1)
0.24	7.816773(−1)	4.596804(−2)	1.026292(−1)	4.276218(−1)	−3.249926(−1)
0.25	8.564756(−1)	3.788970(−2)	1.020919(−1)	4.083675(−1)	−3.062757(−1)
0.26	9.401742(−1)	3.074758(−2)	1.009554(−1)	3.882899(−1)	−2.873345(−1)
0.27	1.034584	2.451038(−2)	9.918741(−2)	3.673608(−1)	−2.681734(−1)
0.28	1.142035	1.914119(−2)	9.675635(−2)	3.455584(−1)	−2.488020(−1)
0.29	1.265563	1.459669(−2)	9.363250(−2)	3.228707(−1)	−2.292382(−1)
0.30	1.409194	1.082643(−2)	8.979020(−2)	2.993007(−1)	−2.095105(−1)
0.31	1.578352	7.772310(−3)	8.521076(−2)	2.748734(−1)	−1.896627(−1)
0.32	1.780492	5.368300(−3)	7.988662(−2)	2.496457(−1)	−1.697591(−1)
0.33	2.026084	3.540656(−3)	7.382694(−2)	2.237180(−1)	−1.498911(−1)
0.34	2.330191	2.208922(−3)	6.706521(−2)	1.972506(−1)	−1.301854(−1)
0.36	3.214870	6.909640(−4)	5.174956(−2)	1.437488(−1)	−9.199922(−2)
0.38	4.810650	1.418663(−4)	3.508803(−2)	9.233692(−2)	−5.724889(−2)
0.40	8.204249	1.468656(−5)	1.925372(−2)	4.813429(−2)	−2.888057(−2)
0.45	1.499293(+2)	1.010623(−11)	2.918750(−4)	6.486111(−4)	−3.567361(−4)
0.50	∞	0.	0.	0.	−0.

Appendix E
THE ψ- AND χ-FUNCTIONS

In Tables E.1-E.5, the number in parentheses following each entry is the power of ten by which that entry must be multiplied. Source: Aikawa, T., *Sci. Rep. Tôhoku Univ.* (I) **54**, 13, 1971.

TABLE E.1

Associated Functions for the Rotating Polytrope of Index $n = 1.5$

ξ	θ	ψ_0	ψ_2	ψ_4	$-\chi_0$	χ_2	$-\chi_4$
0.0	0.10000(+1)	0.00000	0.00000	0.00000	0.00000	0.00000	0.00000
0.3	0.98510	0.14899(−1)	0.89138(−1)	0.80505(−2)	0.48404(−6)	0.60782(−6)	0.27007(−5)
0.6	0.94159	0.58423(−1)	0.34648	0.12649	0.39499(−4)	0.45369(−4)	0.15042(−3)
0.9	0.87285	0.12728	0.73474	0.62166	0.43931(−3)	0.51210(−3)	0.16004(−2)
1.2	0.78398	0.21669	0.12397(+1)	0.18875(+1)	0.23512(−2)	0.27927(−2)	0.84610(−2)
1.5	0.68112	0.32128	0.17877(+1)	0.43866(+1)	0.84443(−2)	0.10259(−1)	0.30216(−1)
1.8	0.57072	0.43600	0.23429(+1)	0.85937(+1)	0.23557(−1)	0.30335(−1)	0.86529(−1)
2.1	0.45880	0.55695	0.28687(+1)	0.14957(+2)	0.55281(−1)	0.71063(−1)	0.19577
2.4	0.35049	0.68196	0.33413(+1)	0.23891(+2)	0.11474	0.15245	0.40257
2.7	0.24966	0.81095	0.37509(+1)	0.35797(+2)	0.21822	0.30031	0.75415
3.0	0.15886	0.94624	0.41025(+1)	0.51132(+2)	0.39137	0.55844	0.13198(+1)
3.3	0.79315(−1)	0.10929(+1)	0.44152(+1)	0.70526(+2)	0.68145	0.10078(+1)	0.22067(+1)
3.6	0.11091(−1)	0.12607(+1)	0.47266(+1)	0.95032(+2)	0.12131(+1)	0.18547(+1)	0.36395(+1)
ξ_1	0.00000	0.12944(+1)	0.47874(+1)	0.10011(+3)	0.13720(+1)	0.21087(+1)	0.40029(+1)

TABLE E.2

Associated Functions for the Rotating Polytrope of Index $n = 2.0$

ξ	θ	ψ_0	ψ_2	ψ_4	$-\chi_0$	χ_2	$-\chi_4$
0.0	0.10000(+1)	0.00000	0.00000	0.00000	0.00000	0.00000	0.00000
0.3	0.98513	0.14867(−1)	0.88855(−1)	0.80342(−2)	0.17129(−5)	0.18013(−5)	0.74299(−5)
0.6	0.94209	0.57935(−1)	0.34226	0.12552	0.11425(−3)	0.11808(−3)	0.43509(−3)
0.9	0.87521	0.12510	0.72471	0.61174	0.12105(−2)	0.12785(−2)	0.45756(−2)
1.2	0.79067	0.21095	0.11885(+1)	0.18396(+1)	0.61402(−2)	0.66741(−2)	0.23396(−1)
1.5	0.69537	0.31022	0.16857(+1)	0.42362(+1)	0.20704(−1)	0.23306(−1)	0.79952(−1)
1.8	0.59582	0.41902	0.21777(+1)	0.82401(+1)	0.53816(−1)	0.63079(−1)	0.21120
2.1	0.49745	0.53548	0.26404(+2)	0.14288(+2)	0.11700	0.14340	0.46721
2.4	0.40421	0.65980	0.30642(+1)	0.22831(+2)	0.22386	0.28770	0.90954
2.7	0.31859	0.79403	0.34519(+1)	0.34367(+2)	0.39017	0.52648	0.16106(+1)
3.0	0.24182	0.94165	0.38150(+1)	0.49492(+2)	0.63490	0.89928	0.26554(+1)
3.3	0.17418	0.11072(+1)	0.41699(+1)	0.68948(+2)	0.98229	0.14581(+1)	0.41458(+1)
3.6	0.11552	0.12961(+1)	0.45360(+1)	0.93675(+2)	0.14651(+1)	0.22725(+1)	0.62076(+1)
3.9	0.64252(−1)	0.15143(+1)	0.49339(+1)	0.12487(+3)	0.21290(+1)	0.34369(+1)	0.90000(+1)
4.2	0.20160(−1)	0.17690(+1)	0.53853(+1)	0.16407(+3)	0.30386(+1)	0.50811(+1)	0.12728(+2)
ξ_1	0.00000	0.19153(+1)	0.56432(+1)	0.18772(+3)	0.36249(+1)	0.61589(+1)	0.15070(+2)

TABLE E.3

Associated Functions for the Rotating Polytrope of Index $n = 2.5$

ξ	θ	ψ_0	ψ_2	ψ_4	$-\chi_0$	χ_2	$-\chi_4$
0.0	0.10000(+1)	0.00000	0.00000	0.00000	0.00000	0.00000	0.00000
0.3	0.98517	0.14834(−1)	0.88575(−1)	0.80182(−2)	0.36658(−5)	0.33956(−5)	0.15186(−4)
0.6	0.94259	0.57465(−1)	0.33818	0.12458	0.22841(−3)	0.21235(−3)	0.89119(−3)
0.9	0.87746	0.12310	0.70695	0.60246	0.23242(−2)	0.22272(−2)	0.91531(−2)
1.2	0.79686	0.20594	0.11428(+1)	0.17968(+1)	0.11233(−1)	0.11205(−1)	0.45231(−1)
1.5	0.70807	0.30116	0.15993(+1)	0.41083(+1)	0.35879(−1)	0.37552(−1)	0.14871
1.8	0.61744	0.40601	0.20447(+1)	0.79538(+1)	0.88109(−1)	0.97352(−1)	0.37762
2.1	0.52964	0.52036	0.24643(+1)	0.13768(+2)	0.18095	0.21191	0.80428
2.4	0.44771	0.64550	0.28568(+1)	0.22027(+2)	0.32747	0.40723	0.15117(+1)
2.7	0.37321	0.78433	0.32297(+1)	0.33280(+2)	0.54075	0.71395	0.25926(+1)
3.0	0.30668	0.94033	0.35947(+1)	0.48174(+2)	0.83455	0.11678(+1)	0.41511(+1)
3.3	0.24793	0.11172(+1)	0.39649(+1)	0.67482(+2)	0.12244(+1)	0.18101(+1)	0.63058(+1)
3.6	0.19642	0.13188(+1)	0.43530(+1)	0.92117(+2)	0.17281(+1)	0.26882(+1)	0.91930(+1)
3.9	0.15139	0.15487(+1)	0.47701(+1)	0.12315(+3)	0.23661(+1)	0.38549(+1)	0.12969(+2)
4.2	0.11203	0.18102(+1)	0.52264(+1)	0.16181(+3)	0.31602(+1)	0.53664(+1)	0.17810(+2)
4.5	0.77554(−1)	0.21067(+1)	0.57300(+1)	0.20950(+3)	0.41307(+1)	0.72774(+1)	0.23913(+2)
4.8	0.47238(−1)	0.24410(+1)	0.62878(+1)	0.26777(+3)	0.52905(+1)	0.96306(+1)	0.31490(+2)
5.1	0.20443(−1)	0.28153(+1)	0.69050(+1)	0.33834(+3)	0.66314(+1)	0.12434(+2)	0.40753(+2)
ξ_1	0.00000	0.31665(+1)	0.74791(+1)	0.40942(+3)	0.78676(+1)	0.15112(+2)	0.50096(+2)

TABLE E.4

Associated Functions for the Rotating Polytrope of Index $n = 3.0$

ξ	θ	ψ_0	ψ_2	ψ_4	$-\chi_0$	χ_2	$-\chi_4$
0.0	0.10000(+1)	0.00000	0.00000	0.00000	0.00000	0.00000	0.00000
0.3	0.98520	0.14802(−1)	0.88297(−1)	0.80022(−2)	0.64727(−5)	0.53131(−5)	0.26644(−4)
0.6	0.94307	0.57011(−1)	0.33422	0.12366	0.38669(−3)	0.32263(−3)	0.15492(−2)
0.9	0.87962	0.12123	0.69034	0.59376	0.37904(−2)	0.32922(−2)	0.15515(−1)
1.2	0.80259	0.20152	0.11019(+1)	0.17582(+1)	0.17530(−1)	0.16046(−1)	0.74284(−1)
1.5	0.71950	0.29364	0.15251(+1)	0.39979(+1)	0.53417(−1)	0.52006(−1)	0.23629
1.8	0.63631	0.39601	0.19349(+1)	0.77159(+1)	0.12518	0.13044	0.58148
2.1	0.55703	0.50935	0.23236(+1)	0.13349(+2)	0.24596	0.27517	0.12043(+1)
2.4	0.48393	0.63593	0.26944(+1)	0.21391(+2)	0.42737	0.51356	0.22097(+1)
2.7	0.41796	0.77878	0.30563(+1)	0.32415(+2)	0.68003	0.87619	0.37137(+1)
3.0	0.35923	0.94108	0.34203(+1)	0.47094(+2)	0.10146(+1)	0.13968(+1)	0.58455(+1)
3.3	0.30735	0.11260(+1)	0.37964(+1)	0.66210(+2)	0.14427(+1)	0.21126(+1)	0.87511(+1)
3.6	0.26169	0.13363(+1)	0.41938(+1)	0.90659(+2)	0.19774(+1)	0.30635(+1)	0.12595(+2)
3.9	0.22156	0.15745(+1)	0.46198(+1)	0.12146(+3)	0.26329(+1)	0.42921(+1)	0.17563(+2)
4.2	0.18623	0.18430(+1)	0.50805(+1)	0.15973(+3)	0.34246(+1)	0.58422(+1)	0.23861(+2)
4.5	0.15507	0.21435(+1)	0.55804(+1)	0.20679(+3)	0.43670(+1)	0.77567(+1)	0.31715(+2)
4.8	0.12835	0.24778(+1)	0.61235(+1)	0.26397(+3)	0.54733(+1)	0.10075(+2)	0.41373(+2)
5.1	0.10296	0.24873(+1)	0.67124(+1)	0.33277(+3)	0.67524(+1)	0.12831(+2)	0.53100(+2)
5.4	0.81065(−1)	0.32530(+1)	0.73493(+1)	0.41483(+3)	0.82070(+1)	0.16046(+2)	0.67175(+2)
5.7	0.61430(−1)	0.36958(+1)	0.80358(+1)	0.51184(+3)	0.98300(+1)	0.19724(+2)	0.83892(+2)
6.0	0.43738(−1)	0.41762(+1)	0.87727(+1)	0.62564(+3)	0.11601(+2)	0.23849(+2)	0.10355(+3)
6.3	0.27723(−1)	0.46945(+1)	0.95603(+1)	0.75811(+3)	0.13482(+2)	0.28373(+2)	0.12645(+3)
6.6	0.13162(−1)	0.52507(+1)	0.10399(+2)	0.91125(+3)	0.15411(+2)	0.33209(+2)	0.15290(+3)
ξ_1	0.00000	0.58378(+1)	0.11277(+2)	0.10851(+4)	0.17281(+2)	0.38170(+2)	0.18285(+3)

TABLE E.5

Associated Functions for the Rotating Polytrope $n = 3.5$

ξ	θ	ψ_0	ψ_2	ψ_4	$-\chi_0$	χ_2	$-\chi_4$
0.0	0.10000(+1)	0.00000	0.00000	0.00000	0.00000	0.00000	0.00000
0.6	0.94177	0.56572(−1)	0.32033	0.12278	0.59465(−3)	0.44252(−3)	0.24444(−2)
1.2	0.80793	0.19761	0.10649(+1)	0.17771(+1)	0.24993(−1)	0.20918(−1)	0.11597
1.8	0.65299	0.38800	0.18426(+1)	0.75144(+1)	0.16464	0.16139	0.82406(+1)
2.4	0.51472	0.62940	0.25628(+1)	0.20868(+2)	0.52585	0.60862	0.30130(+1)
3.0	0.40294	0.94311	0.32779(+1)	0.46819(+2)	0.11865(+1)	0.15995(+1)	0.77755(+1)
3.6	0.31556	0.13509(+1)	0.40571(+1)	0.89349(+2)	0.22230(+1)	0.34092(+1)	0.16482(+2)
4.2	0.24749	0.18694(+1)	0.49485(+1)	0.15784(+3)	0.37301(+1)	0.63453(+1)	0.30849(+2)
4.8	0.19397	0.25107(+1)	0.59811(+1)	0.26082(+3)	0.58157(+1)	0.10730(+2)	0.52972(+2)
5.4	0.15127	0.32829(+1)	0.71716(+1)	0.40906(+3)	0.85829(+1)	0.16871(+2)	0.85313(+2)
6.0	0.11665	0.41915(+1)	0.85296(+1)	0.61500(+3)	0.12105(+2)	0.25020(+2)	0.13069(+3)
6.6	0.88137(−1)	0.52398(+1)	0.10061(+2)	0.89259(+3)	0.16401(+2)	0.35332(+2)	0.19225(+3)
7.2	0.64307(−1)	0.64294(+1)	0.11767(+2)	0.12574(+4)	0.21320(+2)	0.47599(+2)	0.27037(+3)
7.8	0.44119(−1)	0.77603(+1)	0.13648(+2)	0.17262(+4)	0.26948(+2)	0.62347(+2)	0.37794(+3)
8.4	0.26807(−1)	0.92315(+1)	0.15704(+2)	0.23175(+4)	0.32816(+2)	0.78567(+2)	0.50974(+3)
9.0	0.11803(−1)	0.10841(+2)	0.17933(+2)	0.30509(+4)	0.38604(+2)	0.96004(+2)	0.67304(+3)
ξ_1	0.00000	0.12394(+2)	0.20066(+2)	0.38430(+4)	0.43420(+2)	0.11217(+3)	0.84912(+3)

Index of Names

Page numbers in *italics* refer to the Bibliographical Notes, tables, figures, or footnotes.

Index of Subjects

Library of Congress Cataloging in Publication Data

Tassoul, Jean-Louis.
  Theory of rotating stars.

  (Princeton series in astrophysics; 1)
  Includes bibliographical references and index.
  1. Stars—Rotation. I. Title.
QB810.T37   523.8′3   78-51198
ISBN 0-691-08211-1
ISBN 0-691-8214-6 pbk.